MEASURE OF THE MOON

ASTROPHYSICS AND SPACE SCIENCE LIBRARY

A SERIES OF BOOKS ON THE RECENT DEVELOPMENTS
OF SPACE SCIENCE AND OF GENERAL GEOPHYSICS AND ASTROPHYSICS
PUBLISHED IN CONNECTION WITH THE JOURNAL
SPACE SCIENCE REVIEWS

VOLUME 8

MEASURE OF THE MOON

Proceedings of the Second International Conference on Selenodesy and Lunar Topography held in the University of Manchester, England

May 30 – June 4, 1966

Edited by

ZDENĚK KOPAL
Department of Astronomy, University of Manchester

and

CONSTANTINE L. GOUDAS
Boeing Scientific Research Laboratories, Seattle, Wash.

D. REIDEL PUBLISHING COMPANY

DORDRECHT-HOLLAND

ISBN-13:978-94-010-3531-6 e-ISBN-13:978-94-010-3529-3
DOI: 10.1007/978-94-010-3529-3

Reprint of the original edition 1967

PREFACE

After many decades spent in astronomical semi-obscurity, the Moon has of late suddenly emerged to claim renewed interest on the part of the students of astronomy, as well as of other branches of physical science and technology; and the reasons which brought this about are indeed of historical significance.

From time immemorial, astronomy has been debarred from the status of a genuine experimental science by the utter remoteness of the objects of its study. With the exception of meteors – those small freaks of cosmic matter intercepted by the Earth on its perpetual journey through space – the properties of all celestial bodies outside the gravitational confines of our planet could be studied only at a distance: namely, from the effect of attraction exerted by their masses; or from the ciphered message of their light brought to us by nimble-footed photons across the intervening gaps of space.

A dramatic emergence of long-range rockets in the last decade bids fair to bring about a profound change in this situation. On September 13, 1959 – a memorable date in the history of human endeavour – a man-made missile of Russian origin crash-landed on the surface of the Moon in the region of its Mare Imbrium, and thus ended the age-long separation of the Earth and its only natural satellite which lasted not less than $4\frac{1}{2}$ billion years. Moreover, subsequent advances in astronautics have made it a virtual certainty that the space flights made since – at the time of writing, not less than 12 such spacecraft of American as well as Russian origin landed already on different parts of the lunar surface (two of which being soft-landers) – will be followed by others eventually carrying men to the Moon, and thus inaugurating an era of direct exploration of our natural satellite within our lifetime.

It was against this background, and in order to meet the anticipated needs of the knowledge of the Moon's size, shape, topography, and gravitational field for the purpose of circumlunar navigation and surface landings, that a second International Conference on Selenodesy and Lunar Topography met at the University of Manchester in England between May 30 and June 4 of 1966, under the joint sponsorship of its Department of Astronomy and of the Terrestrial Sciences Laboratory, Air Force Cambridge Research Laboratories, Bedford, Mass., U.S.A. This partnership is by no means accidental. A close collaboration between these institutions in the field of selenodetic studies goes back to 1959; and in April 1960 a working group of interested investigators met under the auspices of AFCRL at the Observatoire du Pic-du-Midi, Bagnères de

Bigorre in France, by invitation of its director, Professor Jean Rösch, to plan the work that laid ahead. The aim of the second conference, held at Manchester, England, between May 30–June 4th of this year, was to review the progress made in this field since 1960, to take stock of our present knowledge as well as needs, and to discuss plans for further work by astronomical and other methods on the eve of the exploration by spacecraft operating (or soon to operate) in closer proximity of our satellite.

In order to meet these needs, 42 invited guests (for their list, see pp. XIII–XIV) from 13 countries congregated at the University of Manchester; and in accordance with our aims, the following programme was arranged for the meetings:

May 31 (Chairman: Professor Jean Rösch, University of Paris)

K. Koziel (University of Krakow): 'Physical Librations of the Moon in Longitude';
G. Colombo (University of Padua): 'Cassini's Second and Third Laws';
J. Maslowski (University of Krakow): 'The Moon's Mechanical Ellipticity';
J. Mietelski (University of Krakow): 'The Moon's Rotation Constants and the Coordienast of Mösting A';
M. D. Moustoulas (University of Manchester): 'A Contribution to the Solution for the Moon's Physical Libration in Longitude';
D. H. Eckhardt (Air Force Cambridge Research Laboratories): 'Lunar Physical Libration Theory';
D. W. G. Arthur (University of Arizona): 'Photographic Techniques for the Determination of the Moon's Constants of Rotation'.

June 1 (Chairmen: Professor M. G. J. Minnaert, University of Utrecht, and Dr. R. H. Stoy, Director of the Cape of Good Hope Observatory)

J. Rösch (University of Paris): 'Selenodetic Observations';
R. V. Karandikar (Osmania University): 'Observational Facilities of the Indian National Centre for Astronomy at Hyderabad';*
S. Miyamoto (University of Kyoto): 'New Observatory for Lunar and Planetary Observations in Japan';*
A. H. Samaha (Helwan Observatory): 'Lunar Research at the Kottamia Observatory';
R. H. Stoy (Cape of Good Hope Observatory): 'Facilities for Selenodetic Work at the Royal Observatory on the Cape of Good Hope';*
M. S. Hunt (Air Force Cambridge Research Laboratories): 'Selenodetic Evaluation of Eleven Paris Lunar Photographs';
L. G. Stoddard (Lockheed California Co.): 'Selenodetic Equipment for Observations of Three Annular Eclipses';**
M. Davidson, C. L. Goudas, and Z. Kopal (University of Manchester and Boeing

* On account of the essentially descriptive nature of this communication, not included in this volume.
** In the author's absence presented by D. H. Eckhardt.

Scientific Research Laboratories): 'Lunar Profiles Determined from the Annular Solar Eclipses of 1962 and 1963';*

A. G. Wilson (Douglas Advanced Research Laboratories): 'The Use of the 48-inch Schmidt Telescope of Palomar Mountain for Selenodetic Observations';**

A. Orszag (Ecole Polytechnique, Paris): 'Laser Measurements of the Earth-Moon Distance and for Selenodesy';

R. Julian (Hughes Aircraft Co.): 'Investigation of the Moon's Motion by means of Laser'.

June 2 (Chairmen: Professor Z. Kopal, University of Manchester, and Mr. Mahlon S. Hunt, Air Force Cambridge Research Laboratories)

C. L. Goudas (Boeing Scientific Research Laboratories): 'Harmonic Analysis of the Moon's Shape and Gravitational Field';

J. Hopmann (Vienna University Observatory): 'What can we say about the shape of the Moon?';

R. W. Carder (Aeronautical Chart and Information Center, U.S. Air Force): 'The Selenodetic Work at ACIC';*

D. L. Meyer (Aeronautical Chart and Information Center, U.S. Air Force): 'Observational Uncertainties in the Lunar Control Systems';

A. Nowicki and M. Marchant (U.S. Army Map Service): 'The AMS Control System';

G. A. Mills and M. Davidson (University of Manchester): 'The Manchester Selenodetic Control System';

D. W. G. Arthur (Lunar-Planetary Laboratory, University of Arizona): 'The LPL Programme for Selenodetic Positions';

W. L. Sjogren (Jet Propulsion Laboratory, California Institute of Technology): 'Estimates of Four Topocentric Lunar Radii';†

W. M. Kaula (University of California at Los Angeles): 'Analysis of Satellite Orbit Perturbations to Determine the Lunar Gravitational Field';††

J. Lorell (Jet Propulsion Laboratory, California Institute of Technology): 'Lunar Gravity from Orbiter Tracking Data';

G. Veis (Technical University, Athens): 'A Differential Orbit Improvement Program for Lunar Orbiter'.*†

June 3 (Chairmen: Dr. Th. Weimer, University of Algiers, and Professor K. Koziel, University of Krakow

M. G. J. Minnaert (University of Utrecht): 'Photometric Methods for Determination of the Lunar Relief';

* Presented by C. L. Goudas.

** Communication received after the conclusion of the meetings.

† In the author's absence reported by Dr. J. Lorell.

†† In the author's absence presented by Dr. C. L. Goudas.

*† In the author's absence presented by Prof. G. Colombo.

J. H. Sasser (NASA Manned Space Flight Center, Houston): 'Selenodesy for the APOLLO Project';
Z. Kopal (University of Manchester): 'Overexposed Terminator Photography for Lunar Topographic Work';
T. W. Rackham (University of Manchester): 'Measurements and Reduction of Lunar Relative Altitudes';
P. V. Sudbury (University of Manchester): 'High-Resolution Lunar Photography';
M. T. Jones (University of Manchester): 'Iso-densitometric Aids to Lunar Charting';
A. Rifaat (Helwan Observatory and University of Manchester): 'Iso-densitometric Measurements of Lunar Slopes from the Ranger Photographs';
O. Calame (Observatorie du Pic-du-Midi): 'Determination of the Moon's Shape by the Photometry of its Terminator';
M. G. J. Minnaert (University of Utrecht): 'Concluding Remarks'.

The first paper presented at the commencement of each session was of the nature of an introduction to the entire field to be discussed each day.

While the proceedings of our first 1960 conference at Bagnères were published only for a limited distribution*, the scope of the second conference as embodied in the foregoing programme has made it desirable to make its proceedings available to the scientific community at large by publishing them in book form; and agreement to this end was reached at an early date with the D. Reidel Publishing Co. in Dordrecht, Holland. Not all papers presented at the meetings are, however, included in this volume in full, as its size would otherwise have grown out of practicable limits. This is, in particular, true of several communications presented in the morning session of June 1st, which were concerned with an account of facilities existing at several observatories in different parts of the world of interest to the students of our satellite.

In concluding our task of submitting this volume to the reader, it is a pleasure to express our sincere thanks to all guests, colleagues and friends for all they have done to contribute to the success of our conference. In the first place, may we assure our visitors – many of whom waged long journeys to join us (the total distance travelled by all participants, in excess of 250000 km, amounted to two-thirds of the mean distance to the Moon!) – how much we appreciated their contributions to the programme and discussion. This is true, in particular, to those distinguished guests who agreed to preside at the individual sessions, or prepare the introductory addresses, and whose tact and patience contributed so large a measure to smooth progress of the meetings. The five pleasant days of uninterrupted sunshine which we were privileged to spend together on this occasion will be long remembered at the University of Manchester; and our regret at seeing so many guests depart after so short a sojourn was lessened only by the hope that their recent visit to Manchester may become one of many to which we may look forward in the future.

* As an *Astronomical Contribution from the University of Manchester*, Series III, No. 90 (1961): *ASTIA Document No. AD-277724.*

On June 4th, at the conclusion of the Manchester meetings, a group of the participants departed for France to visit Bagnères de Bigorre – the site of our first lunar conference in April 1960 – and the Observatoire du Pic-du-Midi, which repeatedly contributed history-making records to lunar studies since the days of Bernard Lyot. We are very grateful to Professor Jean Rösch, director of the Observatoire du Pic-du-Midi and chairman of the first conference on lunar topography in 1960, for extending to us an invitation for this pilgrimage.

Secondly, the undersigned would like to express their indebtedness to the authorities of the sponsoring institutions for all support willingly given to our venture – in particular, to Professor Sir William Mansfield Cooper, Vice-Chancellor of the University of Manchester, who opened the conference on May 31st, and to Colonel Robert F. Long, Commanding Officer of the Air Force Cambridge Research Laboratories. We are also grateful to Colonel Currie S. Downie of the Brussels Office of Aerospace Research (USAF) who was responsible for working out the administrative arrangements between AFCRL and the University of Manchester.

At Manchester, a prominent part in the organization of the meetings was taken by Miss Ellen B. Finlay, who served as our finance officer and organizing secretary in charge of all local arrangements; and her sister Jean who willingly assumed the functions of a transport officer; while, in the office, the assistance of Mrs. Barbara Sudbury in keeping all channels of communication operative during the busy week of the meetings has been invaluable. These three young ladies well deserved the tribute paid to them by Professor Hopmann on behalf of the guests at the closing dinner of the conference; and so does Dr. Thomas Rackham, senior member of the Manchester Lunar Group, whose care for the welfare of our guests did not stop till at the summit of Pic-du-Midi!

The editing of the present volume devolved on the undersigned local chairman and Dr. Constantine L. Goudas, who throughout the meetings served as scientific secretary of the conference. In submitting its proceedings to the reader, the editors wish to express their gratitude to all colleagues for the promptness with which they communicated their mss. for the press, and thus ensured an early publication of this volume. In view of the number of the contributions and the variety of their subjects, hope has been given up to attain complete uniformity in notations, or references, throughout the book; but every effort has been made to ensure it within each communication.

Limitations in size have also prevented us from publishing full transcript of all discussions which followed the individual papers; but it is hoped that nothing essential has been omitted from its published text. Contributions to the discussion of invited members of the conference (as listed on pp. XIII–XIV) have been identified whenever possible, by their names. Exceptions have arisen in certain cases where the magnetic tape record renders the true authorship of a remark dubious. In such a case, it has been ascribed to an anonymous 'voice' – as have been all remarks made by unofficial participants of the conference (mainly research students of the University of Manchester) whose names are not listed.

Dr. M. D. Moutsoulas deserves especial thanks on the part of the Editors for the care with which he read the proofs and helped to clarify a number of obscure points in the text.

Last but not least, we wish to thank Professor C. de Jager, Editor of the *Astrophysics and Space Science Library*, for inviting us to publish these conference proceedings in his series; and to the D. Reidel Publishing Company for the care which they have bestowed on the production of the entire volume.

MAHLON S. HUNT
ZDENĚK KOPAL

Joint Conference Chairmen:

ZDENĚK KOPAL
University of Manchester, England

MAHLON S. HUNT
Air Force Cambridge Research Laboratories, Bedford, Mass., U.S.A.

Scientific Secretary:

CONSTANTINE L. GOUDAS
Boeing Scientific Research Laboratories, Seattle, Wash., U.S.A.

Organizing Secretary:

ELLEN B. FINLAY
Department of Astronomy, University of Manchester

Identification Key:

Sitting in front (left to right): Mrs. Sudbury, Miss Finlay, Mrs. Finlay-Reid, Mrs. Rifaat.

Sitting in second row (left to right): Miyamoto, Rösch, Colombo, Koziel, Mlle Calame, Samaha, Nowicki, Hopmann, Minnaert, Stoy, Karandikar.

Standing in third row (left to right): Davidson, Sudbury, Moutsoulas, Masłowski, Orszag, Weimer, Goudas, Eckhardt, Julian, Kopal, Downie, Sasser, Hunt, Chappas, Mietelski, Carder, Lorell, Mills, Arthur, Meyer, Klages, Rackham, Mounir.

LIST OF PARTICIPANTS

Mr. D. W. G. ARTHUR, Lunar and Planetary Laboratory, University of Arizona, Tucson, Ariz., U.S.A.

Mlle. O. CALAME, Faculty of Science, University of Paris, France.

Mr. R. W. CARDER, Aeronautical Chart and Information Center, U.S. Air Force, St. Louis, Mo., U.S.A.

Lt. Col. Walter H. CHAPPAS, ACIC Squadron, Wiesbaden, Germany.

Professor G. COLOMBO, University of Padua, Italy.

Dr. M. E. DAVIDSON, Department of Astronomy, University of Manchester.

Professor C. DE JAGER, Sterrewacht, Utrecht, Holland.

Colonel C. S. DOWNIE, European Office of Aerospace Research, Brussels, Belgium.

Dr. D. H. ECKHARDT, Terrestrial Sciences Laboratory, Air Force Cambridge Research Laboratories, Bedford, Mass., U.S.A.

Dr. C. L. GOUDAS, Mathematics Research Laboratory, Boeing Scientific Research Laboratories, Seattle, Wash., U.S.A.

Mr. M. HAMDY, Department of Astronomy, University of Manchester and Helwan Observatory, Egypt.

Professor J. HOPMANN, University Observatory, Vienna, Austria.

Mr. M. S. HUNT, Terrestrial Sciences Laboratory, Air Force Cambridge Research Laboratories, Bedford, Mass., U.S.A.

Mr. M. T. JONES, Department of Astronomy, University of Manchester.

Dr. R. S. JULIAN, Hughes Aircraft Company, Aerospace Group, Research and Development Division, Culver City, Calif., U.S.A.

Professor R. V. KARANDIKAR, Nizamiah Observatory, Osmania University, Hyderabad, India.

Professor W. M. KAULA, Institute of Geophysics and Planetary Physics, University of California, Los Angeles, Calif., U.S.A.

Mr. L. KLAGES, Aeronautical Chart and Information Center, U.S. Air Force, St. Louis, Mo., U.S.A.

Professor Z. KOPAL, Department of Astronomy, University of Manchester.

Professor K. KOZIEŁ, Astronomical Observatory, University of Krakow, Poland.

Dr. J. LORELL, Jet Propulsion Laboratory, University of California, Pasadena, Calif., U.S.A.

Mr. M. Q. Marchant, Corps of Engineers, Army Map Service, Washington, D.C., U.S.A.

Dr. J. Masłowski, University Observatory, University of Krakow, Poland.

Mr. D. Meyer, Aeronautical Chart and Information Center, U.S. Air Force, St. Louis, Mo., U.S.A.

Dr. J. Mietelski, University Observatory, University of Krakow, Poland.

Dr. G. A. Mills, Department of Astronomy, University of Manchester.

Professor M. G. J. Minnaert, Sterrewacht Sonnenborgh der Rijksuniversiteit, Utrecht, Holland.

Professor S. Miyamoto, Kwasan Observatory, University of Kyoto, Japan.

Dr. M. D. Moutsoulas, Department of Astronomy, University of Manchester and University of Thessaloniki, Greece.

Mr. A. L. Nowicki, Corps of Engineers, Army Map Service, Washington, D.C., U.S.A.

Mr. A. Orszag, Laboratoire de Physique, Ecole Polytechnique, Paris, France.

Dr. T. W. Rackham, Department of Astronomy, University of Manchester.

Mrs. A. Rifaat, Department of Astronomy, University of Manchester.

Professor J. Rösch, Observatoire du Pic-du-Midi, Bagnères de Bigorre, Hautes Pyrenées, France.

Professor A. H. Samaha, Helwan Observatory, Helwan, Cairo, Egypt.

Mr. J. Sasser, Manned Spaceflight Center, NASA, Houston, Tex., U.S.A.

Dr. L. G. Stoddard, Basic Sciences Laboratory, Lockheed California Co., Burbank, Calif., U.S.A.

Dr. R. H. Stoy, Royal Observatory, Cape of Good Hope, Republic of South Africa.

Dr. P. V. Sudbury, Department of Astronomy, University of Manchester.

Professor G. Veis, Athens Technical University, Athens, Greece.

D. Th. Weimer, Astronomical Observatory, University of Algiers, Algeria.

Dr. A. G. Wilson, Douglas Advanced Research Laboratories, Huntington Beach, Calif., U.S.A.

TABLE OF CONTENTS

PART II / SELENODETIC OBSERVATIONS AND MEASUREMENTS

PART III / SHAPE AND GRAVITATIONAL FIELD OF THE MOON

PART I

LIBRATIONS OF THE MOON

RECENT RESEARCHES ON THE DETERMINATION OF THE MOON'S PHYSICAL LIBRATION CONSTANTS

With Special Consideration of Cracow Investigations

KAROL KOZIEŁ*

Although the problem of determination of the physical libration constants has been the object of interest for astronomers since over 150 years, yet various methods applied by various authors yielded up to the present day very divergent values for these constants. Thus, e.g., the values of the two principal unknowns of the libration problem – i.e., of the Moon's mechanical ellipticity f, characterizing the differences of the Moon's principal moments of inertia and of the mean inclination I of the Moon's equatorial plane to the ecliptic – obtained by various investigators for the past few years, are not only discordant among themselves within the limits of their mean errors, but do not even show any tendency of converging to some definitive values. Thus the values of f obtained in recent years ranged from $f=0.50$ (Habibulin, 1958) to $f=0.89$ (Gorynya, 1965). Similarly, the values of I ranged from $I=1°30'54''$ (Shakirov, 1963) to $I=1°33'50''$ (Watts, 1955). If we compare these values of the fundamental physical libration constants with those obtained for these constants for the past 120 years, commencing with the first serious discussion of this topic by Wichmann (1847, 1848), we might come to a discouraging conclusion that the endeavours to solve the problem of the Moon's physical libration have ended in a deadlock and did not advance for 120 years.

The present discussion will show that the problem of determining the Moon's physical libration constants is not as hopeless as it might seem in the light of some papers appearing in this field. On the basis of 3282 observations carried out with a heliometer in 340 evenings, covering the period from 1877 to 1915, a system of constants of the Moon's physical libration, including the coordinates of crater Mösting A, will be determined simultaneously with other constants of the Moon's free libration in longitude in a homogeneous way – which constitutes a distinct step forward in the solution of the problem of the Moon's physical libration. The reductions of the observations just mentioned have been carried out on the basis of a new method of adjustment of heliometric libration observations, and a strict proof of the uniqueness of the solution for f will be given. In conclusion, we shall try to throw some light upon, and discuss critically, the existing discrepancies in the results obtained by various authors in their discussions of the problem of the Moon's physical libration. Finally, from the above discussion we shall also draw conclusions concerning the Moon's principal moments of inertia.

* Department of Theoretical Astronomy and Astronomical Geophysics, Jagellonian University, Cracow, Poland.

In heliometric measurements – which constitute as yet the vast majority of the libration measurements of the Moon – it has been customary to refer the crater Mösting A, situated near the centre of the Moon's disc visible from the Earth, to the Moon's illuminated limb. This is done by measuring the distance s_0 (s observatum) of this crater from the limb in specified position angles p, in order to determine the position of Mösting A as referred to the centre of mass of the lunar body.

From Bessel's time (1839) until recently, the adjustment of the observations of the Moon's libration was performed in two stages: first, the auxiliary unknowns of the problem (i.e., the corrections to the rectangular plane coordinates of crater Mösting A) were found, and only the second adjustment gave the proper unknowns of the problem – i.e., the corrections to the selenocentric coordinates λ, β as well as to the Moon's radius h to the crater Mösting A, and also the corrections to the mean inclination of the lunar equator to the ecliptic I and to the mechanical ellipticity of the Moon

$$f = \frac{B(C-B)}{A(C-A)},$$

where A, B and C denote the principal moments of inertia. From the very beginning this manner of treatment created great difficulties in the choice of weights of the right-hand sides of the observation equations in the second adjustment. These right-hand sides are not independent, as they were obtained from the same equations in the first stage. But the adjustment of such equations in the second stage is incorrect from the point of view of the least-squares method, and no choice of weights can secure here a correct solution. The reason for such a state of affairs was the difficulty in carrying out the very extensive calculations.

In his paper published in the years 1948/49 the writer (Kozieł, 1949c) gave a method for adjustment of heliometric libration series comprising the observation equations of the problem, in which, according to the suggestion of T. Banachiewicz, the so-called triangular cracovian square-roots were used. Such square-roots not only require rather tedious computations – especially when electronic computers are applied – but their use restricts, moreover, the application of the observation equations of the problem to the case of only such observation evenings in which at least three measurements of the crater's distance to the limb were made.

This state of affairs has been changed after the writer developed a new method of adjustment of the heliometric libration observations of the Moon, an outline of which has first been published in 1956 (Kozieł, 1956). Given below is the correct mathematical form of our final observation equations (Kozieł, 1962):

$$\tau \cdot \{\mathrm{d}\xi \;\mathrm{d}\eta \;\mathrm{d}\zeta \;\mathrm{d}I \;\mathrm{d}f_1 \;\mathrm{d}f_2 \;U \;W \;\mathrm{d}R_0\}\,.$$

$$\times \sec\pi'' \begin{Bmatrix} -\sin(p-\pi'') \\ -\cos(p-\pi'') \\ 1 \end{Bmatrix} \begin{Bmatrix} A & B & C & D & E_1 & E_2 & F & G & 0 \\ A' & B' & C' & D' & E_1' & E_2' & F' & G' & 0 \\ 0 & 0 & 0 & 0 & 0 & 0 & 0 & 0 & \dfrac{R_{☾}'}{R_0} \end{Bmatrix} = \{s_0 - s_c\}\,. \qquad (1)$$

The proper unknowns of the problem occurring in Equations (1) are, as has been said above: the corrections to the rectangular coordinates of crater Mösting A: $d\xi$, $d\eta$, $d\zeta$, the correction dI, the correction df decomposed into two unknowns df_1 and df_2 on account of the peculiar point of the physical libration in longitude, then the unknowns are $U = A \sin a$ and $W = A \cos a$, where A denotes the amplitude of the free libration in longitude; a, the phase of this libration; and, finally, the correction to the Moon's mean radius dR_0. The coefficients of Equations (1) contain, moreover, the position angles p of crater Mösting A, and the corresponding angles π'', between the directions from the limb to the disc's centre and to the crater.

The differential coefficients occurring in these equations can be found from our formulae, which were obtained on the basis of our differential formulae of spherical polygonometry (KOZIEŁ, 1949a). The manner of treating the adjustment of the libration observations having on their left-hand sides the proper unknowns of the problem and, on the right-hand sides, the independent quantities $(s_0 - s_c)$, permits us to carry out the adjustment in accordance with the principles of the least-squares method. Moreover, our form of the observation equations enables use to be made of even the so-called quite incomplete observation evenings – i.e., such evenings as, in the extreme case, consist of only one measurement of the illuminated limb of the Moon's disc. A further advantage of observation equations of our type rests on the fact that they allow an introduction, on their right-hand sides, of the corrections to s_0 for the irregularities of the Moon's limb – or any other corrections – and then a parallel adjustment may be carried out without any change on the left-hand sides.

It is interesting that the strictest mathematical form of the observation equations which were given above is also the simplest for programming on an electronic computer.

The method of adjustment of libration observations just described has been applied to four heliometric series: the Strasbourg series (1877–1879) (HARTWIG, 1880), Dorpat series (1884–1885) (KOZIEŁ, 1948a, b, c; 1949b, c, d), the first half of Bamberg series (1890–1912) (NAUMANN, 1939), all of them carried out by E. Hartwig, and the Kazan series (1910–1915) (YAKOVKIN, 1928) by T. Banachiewicz. The reductions of these series have been undertaken by the Department of Theoretical Astronomy and Astronomical Geophysics of the Jagellonian University in Cracow. The comparison of the observations with theory for the Strasbourg and Dorpat series was carried out in Cracow by K. Kozieł, for the first half of Bamberg series by J. Masłowski, and for the Kazan series by J. Mietelski. The four series comprise a total of 3328 individual measurements of the distances s_0 of crater Mösting A from the Moon's illuminated limb performed in 340 evenings. From all this observational material only 46 such individual measurements were rejected, as the corresponding differences $s_0 - s_c$ were more than three times as large as the mean error of one observation, the corrections to the irregularities of the Moon's limb being taken onto account with the aid of Hayn's charts. However, in no case did a whole evening have to be rejected, since, using the type of observation Equations under consideration (1), it is possible to utilise every observation. Thus the final adjustment comprised 3282 observations of the crater Mösting A, made in 340 evenings.

The vast computations of the differential coefficients of observational equations, and of the coefficients of normal equations for the adjustment of the four series of heliometric observations referred to above, were carried out according to K. KOZIEŁ's (1962) formulae by means of the electronic computer "Mercury" at the University of Manchester, which was made possible owing to the kind invitation by Professor Kopal, of the writer of this paper, together with his assistants J. Masłowski and J. Mietelski. The solution of different versions of the normal equations of the problem was carried out later in Cracow.

Preliminary determinations of the unknowns from normal equations for the individual heliometric series have already shown that the unknown f admits of two solutions situated on either side, and not far from, the singular point 0.662. Such being the case, the determination of the unknowns was performed by the use of SCHRUTKA-RECHTENSTAMM's (1955) artifice (*Kunstgriff*), consisting in taking two unknowns df_1 and df_2 instead of the unknown df, and in combining them in such a way as to avoid the nonlinearity of the problem in the vicinity of the singular point.

But the principal concern was the problem of combining the system of equations belonging to different heliometric series for a simultaneous solution. I solved the problem by eliminating the correction to the Moon's mean radius for each individual series before combining all series – i.e., I determined first the numerical value of dR_0 and after substituting it into the corresponding system of normal equations I transferred the value thus obtained to the free terms. Then I combined the system of normal equations belonging to our four series according to their corresponding mean errors of unit weight. The correction to the Moon's mean radius was no longer included among the unknowns, since it is bound to be affected to some extent by irradiation; moreover, this correction depends in some way on the instrument and the observer.

Since 120 years – and especially in the last 60 years – the astronomers concerned with the problem of the Moon's physical libration discussed the question, whether the fundamental unknown of the problem – the Moon's mechanical ellipticity f – is greater or smaller than the value of the singular point. At first, most authors kept obtaining for f values smaller than the critical value; but since the days of Hayn an opinion begun to gain ground that f is greater than 0.662. In 1944 the present author (KOZIEŁ, 1949) has shown that, mathematically, the problem admits two solutions for f – one of them smaller and the other greater than the critical value. But the Dorpat series of Hartwig (36 heliometric evenings) from which I detected this phenomenon (KOZIEŁ, 1948a, b, c; 1949b, c, d), was too limited in scope to decide which of the two values for f constituted the proper solution. SCHRUTKA-RECHTENSTAMM (1955) who reduced a large observational material consisting of 680 heliometric evenings, did not succeed in settling this question; he only stated that the value of f smaller than 0.662 was more probable than the larger one. These difficulties resulted, of course, mainly from the fact that Schrutka applied to his reductions the old theory of adjustment of the libration observations, a theory which from the point of view of the least-squares method was not quite correct. Moreover, he used reductions of the observations made in a not very precise manner by various authors, who reduced the individual libration series.

A new simultaneous solution of the problem has been performed in two versions: without taking into account the free libration in longitude (the amplitude of which is certainly very small) and with account taken of this libration. In both cases, an exact proof has been given that of the two values for f obtained in each solution, the proper solution corresponds to $f<0.662$. The way in which this has been proved I shall illustrate for the case without free libration: after solving the normal equations for five unknowns, which represent jointly our four heliometric series, I obtained two values for f together with the corresponding sums of squares of the residuals

$$\left.\begin{aligned} f &= 0.633, \ [v\,v]_{\min} = 1211.8397\,, \\ f &= 0.720, \ [v\,v]_{\min} = 1238.6667\,. \end{aligned}\right\} \tag{2}$$

The mean error of the unit of weight

$$\mu_0 = \pm \sqrt{\frac{1211.8397}{3282-5}} = \pm\, 0.6081\,, \tag{3}$$

thus, according to the least-squares theory, the expected mean deviation of the sum of squares of the residuals from its minimum

$$\mu^2 = 5\mu_0^2 = 1.849\,. \tag{4}$$

But, according to (4), the difference of $[vv]_{\min}$ occurring in (2) amounts to

$$26.8270 = 14.51\mu^2$$

favouring the smaller sum corresponding to $f=0.633$. Both theory and practice of the least-squares method show that, between the two values for f the one to which corresponds a smaller $[vv]_{\min}$ must be chosen when the difference of the two $[vv]$'s exceeds the value $(3\mu)^2=9\mu^2$. Since, in our case, this disparity is much greater, as the coefficient of μ^2 does not amount to only 9, but 14.5. Therefore, we have proved that the unique solution of the problem is given by

$$f = 0.633\,.$$

Below we give the full set of solutions for the problem of determination of the Moon's physical libration constants, based on the four heliometric series discussed (1877–1915), obtained without taking into account the free libration in longitude (Kozieł, 1964b):

$$\left.\begin{aligned} \lambda &= \quad -5^\circ\ 9'53'' \pm 4''.5\ \text{(m.e.)}, \\ \beta &= \quad -3^\circ 10'49'' \pm 4''.3\,, \\ h &= 932''.85 \qquad \pm 0''.18_5\,, \\ I &= \qquad 1^\circ 32'\ 4'' \pm 7''.0\,, \\ f &= 0.633 \qquad \pm 0.011_3\,; \end{aligned}\right\} \tag{5}$$

λ and β denote the selenocentric longitude and latitude of the crater Mösting A and h its radius vector.

Taking into account for the first time the constants of forced libration together with those of free libration in longitude A, a, simultaneously, I obtained (KOZIEŁ, 1964a, b):

$$\left.\begin{aligned} \lambda &= \quad -5^\circ\ 9'50'' \pm 4''.5 \text{ (m.e.)} \\ \beta &= \quad -3^\circ 10'47'' \pm 4''.4 \\ h &= 932''.98 \qquad \pm 0''.19_1 \\ I &= \qquad 1^\circ 32'\ 1'' \pm 7''.1 \\ f &= 0.633 \qquad \pm 0.011_5 \\ A &= \ 18''.7 \qquad \pm 4''.7 \\ a &= 334^\circ.3 \qquad \pm 15^\circ.7 \text{ (1800.0)} \end{aligned}\right\} \tag{6}$$

The solutions (5) and (6) do not merely represent additional sets of quantities, differing from other similar sets by more than is admissible considering their mean errors. On comparing our set of solutions with those of SCHRUTKA (1955), we find this time an agreement within the limits of the respective mean errors. It should, moreover, be pointed out here that although Schrutka used twice as large observational material, comprising 680 heliometric evenings (1890–1942), yet he obtained larger mean errors of the unknowns than the writer, for reasons already given above.

But now we have, moreover, an independent confirmation of our results. W. J. ECKERT (1965) in his paper 'On the Motions of the Perigee and Node and the Distribution of Mass in the Moon' found on the basis of a discussion of the Moon's orbital motion a value of

$$f = 0.638\,. \tag{7}$$

At the time, Eckert stated that f should lie within the interval (0.60, 0.65). Eckert's result (7), obtained in a way having nothing in common with heliometric observations and with the theory of the Moon's rotation, is an excellent confirmation of our results (5) and (6).

In the light of this comparison, we can state that, at present, there is no more any doubt about the proper value of f as regards its first or even second decimal place; and only the thousandths differ among the determinations quoted above well within the limits of its mean error, which amounts here to ± 0.01.

Thus a determination of f such as, e.g., by GORYNIA (1965), who obtained $f=0.89$ must, in the light of our discussion, be regarded as a failure. It merely remains to find the reasons for such discrepancies between the results of various authors. In the first place attention should be given to the precision of the reduction of observations. As regards heliometric measurements, great care is necessary in considering all kinds of even slight corrections which, at first sight, may seem negligible; and the least-squares method must be applied strictly in accordance with its theoretical foundations.

Besides, many observational "methods" aiming at the determination of libration constants have appeared recently which, for various reasons, abandoned the fundamental ideas of the classical method of BESSEL (1839). Among these belong such ideas as measurements of the position angles of the directions from crater Mösting A to craters situated at the Moon's limb (YAKOVKIN, 1961; GORYNYA, 1965), referring the crater immediately to stars (SHAKIROV, 1963) and other methods. The main disadvantage of the classical method of Bessel is its sensitivity to the irregularities of the lunar limb. But, nowadays these can be eliminated, since we have at our disposal such excellent charts of the Moon's limb as HAYN's (1914), WEIMER's (1952), or WATT's (1963). But in the method of Bessel not only the fundamental unknowns of the libration problem (i.e., f and I) but also the position of crater Mösting A are regarded as unknowns, while some of these new observational methods introduce coordinates of sometimes many craters, as "known from somewhere". Evidently, these coordinates are affected with errors which, in turn, affect the determination of f and of other libration constants. Besides, nowadays it is evident that f cannot be determined from a few dozen, or even from a few hundreds, of observations; thousands of observations are necessary for its determination. And all these new methods are being tested with small numbers of observations. Thus the determination of libration constants from short observation series on the basis of untried methods and, moreover, with introducing into the computation coordinates of many lunar craters as known quantities, must be regarded as a misunderstanding of the true nature of our problem.

In the literature of the subject the constants of the free libration in longitude, i.e., the amplitude A and the phase a were heretofore determined first of all from deviations obtained after adjustment in the libration problem. FREEDLAND (1959), it is true, endeavoured to determine the constants of the free libration in longitude together with other constants; but he did not obtain any definite result. Our determination (6) shows for the first time that, although the amplitude of the free libration in longitude $A = 18''.7$ is quite small – as seen from the Earth, it amounts to less than $0''.1$ – nevertheless it is realistic, because it exceeds three times the value of its mean error. At any rate, the value determined by YAKOVKIN (1952), $A = 55''$ is by no means acceptable.

The results obtained above concerning the determination of f will still be used in our determination of the Moon's principal moments of inertia. When integrating the equations for the Moon's rotation we obtain an identity relating the mechanical ellipticity f with the inclination I and with the Moon's principal moments of inertia A, B, C. Denoting

$$\beta = \frac{C - A}{B} \tag{8}$$

and taking for I the value from (5), we obtain this identity in the form

$$\beta = 0.0006314 - 0.00528 f\beta + 1.00744 f\beta^2 , \tag{9}$$

or taking on the right-hand side in the mean for $\beta = 0.000630$, we get with sufficient accuracy

$$\beta = 0.0006314 - 0.00000293 f . \tag{10}$$

On introducing further the quantities

$$\alpha = \frac{C - B}{A}, \quad \gamma = \frac{B - A}{C} \tag{11}$$

and taking into account that

$$\alpha = \beta f, \quad \gamma = \beta(1 - f), \tag{12}$$

we obtain – on the basis of the values

$$f = 0.633 \quad \text{and} \quad I = 1°32'4''$$

resulting from (5) – the moments of inertia

$$\left.\begin{aligned} \alpha &= \frac{C - B}{A} = 0.0003984, \\ \beta &= \frac{C - A}{B} = 0.0006294, \\ \gamma &= \frac{B - A}{C} = 0.0002310, \end{aligned}\right\} \tag{13}$$

leading to

$$\left.\begin{aligned} B &= 1.0002311\,A, \\ C &= 1.0006295\,A. \end{aligned}\right\} \tag{14}$$

Acknowledgements

My thanks are due to professor Z. Kopal of Manchester University for making it possible to me to carry out the vast adjustment computations on the electronic computer "Mercury" of the Manchester University. I also wish to acknowledge the financial support granted by the Astronomical Institute of the Polish Academy of Sciences for carrying out in Cracow the other extensive computations of our problem.

References

Bessel, F. W.: 1839, 'Über die Bestimmung der Libration des Mondes', *Astron. Nachr.* **16,** 257.

Eckert, W. J.: 1965, 'On the Motions of the Perigee and Node and the Distribution of Mass in the Moon', *Astron. J.* **70,** 787.

Freedland, M. V.: 1959, 'On the Determination of the Moon's Free Libration in Longitude', *Bull. Inst. Theor. Astr.* (Leningrad) **7,** 293.

Gorynya, A. A.: 1965, *Opred. postoj. fisicz. libr.* Nauk. dumka (ser. Astron. i Astrof.), Kiev.

Habibulin, Sh. T.: 1958, 'Fisicz. libr. Luny 9', *Izv. Obs. Engelhardt* **31,** 1.

Hartwig, E.: 1880, *Beitrag zur Bestimmung der physischen Libration des Mondes.* Karlsruhe.

Hayn, F.: 1914, 'Selenographische Koordinaten IV', *Abhandl. Sächs. Ges. Wiss.* **33,** 1.

Kozieł, K.: 1948a, 'The Moon's Libration and Figure, I', *Acta Astr.* (Cracow), a, **4,** 61; 1948b, II, *ibid.* **4,** 83; 1948c, III, *ibid.* **4,** 99.

Kozieł, K.: 1949a, 'The Differential Formulae of Spherical Polygonometry and their Application to Astronomy', *Bull. Acad. Polon. Sci.* Ser. A, **1.**

Kozieł, K.: 1949b, 'The Moon's Libration and Figure, IV', *Acta Astr.* (Cracow), a, **4,** 141; 1949c, V, *ibid.* **4,** 153; 1949d, VI, *ibid.* **4,** 179.

Kozieł, K.: 1956, 'O krakowskim opracowaniu sztrasburskiego szeregu księżycowych obserwacji heliometrycznych Hartwiga', *Postępy Astronomii* (Cracow) **4,** 78.

Kozieł, K.: 1962, 'Libration of the Moon', in *Physics and Astronomy of the Moon* (ed. Z. Kopal). Academic Press, New York and London, pp. 27–59.

Kozieł, K.: 1964a, 'Crater Mösting A as the First-Order Point of Triangulation on the Moon', *Life Sciences and Space Research* II (COSPAR), p. 141.

Kozieł, K.: 1964b, 'Reports on Astronomy', *Trans. I.A.U.* **12A,** 223.
Naumann, H.: 1939, 'Selenographische Koordinaten V', *Abh. Sächs. Akad. Wiss.* **43,** 1.
Schrutka-Rechtenstamm, G.: 1955, 'Zur physischen Libration des Mondes', *S.B. Öst. Akad. Wiss.* **164,** 323.
Shakirov, K. S.: 1964, *Trans. I.A.U.* **12A,** 219.
Watts, C. B.: 1955, 'A new Method of Measuring the Inclination of the Moon's Equator', *Astron. J.* **60,** 443.
Watts, C. B.: 1963, 'The Marginal Zone of the Moon', *Astr. Papers Amer. Eph. Naut. Alm.* **17,** 1.
Weimer, Th.: 1952, *Atlas de profils lunaires.* Obs. de Paris.
Wichmann, M.: 1847, 'Erster Versuch zur Bestimmung der physischen Libration des Mondes', *Astr. Nachr.* **26,** 289; 1848, *ibid.* **27,** 53, 81, 97.
Yakovkin, A. A.: 1928, 'Postajan. fizicz. libr. Luny', *Publ. Obs. Engelhardt* **13,** 1.
Yakovkin, A. A.: 1952, *Trans. I.A.U.* **8,** 231; 1961, *ibid.* **11A,** 161.

CASSINI'S SECOND AND THIRD LAWS

G. COLOMBO*

Dedicated to the Memory of Imre Iszak

1. Introduction

In 1693 G. D. Cassini published the three following empirical laws on the Moon's rotational motion:

(1) The Moon rotates uniformly about its polar axis with a rotational period equal to the mean sidereal period of its orbit about the earth.

(2) The inclination of the Moon's equator to the ecliptic is a constant angle approximately $1^\circ.5$.

(3) The ascending node of the lunar orbit on the ecliptic coincides with the descending node of the lunar equator on the ecliptic.

These laws describe fairly well the main rotational motion of the Moon, as well as the behavior of some of the lunar orbital elements, in particular the effect of solar perturbations on the motion of the nodal line.

Physical libration of the Moon means the departure of the Moon's rotational motion from that defined by CASSINI's (1693) three laws.

In the literature on the subject, Cassini's laws are taken as a reference motion, and the perturbations due to the gravitational torque acting on a rigid body are studied. Naturally the parameters (principal moments of inertia) are adjusted in such a way that the motion can be a dynamical solution, after which the small oscillations about this solution due to the different torques are investigated, taking advantage of the fact that the equation can be simplified if one assumes that the actual motion of the Moon is close to that described by Cassini's laws. To my knowledge no specific investigation has been made of all other possible motions of the Moon permitted by the actual gravitational dynamics, in order to understand the properties of the actual motion.

I will be concerned only with Cassini's second and third laws, since the first one seems to be quite clear and easily understandable if we assume the inertial ellipsoid of the Moon to be triaxial. With this in mind, I have studied the motion for a model representing the real case to an extent that I judged sufficient for the main purpose of understanding the second and third laws. I will show that the second and third laws are independent of the first one, at least qualitatively, in the sense that even if the Moon's inertial ellipsoid were rotationally symmetric, the second and third laws

* Smithsonian Astrophysical Observatory, Cambridge, Mass., and University of Padua, Italy.

could still be satisfied (provided the rotational angular velocity of the Moon is the same) since they represent a motion corresponding to the minimum dissipation of energy by internal friction. For a different rotational angular velocity, only the inclination of the Moon's equator on the ecliptic would change.

This very simple theory applies to any satellite or planet whose nodal line on the invariable plane shifts because of perturbations that may be of different origin (e.g., oblateness of the primary body, a perturbing third body, or both). We will treat the case of Mercury, the case of an aritficial Earth satellite (gravitational torque only), and finally the case of Iapetus since G. D. Cassini was the discoverer of this, for many reasons peculiar, satellite of Saturn.

2. Equations of Motion

We study first the general problem of the motion of the rotational axis of a rigid satellite due to a gravitational torque. We deduce the equation of motion in a closed, very simple form since in our model a reduction to one degree of freedom is possible and also a first integral is easily found. The motion may also be represented in a very simple geometrical form.

Let Σ be a satellite of the primary E; G the center of mass of Σ; $\mathbf{i}, \mathbf{j}, \mathbf{k}$ the unit vectors along the principal axes of the inertia ellipsoid of Σ; and A, B, C the principal moments of inertia ($A<B<C$). A permanent rotation of Σ about C is a stable motion, and internal dissipation minimizes any possible wobbling of the spin axis in relation to the external torque. We assume that only the gravitational torque is present and that the body is rigid. Later on we consider the effects of internal dissipation. The gravitational torque acting on the satellite may be written

$$\mathbf{T} = \frac{3\mu}{r^3}\{(C-B)(\mathbf{r}\cdot\mathbf{j})(\mathbf{r}\cdot\mathbf{k})\,\mathbf{i} + (A-C)(\mathbf{r}\cdot\mathbf{k})(\mathbf{r}\cdot\mathbf{i})\,\mathbf{j} + (B-A)(\mathbf{r}\cdot\mathbf{j})(\mathbf{r}\cdot\mathbf{i})\,\mathbf{k}\}\,. \tag{1}$$

Here μ is the product of the gravitational constant and the mass of E, $\mathbf{r}$ is the unit vector from G to E, and r is the distance GE.

(a) Suppose $A=B$; then the torque reduces to

$$\mathbf{T} = \frac{3\mu}{r^3}(C-A)(\mathbf{r}\cdot\mathbf{k})(\mathbf{r}\times\mathbf{k})\,. \tag{2}$$

(b) Suppose the rotational angular velocity ω_3 is much larger than the orbital angular velocity. We may then consider $\mathbf{r}$ and $\mathbf{k}$ constant during one rotational period; and averaging we find

$$\mathbf{T} = \frac{3\mu}{r^3}\left(C - \frac{A+B}{2}\right)(\mathbf{r}\cdot\mathbf{k})(\mathbf{r}\times\mathbf{k})\,. \tag{3}$$

(c) Suppose the body is moving in such a way that the axis of minimum moment of inertia is nearly pointing to E (stable relative equilibrium position); and suppose that B–A is small compared to C–A.* Thus, neglecting second-order terms, we may write

$$\mathbf{T} = \frac{3\mu}{r^3}(C - A)(\mathbf{r}\cdot\mathbf{k})(\mathbf{r}\times\mathbf{k}). \tag{4}$$

The perturbing torque, being generally small, may be averaged over an orbital period.

Let us call $\boldsymbol{\Omega}$ the unit vector along the ascending node of the orbit of E with respect to G; $\mathbf{n}$ the unit vector normal to the orbital plane; $\mathbf{m}=\mathbf{n}\times\boldsymbol{\Omega}$; and $\mathbf{c}_1$, $\mathbf{c}_2$, $\mathbf{N}$ an inertial reference frame centered at G. We have

$$\left.\begin{aligned} \mathbf{r} &= \cos(\omega + v)\,\boldsymbol{\Omega} + \sin(\omega + v)\,\mathbf{m} \\ r &= \frac{a(1 - e^2)}{1 + e\cos v} \end{aligned}\right\}, \tag{5}$$

where a, e, ω, and v are the semi-major axis, eccentricity, argument of perigee, and true anomaly, respectively, of the orbit of E with respect to G. Now let us consider the dynamical equation (with $\mathbf{K}$ the angular momentum vector):

$$\left.\begin{aligned} \frac{\mathrm{d}\mathbf{K}}{\mathrm{d}t} &= \frac{3\mu}{r^3}(C - A)(\mathbf{r}\cdot\mathbf{k})(\mathbf{r}\times\mathbf{k}) \\ \frac{\Delta\mathbf{K}}{P} &= \frac{1}{P}\int_0^P \frac{3\mu}{r^3}(C - A)(\mathbf{r}\cdot\mathbf{k})(\mathbf{r}\times\mathbf{k})\,\mathrm{d}t \end{aligned}\right\}, \tag{6}$$

where P is the orbital period of E about G. We may also write

$$\frac{\Delta\mathbf{K}}{P} = \frac{1}{P}\int_0^{2\pi} \frac{3\mu}{r^3}(C - A)(\mathbf{r}\cdot\mathbf{k})(\mathbf{r}\times\mathbf{k})\frac{\mathrm{d}v}{\dot{v}}. \tag{7}$$

Since

$$r^2\dot{v} = h = \frac{2\pi a^2(1 - e^2)^{1/2}}{P},$$

we have

$$\frac{1}{\dot{v}} = \frac{Pr^2}{2\pi a^2(1 - e^2)^{1/2}},$$

and consequently

$$\frac{\Delta\mathbf{K}}{P} = \frac{1}{2\pi a^2(1 - e^2)^{1/2}}\int_0^{2\pi} \frac{3\mu}{r}(C - A)(\mathbf{r}\cdot\mathbf{k})(\mathbf{r}\times\mathbf{k})\,\mathrm{d}v. \tag{8}$$

Let us now introduce the unit vector $\mathbf{a}$ from G to the pericenter and define the unit vector $\mathbf{b}$ as $\mathbf{b}=\mathbf{n}\times\mathbf{a}$. We have

$$\mathbf{r} = \mathbf{a}\cos v + \mathbf{b}\sin v,$$

* Naturally in case (c) this expression of the average torque holds only when the motion of the body is such that $\mathbf{r}$ is close to $\mathbf{i}$. This means that the validity of the considerations which follow is restricted in this case to a particular class of motions.

and

$$\frac{\Delta \mathbf{K}}{P} = \frac{3\mu(C-A)}{2\pi a^3(1-e^2)^{3/2}} \int_0^{2\pi} (1+e\cos v)(k_a \cos v + k_b \sin v) \times$$

$$\times [\mathbf{a} \sin v k_n - \mathbf{b} \cos v k_n + \mathbf{n}(k_b \cos v - k_a \sin v)]\, \mathrm{d}v$$

$$= \frac{3\mu(C-A)}{2\pi a^3(1-e^2)^{3/2}} 2\pi \frac{k_n}{2}(k_b \mathbf{a} - k_a \mathbf{b}) \tag{8'}$$

$$= \frac{3}{2}\frac{\mu(C-A)}{a^3(1-e^2)^{3/2}}(\mathbf{k}\cdot\mathbf{n})(\mathbf{k}\times\mathbf{n}) = \frac{3}{2}\frac{n^2}{(1-e^2)^{3/2}}(C-A)(\mathbf{k}\cdot\mathbf{n})(\mathbf{k}\times\mathbf{n}),$$

where n is the mean motion of E.

Now let us write the angular momentum vector

$$\mathbf{K} = A\omega_1 \mathbf{i} + B\omega_2 \mathbf{j} + C\omega_3 \mathbf{k}.$$

As is usual in the treatment of gyrophenomena, we shall now assume as a first approximation that

$$\mathbf{K} = C\omega_3 \mathbf{k}. \tag{9}$$

For the Moon $A \cong B \cong C$, and we may write

$$\mathbf{K} = (A-C)\,\omega_1 \mathbf{i} + (B-C)\,\omega_2 \mathbf{j} + C(\omega_1 \mathbf{i} + \omega_2 \mathbf{j} + \omega_3 \mathbf{k}).$$

Neglecting small terms of the order of 3×10^{-6} of the main term leaves

$$\mathbf{K} = C\omega \mathbf{k}_1.$$

By writing the equation of motion in the form

$$\frac{\mathrm{d}}{\mathrm{d}t}(C\omega \mathbf{k}_1) = \frac{3}{2}\frac{n^2(C-A)}{(1-e^2)^{3/2}}(\mathbf{k}_1\cdot\mathbf{n})(\mathbf{k}_1\times\mathbf{n}), \tag{10}$$

we neglect only very small terms. From now on we shall call $\mathbf{k}_1 = \mathbf{k}$ and go back to the fundamental equation

$$C\omega \frac{\mathrm{d}\mathbf{k}}{\mathrm{d}t} = \frac{3\, n^2(C-A)}{2\,(1-e^2)^{3/2}}(\mathbf{k}\cdot\mathbf{n})(\mathbf{k}\times\mathbf{n}). \tag{11}$$

The derivative $\mathrm{d}\mathbf{k}/\mathrm{d}t$ is naturally evaluated with respect to an invariable system.

It is much simpler to study the motion of $\mathbf{k}$ with respect to a moving reference system, that is, the rotating system $\mathbf{N}, \mathbf{\Omega}, \mathbf{N}\times\mathbf{\Omega} = \mathbf{M}$. Let us call $\dot{\Omega}\mathbf{N}$ the angular velocity of this system around $\mathbf{N}$. The sign of $\dot{\Omega}$ is positive if the nodal line is advancing and negative if (as for a low-inclination Earth satellite* or for the Moon) the nodal line is regressing.

* For an Earth satellite we have $\Delta\Omega/\text{rev} = -3\pi \frac{J^2}{p^2} \cos i$, where $p = a\,(1-e^2)$, $J_2 = 1.083 \times 10^{-3}\, R_0$, and $R_0 \approx 6{,}378$ km.

The preceding equation may be rewritten as

$$C\omega\left(\frac{\mathrm{d}\mathbf{k}}{\mathrm{d}t} + \dot{\Omega}\mathbf{N} \times \mathbf{k}\right) = \frac{3}{2}\frac{n^2 (C - A)}{(1 - e^2)^{3/2}}(\mathbf{k}\cdot\mathbf{n})(\mathbf{k} \times \mathbf{n}).$$

Dividing by $C\omega$ we may write

$$\frac{\mathrm{d}\mathbf{k}}{\mathrm{d}t} = -\dot{\Omega}\mathbf{N} \times \mathbf{k} + \alpha(\mathbf{k}\cdot\mathbf{n})(\mathbf{k} \times \mathbf{n}), \tag{12}$$

where

$$\alpha = \frac{3}{2}\frac{n^2}{(1 - e^2)^{3/2}}\frac{C - A}{C\omega}.$$

Let us now introduce the inclination i of the orbit. We have

$$\mathbf{n} = -\sin i\,\mathbf{M} + \cos i\,\mathbf{N}$$
$$\mathbf{k} = k_x\boldsymbol{\Omega} + k_y\mathbf{M} + k_z\mathbf{N}$$

and the following three equations:

$$\left.\begin{aligned}
\frac{\mathrm{d}k_x}{\mathrm{d}t} &= \dot{\Omega}k_y + \alpha k_y k_z \cos 2i + \frac{\alpha}{2}(k_z^2 - k_y^2)\sin 2i \\
\frac{\mathrm{d}k_y}{\mathrm{d}t} &= -\dot{\Omega}k_x - \alpha k_x k_z \cos^2 i + \frac{\alpha}{2}k_x k_y \sin 2i \\
\frac{\mathrm{d}k_z}{\mathrm{d}t} &= -\frac{\alpha}{2}k_x k_z \sin 2i + \alpha k_x k_y \sin^2 i
\end{aligned}\right\}. \tag{13}$$

The path of the vertex of $\mathbf{k}$ on the unit sphere

$$k_x^2 + k_y^2 + k_z^2 = 1$$

is determined by the quadratic integral

$$\alpha\, k_z^2 \cos^2 i + \alpha\, k_y^2 \sin^2 i - 2\alpha\, k_y k_z \sin i \cos i + 2\dot{\Omega}k_z = H. \tag{14}$$

3. Relative Equilibrium Configuration of the k Axis

Let us first consider the relative equilibrium configuration. It is easily shown that the only equilibrium configurations are in the **M**, **N** plane. That is

$$\left.\begin{aligned}
&k_x^\circ = 0, \\
&\dot{\Omega}k_y^\circ + \alpha k_y^\circ k_z^\circ \cos 2i + \frac{\alpha}{2}(k_z^{\circ 2} - k_y^{\circ 2})\sin 2i = 0
\end{aligned}\right\} \tag{15}$$

Let us put

$$k_y^\circ = \sin\lambda, \qquad k_z^\circ = \cos\lambda. \tag{16}$$

We have

$$2\dot{\Omega}\sin\lambda + \alpha\sin 2(i + \lambda) = 0. \tag{17}$$

This relative equilibrium configuration for $\mathbf{k}$, naturally corresponds to a regular precessional motion of the body, the precessional angle being λ, and the angular velocity of precession being $\dot{\Omega}$. The two Poinsot cones (the body-fixed and the space-fixed) are easily computed.

4. Precessional and Nutational Motion

Equation (14) may be written in the following form

$$\alpha(k_y \sin i - k_z \cos i)^2 + 2\dot{\Omega}k_z = H. \tag{18}$$

It represents a family of parabolas in the **M–N** plane determined by the value of H. The axis of these parabolas is independent of H as is the latus rectum. The equation of the axis is

$$k_y \sin i - k_z \cos i - \cos i \frac{\dot{\Omega}}{\alpha} = 0. \tag{19}$$

Its intersection with the z axis is at $y = -\dot{\Omega}/\alpha$; its inclination is i. The length $2p$ of the latus rectum is

$$2p = \left|\frac{2 \sin i\dot{\Omega}}{\alpha}\right|, \tag{20}$$

The convexity of the parabola is determined by the sign of $\dot{\Omega}/\alpha$; if $\dot{\Omega}/\alpha$ is positive, the convexity of the parabola is directed toward negative y; if $\dot{\Omega}/\alpha$ is negative (as for the Moon), the convexity is directed toward positive y. The *relative* motion is in some cases a pure regular precessional motion but, in many cases, it may be very different from regular precessional motion.

5. The Moon

First consider the Moon case. Since A–B is smaller than C–A and C–B, we may treat the problem of the dynamic of the spin axis assuming we are in case (a) of p. 13. Actually B–A is not negligible, but for motions close to the actual one conditions of case (c) of p. 13 may be considered satisfied. Assuming therefore the unit of time to be the sidereal period of the Moon's orbital motion we have

$$\begin{cases} \Omega = -2.525 \times 10^{-2} \text{ rad/rev}, \\ \alpha = 10.188 \dfrac{C-A}{C} \text{ rad/rev}, \\ i = 5^\circ\!.8'43''. \end{cases}$$

From Equation (17), assuming the Moon's axis of rotation to be in the relative equilibrium configuration, we have

$$10.188 \frac{C-A}{C} = -2.525 \times 10^{-2} \frac{\sin \lambda}{\sin 2(i+\lambda)}.$$

From this relation, using the observed values of λ and i, we have for $(C-A)/C$ the usual value 6.3×10^{-4}.

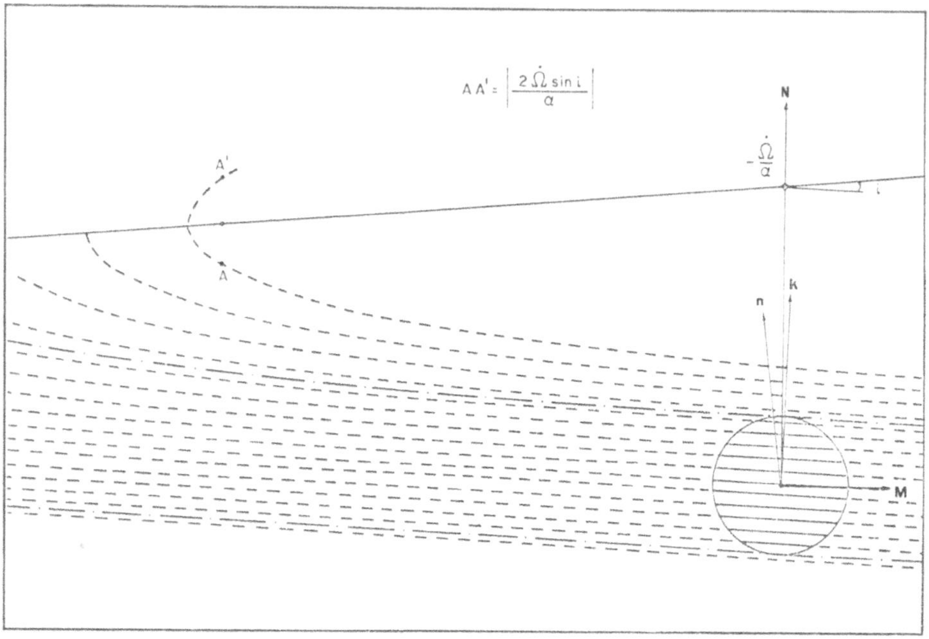

Fig. 1. Paths of the vertex of the unit vector along the spin axis as intersection of parabolic cylinders with the unit sphere (in the case of the Moon).

In Figure 1 we have shown the unit sphere and the projection on the y–z plane of the possible motion of the **k** axis. In this case (**k**, **N** small and also i small) we may very easily compute the angular velocity of the relative precessional motion about this relative equilibrium configuration, that is, the angular velocity of nutation about the absolute precessional motion of the Moon defined by Cassini's second and third laws. Neglecting small quantities, we find for this angular velocity $\dot{\Omega}+\alpha$. The absolute velocity $\mathrm{d}\mathbf{k}/\mathrm{d}t$ will be to a first approximation

$$\frac{\mathrm{d}\mathbf{k}}{\mathrm{d}t} = \dot{\Omega}\mathbf{N} \times \mathbf{k} + \theta_1 \{\cos[\Phi_0 + (\dot{\Omega} + \alpha)\, t]\, \mathbf{c}_1 + \sin[\Phi_0 + (\dot{\Omega} + \alpha)\, t]\, \mathbf{c}_2\},$$

θ_1 and Φ_0 being the amplitude and the phase of the nutation. We may also write

$$\frac{\mathrm{d}\mathbf{k}}{\mathrm{d}t} = [\lambda \cos(\dot{\Omega}t + \Psi_0) + \theta_1 \cos[\Phi_0 + (\dot{\Omega} + \alpha)\, t]]\, \mathbf{c}_1 + $$
$$+ [\lambda \sin(\dot{\Omega}t + \Psi_0) + \theta_1 \sin[\Phi_0 + (\dot{\Omega} + \alpha)\, t]]\, \mathbf{c}_2 .$$

We have consequently

$$\left|\frac{\mathrm{d}\mathbf{k}}{\mathrm{d}t}\right| = [\lambda^2 + \theta_1^2 + 2\lambda\theta_1 \cos(\alpha t + \Psi_0 - \Phi_0)]^{1/2}.$$

For small θ_1, we have the average value of $\mathrm{d}\mathbf{k}/\mathrm{d}t$ roughly equal to λ, but the maximum value is $\lambda+\theta_1$ and the minimum value is $\lambda-\theta_1$.

We may here observe that internal friction produces a dissipation of energy, because both the magnitude of the angular velocity and the instantaneous axis of rotation change. In fact, in a simple way, let us consider two neighboring points P and Q, in the body, and let P' and Q' be the projection of P and Q on the instantaneous axis of rotation about the center of mass. The accelerations of the two points in the motion relative to the center of mass, are

$$-\omega^2 \mathbf{P'P} + \frac{d\boldsymbol{\omega}}{dt} \times \mathbf{OP}; \qquad -\omega^2 \mathbf{Q'Q} + \frac{d\boldsymbol{\omega}}{dt} \times \mathbf{OQ}.$$

When the motion is a uniform rotational motion, the second term in each expression goes to zero and the inertial forces correspond to a static field of forces, and no relative motion occurs. When the angular velocity vector $\boldsymbol{\omega}$ changes, internal forced elastic vibration takes place under the action of internal friction. If the internal friction dissipation is linear with the amplitude of the forced vibration, any regular precession-nutation will dissipate the same amount of energy as in the pure precessional motion; but if the dissipation is an increasing function of the amplitude with a positive second derivative, the energy dissipation will also increase if nutation is present. For any nonrigid body (elastic) any precessional motion will be damped out and a pure rotational motion will finally result. Since here a pure rotational motion is dynamically impossible, the forced precessional motion will correspond to a minimum dissipation of energy by internal friction. We conclude by saying that the second and third laws correspond to a motion that, among all possible dynamical solutions, minimizes the internal dissipation.

6. An Artificial Earth Satellite

We shall consider a 24-hour satellite (case (b) of p. 13) at an inclination of 30°. We suppose the inertial ellipsoid to be rotationally symmetric about the axis of maximum moment of inertia. In addition, we suppose that the magnetic torque is minimized by suitable design. We have

$$\dot{\Omega} = -2.5 \times 10^{-3} \text{ rad/day} \qquad \alpha = 6\pi^2 \frac{C - A}{C\omega} \text{ rad/day}$$

where ω is the spin velocity about the symmetric axis.

Suppose $(C-A)/C = \frac{2}{3}$ and $\omega = 10^4$ rad/day; we then have

$$-\frac{\dot{\Omega}}{\alpha} = 0.7, \left|\frac{2\dot{\Omega}}{\alpha} \sin i\right| = 0.7.$$

The path of the spin axis is represented in Figure 2. We note that there are three stable and one unstable relative equilibrium configurations, as for the case of Mercury. We considered the satellite case in order to give a very clear picture of the behavior of the path of $\mathbf{k}$ on the sphere.

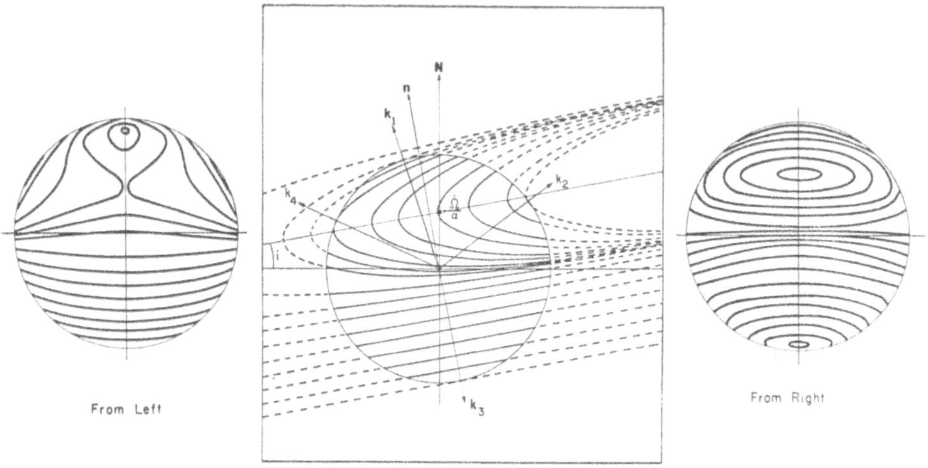

Fig. 2. Projection of the relative paths of the vertex of the unit vector along the spin axis for a particular 24-hour artificial Earth satellite.

7. Mercury

For Mercury, assuming *B–A* negligible with respect to *C–A*, we have (case (a) of p. 13).

$$\left.\begin{aligned} &i \approx 7^\circ \\ &\dot{\Omega} = 6''.2/\text{year} \cong 8.1 \times 10^{-8}\ \text{rad/day} \\ &\alpha = 8.1 \times 10^{-2}\frac{C-A}{C}\text{rad/day} \end{aligned}\right\},$$

and hence the following result:

$\frac{C-A}{C}$	10^{-3}	10^{-4}
$-\frac{\dot{\Omega}}{\alpha}$	-10^{-3}	-10^{-2}
$\frac{2\dot{\Omega}\sin i}{\alpha}$	2.6×10^{-4}	2.6×10^{-3}

The configurations of the possible paths of the rotational axis of Mercury are represented in Figure 3. If Mercury is locked-in as the Moon is, Cassini's laws make it possible to infer the position of the axis of rotation. In this case, three equilibrium configurations are compatible, but for one of them, even if actually stable, the region of stability is very narrow.* Moreover, the argument given above for the Moon needs to be revised from the point of view of the effects of internal dissipation. Remembering that Mercury rotates in a direct fashion, we shall compute only the relative equilibrium

* One of them is close to the invariable plane and recalls the rotational motion of the Uranus system.

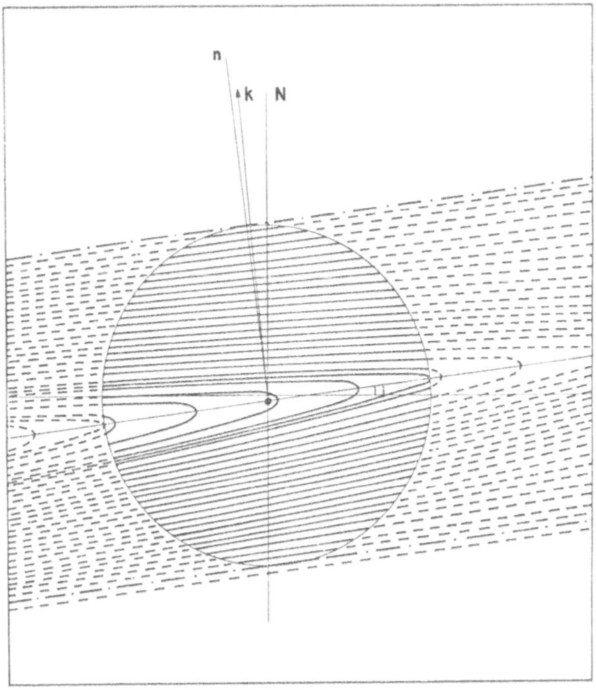

Fig. 3. Paths of the vertex of the unit vector along the spin axis as intersection of parabolic cylinders with the unit sphere (in the case of Mercury).

configuration with **k** very near **n**. In fact, we find for the angle between **n** and **k**, the result 4′28″ with $(C-A)/C=10^{-4}$ and the result 26″8 for $(C-A)/C=10^{-3}$.

Since the precession of the node with respect to an invariable system is at present very slow, we cannot be sure that Mercury has had enough time to settle down into the final configuration in accordance with the first as well as with the second and third of Cassini's laws. Naturally this discussion is true only if A–B is small compared to C–A and C–B since otherwise none of the three cases we have considered at the beginning is consistent with the Mercury case.

There is no need to emphasize the importance of the determination of the orientation of the rotational axis of Mercury. We may determine this orientation within the next few years by obtaining accurate delay-doppler radar maps of Mercury's surface.

8. Iapetus

This very simple theory may be applied to many of the satellites that are in some way locked-in to the primary. Since G. D. Cassini discovered Iapetus (Oct. 1671), we want to end this paper by devoting a few words to this particular inhabitant of the sky. Actually Iapetus is peculiar from two points of view. Revolving with the same face toward Saturn, Iapetus' brightness varies during a rotation by a factor of 6 (WHIPPLE,

1963). On the other hand, the perturbations due to the Sun and the oblateness of Saturn cause the pole of the orbit to describe a circular path about a mean pole that lies in the great circle passing through the pole of Saturn's orbital plane and of Saturn's equator. It takes the pole of Iapetus about 3000 years to describe its curve with a mean radius about 8° (TISSERAND, 1891; JEFFREYS, 1953). Taking for granted all these data, we will find out what the theory gives assuming case (c) or case (a) of p. 13. Thus, we have

$$i = 8^\circ, \quad \dot{\Omega} = \frac{2\pi}{3000}\,\text{years}^{-1}, \quad n \approx \frac{2\pi\,(366)}{79.33}, \quad \omega = n$$

$$\alpha = \frac{3}{2} n \,\frac{C-A}{C} = \frac{3}{2}\,\frac{2\pi(366)}{79.33}\,\frac{C-A}{C} = 2\pi\,(6.9)\frac{C-A}{C},$$

and

$$\frac{\dot{\Omega}}{\alpha} = -\,\frac{1}{3000 \times 6.9}\,\frac{C}{C-A} = -\,4.83 \times 10^{-5}\,\frac{C}{C-A},$$

and

$$\left|\frac{2\dot{\Omega}\sin i}{\alpha}\right| = 1.3 \times 10^{-5}\,\frac{C}{C-A}.$$

Suppose, for instance, that $C/(C-A)=10^4$; we then have

$$\frac{\dot{\Omega}}{\alpha} = -\,0.483, \quad \left|\frac{2\dot{\Omega}\sin i}{\alpha}\right| = 0.13.$$

We examine only the relative equilibrium configuration $\mathbf{k}_1$. As a first approximation we have $\lambda = -2i$. If the body, as suggested by Whipple, is very irregular, we should have $C/(C-A)$ much smaller and $\mathbf{k}_1$ will be close to n.

Acknowledgments

I am grateful to Dr. C. Lundquist and Dr. I. Michelson for suggesting this study and to Dr. I. I. Shapiro and Dr. Lundquist for reading the manuscript. This work was supported in part by NASA NsG 87-60.

References

CASSINI, G. D.: 1693, *Traité de l'origine et du progrès de l'astronomie*. Paris.
JEFFREYS, H.: 1953, *Monthly Notices Roy. Astron. Soc.* **113,** 81.
TISSERAND, F.: 1896, *Traité de Mécanique Céleste* IV, Gauthier-Villars, Paris, p. 91.
WHIPPLE, F. L.: 1963, *Earth, Moon, and Planets*. Harvard University Press, Cambridge, Mass.

THE MOON'S MECHANICAL ELLIPTICITY

Schrutka-Rechtenstamm's and Habibullin's Artifice in the Light of the Reduction of the 1st Part of Bamberg Series

J. MASŁOWSKI*

In a discussion of problems connected with the Moon's physical libration one cannot fail to pay particular attention to the quantity f, the so-called Moon's mechanical ellipticity.

Apart from the undoubtedly interesting question whether or not the values of f determined from heliometric series have any physical meaning, the very existence of f in the physical libration affords in practice many difficulties in the determination of the remaining unknowns of the problem. These difficulties are due, as we know, to the existence of the well-known singularity in the development of the libration component τ and even – as has been shown by MIETELSKI and MASŁOWSKI (1963) – of quite a number of such singularities in the most interesting interval of f values, namely from 0.60 to 0.70.

Many endeavours undertaken by various investigators aiming at a theoretical elimination of the singularity in τ remained unsuccessful to this day, which makes two suggestions of a great practical significance seem very valuable. For these suggestions enable us at least to choose – in accordance with the least squares method – between the two solutions given by heliometric series.

I mean here the suggestions of SCHRUTKA-RECHTENSTAMM (1955) and of HABIBULLIN (1958) concerning the determination of the unknown f. I need not describe the suggestion of Schrutka-Rechtenstamm, because it is well known and, besides, it was already mentioned here by Professor Kozieł.

As for the other one, I should like to make some remarks. In 1958, Habibullin indicated the possibility of the determination of f from two coefficients of the expansion of the physical libration in longitude, $a_{g'}$ and $a_{2\omega}$, after introducing them into the observation equations as the chief unknowns of the problem in place of the unknown df or of Schrutka-Rechtenstamm's unknowns df_1 and df_2.

I should like to emphasize that, mathematically, Habibullin's approach and formulae are correct, but his further steps leading to the determination of the values $a_{g'}$ and $a_{2\omega}$ from the normal equations and thus of the value of f are inappropriate if not erroneous. The reason is the fact that because of the theoretical dependence of $a_{g'}$ and $a_{2\omega}$ on f (the same $a_{g'}$ and $a_{2\omega}$ are dependent on each other) Habibullin maintains that, in order to choose between the two different solutions situated on

* University of Cracow, Poland.

both sides of the singular point 0.662, it is sufficient to solve the normal equations with the unknown $a_{g'}$ only, neglecting the term $a_{2\omega}$ which assumes an indeterminate value for the mentioned point. Evidently, a more precise value of f Habibullin determines from both coefficients $a_{g'}$ and $a_{2\omega}$.

The value for $a_{g'}$ computed from such normal equations should settle – as Habibullin wants – the position of f below or above the singular point (f is computed from $a_{g'}$) without regard for the initial value of f. It is necessary to keep in mind the fact that Habibullin's solution of the normal equations for $a_{g'}$ cannot settle the problem of f; for if we remove from the normal equations the unknown $a_{2\omega}$ possessing the singularity, these equations "will not know" that such singularity exists.

Nevertheless, this very interesting (and valuable) suggestion of Habibullin can be saved by a litle modification which will make it correct according to the basic rules of the least-squares method. For this purpose it is sufficient to bring the normal equations to a quadratic form (QF) in some sense analogous to that of Schrutka-Rechtenstamm on the basis of the unknowns $a_{g'}$, $a_{2\omega}$ and to find its minimum values $[v, v]_{\min}$ ($QF_{\min}$).

Now, I should like to make known the results relative to f, which can be obtained by means of both suggestions, i.e., the suggestion (or artifice) of Schrutka-Rechtenstamm and the suggestion of Habibullin. The results which I shall show presently have been obtained from a new reduction of the 1st part of Hartwig's Bamberg series containing 132 evenings and covering the period 1890–1912 (22 years). The total number of single observations was 1613.

I shall not speak about the details of the reductions of this material because it will soon be printed in the *Acta Astronomica*. I shall mention only that the observation equations and the normal ones have been obtained using the method of Kozieł (1948–49, 1962), which is the best for that purpose. The comparison of the observations with theory was performed in Cracow, but the adjustment of these observations could be performed only using an electronic computer. This was facilitated by professor Kopal, who made the computer "Mercury" available to me. I wish to thank professor Kopal for his great kindness.

Now, the artifice of Schrutka-Rechtenstamm applied to the 1st part of Bamberg series for the initial value $f_0=0.60$ gave as a result of the adjustment the QF shown in Figure 1. We see that the difference between the minimum values of the QF (for f smaller than f_s and for f greater than f_s) is close to zero – much smaller than $9\cdot\mu^2$, particularly for the case when Hayn's lunar profiles were taken into account. Thus, the artifice of Schrutka-Rechtenstamm did not give a unique solution for f. If we take into account the QF obtained with Weimer's lunar profiles or without any profiles at all the result of the adjustment speaks in favour of f situated above the singular point (a deeper minimum).

For the initial value $f_0=0.73$ the QF of Schrutka-Rechtenstamm behaves in a similar manner (Figure 2). The difference between the minimum values of QF is much smaller than $9\cdot\mu^2$, which makes a choice between the two values of f impossible. The artifice of Schrutka-Rechtenstamm applied to the 1st part of Bamberg series either

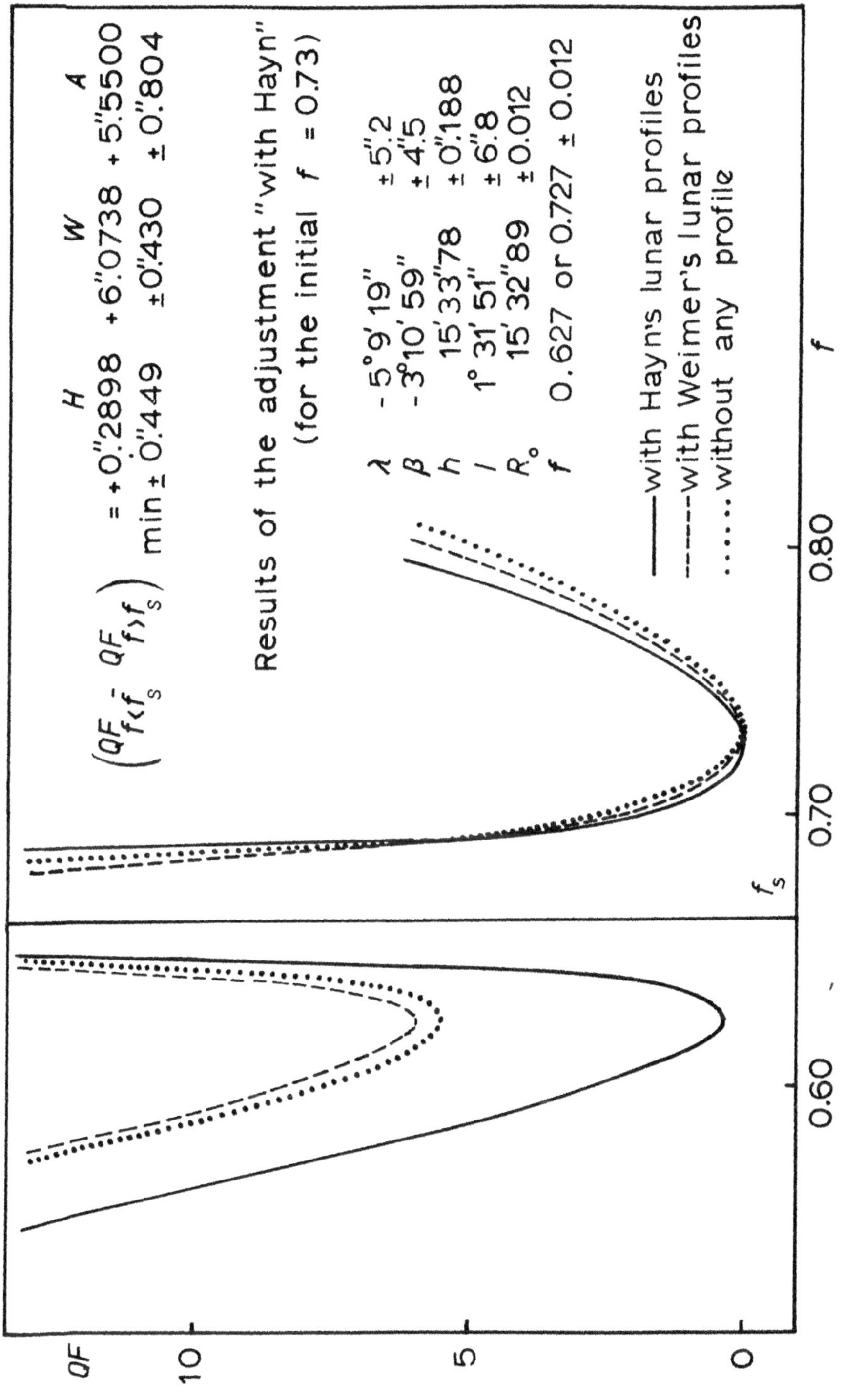

Fig. 1. Diagram of $QF(df_1, df_2)$ minus free term.

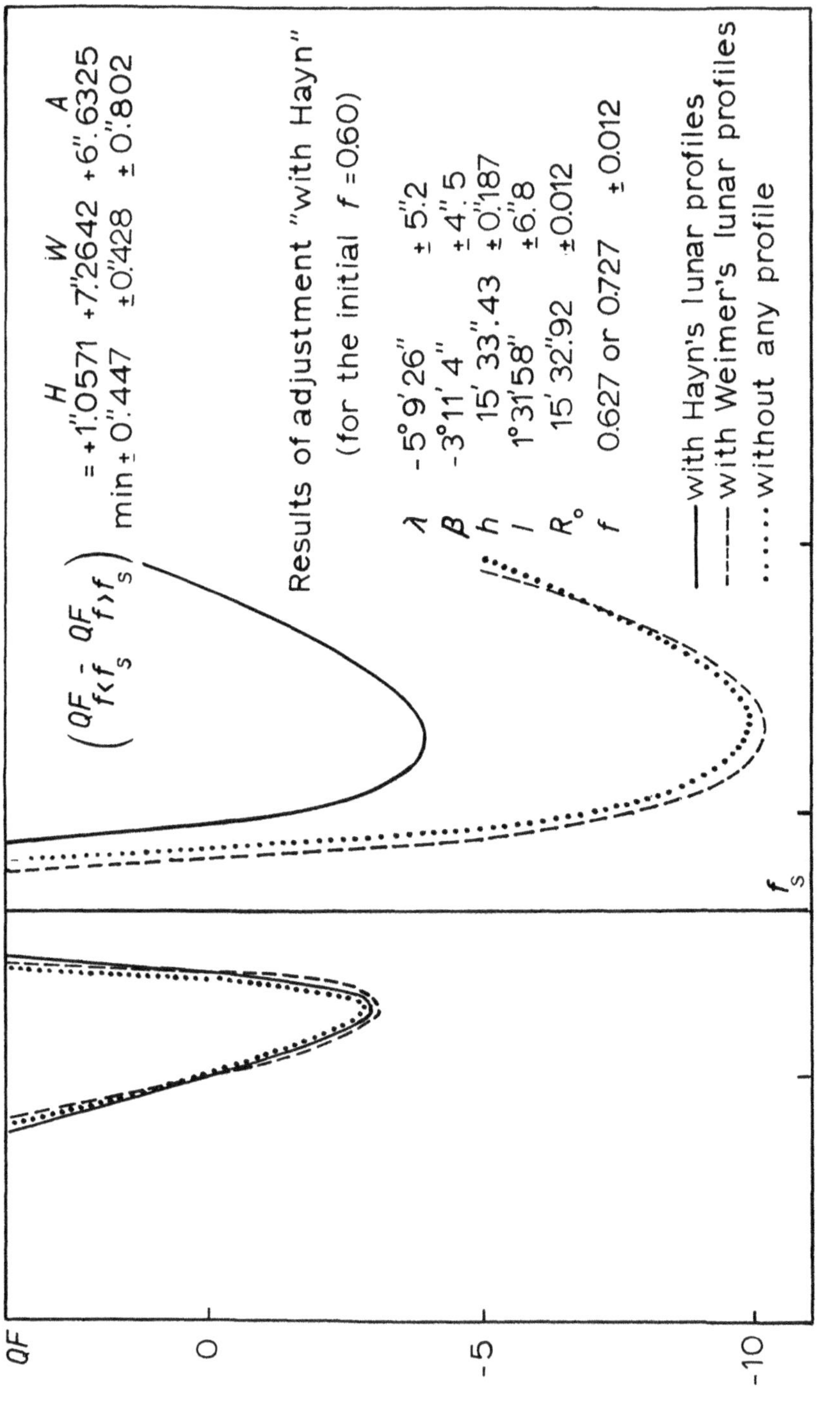

Fig. 2. Diagram of $QF(\mathrm{d}f_1, \mathrm{d}f_2)$ minus free term.

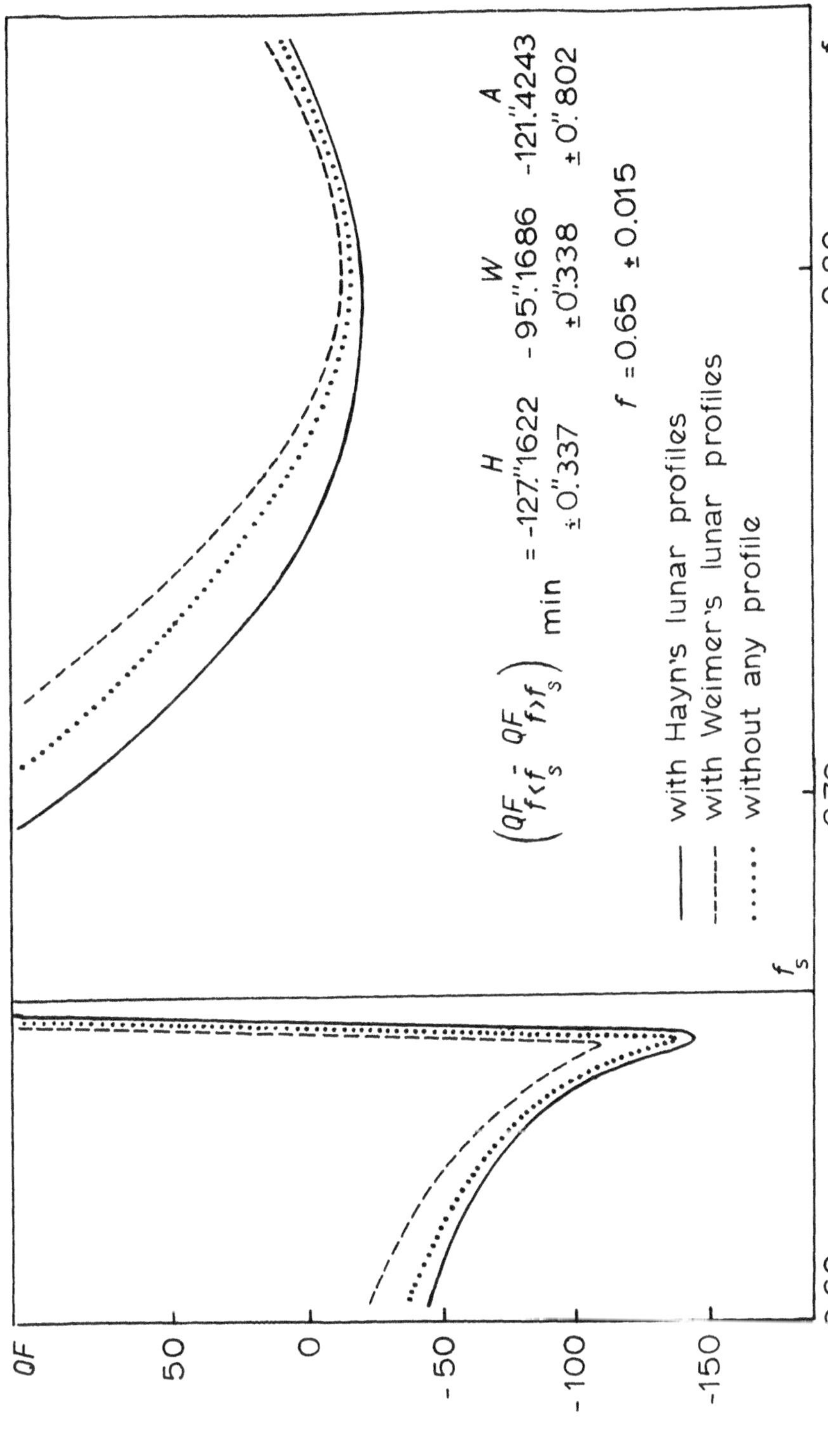

Fig. 3. Diagram of $QF(a_{g'}, a_{2\omega})$ minus free term.

fails to give a unique solution for f or speak in favour of f situated above 0.662.

The modified QF of Habibullin behaves quite differently (Figure 3). As we see from this Figure, the QF has a very deep and certain minimum for f situated below the singular point, though very closely. Since the difference between these two minima is very large (about 127 with Hayn) and is equal to $180 \cdot \mu^2$, the modified artifice of Habibullin enables us definitely to settle the question whether f lies below or above the singular point 0.662. We see that the other minimum is shifted from 0.662 to the very remote region 0.80, which is already unacceptable.

Thus, from the point of view of the $9 \cdot \mu^2$ criterion, the modified artifice of Habibullin gives one and only one unique solution for f (below the singular point).

Besides, I should like to remark that after removing – as Habibullin wants – the coefficient $a_{2\omega}$ as an unknown, I obtained for f the values situated above 0.662, namely: $a_{g'}=67.54$ ($f=0.722$) with Hayn, $a_{g'}=57.78$ ($f=0.759$) with Weimer, $a_{g'}=59.58$ ($f=0.752$) without lunar profiles. This is contradictory to Habibullin's statement that only $a_{g'}$ is decisive for the choice of f.

The results obtained on the basis of the 1st part of Bamberg series by using both suggestions lead to the following conclusions:

(1) The values of f determined from the adjustment of the same heliometric series depend essentially on the method and on the preciseness of the reduction of heliometric observations.

(2) Habibullin's artifice adapted to the artifice of Schrutka-Rechtenstamm is – in my opinion – a much more sensitive and maybe better method of settling the problem than the artifice of Schrutka-Rechtenstamm alone. But Habibullin's artifice is not fit for a determination of f from heliometric series, because it takes into account only two coefficients in the development of forced libration in τ, neglecting all the others.

In the light of what has already been said, only the solution lying below the singular point 0.662, as given in Figure 1, can be accepted, as the result of the adjustment of the 1st part of Hartwig's Bamberg series.

References

Habibullin, Sz. T.: 1958, *Izv. Astron. Observatory Engelhardt*, No. 31.

Kozieł, K.: 1948–49, *Acta Astronomica* **4**, 83.

Kozieł, K.: 1962, in *Physics and Astronomy of the Moon* (ed. by Z. Kopal). Academic Press, New York and London, Chapter 2, p. 27.

Masłowski, J. and Mietelski, J.: 1963, *Acta Astronomica* **13**, 135.

Schrutka-Rechtenstamm, G.: 1955, 'Zur physischen Libration des Mondes', *Berichte Österr. Akad. Wiss., math.-nat. Kl.*, Abt. II **164**, Band 8–10.

THE MOON'S ROTATION CONSTANTS AND THE COORDINATES OF MÖSTING A FROM THE HELIOMETRIC OBSERVATIONS OF BANACHIEWICZ

J. MIETELSKI*

1. Introduction

As a result of an analysis of the five years' (1910–1915) observation series of Banachiewicz I determined the constants of the physical libration and figure of the Moon, i.e. its mechanical ellipticity, the mean radius of its disc, the inclination of its rotation axis to the axis of the ecliptic, and the coordinates of the crater Mösting A, which was observed with Bessel's method. Beside these fundamental parameters I cleared out the subject of Yakovkin's effect and tried to determine the amplitude and the phase of the free libration in longitude.

The final results of my computations were quoted in Draft Reports just before the Hamburg Assembly of the IAU. The original paper containing details concerning the methods of the treatment and extensive numerical data has been given to print in *Acta Astronomica* and will probably appear next year.

My treatment of the Kazan series of Banachiewicz is not the only one, of course. The first full treatment of this series of over one thousand observations has been performed by YAKOVKIN (1928, 1934), and afterwards this material was analyzed in some parts by BELKOVICH (1949) and SCHRUTKA-RECHTENSTAMM (1955).

It could seem superfluous to repeat the enormous work represented by the first investigation by Yakovkin. Knowing, however, the high quality of these observations of Banachiewicz, we have decided that they were worth to be treated anew with a mathematically correct, modern method. Such a point of view will not be surprising to anybody who knows the large differences between the results of treatments of even the same heliometric observations by different methods.

A treatment of a similar problem consists usually, as it is well known, of three general parts: *O* (observatum) – reduced observations, *C* (calculatum) – reproduced observations, and *S* (solution) – least squares procedure, etc.

The treatment by Yakovkin raised a number of objections from the mathematical point of view in its part *S*, which was also systematically incorrect in other libration papers preceding the first paper by KOZIEŁ (1948). With Yakovkin the *C* part was also rather far from being ideal.

Relatively the best was part *O* into which I have introduced in my treatment only

* Department of Theoretical Astronomy and Astronomical Geophysics, University of Cracow, Poland.

three additional, small but unnegligible reductions, i.e. corrections to position angles for the convergence of hour circles, an allowance for nonlinearity of variations of the Moon's topocentric radius, and the question of transgressing the terminator by the observer.

On reproducing the observations of the crater's positions I endeavoured to work with the highest possible precision, and this is why I allowed for Greenwich corrections to tabular places of the Moon and also for corrections to its mean longitude as given by NEWCOMB (1878) and SPENCER JONES (1925).

2. Theoretical Basis

As the theoretical basis for this reproduction of observations I used HAYN's (1914, 1923) expansions for librations in node and in inclination, and that of KOZIEŁ (1948) for libration in longitude. In 1962 the latter's result (as well as Hayn's in this matter) was put into a doubt by MAKOVER (1962), who proposed to treat the problem, not as an ordinary differential equation with constant coefficients, but an inhomogeneous counterpart of Hill's equation which, as he hoped, would enable one to avoid the well-known singularity of τ for f=0.662.

Such a suggestion seemed interesting to Masłowski and myself, and stimulated us to make an attempt of approaching the problem by way of a Mathieu's equation which could be regarded as a reduced Hill's equation.

TABLE I

Equations proposed for τ

$$\frac{d^2\tau}{dx^2} + a\tau = \text{right-hand periodic terms}_{(1)}, \text{ as used by Hayn and others}$$

$$\frac{d^2\tau}{dx^2} + (q^2 + 2q_1 \cos 2x + 2q_2 \cos 4x + \ldots)\,\tau = r{\cdot}h{\cdot}p{\cdot}t_{(2)}, \text{ as suggested by Makover}$$

$$\frac{d^2\tau}{dx^2} + (q^2 + 2q_1 \cos 2x)\,\tau = \text{right-hand periodic terms}_{(3)}, \text{ as regarded by Maslowski and Mietelski,}$$

where a, q^2, q_1, q_2, ... are constants depending on f, and x is linear function of time.

Our common paper on the topic was published in *Acta Astronomica* (MASŁOWSKI and MIETELSKI, 1963). This attempt showed theoretically that the singularity of τ, known since Hayn's days, not only did not vanish; but, on the contrary, a series of new singular points appeared.

However, the practical result of this research has been a little more fortunate for us, because our numerical values of the expansion coefficients proved nearly identical with those of KOZIEŁ (cf. Table 1 in the paper just quoted), which are, as it was said, one of the bases for the reproduction of the crater's positions on the disc in all our Cracow treatments.

3. Brief Description of the Treatment

In reproducing and adjusting the observations of Banachiewicz I used the methods proposed by KOZIEŁ in 1948, 1956 and 1962, which have already been generally discussed here by their author.

The least-squares adjustment of Banachiewicz's series has been performed here in Manchester, on "Mercury" computer, and the possibility to use it had been kindly given me by Professor Kopal. During my stay here I performed my computations with the help of some colleagues of the Department of Astronomy, and that circumstance is emphasized in my paper about to be printed. Also my personal contact with Professor Kozieł was then especially helpful for my work because of his important suggestions.

It is noteworthy that my observation equations have had not one, but three, interchangeable columns of differences $O-C$ on their right-hand sides. The N column did not allow for the lunar profile, since H and W columns allowed for its irregularities as taken from Hayn's charts or from the *Atlas* of WEIMER (1952) respectively.

A really quick procedure of automatic computations gave an opportunity to perform them in several versions, which allowed for different sets of unknowns and started from $f=0.60$ or $f=0.73$ as from the basic reproduction data.

The manner of the transfer between those initial values of the mechanical ellipticity of the Moon has just been explained in the address of Professor Kozieł, who described also his other own ideas concerning some details of the adjustment procedure.

4. Final Remarks

Among the most interesting results of my research I consider the fact that, for the main unknown f characterizing the mass distribution in the Moon as a solid, we obtain by use of SCHRUTKA-RECHTENSTAMM's artifice (1955) the values in the close neighbourhood of $f=0.63$. This result is independent of the adopted initial value for f, and independent of the allowance for the profile of the Moon's disc in $O-C$ terms, as it can be easily seen on Figures 1 and 2.

In our case the difference of 13.5 units of the QF axis ($9\mu^2$, where $\mu^2=m\mu_0^2$) between two corresponding minima of QF_H or QF_W would be sufficient to indicate quite surely the realistic solution. Our differences are:

$$\text{diff.}\,(QF_H)=9.03;\quad \text{diff.}\,(QF_W)=10.54 \text{ (cf. Figure 1)},$$
$$\text{diff.}\,(QF_H)=7.93;\quad \text{diff.}\,(QF_W)=\ 9.50 \text{ (cf. Figure 2)}.$$

Here QF is the usual quadratic function of the unknowns df_1 and df_2. These quantities are explained in the paper of SCHRUTKA-RECHTENSTAMM (1955).

It is also interesting to note that the Yakovkin effect seems to be detectable. It is true that this effect is totally absorbed by the error of the estimated radius, but the allowance for it, as it can be seen in Table II, causes that the mean error of the weight unit to decrease.

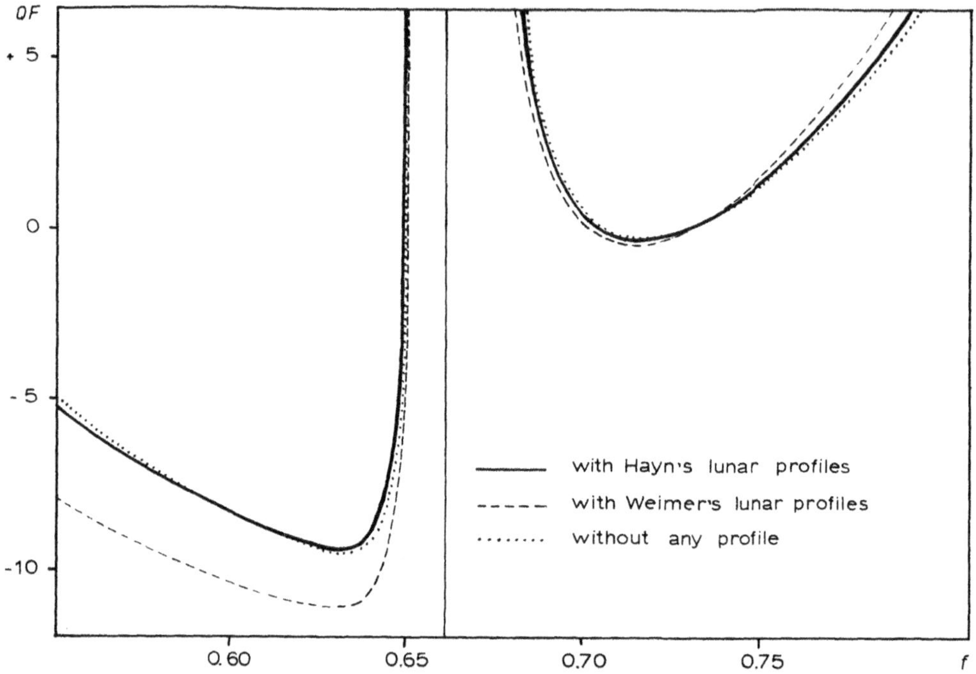

Fig. 1. Diagram of $QF(\mathrm{d}f_1, \mathrm{d}f_2)$ minus free term (for initial $f = 0.73$).

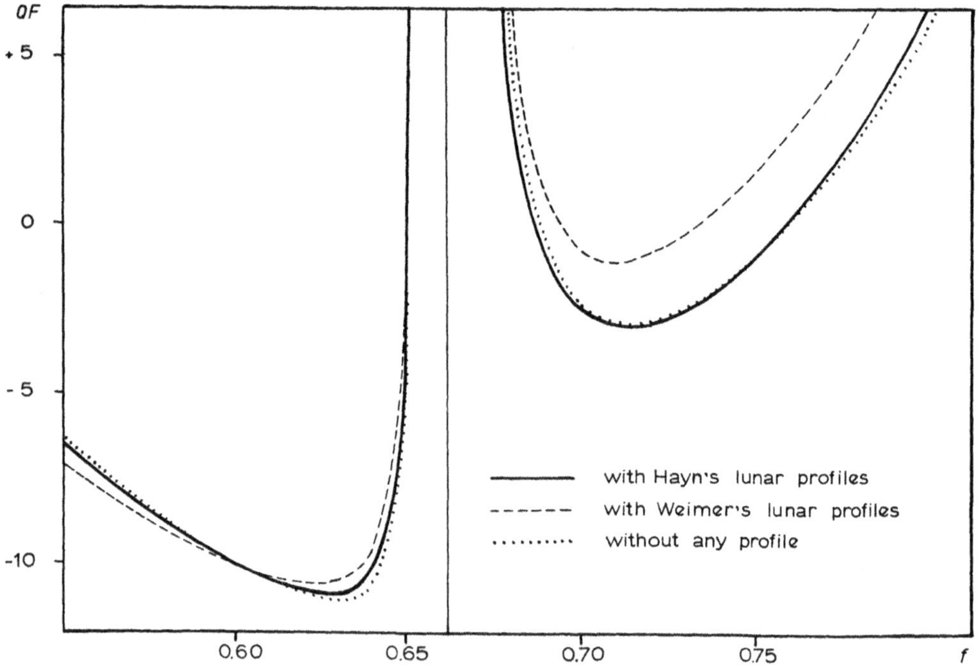

Fig. 2. Diagram of $QF(\mathrm{d}f_1, \mathrm{d}f_2)$ minus free term (for initial $f = 0.60$).

TABLE II

The Yakovkin Effect ($dr_0 = x + y\beta$) in Banachiewicz's Observations

Author	m.e. of x	y	m.e. of y
YAKOVKIN (1934) (Hayn's charts)	± 0″.04	0″.02	± 0″.01
MIETELSKI (1963) (3rd s., Hayn)	± 0″.019	0″.018	± 0″.004
(3 s., Weimer)	± 0″.021	0″.034	± 0″.004
(3 s., no profile)	± 0″.035	0″.030	± 0″.007

Mean errors of the weight unit (one observation)
(the first lines without, the second ones with, the Yakovkin effect)

Solution 2

H	*W*	*N*
± 0″.4840	± 0″.5251	± 0″.8768
0″.4793	0″.5091	0″.8697

Solution 3

H	*W*	*N*
0″.4755	0″.5168	0″.8719
0″.4709	0″.5007	0″.8649

Solution 4

H	*W*	*N*
0″.4754	0″.5146	0″.8714
0″.4712	0″.5003	0″.8652

Ultimately, the results obtained here disprove the belief, held in some papers, that the Yakovkin effect is present in Hayn's charts only. The values of mean errors speak in favour of the charts of Hayn rather than of WEIMER's *Atlas* (1952) as giving lunar profiles. This result, however, is not a definitive quality criterion, because in Masłowski's investigation the situation is just opposite.

A comparison of other results obtained from the same material with ours can create a rather negative impression of some subjectivity. Some differences can easily be explained by slightly different initial conditions (e.g. the presence or absence of the corrections to the Moon's mean longitude); but there are others, an explanation of which is to be found in different manners of the theoretical reduction and also of the least-squares adjustments in different treatments. These "methodical" differences in results demonstrate the extent to which the latter are sensitive to the adopted method of reduction, for we can certainly assume that every author made his computations with care and used the method best in his opinion. At present the method of Kozieł is generally regarded as giving reliable results and this method was used in my treatment.

As the final results of this treatment of Banachiewicz's series we present the solution which can be compared here with the results of the other authors (see Table III).

It is to be noted that the estimates of the free libration here are only tentative. In fact I think my results are not affected by methodological errors and their imperfection can lie only in a relatively limited material (i.e., 1000 observations) and the short time-interval (five years) covered by them.

TABLE III
Comparison of Final Results

	YAKOVKIN (1928)	BELKOVICH (1949), mean	SCHRUTKA-RECHTENSTAMM (1955), 1st sol.	MIETELSKI (1963), 6 − 1–H
λ	− 5°10′19″	− 5°10′58″	− 5°10′26″	− 5°10′ 5″
m.e.	± 11	± 9	± 11	± 6.2
β	− 3°10′56″	− 3°11′20″	− 3°11′ 1″	− 3°10′52″
m.e.	± 10	± 10	± 10	± 6.3
h	15′35″.5	15′32″.0	15′31″.58	15′31″.82
m.e.	± 0.5	± 0.5	± 0.49	± 0.28
I	1°32′2″	1°33′51″	1°32′10″	1°32′37″
m.e.	± 17	± 17	± 17	± 10.9
R	15′32″.68	—	—	15′32″.834
m.e.	± 0.040			± 0.0193
f	0.74	0.70	0.633	0.628
m.e.	± 0.03	± 0.03	+ .008, − .014	± 0.0191

The free libration in longitude

	Amplitudes	Phases (1800.0)
Mietelski (H, W, N)	13 ± 9″, 29 ± 9″, 26 ± 15″	19° ± 36°, 18° ± 17°, 11° ± 34°
Yakovkin (H)	35 ± 13″	mutually contradictory!
Belkovich (H)	40 ± 17″	

As was said already by Professor Kozieł, it is almost certain that definitive results in the field of lunar librations, as derived from heliometric or photographic observations, can be obtained only by a full treatment of the entire ensemble of all available observations; and these, fortunately, are available to us in sufficient numbers.

References

BELKOVICH, I. V.: 1949, *Izv. Engelhardt Obs. Kazan*, No. 24.
HAYN, F.: 1914, 'Selenographische Koordinaten', *Abh. König. Sächs. Gesell. Wiss.*, **32**, 1–113.
HAYN, F.: 1923, 'Die Rotation des Mondes', in *Enzykl. Math. Wiss.* **6**, pp. 1020–1043.
KOZIEŁ, K.: 1948, *Acta Astr. Cracoviae* (a) **4**, 61–139.
KOZIEŁ, K.: 1956, *Postepy Astr.*, **4**, 80.
KOZIEŁ, K.: 1962, in *Physics and Astronomy of the Moon* (ed. by Z. Kopal). Academic Press, New York and London, Chapter 2, pp. 27–59.
MAKOVER, S. G.: 1962, *Bull. Inst. Theor. Astr. Leningrad* **8**, 249.
MASŁOWSKI, J. and MIETELSKI, J.: 1963, *Acta Astr. Cracoviae* **13**, 135.
MIETELSKI, J.: 1963, 'Wyznaczenie stalych libracji fizycznej i figury Ksiezyca na podst. kazanskiego szeregu obserwacji heliometr. T. Banachiewicza' (unpublished MS).
NEWCOMB, S.: 1878, U.S. Naval Observ. Washington, Observations 1875, Appendix 2.
SCHRUTKA-RECHTENSTAMM, G. VON: 1955, *Mitt. Univ. Sternw. Wien* **8**, 151.
SPENCER JONES, H.: 1925, *Mon. Not. Roy. Astr. Soc.*, **85**, 11.
WEIMER, Th.: 1952, *Atlas des Profils Lunaires*. Publ. Observ. de Paris.
YAKOVKIN, A. A.: 1928, *Izv. Engelhardt Obs. Kazan*, No. 13.
YAKOVKIN, A. A.: 1934, *Bull. Engelhardt Obs. Kazan*, No. 4.

A CONTRIBUTION TO THE STUDY OF THE MOON'S PHYSICAL LIBRATION IN LONGITUDE

M. D. MOUTSOULAS*

The equation of the physical libration in longitude, as is only too well known, can be treated independently of the first two equations and, in particular, be placed in the form

$$\ddot{\tau} + A(g, g', \omega, \omega')\tau = B(g, g', \omega, \omega') \tag{1}$$

where g is the mean anomaly of the Moon, g' the mean anomaly of the Sun, ω the distance of the Moon's perigee from the ascending node of its orbit and, ω' the distance of the Sun's perigee from the ascending node of the orbit of the Moon. The functions A and B depend only on the time implicitly and can be expanded in a Fourier series in terms of multiples of four different frequencies. This equation is true if τ is small and this is the case for the Moon.

The general solution of Equation (1) can be found once we have obtained the general solution of the homogeneous equation corresponding to Equation (1).

If we put zero on the right-hand side of Equation (1) and assume that τ_1, τ_2 are two independent solutions of the resulting homogeneous equation, then its general solution is given by the expression

$$\tau = V_1\tau_1 + V_2\tau_2 . \tag{2}$$

If we assume that the four independent frequencies appearing in the expression of A are commensurable, then we can express the function A as a Fourier expansion in terms of only one frequency. In such a case the equation of the physical libration in longitude takes the form of Hill's equation. We have tried this, considering a period equal to the lowest common multiple of the contributing frequencies, but the terms of the Fourier series were decreasing very slowly and so the approximate solution of the corresponding Hill's equation could not be accepted as satisfactory.

Nevertheless, we can approach a quite satisfactory solution in the case where ω is very small, i.e., when the lunar perigee is very close to the node of the lunar orbit. Then, the coefficient of τ is periodic with period $l = 0.07544$ and the homogeneous equation takes the form

$$\frac{d^2\tau}{dz^2} + (p_0 + p_1 \cos 2z + p_2 \cos 4z + p_3 \cos 6z + \cdots)\,\tau = 0 \tag{3}$$

where $z = (\pi/l)\,t$.

* Department of Astronomy, University of Manchester.

TABLE I

γ	p_0	p_1	p_2	p_3	p_4	p_5	p_6
0.00014	0.00166853747	0.00017181597	0.00000094274	0.00000000015	0.00000000008	0.00000000005	0.00000000003
0.00016	0.00190689997	0.00019636111	0.00000107742	0.00000000017	0.00000000009	0.00000000006	0.00000000004
0.00018	0.00214526247	0.00022090625	0.00000121210	0.00000000019	0.00000000010	0.00000000006	0.00000000004
0.00020	0.00238362496	0.00024545138	0.00000134677	0.00000000021	0.00000000011	0.00000000007	0.00000000005
0.00022	0.00262198846	0.00026999652	0.00000148145	0.00000000023	0.00000000012	0.00000000008	0.00000000005
0.00024	0.00286034996	0.00029454166	0.00000161613	0.00000000025	0.00000000013	0.00000000008	0.00000000006
0.00026	0.00309871245	0.00031908680	0.00000175080	0.00000000027	0.00000000014	0.00000000009	0.00000000006

As we do not know, at the moment, the true value of γ and we suppose that it is probably near the value $\gamma = 0.00020$, we give the values of $p_0, p_1, p_2 \ldots$ for $\gamma = 0.00014$, 0.00016, 0.00018, 0.00020, 0.00022, 0.00024, 0.00026 (Table I).

Defining

$$q_n = q_{-n} = \frac{p_n}{2} \quad (\text{for } n \neq 0), \quad \text{and} \quad q_0 = p_0$$

we assume

$$\tau = e^{\mu z} \sum_{n=-\infty}^{\infty} b_n e^{2niz} \tag{4}$$

as a solution of Equation (3).

The quantity μ is determined by the relation

$$\sin^2(\tfrac{1}{2}\pi i\mu) = \Delta_{(0)} \sin^2(\tfrac{1}{2}\pi i \sqrt{q_0})$$

where

$$\Delta_{(0)} = \begin{vmatrix} \cdots & \cdot & \cdot & \cdot & \cdot & \cdot & \cdots \\ \cdots & \cdot & \cdot & \cdot & \cdot & \cdot & \cdots \\ \cdots & 1 & \frac{q_1}{q_0 - 16} & \frac{q_2}{q_0 - 16} & \frac{q_3}{q_0 - 16} & \frac{q_4}{q_0 - 16} & \cdots \\ \cdots & \frac{q_1}{q_0 - 4} & 1 & \frac{q_1}{q_0 - 4} & \frac{q_2}{q_0 - 4} & \frac{q_3}{q_0 - 4} & \cdots \\ \cdots & \frac{q_2}{q_0} & \frac{q_1}{q_0} & 1 & \frac{q_1}{q_0} & \frac{q_2}{q_0} & \cdots \\ \cdots & \frac{q_3}{q_0 - 4} & \frac{q_2}{q_0 - 4} & \frac{q_1}{q_0 - 4} & 1 & \frac{q_1}{q_0 - 4} & \cdots \\ \cdots & \frac{q_4}{q_0 - 16} & \frac{q_3}{q_0 - 16} & \frac{q_2}{q_0 - 16} & \frac{q_1}{q_0 - 16} & 1 & \cdots \\ \cdots & \cdot & \cdot & \cdot & \cdot & \cdot & \cdots \\ \cdots & \cdot & \cdot & \cdot & \cdot & \cdot & \cdots \end{vmatrix}$$

Substituting μ in the system

$$(\mu + 2ni)^2 b_n + \sum_{m=-\infty}^{\infty} q_m b_{n-m} = 0$$

$$(n = 0, \pm 1, \pm 2, \ldots)$$

we determine the coefficients b_n in terms of b_0. As the values of $q_3, q_4, \ldots$ are small enough compared with q_0, an approximate solution may be obtained by neglecting all b except $b_{-3}, b_{-2}, b_{-1}, b_0, b_1, b_2, b_3$. Table II gives for various values of γ the values of μ and the coefficients b.

The general solution of the non-homogeneous equation is given by Equation (2), where

$$V_1 = -\int \frac{\tau_2}{W(\tau_1, \tau_2)} B_{(t)}\, dt,$$

$$V_2 = \int \frac{\tau_1}{W(\tau_1, \tau_2)} B_{(t)}\, dt.$$

TABLE II

γ	μ	$b_{-3} \times 10^{11}$	$b_{-2} \times 10^{11}$	$b_{-1} \times 10^{7}$	$b_1 \times 10^{7}$	$b_2 \times 10^{7}$	$b_3 \times 10^{11}$
	$0.040848i$	0.191816	0.234845	−0.134934	0.133513	0.137745	0.006024
0.00014	$-0.040848i$	0.197110	0.244715	−0.141183	0.128449	0.138354	0.000653
	$1.959152i$	0.110500	0.059725	−0.046045	0.171482	0.176976	0.194467
	$-1.959152i$	0.489397	0.550350	0.203785	0.004827	−0.091807	−0.134421
	$0.043668i$	0.219012	0.267474	−0.154004	0.152800	0.157434	0.007650
0.00016	$-0.043668i$	0.225481	0.279522	−0.161643	0.146614	0.158181	0.001073
	$1.956332i$	0.126375	0.067816	−0.052819	0.196440	0.202456	0.223285
	$-1.956332i$	0.489397	0.629634	0.233157	0.005513	−0.105138	−0.153560
	$0.046317i$	0.246171	0.299901	−0.173007	0.172125	0.177100	0.009452
0.00018	$-0.046317i$	0.253890	0.314265	−0.182123	0.164742	0.177990	0.001589
	$1.953683i$	0.142267	0.075779	−0.059626	0.221494	0.227969	0.252395
	$-1.953683i$	0.549870	0.709087	0.262406	0.006223	−0.118432	−0.172702
	$0.048822i$	0.273295	0.332129	−0.192000	0.191486	0.196788	0.011431
0.00020	$-0.048822i$	0.282336	0.348937	−0.202680	0.182839	0.197834	0.002203
	$1.951178i$	0.158174	0.083610	−0.066469	0.246614	0.253516	0.281803
	$-1.951178i$	0.610232	0.788715	0.291835	0.006912	−0.131824	−0.191852
	$0.051205i$	0.300386	0.364162	−0.210932	0.210884	0.216456	0.013583
0.00022	$-0.051205i$	0.310816	0.383537	−0.223255	0.200906	0.217663	0.002916
	$1.948795i$	0.174095	0.091309	−0.073342	0.271825	0.279095	0.311513
	$-1.948795i$	0.670490	0.868513	0.321143	0.007625	−0.145177	−0.211011
	$0.053482i$	0.327446	0.396002	−0.229859	0.230318	0.236150	0.015909
0.00024	$-0.053482i$	0.339330	0.418059	−0.243907	0.218949	0.237529	0.003730
	$1.946518i$	0.190031	0.098872	−0.080249	0.297098	0.304709	0.341528
	$-1.946518i$	0.730648	0.948488	0.350631	0.008317	−0.158629	−0.230183
	$0.055666i$	0.354475	0.427654	−0.248729	0.249782	0.255820	0.018406
0.00026	$-0.055666i$	0.367875	0.452503	−0.264571	0.236961	0.257376	0.004646
	$1.944334i$	0.205980	0.106299	−0.087182	0.322452	0.330349	0.371852
	$-1.944334i$	0.790710	1.028630	0.379992	0.009033	−0.172036	−0.249369

$W(\tau_1, \tau_2)$ is the Wronskian matrix

$$W(\tau_1, \tau_2) = \begin{vmatrix} \tau_1 & \tau_2 \\ \tau_1' & \tau_2' \end{vmatrix}.$$

The values of the indefinite integrals V_1 and V_2 depend on the time t and the initial time t_0. The general solution has a singularity, if there is a time t_1 between t_0 and t such that $W(t_1)=0$. This singularity can be removed if V_1 and V_2 are finite, i.e., if the improper integrals converge, otherwise it cannot be removed. In such a case the solution becomes very large and hence the original linearized Equation (1) is no longer valid. It is important to note that the singularity depends only on the function $A(t)$ and not the function $B(t)$, i.e., the action of the disturbing body.

But, in any case, whether we find a better approximation to the solution for a special case, as we have done, or whether we try to find a formula for the general case consisting of a long series of various terms, as previous investigators have done, any suggestion on the position or even the existence of singularities in the solution of the problem is premature as long as this solution is not the real mathematical one, but is based on Hayn's approximating formula, derived after we have made many simplifications and neglected many terms.

In addition, one can independently test the behaviour of the function $\tau = \tau(f, t)$ by going directly to the exact equation (i.e., before linearization) and study the corresponding equation of sensitivity of this function. This equation can be found by partial differentiation of the third equation of Euler with respect to f. The particular form of this equation will be

$$\frac{\mathrm{d}^2}{\mathrm{d}t^2}\left(\frac{\partial\tau}{\partial f}\right) = q(t, f, \tau, \partial\tau/\partial t).$$

The initial conditions of this equation are:

$$t = t_0, \quad \tau = \tau_0, \quad \frac{\partial\tau}{\partial f} = 0.$$

We hope that in the near future we will able be to present the conclusion of this investigation which is now in progress.

References

HAYN, F.: 1902, 'Selenographische Koordinaten', *Abh. König. Sächs. Gesell. Wiss.* **27**, 861.

KOPAL, Z.: 1966, *An Introduction to the Study of the Moon*, pp. 24–40.

MAKOVER, S. G.: 1962, *Bull. Inst. Theor. Astron. Leningrad* **8**, 249.

WHITTAKER, E. T. and WATSON, G. N.: 1962, *A Course of Modern Analysis*. 4th ed., Cambridge University Press, Cambridge, pp. 413–417.

LUNAR PHYSICAL LIBRATION THEORY

DONALD H. ECKHARDT*

1. Introduction

Like many problems of dynamical astronomy, the integration of the three Euler dynamical equations, which describe the forced physical librations of the Moon, is a very onerous task. The differential equations are complicated, mutually coupled and non-linear. The elaborate numerical treatment of HAYN (1902) furnished solutions which remain authoritative to this day except for the modifications of KOZIEŁ (1948) who replaced Hansen's lunar theory with Brown's theory.

The solutions of Hayn and Kozieł, as painstaking as they were, would have been unattainable were it not for their use of a number of clever approximations and linearizations which were devised by Hayn. The approximations restricted the precision of the solutions and the linearizations tended to degrade their accuracies, especially in the neighborhoods of solution singularities. Now, fortunately, these limitations can be ameliorated through the use of a digital computer. The computer can be programmed to perform all the relevant numerical algorithms which are required for solutions of any reasonable precision and the solutions can be generated iteratively to allow for the non-linearities and the coupling between the equations. Because the labor-saving approximations and linearizations of Hayn are not needed for a computer solution, the computer programming is relatively simple.

2. Cassini's Laws and their Perturbations

The Moon's rotation is fairly accurately described by three empirical "laws" published in 1693 by G. D. Cassini:

(1) The Moon rotates uniformly about its polar axis with a rotational period equal to the mean sidereal period of its orbit about the Earth.

(2) The inclination of the Moon's equator to the ecliptic is a small constant angle, approximately $I = 1\frac{1}{2}°$.

(3) The ascending node of the lunar orbit on the ecliptic coincides with the descending node of the lunar equator on the ecliptic.

The Moon's optical librations are computed by adopting an inclination I in

* Air Force Cambridge Research Laboratories, Bedford, Mass., U.S.A.

Cassini's laws which are then assumed to be exact and used together with lunar theory.

Rotations which are observed from the Earth as optical librations of the Moon would appear from the Moon as a wandering of the apparent direction of the Earth. This apparent motion of the Earth in the triaxial gravitational field of the Moon induces small oscillations of the Moon about its center of gravity which are the forced physical librations.

Even without optical librations, the triaxial Moon could theoretically oscillate subject to the Earth's restoring torque much like a pendulum. Indeed, even without the Earth, the Moon could wobble to maintain constant its angular momentum. The components of these free physical librations can occur only at discrete frequencies which are weakly dependent on their amplitudes because the mechanical system is nonlinear. The free physical librations are required to satisfy initial conditions of the pertinent Euler dynamical equations, and if they ever did exist, they would be dampened through tidal dissipation of energy on the Moon. They have never been detected satisfactorily and, for the purpose of this paper, the free physical librations are assumed to be negligible.

A Cartesian coordinate system for the Moon may be defined, whose axes are assumed to be the same as those of the Moon's principal moments of inertia. Axis 1 is the mean direction of the Earth; Axis 3 is in the direction of the north lunar pole; and Axis 2 is orthogonal to 1 and 3 to complete a right-handed system. The orientation of these axes with respect to a fixed ecliptic system is given by three Euler angles, ϑ, ψ, and φ. ϑ is the inclination of the lunar equator to the ecliptic; ψ is the longitude of the descending node of the lunar equator; and φ is the angle between the descending node and Axis 1. If Cassini's laws were exact, one would have

$$\vartheta = I,\ \psi = \Omega,\ \varphi + \psi = \pi + L$$

where L is the mean longitude of the Moon and Ω is the mean longitude of its ascending node. Since Cassini's laws are not exact, one must consider the perturbed Euler angles

$$\vartheta = I + \rho,\ \psi = \Omega + \sigma,\ \varphi + \psi = \pi + L + \tau$$

where ρ, σ, and τ are the perturbing terms of the physical librations.

The Euler angles are related to the angular rotational velocities of the Moon about its 1, 2, 3 axes, ω_1, ω_2, and ω_3 as follows:

$$\begin{aligned} \omega_1 &= -\dot{\psi} \sin\vartheta \sin\varphi - \dot{\vartheta} \cos\varphi \\ \omega_2 &= -\dot{\psi} \sin\vartheta \cos\varphi + \dot{\vartheta} \sin\varphi \\ \omega_3 &= \dot{\psi} \cos\vartheta + \dot{\varphi}\,. \end{aligned}$$

3. Derivation of the Dynamical Equations

Let $r_i = ru_i$ locate the center of mass of the Earth, a distance r from the origin and let $\rho_i = \rho\xi_i$ locate a point within the Moon, a distance ρ from the origin. The gravitational potential per unit mass in the field of the Moon at r_i is

$$U = -(G/r)\left[\int_{☾} \mathrm{d}M + \sum_{n=2}^{\infty} (R/r)^n \int_{☾} (\rho/R)^n P_n(\cos A)\,\mathrm{d}M\right]$$

where G is the gravitation constant; R is the mean radius of the Moon; P_n is the Legendre function of degree n; A is the angle between vectors r_i and ρ_i; and the integrals are over the mass, M, of the Moon.

At $r=a$, the mean range of the Earth, $R/r=R/a=1/221$, so even if all degree harmonics have the same magnitude on the Moon's surface, at the Earth their magnitudes decrease as $(1/221)^n$. A good approximation to the potential at the Earth, then, is provided by truncating all third or higher degree harmonics.

Expanding P_2 in terms of $\cos A = u_i\xi_i$, the truncated potential may be written as

$$U = -GM/r - (G/r^3)\left[\tfrac{3}{2}\int_{☾} [(u_i\rho_i)^2 - \rho_i^2]\,\mathrm{d}M + \int_{☾} \rho_i^2\,\mathrm{d}M\right].$$

Since the moment of inertia tensor is

$$C_{ij} = \int_{☾} (\rho_k^2\,\delta_{ij} - \rho_i\rho_j)\,\mathrm{d}M$$

where δ_{ij} is the Kronecker delta, the potential may be reformulated as

$$U = -GM/r + (G/2r^3)\,[3u_i\,C_{ij}\,u_j - C_{ii}]\,. \tag{1}$$

Because the Moon is not rigid, its moments of inertia are not constant. Assume that C_{ij} represents only the moment of inertia tensor which the Moon would have if it were non-rotating and removed from the gravitational fields of other bodies. The Moon is deformed from this state principally by forces derived from its centrifugal potential, V, and from the second degree earth induced tidal potential, W. To account for this deformation, the potential, U, must be replaced by

$$U^* = U + k(R/r)^5\,(V + W) \tag{2}$$

where k is the potential disturbance Love number.

The centrifugal and tidal potentials at x_i are

$$V = -\tfrac{1}{2}\,[\varepsilon_{ijk}\,\omega_j\,x_k]^2 = \tfrac{1}{2}x_i\,[\omega_i\,\omega_j - \delta_{ij}\,\omega_k^2]\,x_j \tag{3}$$

$$W = -(Gm/2r^3)\,x_i\,[3u_i\,u_j - \delta_{ij}]\,x_j \tag{4}$$

where ε_{ijk} is the alternating tensor and m is the mass of the Earth. Using Equations (1)–(4), the potential may be written

$$U^* = -(GM/r) + (G/2r^3)\,[3u_i\,C_{ij}^*u_j - C_{ii}^*] - (kR^5\,\omega_i^2/3r^3) \tag{5}$$

with

$$C_{ij}^* = C_{ij} + (kR^5/3G)\,\omega_i\omega_j - (km\,R^5/r^3)\,u_i\,u_j\,. \tag{6}$$

For a small rotation, $\delta\varphi_l$, of the Moon, the direction cosines to the Earth change by

$$\delta u_j = \varepsilon_{jkl} u_k \delta\varphi_l$$

and, therefore,

$$\frac{\partial u_j}{\partial \varphi_l} = \varepsilon_{jkl} u_k .$$

The torque exerted on the Moon is

$$N_l = -m \frac{\partial U^*}{\partial \varphi_l} = -m \frac{\partial U^*}{\partial u_j} \varepsilon_{jkl} u_k = -(3Gm/r^3)\, u_i C^*_{ij}\, \varepsilon_{jkl}\, u_k . \tag{7}$$

Setting $a = r$ and $\dot{L} = |\omega_i|$, and using Kepler's third law, which is approximately $1/G = m/(\dot{L}^2 a^3)$, the maximum components of the three terms in Equation (6) for C^*_{ij} are, respectively, about $(2/5)\, MR^2$, $\mathrm{kmR}^2 (R/a)^3$, $\mathrm{kmR}^2 (R/a)^3$. The ratios of the deformation components in $C^*_{ij} - C_{ij}$ to the principal components of C_{ij} are no more than about $(5/2)\,(k/\kappa)\,(R/a)^3$, where $\kappa = M/m$.

The Love number, k, has been calculated for a variety of hypothetical lunar models by HARRISON (1963) and KAULA (cited by Harrison). Its most likely value is $k = 0.02$, and a realistic extreme is $k = 0.05$. Using the latter Love number and $\kappa = 1/81.30$ and $(R/a) = 1/221$, the ratios of deformation to principal moment of inertia components are no more than about 10^{-6}. They are negligible, so C^*_{ij} in Equation (7) may be replaced by C_{ij}.

In an inertial system, the time derivative of the angular momentum is equal to the torque, so

$$\frac{\mathrm{d}}{\mathrm{d}t}(C_{ij}\omega_j) = C_{ij}\dot{\omega}_j + \varepsilon_{ijk}\,\omega_j\, C_{kl}\,\omega_l = N_i = (3Gm/r^3)\,\varepsilon_{ikj}\, u_k\, C_{jl}\, u_l \tag{8}$$

where $\mathrm{d}/\mathrm{d}t$ is the operator for a time derivative in the inertial system and the dot derivative is with respect to time in the lunar coordinate system. For $i = 1, 2, 3$, these are the Euler dynamical equations.

The coordinate system has been defined such that the products of inertia, that is C_{ij} for $i \neq j$, are zero. Set

$$\alpha = (C_{33} - C_{22})/C_{11},\ \beta = (C_{33} - C_{11})/C_{22},\ \gamma = (C_{22} - C_{11})/C_{33} .$$

Because $\alpha - \beta + \gamma - \alpha\beta\gamma = 0$, only two of these quantities are independent and need be known to integrate the transformed dynamical equations

$$\dot{\omega}_1 + \alpha\omega_2\omega_3 = \alpha q^2 (a/r)^3\, u_2 u_3 \tag{9a}$$

$$\dot{\omega}_2 - \beta\omega_3\omega_1 = -\beta q^2 (a/r)^3\, u_3 u_1 \tag{9b}$$

$$\dot{\omega}_3 + \gamma\omega_1\omega_2 = \gamma q^2 (a/r)^3\, u_1 u_2 \tag{9c}$$

where $q^2 = 3Gm/a^3$. For the perturbed lunar orbit, Kepler's third law is $G(m + M) = \dot{L}^2 [a/0.99909]^3$ (BROWN, 1899), so $q^2 = 3\lambda\dot{L}^2/(1 + \kappa)$ where $\lambda = 1.0027$. The factor, λ, was first used by JEFFREYS (1961). Taking $\kappa = 1/81.30$ one obtains $q^2 = 0.00398081$ per day^2.

At the Moon, let v be the longitude of the Earth from the descending node of the Moon's equator and let $-B_☾$ be the latitude of the Earth. At the Earth, $v+\sigma$ is the longitude of the Moon from the descending node of its orbit on the ecliptic, and $B_☾$ is the latitude of the Moon. The direction cosines of the Earth from the Moon are

$$\begin{bmatrix} u_1 \\ u_2 \\ u_3 \end{bmatrix} = \begin{bmatrix} \cos\varphi & \sin\varphi & 0 \\ -\sin\varphi & \cos\varphi & 0 \\ 0 & 0 & 1 \end{bmatrix} \begin{bmatrix} 1 & 0 & 0 \\ 0 & \cos\vartheta & -\sin\vartheta \\ 0 & \sin\vartheta & \cos\vartheta \end{bmatrix} \begin{bmatrix} \cos B_☾ \cos v \\ \cos B_☾ \sin v \\ -\sin B_☾ \end{bmatrix}$$

$$= \cos B_☾ \begin{bmatrix} \sin^2\tfrac{1}{2}\vartheta \cos(v+\varphi) + \cos^2\tfrac{1}{2}\vartheta \cos(v-\varphi) + \sin\vartheta \sin\varphi \tan B_☾ \\ -\sin^2\tfrac{1}{2}\vartheta \sin(v+\varphi) + \cos^2\tfrac{1}{2}\vartheta \sin(v-\varphi) + \sin\vartheta \cos\varphi \tan B_☾ \\ \sin\vartheta \sin v - \cos\vartheta \tan B_☾ \end{bmatrix}.$$

The direction cosines of the pole of the ecliptic are $(B_☾ = -\tfrac{1}{2}\pi)$

$$\begin{bmatrix} p_1 \\ p_2 \\ p_3 \end{bmatrix} = \begin{bmatrix} -\sin\varphi \sin\vartheta \\ -\cos\varphi \sin\vartheta \\ \cos\vartheta \end{bmatrix}.$$

Because the pole of the ecliptic remains fixed in an inertial system,

$$\frac{dp_i}{dt} = \dot{p}_i + \varepsilon_{ijk}\omega_j p_k = 0.$$

For $i=1$ and 2, the following equations are derived:

$$\begin{aligned} \omega_1 &= (\dot{p}_2 + p_1\omega_3)/p_3 \\ \omega_2 &= (-\dot{p}_1 + p_2\omega_3)/p_3 \\ \dot{\omega}_1 &= (\ddot{p}_2 + \dot{\omega}_3 p_1 + \omega_3\dot{p}_1 - \omega_1\dot{p}_3)/p_3 \\ \dot{\omega}_2 &= (-\ddot{p}_1 + \dot{\omega}_3 p_2 + \omega_3\dot{p}_2 - \omega_2\dot{p}_3)/p_3 . \end{aligned}$$

Using these equations, (9a) and (9b) may be reformulated as follows:

$$\ddot{p}_1 + \beta\omega_3^2 p_1 - (1-\beta)\,\omega_3\dot{p}_2 - \dot{\omega}_3 p_2 + \omega_1\dot{p}_3 = p_3\beta q^2 (a/r)^3 u_3 u_1 \tag{10a}$$

$$\ddot{p}_2 + \alpha\omega_3^2 p_2 + (1-\alpha)\,\omega_3\dot{p}_1 + \dot{\omega}_3 p_1 - \omega_2\dot{p}_3 = p_3\alpha q^2 (a/r)^3 u_2 u_3 . \tag{10b}$$

From the equation for ω_3 in terms of the Euler angles,

$$\dot{\omega}_3 = \ddot{\tau} - 2\ddot{\sigma}\sin^2\tfrac{1}{2}\vartheta - (\dot{\Omega} + \dot{\sigma})\,\dot{\rho}\sin\vartheta$$

which, combined with (9c), yields

$$\ddot{\tau} + \gamma\omega_1\omega_2 - 2\ddot{\sigma}\sin^2\tfrac{1}{2}\vartheta + (\dot{\Omega} + \dot{\sigma})\,\dot{\rho}\sin\vartheta = \gamma q^2 (a/r)^3 u_1 u_2 . \tag{10c}$$

Equation (10c) represents the Moon's rotation about its polar axis, and Equation (10a, b), represent the motion of the axis.

4. Approximate Solution of the Dynamical Equations

Before proceeding with the solution of Equations (10a, b, c), their first order solutions will be reviewed. Let $L_☾$ be the true longitude of the moon; let $s = \Sigma_i H_i \sin h_i$ be the center equation and inequalities in longitude; let $F=L-\Omega$ be the principal argument of latitude; and let i be the inclination of the Moon's orbit to the ecliptic. Then

$$\begin{aligned} L_☾ &= L + s \\ v &= \pi + L_☾ - \Omega - \sigma \\ \varphi &= \pi + L + \tau - \Omega - \sigma \\ v - \varphi &= s - \tau . \end{aligned}$$

To the first order, set

$$\begin{aligned} \omega_3 &= \dot{L} \\ r &= a \\ q^2 &= 3\dot{L}^2 \\ B_☾ &= \sin i \sin F \\ \vartheta &= I . \end{aligned}$$

Dropping all quadratic terms in $B_☾$, I, s, τ and p_1,

$$\begin{aligned} p_3 &= 1 \\ \dot{p}_3 &= 0 \\ u_1 &= \cos(v - \varphi) = 1 \\ u_2 &= \sin(v - \varphi) = s - \tau \\ u_3 &= \sin\vartheta \sin\varphi - B_☾ = - p_1 - B_☾ \\ u_1 u_2 &= s - \tau \\ u_2 u_3 &= 0 \\ u_1 u_3 &= - p_1 - B_☾ . \end{aligned}$$

Equations (10a, b, c) become, respectively,

$$\ddot{p}_1 + 4\beta\dot{L}^2 p_1 - (1 - \beta)\,\dot{L}\dot{p}_2 = - 3\beta\dot{L}^2 \sin i \sin F \tag{11a}$$

$$\ddot{p}_2 + \alpha\dot{L}^2 p_2 + (1 - \alpha)\,\dot{L}\dot{p}_1 = 0 \tag{11b}$$

$$\ddot{\tau} + 3\gamma\dot{L}^2\tau = 3\gamma\dot{L}^2\, \Sigma_i H_i \sin h_i \tag{11c}$$

and, with $\dot{F}/\dot{L} = 1 + \mu$, the particular solutions are

$$\begin{aligned} p_1 &= 3\beta \sin i \sin F/(2\mu - 3\beta) \\ p_2 &= 3\beta \sin i \cos F/(2\mu - 3\beta) \\ \tau &= \Sigma_i (1 - h_i^2/3\gamma\dot{L}^2)^{-1} H_i \sin h_i . \end{aligned}$$

By Cassini's laws, the amplitude of p_1 and p_2 is I. Therefore, to the first order

$$\beta = \tfrac{2}{3}\mu I/(I + \sin i) = 6.2 \times 10^{-4} . \tag{12}$$

Dissipative terms have been omitted from the dynamical equations, so the homogeneous solutions of Equations (11a, b, c) represent undamped free librations. Physically there is damping which should be considered to estimate the possible amplitudes, if any, of the free librations. Nevertheless, if there are free librations in the Moon's wobble, they are of the form

$$p_1 = c_1 \sin \nu t,\; p_2 = c_2 \cos \nu t$$

where c_1 and c_2 are small, perhaps negligible, amplitudes and ν is the angular velocity. Homogeneous solutions to Equations (11a, b) exist, then, only if

$$\begin{vmatrix} 4\beta\dot{L}^2 - \nu^2 & (1-\beta)\,\dot{L}\nu \\ (1-\alpha)\,\dot{L}\nu & \alpha\dot{L}^2 - \nu^2 \end{vmatrix} = 0 .$$

That is, with $\eta = (\nu/\dot{L})^2$

$$\eta^2 - (1 + 3\beta + \alpha\beta)\,\eta + 4\alpha\beta = 0 .$$

The sum of the roots of this equation is approximately

$$\eta_1 + \eta_2 = 1 + 3\beta$$

and the product of the roots is

$$\eta_1\eta_2 = 4\alpha\beta .$$

Because α and β are very small quantities, $\eta_1\eta_2 \ll \eta_1 + \eta_2$, so one root, η_1, say, is much larger than the other and

$$\eta_1 = 1 + 3\beta$$
$$\eta_2 = 4\alpha\beta/\eta_1 \approx 4\alpha\beta .$$

The angular frequencies for the free librations in wobble are

$$\nu_1 = \dot{L}\sqrt{\eta_1} \approx (1 + \tfrac{3}{2}\beta)\,\dot{L} \qquad (13a)$$

and

$$\nu_2 = \dot{L}\sqrt{\eta_2} \approx 2\sqrt{\alpha\beta}\,\dot{L} . \qquad (13b)$$

If there is a free libration in the longitude of the Moon, its angular frequency occurs approximately at the singularity of the particular solution of Equation (11c),

$$\nu_3 \approx \sqrt{3\gamma}\,\dot{L} \qquad (13c)$$

Equations (12), (13b), and (13c) imply, respectively, that $\beta > 0$, $\alpha\beta > 0$, and $\gamma > 0$; in other words, $C_{33} > C_{22} > C_{11}$.

5. The Direct Effect of the Sun

The attraction of the Sun directly on the Moon contributes to the physical librations of the Moon. The solar effect is evaluated by replacing $q^2 (a/r)^3\, u_i u_j$ on the right-hand sides of Equations (9a, b, c) by $(q')^2\, (a'/r')^3\, u_i' u_j'$.

The prime denotes that the symbol is referred to the Sun rather than the Earth.

Because $(q')^2 = (m'/m)\,(a/a')^3\, q^2/\lambda = 0.005649\, q^2$, the direct effect of the Sun on the physical librations of the Moon is much less than that of the Earth. A first order development, paralleling the development for the Earth, is sufficient.

Corresponding to v and $B_☾$ are

$$v' = L' + s' - \Omega - \sigma$$

and $B_☾' = (a/a')\, B_☾ \approx 0$.

Here L' is the mean longitude of the Sun from the Earth and $s' = s_1' + s_2'$ is composed of two parts: s_1' is the center equation of the Sun, and s_2' is the contribution to

v' caused by the motion of the Moon about the Earth. For this development, no distinction is required between the center of mass of the Earth and the Earth-Moon barycenter.

To the first order in the eccentricities, e and e', of the lunar and solar orbits about the Earth, and in the ratio a/a',

$$\begin{aligned} s_1' &= 2e' \sin l' \\ s_2' &= -(a/a') \sin D \\ a'/r' &= 1 + e' \cos l' + (a/a') \cos D \end{aligned}$$

where l' is the mean anomaly of the Sun and D is half the argument of the variation. Then

$$\begin{aligned} v' - \varphi &= \pi + s' - \tau - D \\ 2v' - \varphi &= \pi + F - 2D + 2s' - \tau - \sigma \\ u_1' &= \cos(v' - \varphi) \\ u_2' &= \sin(v' - \varphi) \\ u_3' &= I \sin v' \\ u_1'u_2' &= \tfrac{1}{2}\sin(2v' - 2\varphi) \approx -\tfrac{1}{2}\sin 2D + s'\cos 2D = -\tfrac{1}{2}\sin 2D + \tfrac{1}{2}(a/a')\sin D - \\ &\quad - \tfrac{1}{2}(a/a')\sin 3D + e' \sin(l' + 2D) + e' \sin(l' - 2D) \\ u_2'u_3' &= \tfrac{1}{2}I[-\cos(2v' - \varphi) + \cos\varphi] \approx \tfrac{1}{2}I[\cos(F - 2D) - \cos F] \\ u_3'u_1' &= \tfrac{1}{2}I[\sin(2v' - \varphi) + \sin\varphi] \approx -\tfrac{1}{2}I[\sin(F - 2D) + \sin F] \\ (a'/r')^3 &= 1 + 3e' \cos l' + 3(a/a') \cos D \end{aligned}$$

$$\begin{aligned} (a'/r')^3 u_1'u_2' = -\tfrac{1}{2}\sin 2D - \tfrac{1}{4}(a/a')[5\sin 3D + \sin D] + \\ + \tfrac{1}{4}e'[7\sin(l' - 2D) + \sin(l' + 2D)] \end{aligned}$$

$$(a'/r')^3 u_2'u_3' = \tfrac{1}{2}I[\cos(F - 2D) - \cos F]$$

$$(a'/r') u_3'u_1' = -\tfrac{1}{2}I[\sin(F - 2D) + \sin F]$$

These terms, when multiplied by $(q')^2/q^2 = 0.005649$, may be added to the developments of the terms on the right-hand sides of Equations (10a, b, c), following the factor q^2. For these and higher order solar terms, the solutions of the linearized equation – with the left-hand sides of the form in Equations (11a, b, c) – show that the direct solar terms contribute very little to the physical librations of the Moon.

6. Solutions of the Dynamical Equations

Equations (10a, b, c) are to be solved. When taken together, they are of the form

$$RX = Y \tag{14}$$

in which R is an operator; X is the solution vector desired; and Y is the vector that represents the right-hand side of the equations and which depends on X. If one could find the inverse of R, then the solution would be simply

$$X = R^{-1} Y .$$

Such a direct solution is not possible. On the other hand, Equation (14) may be transformed to the form

$$TX = Y + (T - R)\,X$$

in which T is the operator of Equations (11a, b, c). Except at the singularities of the free physical librations for this operator, its inverse exists, and a suitable iterative process to solve Equation (14) is afforded by

$$X_{n+1} = X_n + T^{-1}(Y_n - RX_n) \tag{15}$$

in which the subscript identifies the solution cycle. If this process eventually converges such that $X_{n+1} = X_n$, then $Y_n - RX_n = 0$, and a solution is obtained. Essentially, this is the technique which has been used in iterative solutions of the Moon's forced physical librations.

Equation (15) represents a complex process which, fortunately, can be automated. The first order generation of the solar contributions to the vector Y_1, which is presented in the previous section, is only a very small part of the total process. Nevertheless it is illustrative of some of the manipulations required, such as the addition and multiplication of series of trigonometric terms.

In the generation of the solar contributions, each trigonometric term consists of a symbolic coefficient and a symbolic argument. The dominant terms in Y_1 are the earth contributions which are derived from spherical coordinates of the Moon from the Earth. These coordinates, in turn, are derived by lunar theory as series of trigonometric terms with numerical coefficients and symbolic arguments. This is the form of the series which the IBM 7044 and 7090 computers have been programmed to handle.

Essentially, the computer programs for the IBM 7044 and 7090 are the same. They yield solutions of the forced physical librations of the Moon by an iterative process of the form of Equation (15). Each program includes subprograms to perform the necessary numerical-symbolic algorithms for successively updating $Y - RX_n$ in Equation (15). There are, for example, subprograms to differentiate or scale a series of trigonometric terms or to add, subtract, or multiply two such series. Each series contains up to 200 trigonometric terms, and each term consists of a numerical coefficient and an argument which is of the form

$$il + jl' + kF + mD$$

for which l, l', F and D are the Delaunay arguments and i, j, k, and m are integers with the program restriction that

$$0 \leq i \leq 7, -3 \leq j \leq 4, -3 \leq k \leq 4, -7 \leq m \leq 8.$$

Results of the algorithm subprograms are accumulated into a special area of computer storage with a capacity for 8192 terms. Where required, the contents of this buffer area are stored into a 200 term storage block, truncating only insignificant terms. The truncation level is generally 10^{-7}, but it is automatically increased if necessary to limit each series to no more than 200 terms. The precision of each block storage is monitored and it can be shown that the precision is degraded only for a few series, all of which are of little weight, and the overall precision is consistent with a 10^{-7} truncation level for each block storage operation.

Solutions are generated for various combinations of the parameters β and γ. The angular velocities of all fundamental arguments and the series for the solar terms in longitude, s, and in sine parallax, which is proportional to a/r, are taken from ECKERT *et al.* (1954). The series for the solar terms in latitude, $B_{☾}$, is taken directly from BROWN (1905). The longitude (except for the inclusion of Brown's term #268) and latitude series are truncated at 0".100 and the sine parallax series is truncated at 0".0150.

From the solutions for p_1 and p_2, ρ, $I\sigma$ and I are updated after each iteration. To a sufficient accuracy

$$p_1^2 + p_2^2 = \vartheta^2 = I^2 + 2I\rho$$

Thus

$$\rho(\text{updated}) = \text{variable part of } (p_1^2 + p_2^2)/2I$$
$$I(\text{updated}) = \tfrac{1}{2}I + \text{constant part of } (p_1^2 + p_2^2)/2I$$

Also, to a sufficient accuracy,

$$p_1 \cos F - p_2 \sin F = - \vartheta \sin(\varphi - F) = \vartheta \sin(\tau - \sigma) = I\tau - I\sigma - \rho\sigma$$

Thus

$$I\sigma\,(\text{updated}) = I\tau - p_1 \cos F + p_2 \sin F - (\rho/I)\, I\sigma\,.$$

Iterations are terminated after the change in each of the coefficients of τ, ρ, and $I\sigma$ is less than 0".1. For initial values $I = 1°32'50''$, τ, ρ, $\sigma = 0$, $p_1 = I \sin F$, and $p_2 = I \cos F$; this usually occurs after three to five cycles, depending on the proximity of singularity in the solution of τ.

The true values of β and γ are probably near $\beta = 0.00063$ and $\gamma = 0.00020$, so one series of solutions has been generated for $\beta = 0.00063$ and γ ranging between $\gamma =$ 0.00010 and $\gamma = 0.00032$, and another series of solutions has been generated for $\gamma = 0.00020$ and β ranging between $\beta = 0.00060$ and $\beta = 0.00066$. These solutions have been published in the *Astronomical Journal* (ECKHARDT, 1965). In subsequent runs, Brown's longitude term #182, whose argument is $2F - 2l$ (twice the distance of the Moon's mean perigee from the mean ascending node of the Moon's orbit), has been modified to that of the recent solutions of the main problem of lunar theory by W. J. Eckert and H. F. Smith, Jr. (ECKERT, 1964). The change in the coefficient of this term from $-1''.298$ to $-1''.370$ effects a virtually uniform five percent diminution in the magnitudes of the coefficients of the $2F - 2l$ term in τ. Other terms are negligibly affected.

The $2F - 2l$ (2ω in Hansen's notation) term in τ has a singularity which falls between $\gamma = 0.00020$ and $\gamma = 0.00022$. Assuming that the coefficient behaves as a simple pole in the neighborhood of the singularity, it is found that the singularity falls at $\gamma = 0.000213$ which corresponds to a mechanical ellipticity of $f = \alpha/\beta = 0.662$, in agreement with Hayn and Kozieł.

From the solutions for I for parameter ranges cited, the following equation is derived

$$I = -1622'' - 4''.8 \times 10^4 \gamma + 11''.4 \times 10^6 \beta$$

This equation has been used, along with the equation $\gamma = \beta(1-f)$, to determine the values of β and γ required to generate solutions which can be directly compared with the solutions of Hayn and Kozieł for $I = 5540''$ and $f = 0.5$, 0.6, 0.7 and 0.8. The comparison shows that the computer solutions agree with those of Hayn and Kozieł, usually within a few tenths of a second of arc to about a second at most for all coefficients, except for the τ coefficient of the term whose argument is $2F-2l$. The solutions for this coefficient by the computer and the careful calculations of Kozieł are contrasted in the following tabulation:

$I=5540''$	$f=0.5$	$f=0.6$	$f=0.7$	$f=0.8$
Computer	$-5''.3$	$-10''.7$	$12''.5$	$2''.2$
Kozieł	$-7''.5$	$-15''.7$	$19''.3$	$3''.5$

The conclusions that can be drawn from this comparison are that the approximations introduced by Hayn do not reduce the precision of the physical libration solutions by more than one second of arc but that the linearizations of Hayn do lead to inaccurate solutions for the $2F-2l$ term in τ.

Kozieł (1964) has derived the physical libration constants, $I = 5521''$ and $f = 0.633$; the corresponding moment of inertia ratios are $\beta = 0.00006276$ and $\gamma = 0.0002303$.

TABLE I

Physical Librations for $\beta = 0.0006276$ and $\gamma = 0.0002303$

	Argument l	l'	F	D	Coefficient
τ, sine terms	0	0	2	−2	$1''.7$
	0	1	0	0	$91''.7$
	1	−1	0	−1	$-1''.2$
	1	0	0	−2	$4''.2$
	1	0	0	−1	$-3''.5$
	1	0	0	0	$-16''.9$
	2	−1	0	−2	$1''.0$
	2	0	−2	0	$20''.1$
	2	2	2	−2	$10''.0$
$I\sigma$, sine terms	0	0	2	−2	$-3''.2$
	0	0	2	0	$-10''.6$
	1	0	−2	0	$-23''.8$
	1	0	0	−2	$2''.5$
	1	0	0	0	$-100''.7$
ρ, cosine terms	0	0	2	−2	$-3''.2$
	0	0	2	0	$-11''.0$
	1	0	−2	0	$23''.9$
	1	0	0	−2	$-1''.9$
	1	0	0	0	$-98''.5$
I, constant term	–	–	–	–	$5520''.6$

Table I presents the final solution for τ, $I\sigma$, ρ and I using these parameters and the Eckert-Smith modification of Brown's term #182. Terms whose coefficients are less than 1″.0 are omitted. One should note that Kozieł's physical libration constants were derived from heliometer measurements of the Moon using a different libration theory. His constants would be slightly modified if the current theory had been used; so Table I is more an illustration of a typical solution than an authoritative solution based on observational reductions.

The physical librations induced by planetary terms and additive terms to the fundamental arguments of lunar theory are not included in the computer solutions. The influence of these terms on the libration in longitude may be seen, however, using the first-order solution for τ. Except near $h_i^2 = 3\gamma \dot{L}^2$ and where $3\dot{L}^2 \gg h_i^2$, the amplitude of the forced libration is considerably less than the amplitude of the inducing term. The largest planetary term in longitude has an amplitude of 1″.137, but its period is near a month so $h_i^2 \approx \dot{L}^2 \gg 3\gamma \dot{L}^2$ and its effect is negligible. By inspection of the amplitudes and periods of all Brown's planetary terms in longitude, it is evident that none has a significant effect on forced librations. All of the significant additive terms to the mean longitude of the Moon have long enough periods such that they induce librations of approximately the same magnitudes and thus cancel out the equivalent optical librations. For instance, Brown's term #1375 is the additive term to the longitude 7″.261 $\sin \Omega$. It causes the optical libration term 7″.3 $\sin \Omega$ which, for $\gamma = 0.0002$, is effectively cancelled by the physical libration term $-7''.5 \sin \Omega$. The term can be safely ignored provided that it is omitted in the generation of the lunar ephemeris from which the optical librations are derived.

References

Brown, E. W.: 1899, *Mem. Roy. Astron. Soc.* **53**, 39.

Brown, E. W.: 1905, *Mem. Roy. Astron. Soc.* **57**, 129.

Eckert, W. J.: 1964, *Trans. IAU* **12B**.

Eckert, W. J., Jones, R., and Clark, H. K.: 1954, *Improved Lunar Ephemeris 1952–1959*. Washington, p. 283.

Eckhardt, D. H.: 1965, *Astron. J.* **70**, 466.

Harrison, J. C.: 1963, *J. Geophys. Res.* **68**, 4269.

Hayn, F.: 1902, *Abhandl. Sächs. Ges. Wiss.* **27**, 861.

Jeffreys, H.: 1961, *Monthly Notices Roy. Astron. Soc.* **122**, 421.

Kozieł, K.: 1948, *Acta Astron.* **4**, 61.

Kozieł, K.: 1964, *Trans. IAU* **12A**, 233.

PHOTOGRAPHIC TECHNIQUES FOR THE DETERMINATION OF THE MOON'S CONSTANTS OF ROTATION*

D. W. G. ARTHUR**

Abstract. To date all primary selenodetic results rest upon heliometer measures. The heliometer can be replaced by the long focus photographic refractor, provided that the problems of scale and orientation can be surmounted. The paper outlines the use of ordinary and sinusoidal star trails for the solution of these problems. The analysis of the measures can follow that of classical selenodesy, involving the lunar limb, but the lunar photograph permits a better approach. In this no reference is made to the limb and the weakness of the classical method, that is, the vagueness of the center of figure, is avoided. Instead the selenodetic coordinates are referred to a visible point on the lunar surface as origin.

1. General

The LPL approach to the problem of the determination of the Moon's constants of rotation is based on measures on scaled and oriented photographs. This use of photographs permits a larger number of measurements than is possible with the heliometer, and enables us to bypass the inherent accuracy limitation of the Bessel-Wichmann method, namely, the use of the vague *center of figure* as the origin of the selenodetic coordinate system. The classical method is forced into this by the measures themselves, since these refer the points of the limb to the fundamental point Mösting A. In consequence, the reductions must be based on the doubtful assumption that there is some unique point in the interior of the Moon which is related to the arcs of the limb. Refinements of this assumption, such as occur in the works of Watt, are even more arbitrary in nature.

In our scheme this source of uncertainty is avoided by taking Mösting A itself as the origin of selenodetic coordinates. On the plate we take the parallel step of referring the plate coordinates to the image of Mösting A. The observation equations then state the instantaneous values of these reduced plate coordinates for a number of well defined points. Thus, the reductions do not determine the barycentric coordinates of Mösting A; indeed, these are assigned approximate values. Instead, the selenodetic coordinates, measured from Mösting A, are determined for the other measured points.

The technique demands lunar plates with well determined orientations and scales. The orientations are found from star-trailed lunar photographs, while the scales are

* Research supported by the Air Force Cambridge Research Laboratories, U.S. Air Force, under Contract AF 19(628)-4332.

** University of Arizona, Lunar and Planetary Laboratory, Tucson, Ariz., U.S.A.

found from sinusoidal star trails registered by an oscillating plateholder. These techniques are now described in detail.

2. The Determination of Orientation from Star Trails

Star-trailed lunar photographs were obtained in the period 1959–1962 using the 40-in. refractor of the Yerkes Observatory. Each such photograph is a double exposure in which the lunar image is recorded first. The telescope is then arrested, and if necessary, moved up to 2° in declination to cover the transit of a star. This is monitored using the 6-in. long focus finder. Both exposures are carefully timed, the time for the star trail corresponding to the center of the field of the finder which has already been collimated with the main instrument.

In the measures and reductions we take into account the curvature of the trail and the difference between the refractive compressions in the two exposures. Since the trail is curved, some particular part of it must be identified as defining the east-west direction at the center of the disk. We assume that the relevant point is the foot of the perpendicular from the center of face, that is, the point on the trail and the Moon's hour circle. The direction actually accepted is the tangent at this point.

To simplify the presentation, assume that the plate is observed with the star trail aligned closely parallel to the x-axis of the machine. In practice this situation is reached computationally by a preliminary rotation of the measures. This step makes dy/dx for the trail very small, so that it can be ignored in the expression for the curvature. The latter cannot be reliably determined from the measures on the trail itself since this is appreciably distorted by seeing movements. Instead we impose a curvature of $\tan\delta^*/\mathscr{F}$, where δ^* is the declination of the star and $\mathscr{F}$ is the approximate focal length. Thus, if we write the equation of the (hyperbolic) trail as

$$y = a + bx + cx^2 \tag{1}$$

which is quite precise enough for our purpose, we have, when dy/dx is neglected in the curvature,

$$c = \tan \delta^*/2F \tag{2}$$

The observation equation to determine a and b is then

$$a + bx = y - cx^2 = l \tag{3}$$

For n points on the trail these give the normals

$$\left.\begin{aligned} na + b\Sigma x &= \Sigma l \\ a\Sigma x + b\Sigma x^2 &= \Sigma lx \end{aligned}\right\} \tag{4}$$

from which b is found. In general the slope is

$$\frac{du}{dx} = b + 2cx$$

and if the approximate coordinates of the center of face are (h, k), the relevant slope is

$$\tan\theta = b + 2ch \tag{5}$$

The approximate center of face (h, k) is found from the observation equation

$$(x - h)^2 + (y - k)^2 = r^2 \tag{6}$$

as applied to the measured points of the bright limb. The solution is not rigorous, since it minimizes the amount

$$\Sigma[(x - h)^2 + (y - k)^2 - r^2]$$

but this lack of rigor is not important at this stage, since the coordinates (x, y) are themselves affected by refraction.

Equation (5) defines the refraction-affected east-west direction in the star-trail exposure and requires a small correction to adapt it for use in the lunar exposure. Let λ and μ be the vertical and horizontal refraction compressions in the lunar exposure, and let λ^*, μ^* be the corresponding values in the stellar exposure. These are easily found from the zenith distances, and their derivations need not be detailed here. We also require the ratios

$$\mu = v/\lambda \quad \text{and} \quad \mu^* = v^*/\lambda^* \tag{7}$$

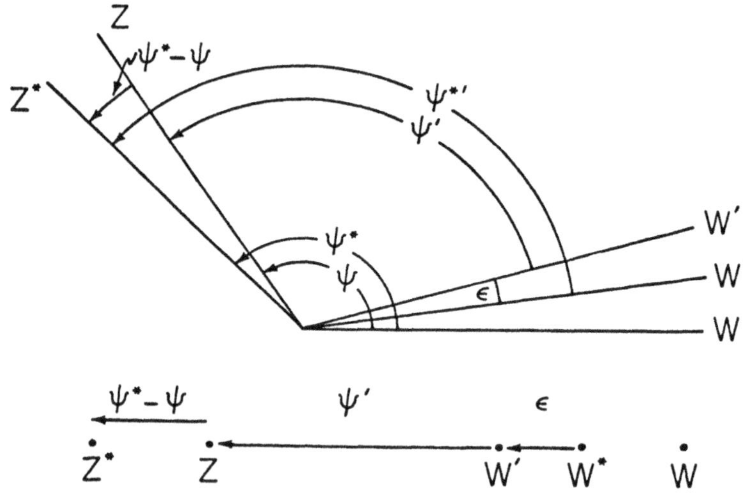

Fig. 1.

In Figure 1 let OW represent the refraction-free direction of the east-west line. *This is assumed to be identical in both pictures.* Let OZ and OZ^* represent the directions of the zenith in the refraction-free lunar and stellar exposures. Clearly they are also the directions of the zeniths in the refracted pictures, so that

$$\left.\begin{aligned} WOZ &= 90^\circ + Q = \psi \\ WOZ^* &= 90^\circ + Q^* = \psi^* \end{aligned}\right\} \tag{8}$$

where Q and Q^* are the parallactic angles without refraction for the Moon and star. Also

$$ZOZ^* = \psi^* - \psi = Q^* - Q \tag{9}$$

before and after refraction. Let the refraction in the lunar and stellar pictures displace W to W' and W^* respectively, and let

$$\begin{aligned} W'OZ &= \psi', \\ W^*OZ^* &= \psi^{*\prime}. \end{aligned}$$

Then ψ' and $\psi^{*\prime}$ are the values of ψ and ψ^* affected by refraction, and are found from

$$\left.\begin{aligned} \tan \psi' &= \mu \tan \psi \\ \tan \psi^{*\prime} &= \mu^* \tan \psi^* \end{aligned}\right\} \tag{10}$$

Now let ε be the counterclockwise rotation from the star-trail direction OW^* to the refraction-affected east-west direction OW' in the lunar exposure. From Figure 1,

$$\varepsilon + \psi' + (\psi^* - \psi) = \psi^{*\prime},$$

or

$$\varepsilon = (\psi^{*\prime} - \psi') - (\psi^* - \psi) \tag{11}$$

The original measures of the selenodetic points can now be reduced. We remember that the rotation from the x-axis to the relevant tangent to the trail was found as θ, while the refraction-affected east-west direction in the lunar image makes an angle ψ' with the vertical. Hence, the (x, y) system must be rotated counterclockwise through the angle

$$\xi = \theta + \varepsilon + \psi' - 90^\circ \tag{12}$$

to bring the y-axis into the vertical. Thus for the selenodetic points and points of the limb, the refraction-free horizontal and vertical coordinates are

$$\left.\begin{aligned} u &= \frac{1}{v}(x \cos \xi + y \sin \xi) \\ v &= \frac{1}{\lambda}(y \cos \xi - x \sin \xi) \end{aligned}\right\} \tag{13}$$

Rotating back through the refraction-free parallactic angle Q, the refraction-free coordinates in an equatorial system are

$$\left.\begin{aligned} x_e &= u \cos Q - v \sin Q \\ y_e &= v \cos Q + u \sin Q \end{aligned}\right\} \tag{14}$$

This follows our computations far enough for present purposes. In practice we redetermine the center of face rigorously and reduce (x_e, y_e) to the correct center of face, so far as this can be found from the bright limb.

3. The Scale Problem

Photographs taken with long focus refractors cannot be fully exploited in primary selenodetic work without a knowledge of the precise focal length. Since I have some reservations regarding the appropriateness of the usual approach, via the Pleiades arcs, the LPL technique will depend on star-trail photography with the telescope stationary. The plate is oscillated in declination through a total distance of 1.5 mm with a period of one solar second. This technique permits determinations of focal length immediately before and after the lunar exposure.

The apparatus consists of a dummy 8 × 10″ plateholder whose external dimensions are exactly those of the normal 8 × 10″ plateholders used for lunar photography. The former holds a 1 × 10″ plate in exactly the same focal plane as the 8 × 10″ lunar plates, but contains internal slides on which the 1 × 8″ plateholder moves. A synchronous motor imparts a periodic motion in declination to the latter with the stated amplitude and frequency.

The focal length is determined from the wavelength of the sinusoidal star trail. Figure 2 shows the points of the trail that are measured in the comparator. The points *AAA*... and *BBB*... are measured in y merely to align the x-axis of the measures along the trail and to fix the axis *XX* of the sinusoid. The x-measures on the points 1, 2, 3... then permit the derivation of a regression of x on the number of half-waves, and thus a least squares estimate of the wavelength. The focal length is connected to this and to the declination of the star in an elementary way which need not be detailed

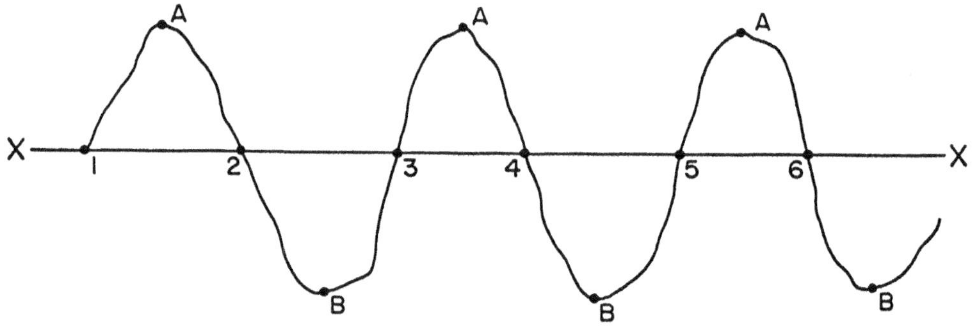

Fig. 2.

here. It is rather early to estimate the precision of the technique, but on good quality star-trailed lunar photographs, the seeing deviations have standard values of 30 microns. If similar dispersions hold for the sinusoidal trails, then the focal length of a large refractor can be determined within a few tenths of a millimeter. Naturally the precision depends on the seeing, so that good lunar plates have precise focal lengths, whereas inferior plates have rather poorly determined focal lengths.

The technique should provide a series of plates with known focal lengths. The problem then arises of transferring scale to other lunar plates. This can be done but only when the plates with known and unknown focal lengths have similar librations, say within one degree in longitude and latitude. Also the same selenodetic points must be measured on both plates, and these must be disposed roughly along a circle at a constant distance from the center of face. For each of the two plates we may compute a *selenodetic* focal length as follows. Let (x, y) be the photographic coordinates of the selenodetic point corrected for refraction, and let (E, F, G) be the solid selenodetic coordinates. Then the instantaneous solid coordinates are (X, Y, Z) where

$$\begin{pmatrix} X \\ Y \\ Z \end{pmatrix} = \begin{pmatrix} a, b, c \\ e, f, g \\ i, j, k \end{pmatrix} \cdot \begin{pmatrix} E \\ F \\ G \end{pmatrix} \tag{15}$$

in which the elements of the libration matrix on the right have the usual values

$$\begin{aligned} a &= \cos l', \\ b &= 0, \\ &\cdots\cdots\cdots\cdots \\ k &= \cos l' \cos b'. \end{aligned}$$

If s' is the augmented semidiameter, the coordinates of the projections on the plane of the limb are

$$\left.\begin{aligned} X' &= X/(1 - Z \sin s') \\ Y' &= Y/(1 - Z \sin s') \end{aligned}\right\} \tag{16}$$

The refraction-free photographic coordinates are connected to these by

$$\left.\begin{aligned} x &= \mathscr{F} \sin s'(X' \cos \theta - Y' \sin \theta) \\ y &= \mathscr{F} \sin s'(Y' \cos \theta + X' \sin \theta) \end{aligned}\right\} \tag{17}$$

where $\mathscr{F}$ is the focal length and θ is an unknown rotation. A least squares value for $\mathscr{F}$ can be derived for both plates, that is, the plate with known focal length and the plate with unknown focal length.

This selenodetic focal length is affected by errors in the values (E, F, G), errors in the assumed librations, and errors in the measures. The errors from the librations are very small. The dominant errors come from E, F and G. The latter cause highly correlated errors in the selenodetic focal lengths $\mathscr{F}$ and $\mathscr{F}_0$ for the two plates, so that we can write, with some precision,

$$\frac{\mathscr{F}}{\mathscr{F}_0} = \frac{L}{L_0}$$

where L and L_0 are the true focal lengths. Hence, the focal length may be transferred by

$$L = L_0 \cdot \frac{\mathscr{F}}{\mathscr{F}_0} \tag{18}$$

The error propagation in this process is extremely complex, and in any case, representative patterns of controls must be used to compute the dispersion in the computed true focal length L. In some simulated transfers using 10 points with errors of 0.001

in E and F, the errors of L were about one part in 10^6, the errors of measurement being ignored.

Thus it appears that scale can be transferred with an accuracy quite commensurate with the precision of the measures on the photographs.

4. The Reduction for the Constants of Rotation

The LPL method for the determination of the constants of rotation is based on a new approach to the geometry of the problem, on Eckhardt's numerical integration of the equations of rotation, and on the use of scaled and oriented photographs.

Eckhardt's results are represented by tables that are interpolated by the high-speed program to determine the values of the coefficients of the expansions of the physical librations ρ, σ, τ, or rather, ρ, $I\sigma$ and τ.

From this point the treatment deviates from the usual in that it makes no use of the differential corrections which are seen in Hayn and his successors. The total librations (l'', b'') corrected for physical libration and the place of the telescope are computed rigorously from

$$\begin{aligned} \cos(☾ + l'' - \Omega + \tau - \sigma)\cos b'' &= \cos(\lambda' - \Omega - N - \sigma)\cos\beta', \\ \sin(☾ + l'' - \Omega + \tau - \sigma)\cos b'' &= \sin(\lambda' - \Omega - N - \sigma)\cos\beta'\cos(I + \rho) \\ &\quad - \sin\beta'\sin(I + \rho), \\ \sin b'' &= -\sin(\lambda' - \Omega - N - \sigma)\cos\beta'\sin(I + \rho) \\ &\quad - \sin\beta'\cos(I + \rho). \end{aligned} \tag{19}$$

These are nothing more than immediate generalizations of the rigorous formulas for the optical librations to which they degenerate when ρ, σ and τ are equated to zero. In these ☾ = Moon's mean longitude, Ω = longitude of mean ascending lunar node, N = nutation in longitude, λ' = Moon's topocentric longitude, β' = Moon's topocentric latitude, ρ = physical libration in inclination of Moon's equator, σ = physical libration in node of Moon's equator, τ = physical libration in longitude, and I = mean inclination of Moon's equator.

The computation of C'', the instantaneous position angle of the Moon's axis, corrected for physical libration and observer's position, requires the computation of the auxiliaries i', Δ' and Ω'', which bear the same relations to the *true* lunar equator as the ephemeris values i, Δ and Ω' bear to the *mean* lunar equator. Once more, replacing I by $I+\rho$, Ω by $\Omega+\sigma$, and ☾ by ☾$+\tau$, we have

$$\left.\begin{aligned} \sin\Delta'\sin i' &= -\sin\varepsilon\sin(\Omega + \sigma + N), \\ \cos\Delta'\sin i' &= +\sin(I + \rho)\cos\varepsilon \\ &\quad - \cos(I + \rho)\sin\varepsilon\cos(\Omega + \sigma + N) \\ \cos i' &= \cos(I + \rho)\cos\varepsilon + \sin(I + \rho)\sin\varepsilon\cos(\Omega + \sigma + N). \end{aligned}\right\} \tag{20}$$

and

$$\left.\begin{aligned} \sin\Omega''\sin i' &= -\sin(I + \rho)\sin(\Omega + \sigma + N), \\ \cos\Omega''\sin i' &= +\cos(I + \rho)\sin\varepsilon - \sin(I + \rho)\cos\varepsilon\cos(\Omega + \sigma + N). \end{aligned}\right\} \tag{21}$$

In these ε is the true obliquity. The value of C'' is then computed from either

$$\sin C'' = \sin i' \cos(l'' + \Delta' + (- \Omega + \tau - \sigma) \sec \delta',$$

or

$$\sin C'' = - \sin i' \cos(\alpha' - \Omega'') \sec b''. \qquad (22)$$

The orientation of the Moon is completely specified for the time of photography by l'', b'', and C''.

At this point the treatment again departs from the classical method. Instead of evaluating the selenodetic coordinates of one fundamental point Mösting A, we determine the differences in the selenodetic coordinates between this point and a number of others. In effect we make Mösting A the origin of our selenodetic system. Let (U, V, W) be the approximate barycentric coordinates of Mösting A in the selenodetic system. We identify these with one or other of the determinations of Kozieł, Schrutka-Rechtenstamm, etc., which are already available. The exact choice and its associated errors are not important. Let (u, v, w) be the selenodetic coordinates with respect to Mösting A. Then the approximate selenodetic coordinates with respect to the centroid as origin are

$$\left.\begin{aligned} E &= U + u \\ F &= V + v \\ G &= W + w \end{aligned}\right\}. \qquad (23)$$

Assuming values for I, f and (u, v, w), the photographic coordinates are computed as follows. The instantaneous coordinates are computed as in (15), and a set of photographic coordinates oriented with the y-axis along the first lunar meridian is found from

$$\left.\begin{aligned} x' &= LX \sin s'/(1 - Z \sin s') \\ y' &= LY \sin s'/(1 - Z \sin s') \end{aligned}\right\} \qquad (24)$$

The photographic coordinates, oriented in an equatorial system with the y-axis along the hour circle through the center of face, are then

$$\left.\begin{aligned} x'' &= x' \cos C'' - y' \sin C'' \\ y'' &= y' \cos C'' + x' \sin C'' \end{aligned}\right\} \qquad (25)$$

Finally, these are reduced to the image of Mösting A as origin on the plate, i.e.,

$$\left.\begin{aligned} x &= x'' - x''_m \\ y &= y'' - y''_m \end{aligned}\right\} \qquad (26)$$

in which (x''_m, y''_m) are the values of (x'', y'') at Mösting A. The last can now be compared with the values determined from the star-trailed plates. In general, the observed and computed values will not agree. The required observation equations are, for each point on each plate,

$$\begin{aligned} \frac{\partial x}{\partial f}\delta f + \frac{\partial x}{\partial I}\delta I + \frac{\partial x}{\partial u}\delta u + \frac{\partial x}{\partial v}\delta v + \frac{\partial x}{\partial w}\delta w &= x_0 - x_c \\ \frac{\partial y}{\partial f}\delta f + \frac{\partial y}{\partial I}\delta I + \frac{\partial y}{\partial u}\delta u + \frac{\partial y}{\partial v}\delta v + \frac{\partial y}{\partial w}\delta w &= y_0 - y_c \end{aligned} \qquad (27)$$

In these we distinguish between (x_0, y_0), the observed values, and (x_c, y_c), the computed values. The increments are the corrections to I, f, u, v, and w.

The formal differentiation to obtain the partial derivatives is slippery and has not always been performed correctly. It is avoided here by varying one unknown at a time and repeating the sequence (19), (20), (21), (22), (23), (15), (24), (25), and (26) as many times as there are unknowns. For example, if we compute x for the set (f, I, u, v, w) and the set $(f+\delta f, I, u, v, w)$, and obtain the values x_1 and x_2 respectively, then

$$\frac{\partial x}{\partial f} \approx \frac{x_2 - x_1}{\delta f}$$

is obtained numerically. The amount of computation is large, but to compensate for this the program is compact.

One point should not be overlooked in forming the normals from (27). The subtraction in (26) introduces algebraic correlation into the observed quantities. Fortunately, this is of a simple nature. Assuming for the present that x'' and y'' are independent and of uniform variance, then the covariance between any two x-values on the same plate is

$$\mu_{ik} = \sigma^2 \tag{28}$$

while the variance of an x (or y) is

$$\mu_{ii} = 2\sigma^2 \tag{29}$$

where σ^2 is the variance of an observation of unit weight. Thus the correlation matrix (which is the covariance matrix normalized by σ^2) for the measures may be written as the compartmented matrix

$$Q = \left[\begin{array}{c:c:c:c} Q_1 & & & \\ \hdashline & Q_2 & & \\ \hdashline & & & \\ \hdashline & & & Q_m \end{array}\right] \tag{30}$$

in which the zero submatrices represent the independence between plates. Assuming for each plate that the measures are ordered in the sequence $x_1, y_1, x_2, y_2 \ldots$, then each submatrix on the diagonal has the structure

$$Q_i = \begin{bmatrix} 2, & 0, & 1, & 0, & 1, & 0, & 1, & 0, & \cdot & \cdot & \cdot \\ 0, & 2, & 0, & 1, & 0, & 1, & 0, & 1, & \cdot & \cdot & \cdot \\ 1, & 0, & 2, & 0, & 1, & 0, & 1, & 0, & \cdot & \cdot & \cdot \\ 0, & 1, & 0, & 2, & 0, & 1, & 0, & 1, & \cdot & \cdot & \cdot \\ \cdot & \cdot & \cdot & \cdot & \cdot & \cdot & \cdot & \cdot & \cdot & \cdot & \cdot \\ \cdot & \cdot & \cdot & \cdot & \cdot & \cdot & \cdot & \cdot & \cdot & \cdot & \cdot \end{bmatrix} \tag{31}$$

To invert this we introduce the lower triangular matrix L such that

$$LL^{\mathrm{T}} = Q_i .$$

Then L is easily calculated as

$$L = \begin{bmatrix} 2^{\frac{1}{2}} & & & & \\ 0, & 2^{\frac{1}{2}} & & & \\ 2^{-\frac{1}{2}}, & 0, & 2^{\frac{1}{2}} & & \\ 0, & 2^{-\frac{1}{2}}, & 0, & 2^{\frac{1}{2}} & \\ \cdots & \cdots & \cdots & \cdots & \\ \cdots & \cdots & \cdots & \cdots & \end{bmatrix} \tag{32}$$

If we write

$$\begin{aligned} s_1 &= \sqrt{2}, \\ s_2 &= -2s_1, \\ s_3 &= -2(s_1 + s_2), \\ s_4 &= -2(s_1 + s_2 + s_3), \\ &\text{etc.,} \end{aligned} \tag{33}$$

the inverse of L is computed as the triangular banded matrix

$$L^{-1} = \begin{bmatrix} s_1, & & & & \\ 0, & s_1, & & & \\ s_2, & 0, & s_1, & & \\ 0, & s_2, & 0, & s_1, & \\ s_3, & 0, & s_2, & 0, & s_1 \\ \cdots & \cdots & \cdots & \cdots & \cdots \\ \cdots & \cdots & \cdots & \cdots & \cdots \end{bmatrix} \tag{34}$$

The inverse of Q_i is then

$$Q_i^{-1} = (L^{-1})^{\mathrm{T}} L^{-1} \tag{35}$$

and the inverse of Q is the generalized weight matrix G,

$$G = Q^{-1} = \begin{bmatrix} Q_1^{-1} & & & \\ & Q_2^{-1} & & \\ & & \ddots & \\ & & & Q_m^{-1} \end{bmatrix} \tag{36}$$

Thus, the inversion of Q reduces to the computation of the terms of a series followed by a matrix multiplication.

If the observations are written in the matrix form

$$D\Delta = X ,$$

where D is the rectangular matrix of derivatives, Δ is the column matrix of corrections to the unknowns, and X is the column matrix of the right-hand sides of (27), the normals, taking account of correlation, are

$$D^{\mathsf{T}}GD\Delta = D^{\mathsf{T}}GX \tag{37}$$

The corrections are applied to f, I, u, v and w, new coefficients in the expansions for ρ, $I\sigma$ and τ computed, and a second iteration commenced. Since we know f and I within close limits and since u, v and w are also known for the points we propose to use, two or three iterations should suffice.

At present some 45 points additional to Mösting A are being observed, but experience should reduce this number to about 4 or 5, as representing the points which are most often visible and measurable. With 5 points the number of unknowns is 17, which is not too large for modern computers.

DISCUSSION

HOPMANN: I wish to make a comment regarding the period of the libration phenomena. We have indications that the solution of the Euler equations may have terms of periods in the neighborhood of five to twenty years. This means that if we wish to establish the amplitudes of these terms it is necessary to cover with observations at least one period, or even better, a few periods. For this reason we have to make use of the old heliometric observations in the way professor Kozieł did. – Another thing that comes to my mind is the outcome of professor Kozieł's work over more than three thousand observations made during the span of half a century. Since all this material has resulted in an error of 0.″25 for the height of Mösting A, I believe that we cannot expect much more for other points measured only a few times on photographs taken only recently.

KOPAL: Mr. Chairman, I should like to make a couple of comments which I shall try to shape in a deliberately provocative manner. I should like to speak against the heliometers. It seems to me that for the past 100 years or so, great effort has been spent in observations made with instruments whose optical resolving power was very limited. If I am not wrong, the largest aperture of any heliometer used so far was 8 inches; and this aperture, on grounds of geometrical optics alone, cannot resolve angular separations smaller than one second of arc. Now the errors of the results, based on an analysis of such observations at the hands of Professor Kozieł and his associates, proved to be between one and two tenths of a second of arc. This, however, clearly falls into the category of fishing for information inside a diffraction pattern of the observed images. – There is, in principle, nothing objectionable to such a strategy. Positional astronomers do it often – e.g., invariably so in measuring the stellar parallaxes. But there the situation is essentially different from one arising in selenodesy. When you set the crosswires at the image of a star on an astrometric plate, you measure the position of a symmetrical image; moreover, the geometrical relation between the actual object and its diffraction image as formed in the focal plane of the telescope (i.e., a light point to a system of Airy's rings) is simple and well known. Its knowledge assures us that the optical center of the apparent image formed on the plate should coincide (within the limits of the errors of measurements) with the position of the actual stellar disc – even though the latter may be incomparably smaller than its often overexposed image. However, for lunar features too small to be resolved by the telescope employed the relationship between the real and diffraction image (which should enable us to localize its true position on the plate) is necessarily unknown. Therefore, it seems to me that to make observations which individually cannot be more precise than to one second of arc, and draw deductions from them within uncertainties of the order of one-tenth of a second of arc implies a tacit assumption that the errors of the results can be diminished at will by dividing them by a square-root of a sufficiently large number of underlying observations – an assumption which clearly implies that the observational errors reflected in the errors of the final results are accidental, and not systematic. This is an assumption which I wish to question; for it is not obvious to me that it must be true. – I should like to make another comment in connection with the free libration mentioned by professor Kozieł. What he (and all his predecessors) actually mean by free libration is a particular integral of the equations of motion, resulting if we expand the disturbing function and retain the term which is linear in the dependent variable. But a physicist may say that this is no free libration at all; because it selects only a particular term from the disturbing function caused by an external body; whereas the Moon can librate freely about its center of gravity only in an absence of any external body, with periods depending solely on its shape and ratio of the differences of its principal moments of inertia. If we adopt their values as known to-day and compute what the periods of the really free librations should be, they come out close to 270 years, or a little less than three centuries. This is a very long time in comparison with all forced librations caused by the Earth, and much too long for any vestige of them to be isolated from the existing observations. But I wish to make it clear that the term "free libration" as used by professor Kozieł and all his predecessors in this connection is only mathematically (but not physically) correct.

RÖSCH: Although I do not wish to kill the heliometer altogether, I must say that I agree with professor Kopal absolutely on all points – especially as regards the limitations inherent in searching for information inside a diffraction pattern. The situation is, I am afraid, even worse; because (unlike for an ordinary refractor) in the case of a heliometer we are dealing with diffraction patterns of two semi-circular halves of an objective; and the result may be an asymmetric diffraction pattern. High accuracy in heliometric observations would, therefore, come as a great surprise for me.

ARTHUR: I believe the main effect of the division of the lens will be large errors not in distances, but in position angles. – May I take a minute to make a comment on the difference between resolution and measurability of a plate. We speak about a resolution of 0".25 of arc, but we should remember that measurability can be three or even four times as good in the usual circumstances.

KOPAL: As long as we are unable to relate the observed diffraction pattern to the true image in a unique manner, mere measurability is irrelevant to the amount of information which we can hope to obtain.

ARTHUR: My argument against the heliometer is that it brings into the picture the center of figure which, as far as I am concerned, is of statistical nature and affects the results of the observations.

GOUDAS: I wish to ask professor Kozieł if the singularity encountered in the solution of the libration in longitude is an essential one, or is it a product of the method employed in tackling this problem?

KOZIEŁ: It is a product of the method used and the most important one is for the value 0.662 of the mechanical ellipticity f.

VOICE: The term *mechanical ellipticity* is incorrectly engaged in this context. The constants α, β, γ qualify to be called mechanical ellipticities whereas the constant f is the ratio of the ellipticities α and β.

GOUDAS: To come back to the previous point, it can be implied that in the iterative process that leads to the solution for f the initial value f_0 of this constant will play an essential rôle in the final outcome since we will be unable to cross the singularity. Thus if the true value of f is on the wrong side of the singularity with respect to the initial value f_0 of f, then we are liable to either not converge to a solution or converge to one that is not the true, but another spurious one.

MOUTSOULAS: The appearance of singularities in our problem is the result of the technique of approximation we use (i.e., the linearization). Drs. Masłowski and Mietelski have found that by retaining terms which are usually dropped in the third Euler equation, the number of singularities are not three, as Hayn has suggested, but many more. These singularities have no real connection with the physicall problem.

KOZIEŁ: In my opinion, the problem of the singularities is relatively minor; the main issue being the adjustment of the observations.

MOUTSOULAS: The seriousness of the singularity or adjustment is a matter of opinion, and I would rather check first the validity of the solution of the Euler equation and the effect of the singularities upon the solution, and then look into the data with, of course, an equal concern.

COLOMBO: I wish to express my agreement with the view that the singularity is a product of the linearization. The precise equations of Euler cannot have such singularities.

GOUDAS: In view of the small size of the physical libration, linearization of the circular functions cannot be placed under suspicion as a cause of the appearance of singularities. Using the pendulum as an analogy, I wish to advance the hypothesis that the singularities arise because of an incomplete coupling of the physical libration in longitude with the orbital motion of the Moon. This results in impossible configurations – such as having the least axis of inertia of the Moon at the instant of zero libration in longitude pointing in a direction that does not pass through the center of mass of the Earth.

KOZIEŁ: A short comment. Dr. Eckert from New York has found the same value of f by use of a technique which entails no singularity.

ARTHUR: But then he obtained a "hollow Moon"!

MINNAERT: Mr. Chairman, I should like to ask professor Kopal about the 270-year period of free libration. Under which circumstances would this libration appear if the Earth were to be removed?

KOPAL: The reason for the presence or absence of a free libration is connected with the rigidity and freedom from damping of the respective body. It should be kept in mind that free librations are characteristic of solid bodies alone; a plastic or visco-elastic Moon could not exhibit them for astronomically long periods of the time; because even if excited, they would be damped out in a relatively short time by dissipation.

MINNAERT: You speak about the excitation of such a free libration, but the possible sources of excitation must be known to us. It should be possible to compute if they existed or not.

KOPAL: The causes which could have excited truly free libration may go back to the initial conditions in the distant past; and these conditions cannot be reconstructed unambiguously today. In particular, the amplitude of the truly free libration can in no way be predicted. However, its period depends only on the shape and internal structure of the Moon as reflected through its moments of inertia.

LORELL: The experiments with the Lunar Orbiter which will soon commence will provide new data, from which we should be able to determine such constants as β or f with an accuracy far surpassing ground-based astronomical determinations; and thus these constants could be eliminated from more general solutions.

HOPMANN: During the recent COSPAR meetings in Vienna I asked Mrs. Massevich about the outcome of the tracking of Luna 9; and she told me that they were consistent with an assumption the Moon to be largely homogeneous.

SASSER: I should like to ask professor Kopal whether a better understanding of the lunar librations could be expected from stellar photography carried out on the lunar surface?

KOPAL: Once we get to the Moon, of course, entirely new vistas and undreamt-of possibilities will open up for astronomical observations. In particular, once we can install and operate zenith telescopes on the Moon, the results obtained with their aid will relegate very swiftly all that we have been able to do from the distance of the Earth into obsolescence. This is likely to happen, I suppose, in the next 10 or 15 years, and will place the heliometers really in the musea.

KOZIEŁ: Regarding the paper by Dr. Eckhardt, I wish to ask what kind of approximations and assumptions you have made. The theory you have thus far presented, although in a different form, is given by Laplace.

ECKHARDT: Well, the approximation was a truncation level of 10^{-7}, which implies that I dropped terms amounting to 10^{-7}. As I already mentioned, I have also assumed that the Moon is rigid and all harmonics beyond the second are dropped. These are my main two assumptions, but I think it is possible to find some more if you think carefully. I have also carried one solar term.

ARTHUR: Do you use Brown's theory instead of Brown's Tables for the motion of the Moon?

ECKHARDT: I use Brown's theory, as given recently by Eckert and Smith. They have re-done Brown's work and presented the latitude terms in a more usable form than the Tables.

ARTHUR: What happened to the empirical term of 7″ in the longitude? Professor Kozieł has used it.

KOZIEŁ: This term is not empirical and does not appear in the longitude, but in the node. It is interesting that the electronic computer is finding the same results as the previous work.

PART II

SELENODETIC OBSERVATIONS AND MEASUREMENTS

SELENODETIC MEASUREMENTS — AN INTRODUCTION

JEAN RÖSCH*

This introductory address will cover the field of general observational methods applied in selenodesy to date. I do not plan to discuss details or results, but rather to give an outline of the various methods which can be used to determine the shape of the so-called selenoid. The interpretation of the results will be discussed in another session; and so will the details of the structures on the surface, which will be the subject of the last session on topography.

I said "so-called" selenoid. This word has been borrowed from Geodesy and it is appropriate to discuss right from the beginning the concept attached to it. The idea of the geoid just comes naturally, because on Earth we have oceans which represent a part of an equipotential surface. So it is quite natural to imagine a fictitious continuity of the oceans below the continents and come to the idea of the geoid. To this fictitious surface we then refer all the other features of the surface of the Earth. In the case of the Moon, as I mentioned yesterday, we can expect no water and no oceans; so the idea of the selenoid is less obvious. Thus our task in starting up Selenodesy is not to determine the shape of a fictitious surface, but simply to establish a number of reference points on the surface of the Moon to which we can refer the positions of other points. So we do not speak about the shape of the selenoid, but rather the coordinates of a number of reference points taken in convenient places on the Moon. If, after that, we succeed in expressing in simple terms an algebraic surface on which all these reference points lie, we could, of course, call it a *selenoid.*

Thus, the result of the observations will be absolute coordinates of well-identifiable points and as such coordinates we can take, for instance, the latitude, longitude, and radius vector to the center of mass of the Moon, which will be taken as the origin of coordinates. Then, the directions of the axes to which we refer longitude and latitude can, in principle, be chosen arbitrarily and such as to suit the convenience of the measurements and description.

A few words should be added here explaining the reasons why we need this system of reference. First, we want to be able to describe the geometrical shape of all parts of the Moon. Selenodesy will give us a number of reference points, and topography, later on, will connect details to these reference points, just as we do on the Earth. Of course, it is important in itself to have accurate positions for all the purposes of future lunar exploration.

* Université de Paris and Observatoire du Pic-du-Midi, France.

Once we have this system of reference points, topographic methods can yield contour maps. From these maps, important results may emerge concerning such problems as the equilibrium and internal structure of the Moon. Of course, one must be careful in this interpretation and keep in mind the fact that the external shape of the Moon is not uniquely related with the internal structure or the moments of inertia of the Moon. The link between the outside form and gravity properties rests on the density distribution, and so one cannot draw conclusions only from a knowledge of the outer shape.

Let us now consider the different methods that are used to determine the absolute coordinates of a point on the surface of the Moon. It happens that none of them can give accurate coordinates of points all over the surface because we always see almost the same view of the Moon at all times. Methods exist which are suitable for the limb part of the apparent disc or its central part. On the central part the angular coordinates can be determined with a better accuracy than the distance to the center of gravity, whereas for the marginal zone (and on the limb itself) we can determine with some accuracy, at least in principle, the distance to the center of gravity (provided that the position of this center is known) and with a low accuracy the angular distance to the front point of the Moon.

We shall divide the methods in several groups, depending on whether we deal with the regions near the limb or those near the center, keeping in mind a remark we heard yesterday that the heliometric measurements run into some trouble since they involve the distance of Mösting A to the limb, the shape of which is not known with sufficient accuracy. Well, one could imagine that the distance of the projection of the center of gravity to the observed point of the limb, together with Mösting A in the heliometer, enters the problem as another unknown in determining not only the libration, but also the contour of the limb. As we saw yesterday, the reduction of heliometric observations involves a tremendous amount of work; and if we introduce one more unknown for each observation, it will be even more so. What one can do, is instead of introducing one more unknown per observation, to assume that the limb is not circular and its form can be defined by, let us say, one or two more parameters. The task is then increased – but not too much. For instance, Yakovkin has found the profile of the Moon to be something like half a circle for the northern hemisphere and half an ellipse for the southern one. Professor Kozieł has shown that by different reduction this assumption can be avoided. Anyhow, the heliometric results can provide a shape of the limb different from a circle, but this would be only first approximation and would not give a real form for the limb valid for any libration.

All this is complicated by the fact that on the limb one observes mountains and the appearance of these mountains changes with the libration. This is why a number of scientists have worked hard in determining the shape of the limb for various librations. I should refer here to the works of Weimer and of Watts, who have produced contour maps of the regions of the Moon which may appear at the limb for all librations.

In order to determine the complete shape of the marginal zone, we have to establish continuity all around the disc; but we never see a complete limb of the Moon unless the phase is exactly zero or 180°, which corresponds to an eclipse. The observation of an uneclipsed full Moon is made when the Moon is out of the shadow cone and the phase is not exactly zero. Therefore, on one half of the disc we shall see the apparent profile whereas on the other half the terminator.

Observations have been made by Watts in the United States and by Potter at Pulkovo, using full-moon phases so close to the shadow cone that the defect of phase could be neglected. The results of the measurements by Watts as well as by Potter, reveal that the apparent profile of the Moon is slightly elongated in a direction about 160° from the direction of the poles, with an excess of the major axis over the minor one of 0".6 according to Watts and 1".5 according to Potter – an agreement not too bad when taking into account the difficulty of such measurements.

Obviously, we also can study the profile of the Moon when there is a solar eclipse, and especially an annular eclipse, by comparing it with the profile of the Sun. In doing so we can assume that the solar profile is perfectly circular. Indeed, because the rotation of the Sun around its axis is very slow, we have reason to believe that the departure from the spherical form is quite small, and certainly negligible as compared to the departure of the Moon from a sphere. Such studies have been conducted by a group from the United States together with the University of Manchester during the eclipses of 1962 and 1963. The reduction of these eclipses met with some trouble because of photographic effects. When the width of the ring is not constant all around the disc – and this is usually the case unless the observation is made from a point on the line of central eclipse and exactly at the time when the eclipse is central – then photographic diffusion produces a spurious spread which varies all around the disc, the correction of which is not easy. However, the reduction of these eclipses has given results which are quite consistent with those by Watts and Potter. I hope that some of the investigators here present will tell us more about it in due course.

I have suggested some years ago that in such a case one should also use a photoelectric scanning method similar to the one used in order to determine the diameter of Mercury during its transit over the dolar disc. If you scan the ring with a radial slit rotating around the center of the disc of the Moon and you measure the photoelectric current, two important features emerge. First, the blurring of the image by the atmospheric turbulence will not matter, because even if the image is blurred, the total amount of light remains roughly the same.

Second, any vibration or motion of the instrument does not matter because it would interfere only with the harmonics in the profile of the Moon of very high order. During the annular eclipse in 1965 a photoelectric method was used and Dr. Eckhardt will speak on this subject this afternoon. For the eclipse of May 20, 1966 (in Greece), the difference between the computed apparent diameters of the Sun and the Moon was so small (smaller than the amount I mentioned a moment ago of 1" difference between the longer and shorter axis of the Moon), that it did not offer a good opportunity for observation because very high-order harmonics would have entered.

One has to take into account also the fact that the brightness on the very limb of the Sun may change from point to point if there are some bright faculae close to the limb. But one can imagine some kind of calibration of the region of the Sun close to the limb, just before or just after the eclipse.

Another method of studying the shape of the limb is to use the occultation of stars. This is again something like the heliometric method in the sense that we have as one of the unknowns the distance from the limb to the projection of center of gravity of the Moon for a point of the limb – the point where the star disappears. I say disappears because it is well known that it is easier to observe the event when the star disappears than when it reappears.

One can certainly obtain accurate results in this way, inasmuch as some photoelectric techniques have been set up to observe the occultations and even to study the decrease of brightness of the star close to the limb, and thus to determine its angular diameter. But this would give only one point of the limb. It would be fruitful to use the Watts charts of the limb in reducing occultation observations because this would give a connection between the limb and the center of gravity of the Moon whose position is well known from the orbital motion about the Earth.

Let us now consider the methods more applicable to the central regions of the Moon. We can take advantage of the phenomena of libration, due to which we do not always see the Moon from the same angle. Unfortunately, the total amount of this libration (mainly optical) is not large enough. Geometrically, the problem is simple in the sense that we have a number of directions connecting the center of gravity of the Moon to reference points on the surface, which we observe from different points of view because of the libration. Writing that the direction joining the optical center of the objective to the image on the plate is a vector parallel to the vector joining the observational point to the point on the Moon gives two conditions. If we have m points observed n times, this gives $2mn$ equations. There are first $3m$ unknowns for the three coordinates of each point, and $6n$ unknowns defining the position of the plate; but one plate can be taken as a reference, so that $6n$ reduces to $6(n-1)$; there is one additional quantity which can be eliminated: and this is the distance of the Earth to the Moon. Finally in order to have sufficient equations the condition $2mn \geq 3m + 6n - 7$ must be fulfilled. There is a trivial solution, $m = n = 1$. The next one is $n = 2$ and $m = 5$, which means determination of 5 points on 2 plates. Obviously, only 5 points on 2 plates will not be enough in order to eliminate the accidental errors. One has to accumulate many points measured at different librations. I shall not go into the details of all this, because a number of participants here present have been involved in such work and will speak in the discussion. I will only mention that the accuracy of this method diminishes from the center to the limb because the projection of the change in position due to libration becomes smaller when you approach the limb.

Another possibility is to use the stereoscopic methods. I have practiced stereoscopy in the laboratory years ago, and I think that measuring by stereoscopy details on the Moon would give rather unreliable results because of the change of the shadows, etc. However, some work has been done in this respect, and I think that for the front part of the disc, and to connect limited regions of the Moon to some reference points, the stereoscopic method could yield good results.

Concerning the regions intermediate between the center and the limb, I would like to mention the method suggested years ago by Ritter, to derive the shape of the so-called selenoid from the shape of the terminator. For a perfect sphere, the terminator would be a perfect ellipse. If the terminator is not an ellipse, this means that the selenoid is not a sphere. Ritter measured indeed the Moon by this method, "using a very sharp pencil to draw the terminator". Unfortunately, he had just forgotten that the tilt of the normal to the surface on the selenocentric radius may not be zero, as Yakovkin remarked later on. This method is being explored at the present time by Mlle Calame. There are not only geometrical problems, but also the problem of determining the position of the true terminator. If you photograph the Moon without any precaution, because the brightness decreases close to the terminator, it may fall below the photographic threshold before you get the true terminator. One solution is to over-expose, but it may be dangerous. We are using another method which consists in pre-exposing the plate, so as to get above the threshold; it seems to work in a significant manner. I should mention that this method is useful when applied far from the center, because at the center the terminator will appear as a straight line no matter what the form of this section of the Moon is. It would be very difficult also to apply it too close to the limb. This is why I call it an intermediate method.

Another method which has some bearing with that one has been used by Professor Hopmann, who measured the position of the tops of the peaks which appear outside the terminator at the moment when there is just the first light or the last light on the peak. I am sure Professor Hopmann will give all details to those interested.

Now I would like to turn to a completely different domain of selenodesy, namely that of laser echoes. I do not like the term *optical radar* and will use instead the term *optical echo*. A laser can produce a very well collimated beam and very short light pulse; besides, it can constitute a very monochromatic and coherent radiation. I shall not enter into the technical details of what can be done with a laser beam to study the shape of the Moon because this afternoon Dr. Julian and Mr. Orszag will give us some information about the plans in this field in the United States and in France. So I shall speak only about the general situation of the problem and try to see what can be done.

We can write the fundamental equation of the problem, which will give us the opportunity to discuss the various features of this method. The principle is that we send a sharp light pulse of a duration of some tens of nanoseconds (a few units of 10^{-8} seconds) and measure the time 2τ between the transmission of the signal and its return from the Moon. The fundamental equation is

$$c\tau = ak_a(t) - \rho k_\rho(t) - rk_z(t),$$

where c is the speed of light, and a the semi-major axis of the orbit of the Moon which is multiplied by a function $k_a(t)$. This first term represents the distance from the center of gravity of the Earth to the center of gravity of the Moon. We subtract ρ, the selenocentric distance of the impact point on the Moon, multiplied again by $k_\rho(t)$. And finally, we subtract r, the equatorial radius of the Earth, multiplied by $k_r(t)$. The functions $k_a(t)$, $k_\rho(t)$ and $k_r(t)$ depend on angular arguments that define in some judiciously chosen frame of reference the center-to-center direction of the Earth-Moon line, the direction of the center-to-spot on the Moon line and the direction of the center-to-observing site on the surface of the Earth line, respectively. What can we get from this equation, and what is the accuracy of the various terms? Let us think about c first. At present the relative accuracy of c is about 3 millionths. We may consider that this is not accurate enough to help us in determining the lengths a, ρ, and r. But if we take as unit of length the distance travelled by the light in one second of ephemeris time, then any change in the value of c as measured in kilometers per second will not matter. Thus we could get a homogeneous system and avoid all worry about the value of c, inasmuch as the value of r is derived from laser measurements of geodetic basis, thus depending too on the velocity of light.

The accuracy of τ as measured by the laser echo may be of the order of 10^{-7} in relative value and still the technology of lasers may be improved in the future in the sense of shortening the pulses.

On the right-hand side of the equation the errors in the angular terms are irrelevant; $k_a(t)$ is derived from the theory of the orbital motion of the Moon. The principal term, apart from a constant, contains the eccentricity of the Moon, e, which is well known. So it turns out that if we accumulate the observations for various values of t, this is tantamount to an elimination of the constant term and measurement of a, the semi-major axis of the Moon, multiplied by e. Of course, because the mean eccentricity of the orbit of the Moon is only one part in 22, we need higher accuracy. But if the accuracy of the laser measurements is high enough, then we can get for the semi-major axis a better value than the accepted one, on the basis of the fact that the eccentricity of the orbit of the Moon is not zero.

$k_\rho(t)$ has no importance if we work close to the center of the disc because then it does not change very much; even if we do not know exactly the parameters of the libration, it will be very close to one and will vary like a cosine, so that its effect on the accuracy will be quite small. Thus for the frontal point of the Moon ρ can be computed by the least squares method because it may be considered as a constant and $k_\rho(t)$ is known with sufficient accuracy.

Then there remains the geocentric position of the observatory; this means that from these laser observations we can also get information about the shape of the geoid. There are plans, indeed, to use these laser experiments for geodesy. As this is out of the scope of this conference, I shall not enter into details.

There is another point which I would like to mention in connection with the question, raised yesterday, of the measurement of libration by the use of laser echoes. In spite of the fact that the monochromatic nature of the signal is assured, I do not think that measurements by Doppler effect would be worth a laser experiment, because measuring the frequencies in the optical range is not as accurate as in the radio range, except perhaps in laboratory experiments with gas lasers, and because libration is a slow phenomenon. But if we have a reflector away from the center of the disc, then the distance to the observer would change enough, and I think that this is the way in which the libration could be studied by a laser echo over a certain range of time; not by the measurement of a wavelength, but by the change in the range of the reflector.

One of the important advantages of the laser method is that it can be applied to the regions located off the center of the apparent lunar disc. This offers an opportunity to attain with good accuracy the longitude, latitude, and range at the same time, because the distances will be referred to the center of gravity of the Moon.

LUNAR RESEARCH AT THE KOTTAMIA OBSERVATORY

A. H. SAMAHA*

The Kottamia Observatory

The climate of Egypt makes it a very favourable country for astronomical observations and an ideal place for a large telescope. It is only recently, however, that the erection of such a telescope has become possible. Since 1905 planetary and stellar observations had been dependent upon a 30-inch, *f*/5 telescope presented to Helwan Observatory by Mr. Reynolds, a past treasurer of the Royal Astronomical Society in London. Valuable work was performed with this instrument, but the Egyptian astronomers realised that its capabilities were strictly limited.

To answer the need for more extreme observations a contract was signed by the Egyptian Government with Messrs. Grubb Parsons of Newcastle-upon-Tyne, England in 1949, for a 74-inch reflecting telescope. Unfortunately, many difficulties, local and technical, delayed the erection of the instrument by several years, but I am pleased to be able to say that it was finally handed over for the use of the Egyptian astronomers in May 1964.

The Observing Site

The site was chosen after repeated discussion regarding such factors as: astronomical seeing, living conditions, electricity supply, accessibility, water supply and industrial smoke. The final position selected is on a plateau at an altitude of 476 m above sea level and 22 km south of the midpoint of the Cairo-Suez road. The latitude of the observatory is almost the same as that of the older Helwan Observatory, being 29°55′9″ North and the longitude is 31°49′5″ East. Access is only possible by means of a specially constructed road from the main Cairo–Suez thoroughfare. The site is supplied with food and water by regular shipment from Cairo. Power is supplied by two 25 KVA diesel generators providing, after transformation, a mains voltage of 220 volts at 50 cycles/sec. and three-phase. Living quarters are provided for up to 15 astronomers, with separate quarters for the workers. Laboratory and workshop space are ample. (Figures 1 and 3.)

* Helwan Observatory, Cairo, Egypt.

Fig. 1. Living quarters at Kottamia Observatory.

The Telescope

The telescope is of conventional design, having a parabolic mirror with secondaries to give Newtonian, Cassegrain and Coudé foci, and is one of several of 74-inch aperture constructed by Grubb Parsons. The mounting consists of north and south piers supporting a polar axis with the tube offset on a declination axis. (Figure 2.)

The 74-inch main mirror is cast from low-expansion glass of focal length 360 inches (*f*/4.9). The edge thickness is 12 inches.

Measurements of the curvature by Professor Kienle, using the Gaviola method, indicate that the wavefront does not deviate from the theory by more than one-twelfth of a wavelength – a better result than called for in the contract. Zonal errors do not exceed ± 20 microns and astigmatism is negligible within the errors of measurement.

All the secondary mirrors are made from fused quartz. The Cassegrain, which has a clear aperture of 19 inches, is of hyperbolic section and works at a focal ratio of *f*/18, thus giving a scale of approximately 6 seconds of arc/mm. A second hyperbolic mirror similar to the Cassegrain has a focal ratio of *f*/28-9 and, in combination with two planes of 13.5 inches and 9.5 inches respectively, gives a Coudé focus at which the scale is 3.8 sec/mm.

A Ross-type Coma Corrector has been supplied to cover a 16 cm field at the Newtonian focus.

Two refractors with achromatic doublets of 6 inches aperture and 60 inches focal length, with 2-inch Kellner eyepieces, are used as finders.

Fig. 2. The 74-inch reflector of Kottamia Observatory.

The telescope is supplied with a control console on which are located the switches for the major telescope motions. Buttons for quick motion at a speed of 45°/min are provided in both Right Ascension and Declination, and are supplemented by setting motions of 15 minutes of arc/minute of time. The handsets have a guiding and a setting motion, the former of which is 15 seconds of arc/minute of time. Trailing speeds of $\pm 1\%$ to $\pm 10\%$ of sidereal rate are available.

Auxiliary Equipment

The Newtonian plate carriage is mounted on a revolving breech piece with manual focussing gear. It carries two guiding microscopes, each of which can be traversed over x, p and y coordinates by a rack and pinion gear. Plate sizes of $3\frac{1}{4} \times 4\frac{1}{4}$ inches and 16×16 cm are accommodated in holders kinematically mounted on the carriage.

A three-prism Cassegrain Spectrograph supplied with two cameras has been constructed by Hilger and Watts. The prisms are of light flint glass with a transmission down to 3650 Å, and with the remainder of the optics they are inside a thermostatically controlled insulated jacket. The smaller camera, which is of focal ratio *f*/2.5, gives a dispersion of 100 Å/mm at 4800 Å, and the larger, of focal ratio *f*/6, gives 48 Å/mm at the same wavelength. The spectral range is from 3600 to 8000 Å and the slit width is variable between 5 and 150 microns. Comparison spectra can be impressed either from an arc or a discharge tube, and a viewing telescope with a field of 4 minutes of arc serves as a finder.

The Coudé spectrograph was also constructed by Hilger and Watts. It is mounted inside a horizontal box girder frame covered with suitable panelling. The slit width is variable between 10 and 500 microns and carries a prism for exposing the plate to comparison spectra either from an arc or a neon discharge lamp. The collimator mirror of focal length of 274 cm sends a parallel beam to a grating ruled with 600 lines per mm over an area of 104 × 101 mm. The grating is blazed in the first order. A mirror and Schmidt corrector plate serve as a short camera of focal length 76 cm giving a dispersion of 20 Å/mm on either plate or film. Spectra on plates are obtained using a field flattening lens carried in the plate holder; another plate holder, which is at the focal plane is used to secure spectra on 35 mm film without using a field-flattening lens over a range of 3600 to 8000 Å. A long camera of focal length 274 cm (*f*/27) also has a curved plate holder and is used without a correcting lens. The dispersion with this camera in use is 5 Å/mm over a spectral range of 2900 to 8000 Å.

The Dome

The dome and building are specially constructed to reduce internal temperature variation. The dome is 18 meters in diameter with an outer shell of galvanised steel plate painted externally with aluminium (Figure 3). The shell is lined with a 5 cm thick insulating material (stramit) and an inner space is formed by a 4 mm lining of building board known as "Intermit". The inner space is ventilated to carry away heat due to insulation. The lower edge of the dome dips into a channel of water to prevent the entry of dust and heat to the interior.

Fig. 3. The dome of the 74-inch reflector at Kottamia.

The building itself is double-walled and passes horizontal concrete fins which act as solar screens to prevent heating of the walls by direct sunlight. As a result of these precautions the diurnal temperature variation at the dome floor does not exceed 3 °C even under conditions of strong sunshine. The dome can be rotated at a speed of one revolution in six minutes. The shutters are in two sections, an upper and a lower, capable of moving at 75 m/minute. The opening of the dome is five meters wide and extends 2.5 meters beyond the zenith.

Telescope Performance

Professor Kienle reported that Hartmann tests at the Newtonian focus show that the mirror working at full aperture collects more than 80% of the light into a circle of 0.8 seconds of arc diameter, and if the aperture is reduced to 70 inches, then 95% of the light falls within this diameter.

Aluminiser

An aluminiser has been purchased from Edwards, England, and is due to arrive on site in September 1966. It is anticipated that after re-aluminisation of the mirror a gain of at least half a magnitude will be achieved.

Meteorological Conditions

After only two years of observational experience at the Kottamia site, yearly weather variation makes it difficult, as yet, to draw up a general scheme of good conditions for the observatory. Nevertheless, a few general ideas will be given.

January and February is the most unfavourable period. The sky is frequently completely overcast, but spectrographic and photographic observations are possible on 60–80% of the nights. For photoelectric observations, where weather conditions are far more stringent, the period is unsuitable, only a few sufficiently clear nights occurring each month.

During March and April the sky is clear, but Khamassin conditions, a steady wind from the South or South-West, lower the atmospheric transparency and frequently make observations difficult. Up to 50% of the nights may be completely clear. However, during the summer months from June to September the conditions are very good, about 75% of the nights being completely clear and none being more than 50% overcast.

Observations of the atmospheric transparency between 4000 and 8000 Å are available for up to 20 nights between January and September from the records of Mr. K. Peacock of Manchester University. The extinction coefficient varies from 0.45 magnitude per zenith air mass at 4035 Å to 0.09 at 7922 Å. The variation of the extinction coefficients observed over 18 nights is up to 0.04 magnitude per zenith air mass in the violet to 0.025 in the infrared.

The "seeing" at Kottamia has frequently been described as good. The late Dr. A. R. Hogg estimated the quality of seeing from the appearance of a 7th-magnitude star under a magnification of 1000 at the Newtonian focus. About 50% of the estimates were in the range of 3–5 seconds of arc, and on occasion estimates of 2 seconds of arc were obtained. Sir Richard Woolley reports that on 90% of the nights on which he has performed observations at Kottamia the seeing has been fair to good, the image at the Cassegrain focus rarely more than 2 seconds of arc in diameter. Photoelectric observations by the Manchester University Lunar Group indicate that a stellar diameter of 2–3 seconds of arc is a reasonable estimate. Lunar photography started by Dr. Rackham from Manchester University and by Mr. M. Hamdy, Helwan Observatory, under the direction of Professor Z. Kopal, confirm the frequent occurrence of seeing conditions which can only be described as excellent. Many photographs obtained are of a quality equal to those obtained from the high altitude Pic-du-Midi Observatory, which is renowned for the quality of its atmospheric conditions.

I should like to devote the remainder of my paper to a description of the lunar observations that have been performed up to the present day with this telescope. The observations fall into three classes; radiometric, photoelectric and photographic. The first lunar observations to be made fall into the radiometric section.

Lunar Infrared Research

In December 1964, Drs. Shorthill and Saari and Mr. Gilbert Bruce of the Boeing Scientific Research Laboratories arrived at the Observatory to study the thermal properties of the lunar surface during the course of an eclipse. With them they brought several tons of complicated scientific equipment necessary for the detection and rapid recording of the radiation. This apparatus included a rapid scanning device capable of covering the entire lunar disk in 16 minutes with a spatial resolution of 10″ of arc at the Newtonian focus. The radiation from the Moon was continuously recorded on magnetic tape in the wavelength band 10 to 12 microns registered by a Mercury-doped Germanium detector and at 4450 Å using a photoelectric detector. The excellent weather on the night of the eclipse, December 19th, 1964, permitted four lunar scans during the first penumbral phase, three during totality and one during the second penumbral phase. Measurements were possible down to a zenith angle of 85°. Cooling curves can be constructed for any 10″ element of the Moon.

The results obtained were, to say the least, unexpected. From previous work by the observers it was expected that the major ray craters would exhibit anomalous cooling. Indeed these craters were found to be warmer than their surroundings during totality. Unexpected, however, was the discovery of hundreds of other localized anomalies or "hot spots". In addition, thermal enhancements existed in all or parts of some maria as can be seen from a published photograph which shows both the infrared and visible scans (Figure 4), the former being taken during totality and the latter at a phase angle of 2°/6′, just outside the penumbra. Maria Tranquilitatis,

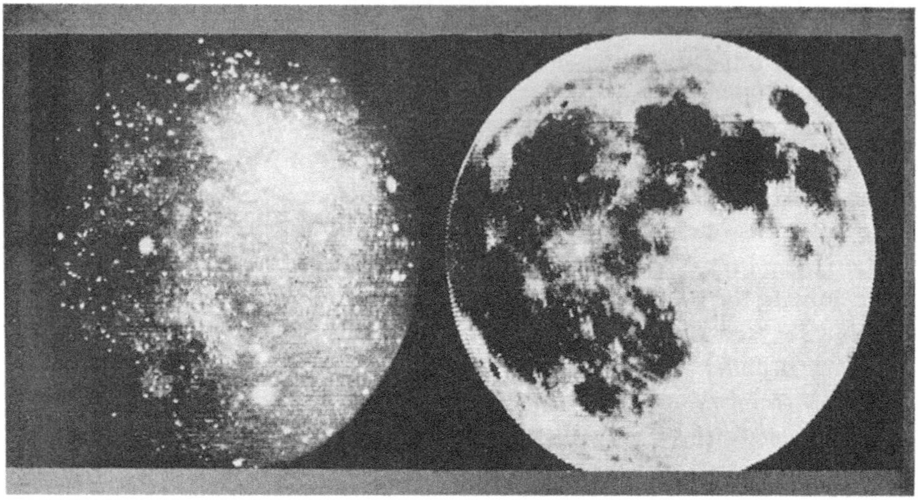

Fig. 4. Visual and infrared scans of the Moon before (right) and during (left) the total eclipse of the Moon on Dec. 19, 1964, made by Drs. Shorthill and Saari at Kottamia Observatory.

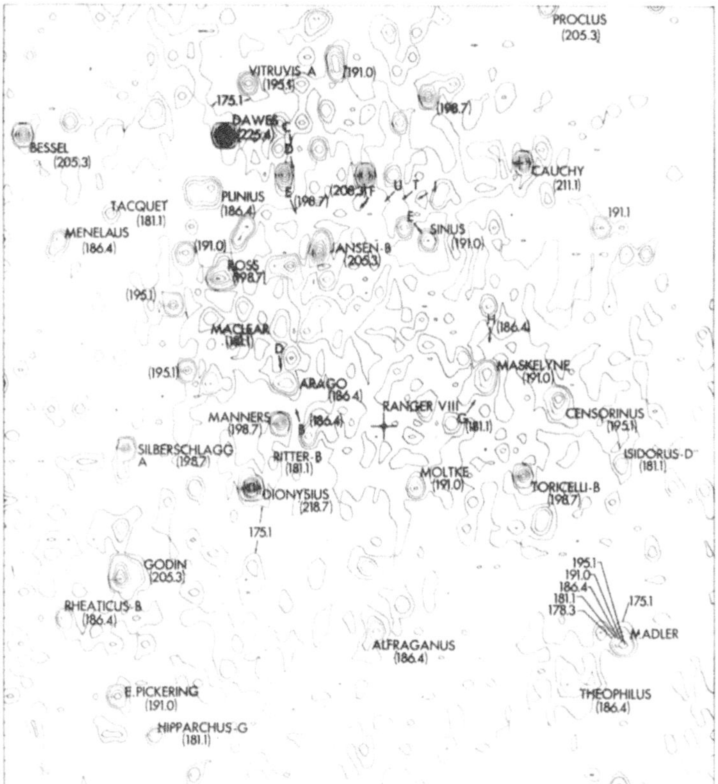

Fig. 5. Thermal contours during the total eclipse of the Moon on Dec. 19, 1964, in the region of Mare Tranquilitatis, measured by Drs. Shorthill and Saari of the Boeing Scientific Research Laboratories at the Kottamia Observatory.

Serenitatis, Foecunditatis and Crisium were found to be thermally enhanced during totality. Contour maps (Figure 5) show in more detail the thermal anomalies. Many of the 1000 discovered hot spots correspond to craters, but some have no corresponding visible feature. Approximately 65% of the hot spots appear in the plains and 35% appear in the uplands. Mare Tranquilitatis has an abundance of hot spots compared with other areas.

By reconstruction of contours of brightness temperatures Drs. Shorthill and Saari have produced a thermal map of the Moon at the time of egress immediately before entry into the penumbra. On the photograph (Figure 6) the limb darkening in the infrared is clearly visible and the large craters and major mountain ranges can be clearly distinguished. This is caused by differential heating of the crater walls, and since the Moon was full, it is not a shadow effect.

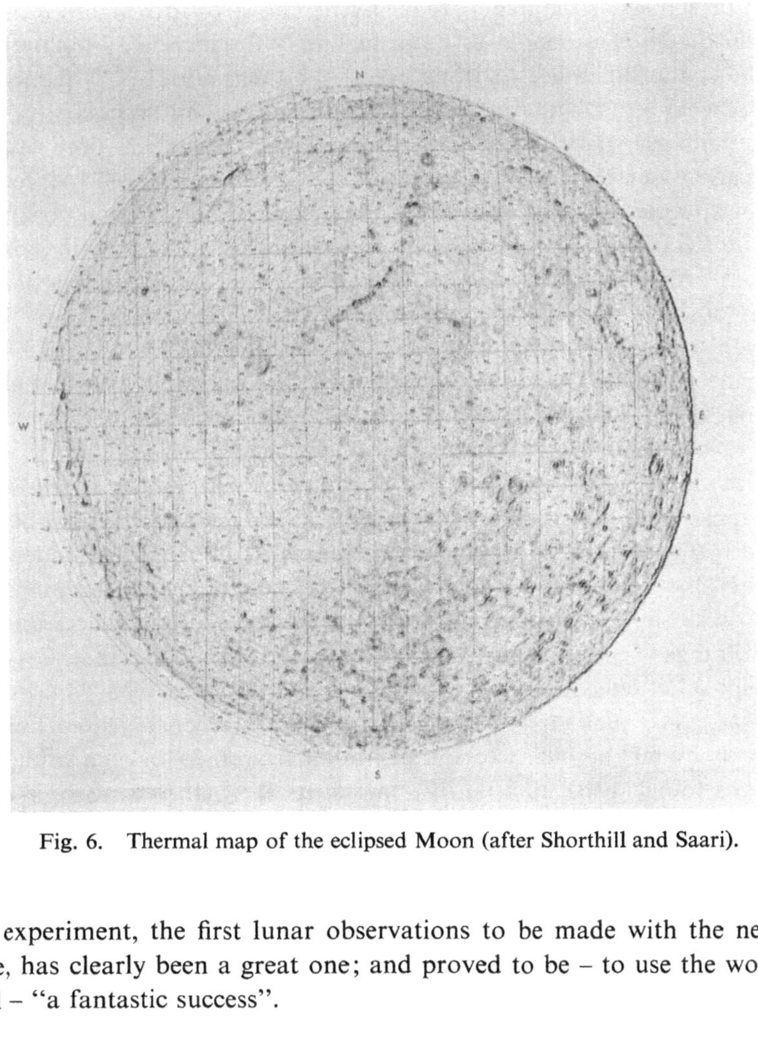

Fig. 6. Thermal map of the eclipsed Moon (after Shorthill and Saari).

This experiment, the first lunar observations to be made with the new 74-inch telescope, has clearly been a great one; and proved to be – to use the words of Dr. Shorthill – "a fantastic success".

Photoelectric Studies

Immediately after the departure of the astronomers from the Boeing Scientific Research Laboratory, Mr. Peacock of the Manchester University, Astronomy Department, began an extensive programme of lunar photometric studies. At various periods in the subsequent year he performed numerous observations, and the programme is at present being continued by Mr. Joseph Mikhail of the Helwan Observatory.

His photoelectric photometer, which was constructed in the Manchester University Astronomy Department workshop, was specially designed for lunar studies. It is a three-beam instrument capable of working in three separate wavelength bands simultaneously, and filter slides provide a total of six colours. Simultaneous recording of the separate intensities in the bands used for estimation of the colour indices is necessary to avoid guidance errors causing the pinhole to drift across a lunar region of different brightness as the filters are changed. The error involved in recording the colours individually is estimated to be from 1 to 3%, which is quite appreciable in terms of the colour differences existing across the lunar surface. As the telescope is not provided with a lunar rate drive in declination, a second necessity is a means of continuous guidance whilst a reading is taken. This problem has been overcome in the photometer by employing a thin transparent plastic film to reflect a few percent of the beam into the guidance optics. The film is left permanently in position.

The photoelectric programme included measurements of the brightness of selected lunar regions with their phase angle, a search for colour variation with phase angle, and a study of the general colour distribution over the lunar surface.

The results obtained indicate that the brightness of a small area of the lunar disk increases almost linearly from the point at which the terminator crosses the region to within about 10° of full Moon, after which the brightness rises more rapidly. Light curves are provided for 23 lunar features.

The colour index versus phase angle measurements tend to show a reddening with the phase angle. The colour index $m_{7922}-m_{4765}$ shows a reddening of approximately 0.0019 m per degree of phase angle. Within the limits of error the reddening is the same for the 23 areas observed. Measurements of over 200 areas distributed over the entire lunar disk show the range and distribution of the colour indices on the lunar surface. The range of colour index is found to be small. The colour index $m_{7922}-m_{4765}$ displays a spread of only 0.12 m for almost all the regions observed, whilst the scatter for $m_{7922}-m_{6692}$ is almost within the experimental error of ± 0.01 magnitude. Aristarchus is by far the bluest region of the lunar surface and the whole of Mare Tranquilitatis found to be unusual in terms of its thermal properties, is the bluest region of the Moon of any extent. At the other end of the scale the reddest regions are the North-East corner of Mare Imbrium and Mare Frigoris.

The same instrument was used by Mr. G. Roberts of Manchester University during August 1965 to search for the time variations of the luminescent component of moonlight. His results did not, however, reveal any significant colour variations that could be attributable to luminescence.

Lunar Photography

Under the direction of Professor Z. Kopal, a considerable amount of lunar photography has been performed by Mr. M. Hamdy of Helwan Observatory with the Kottamia telescope. The equipment used includes two cameras for use at the Cassegrain focus, one of which can be attached to the side of the spectrograph, making photographic and spectrographic observations possible in the same night. The photographic exposure can be varied between $\frac{1}{64}$ second and 8 seconds, and exposures can be taken in rapid sequence. The film is loaded into a magazine. One magazine holds up to 60 frames of 70 mm film. A guidance system permits selection of the times of best seeing for taking the photographs.

The excellent seeing enabled some first-class negatives to be obtained (Figure 7). Up to the present date 2000 exposures have been made at various lunar phase angles, and amongst them are some of the best photographs of the lunar surface ever taken. The photographs are being stored and catalogued in Manchester, and use will be made of them by Mr. Hamdy for height determination of features of the lunar surface using the shadow technique.

Photoelectric observations of the light phase-curve of the Moon in different wave-

Fig. 7. Hipparchus and Albategnius (upper right) region of the lunar surface, photographed in the Cassegrain focus of the 74-inch reflector at Kottamia (North right; South left).

length ranges will be studied by Mr. Joseph. For that purpose a new instrument is under construction at Helwan Observatory. It will include two tubes of V shape, magnesium oxide as a diffuser and the ordinary photoelectric equipment. In this connection the lunar opposition phenomena will be investigated.

This concludes my brief summary of the lunar observations accomplished at the Kottamia Observatory and, as you have seen, the site is already making a name for itself in the field of astronomy.

Our lunar group in Egypt now includes three students working for advanced degrees and the amount of equipment at their disposal is becoming quite considerable.

Therefore, I think that in the future we can confidently expect many more vigorous contributions towards an understanding of the nature of the surface of our nearest celestial neighbour from Kottamia.

Availability

The telescope has been in use in joint programmes with Greenwich astronomers for the determination of radial velocities of variable stars of the SR type, RR Lyrae, W Virginis and MKR Standards.

Observations of spectroscopic binaries, eclipsing binaries and galactic cap stars have been commenced. In all, about 2000 stellar spectra have been taken since the erection of the telescope two years ago.

Photographic and spectrophotometric observations of the Moon in cooperation with astronomers from Manchester University have been carried out regularly.

A joint programme of photoelectric stellar photometry is in preparation with the Swiss astronomers at Basel, Bern and Geneva.

Whenever possible the telescope will be available for other joint programmes proposed by foreign astronomers, pending the approval of the Observatory Board.

A SELENODETIC EVALUATION OF ELEVEN PARIS LUNAR PHOTOGRAPHS

MAHLON S. HUNT*

Abstract. A selection of eleven original photographic plates from the 1894–1909 Paris Observatory collection of lunar photographs was used to acquire 1304 selenodetic measures of 77 lunar features by two investigators. The RMS residual of all measures on all plates for the two investigators was 24 microns and there was no correlation between feature diameters and their reproducibilities. Results are provided as direction cosines of all lunar features corrected for differential refraction.

Introduction

Collections of lunar photographs of possible potential use in selenodetic studies exist usually at observatories. Such collections include lunar photographic plates containing nearly the full disk of the Moon's visible face and are available in original form in sufficient quantity and quality to make a selenodetic sampling of data. The photographs are acquired through the use of long-focus, medium to large aperture terrestrial telescopes. Most observatories can be expected to require that their original photographs be retained and used at the observatory. This imposes the condition that the selection and measuring of original photographs be performed at the observatory and requires that a suitable measuring instrument such as a linear comparator of high precision be available for use at the observatory. To insure success it is also desirable to have an interested astronomer on the observatory staff to participate in the technical investigation.

The purpose of this study is to evaluate selected photographic plates representative of an older collection of lunar photographs expected to be useful to selenodesy. The eleven plates used in this evaluation were selected from the renown Paris Observatory collection (Loewy and Puiseux) of over 2000 lunar photographs acquired from 1894 to 1909 with the 60 cm aperture, 18.22 m focal length "grand équatorial coudé" refractor. This collection and conditions at the observatory met the necessary criteria discussed above for such an evaluation.

Before plate measuring could begin the linear comparator available at the observatory for use in this evaluation had to be modified. The thicker emulsions of the older Paris photographs made it necessary to increase the intensity of the comparator's illumination system to near the manufacturer's maximum recommended limit for thermal considerations. A microscope attachment was provided for the

* Air Force Cambridge Research Laboratories, L. G. Hanscom Field, Bedford, Mass., U.S.A.

Kopal and Goudas (eds.), Measure of the Moon.

comparator to serve as an additional aid in the identification and measurement of selected lunar features. Finally, a temperature control unit for the measuring room was provided to insure temperature stability during the measuring period.

Plate Data

Eleven original photographic plates were selected for measurement from the total photography collection. They represent the best quality photographs in the collection which meet the following plate selection criteria. Plates were selected with as large a range of librations as possible to be useful to selenodesy. Plates were selected near full moon to include as much of the visible face as possible. Figure 1 shows the librational distribution of the 11 selected plates. Five plates were selected before full moon and six plates were selected after full moon.

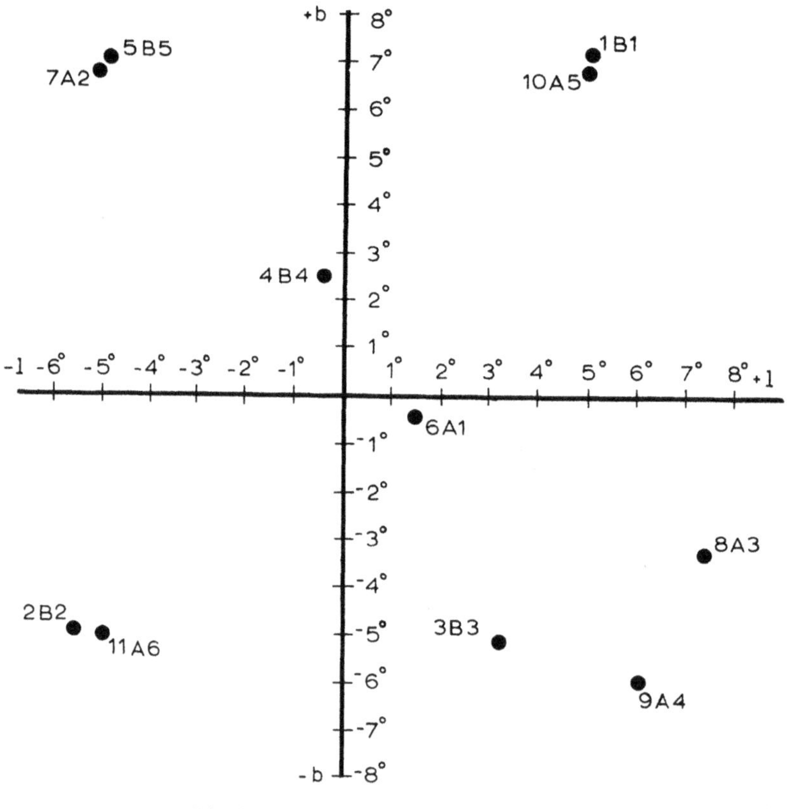

Fig. 1. Distribution of Plate Librations.

Table I provides plate information compiled for the eleven measured plates from observatory records, British ephemeris data and French meteorological records for the dates of the observations.

TABLE I

Plate Information

AFCRL Plate Nos.*	Date		l	b	π	UT	LST
1-B1	16.37002	Nov. 1907	+ 5°.1	+ 7°.3	57′10″	08h.880	00h41m19s
2-B2	10.38612	Dec. 1894	− 5 .7	− 4 .9	58 53	09 .267	02 43 45
3-B3	14.38993	Nov. 1899	+ 3 .3	− 5 .2	59 02	09 .358	01 05 55
4-B4	25.32928	Apr. 1904	− 0 .4	+ 2 .6	59 49	07 .903	10 17 09
5-B5	20.41471	Dec. 1904	− 4 .9	+ 7 .1	56 40	09 .953	04 02 46
6-A1	02.60977	Sep. 1901	+ 1 .6	− 0 .4	59 50	14 .635	01 33 38
7-A2	26.54238	Oct. 1904	− 5 .1	+ 7 .0	55 53	13 .017	03 30 17
8-A3	22.57578	Sep. 1899	+ 7 .5	− 3 .3	58 09	13 .818	02 05 18
9-A4	02.52353	Sep. 1898	+ 6 .1	− 5 .9	57 46	12 .565	23 31 58
10-A5	06.53368	Jan. 1901	+ 5 .0	+ 6 .8	54 15	12 .807	08 01 24
11-A6	01.52669	Mar. 1907	− 5 .0	− 5 .1	57 24	12 .640	11 22 38

AFCRL Plate Nos.*	ε	α	δ	☾	Ω	☾age	Exp.	T	P
1-B1	23°27′05″	01h06m39s	01°32′50″	10°.3683	106°.8654	11	0s.5	03 °C	765 mm
2-B2	23 27 19	03 17 12	22 28 40	58 .7283	356 .9663	13	1 .0	00	764
3-B3	23 27 07	01 05 13	11 52 40	16 .2920	261 .6488	11	0 .8	03	766
4-B4	23 26 57	10 43 55	06 24 20	160 .7853	175 .7077	10	0 .8	09	761
5-B5	23 26 57	04 19 52	16 23 20	71 .0699	163 .0471	13	0 .7	00	769
6-A1	23 27 02	02 19 20	14 41 50	36 .0815	226 .8464	19	0 .7	10	757
7-A2	23 26 57	04 07 11	15 51 40	68 .0502	165 .9528	17	0 .5	07	764
8-A3	23 27 08	02 57 02	19 58 30	40 .3918	265 .4455	17	0 .7	06	760
9-A4	23 27 12	00 39 43	09 49 00	6 .7905	284 .8500	16	1 .0	11	762
10-A5	23 27 03	08 47 36	12 53 20	125 .9201	239 .5065	16	0 .4	−10	760
11-A6	23 27 05	12 13 31	03 24 20	186 .8390	120 .6251	17	0 .9	02	767

* B = Before full moon.
* A = After full moon.

An explanation of the notations used in Table I is as follows: l, b = topocentric librations in longitude and latitude, π = equatorial horizontal parallax, UT = universal time, LST = local sidereal time at Paris Observatory, ε = obliquity of ecliptic, α = right ascension, δ = declination, ☾ = Moon's mean anomaly, Ω = mean longitude of the ascending node of the Moon's orbit on the ecliptic, ☾$_{age}$ = age of the Moon to the nearest day, Exp. time = plate exposure time in seconds, T, P = temperature and pressure estimated from meteorological records for the day of the observation at the Eiffel Tower near Paris Observatory.

Measured Data

From an original selection of 82 lunar features, a final selection of 77 features was made representing 1304 measures by two investigators on 11 photographic plates.

Appendix A lists the measured features used in this investigation. Conventional I.A.U. feature notation after BLAGG and MÜLLER (1935) is used for 73 features. Four features (AFCRL, Nos. 79–82) were added from the *USAF Lunar Photographic Atlas* and supplements.

The criteria used for the final selection of the 77 measured features were as follows. Each lunar feature appeared on at least three photographic plates and was measured independently by two investigators. In selenodetic photogrammetry where the field is restricted to about $\frac{1}{2}$ degree, this narrow field causes the lunar photography to be nearly orthographic in projection. One minimum condition for a statistical non-trivial solution requires that each feature used be measured on three or more plates.

If the reproducibility between two sets of independent measures for each lunar feature was greater than 50 microns, the feature was rejected. The number of lunar features used should be distributed over the Moon's visible face. Figure 2 shows the distribution pattern of the finally selected 77 lunar features and is reasonably balanced.

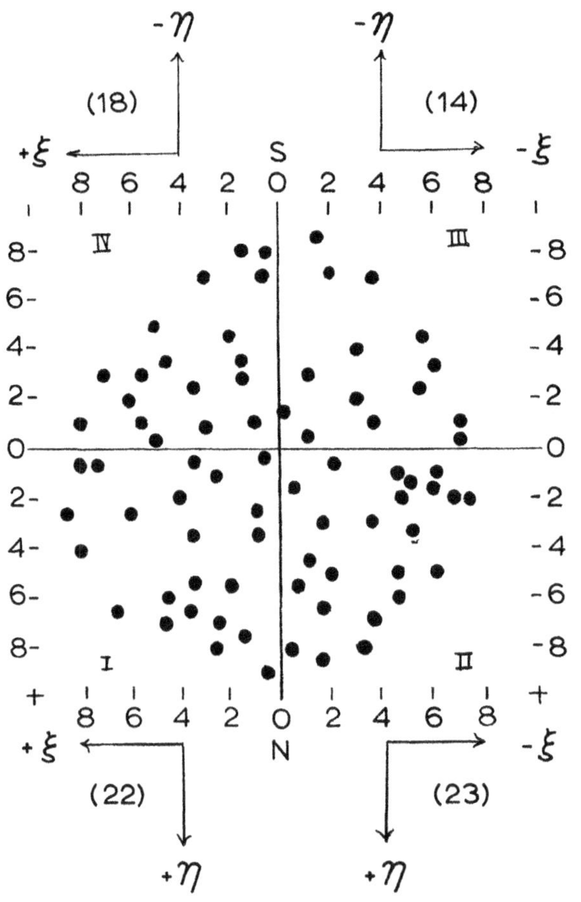

Fig. 2. Lunar Feature Distribution.

Selenodetic Measures

Enlarged photographic positives were made from the original plates as an aid to the identification and selection of lunar features. Selected features were annotated in ink on the non-emulsion side of the original plates and were placed emulsion side up in the modified Mann linear comparator for measurement. All eleven plates were measured separately by T. Weimer and M. Hunt; thus, two independent sets of measures for each feature on each plate were acquired. Ordinate and abscissa plate measures were manually recorded after using a comparator microscope attachment to locate crater positions.

Plates were oriented in the linear comparator by setting on a pinpoint hole near the geometrical center of the image. A second setting was made on an orientation cross scribed on the side of the plate outside the lunar image. Settings were made on orientation marks before and after measures and varied only 1–2 microns for all plates indicating little or no movement of the plate in the comparator during the time of measure. Plates were not rotated 180° in the comparator and remeasured; however, the eyepiece was rotated. Time did not permit the acquisition of two additional sets of measurements and they were not considered completely essential for this evaluation. Temperature readings were taken before and after each plate measure. Temperature did not vary more than 1 °C during each plate measure and not more than 5 °C during the course of measuring all eleven plates covering many days.

Evaluation of Measures

A "boxing" technique was used by both investigators in measuring each lunar feature. Each coordinate is the mean of two measurements, the comparator fiducial mark being set on the rim of both sides of the feature. Each rim measurement is the mean of at least two sequential comparator settings, with the eyepiece oriented at 0° and 180°. The geometrical center of each image was identified and measured and the origin or principal point is assumed to be this center for each plate coordinate system.

Since the two sets of plate measures were in different frames of reference, they were adjusted to a common coordinate system. After referring the measures to their respective image centroids, a least squares criterion was used for each plate to rotate one investigator's set of measurements to best fit the other's. Three estimated parameters, two translation parameters and one angular rotation parameter, were used in these adjustments. Residuals were checked for discrepancies and then the original measured data were checked for transcription errors to resolve discrepancies. Residuals greater than 50 microns or about one kilometer, representing the total distance between the same feature in two measurements were arbitrarily rejected.

Appendix B provides the x and y adjusted plate coordinates of two investigators to a common system. The differences in microns between the adjusted measures in x and y for each measured feature on each plate are shown in the δx and δy columns.

TABLE II

RMS Residuals of Each Plate

AFCRL Plate Nos.*	No. of Measures	$R_p(\mu)$	AFCRL Plate Nos.*	No. of Measures	$R_p(\mu)$
1-B1	62	22	7-A2	63	25
2-B2	60	26	8-A3	52	25
3-B3	68	18	9-A4	63	24
4-B4	54	20	10-A5	66	26
5-B5	55	22	11-A6	56	27
6-A1	53	22			
			Total	652	24 ~ 480 m

* B = before full moon plates.
A = after full moon plates.

TABLE III

RMS Residuals of Each Measured Feature

AFCRL Feature Nos.	No. of Plates	$R_f(\mu)$	AFCRL Feature Nos.	No. of Plates	$R_f(\mu)$	AFCRL Feature Nos.	No. of Plates	$R_f(\mu)$
1	4	25	29	7	27	57	10	23
2	7	25	30	11	21	58	11	29
3	7	13	31	3	15	59	7	38
4	7	23	32	9	25	60	9	28
5	6	25	34	9	19	61	7	25
6	11	23	36	6	25	62	5	18
7	9	15	37	9	29	63	8	30
8	10	30	38	10	18	64	8	22
9	9	15	39	11	17	65	7	33
10	10	30	40	10	28	66	11	26
11	11	25	41	4	27	68	8	28
12	10	22	42	11	22	69	9	21
13	9	20	43	3	19	70	9	31
14	4	13	44	11	18	71	10	24
15	8	23	45	9	20	72	8	22
17	9	22	46	10	21	73	7	30
18	8	25	47	9	25	74	9	18
19	9	31	48	10	24	75	11	21
20	9	21	49	9	17	76	8	21
21	7	28	50	10	28	77	7	26
22	9	25	51	10	17	78	8	22
23	6	19	52	9	23	79	7	21
25	10	23	53	8	22	80	6	26
26	11	19	54	10	24	81	4	25
27	10	20	55	10	23	82	8	17
28	11	18	56	11	21			

The average plate scale at mean lunar distance with an 18.22 meter focal length is about 1 micron, μ, on the plate representing 20 meters on the Moon. The last column provides the magnitudes of the differences of each measured lunar feature. These computed residual magnitudes, R_m, ranged from 1 to 50 microns with a mode of 14 microns.

The RMS residual of measures for each plate was computed and it ranged from 18 to 27 microns, as shown in Table II. The RMS residual of all measures on all plates was computed to be 24 microns or about 480 meters on the Moon. This means that the contribution to the lunar coordinate error of an eleven plate photogrammetric solution due to the reproducibility between the two sets of measures used in this evaluation can be no less than

$$\frac{\text{RMS residual}/\sqrt{2}}{\sqrt{11}} \times \frac{20\ \text{m}}{1\,\mu} \approx 100 \text{ meters}$$

The RMS residual of each measured feature was computed and is shown in Table III. These residuals ranged from 13 to 38 microns. It might be expected that

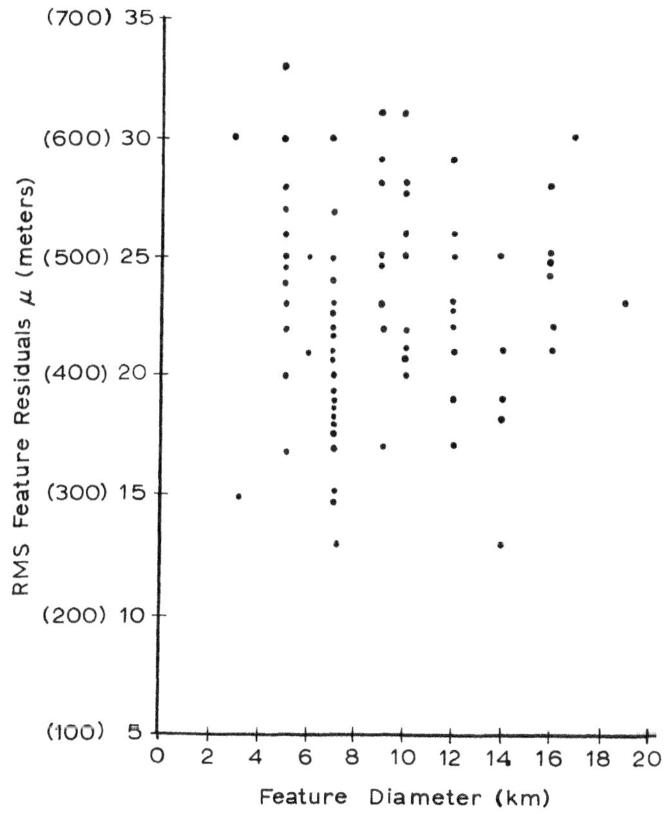

Fig. 3. Reproducibility of Measures.

smaller diameter lunar features would have smaller RMS residuals for reproducibility of feature measurements. However, Figure 3 shows the ordinate values of feature residuals to be generally in normal distribution and does not support this assumption. Thus feature diameter does not affect reproducibility of measures in this evaluation using a "boxing" technique.

Plate Orientation Geometry

Figure 4 depicts the plate orientation geometry used to show the projection of astronomical directions on the plate coordinate system with the emulsion side up.

R is the angle from the plate *x*-axis to the lunar north pole projection on the plate, $N_{☾}$.

S is the angle from the plate *x*-axis to the observers zenith projection, *Z*, on the plate and is required in differential refraction corrections.

P_G is the angle from the plate *x*-axis to a feature center, *G*, on the Moon.

Π_G is the difference between the position angle of the north lunar pole, $N_{☾}$, and the position angle of an orientation feature center, *G*, on the Moon.

C' is the topocentric position angle of the Moon between the Earth's north pole, $N_{\oplus}$, and the north lunar pole, $N_{☾}$.

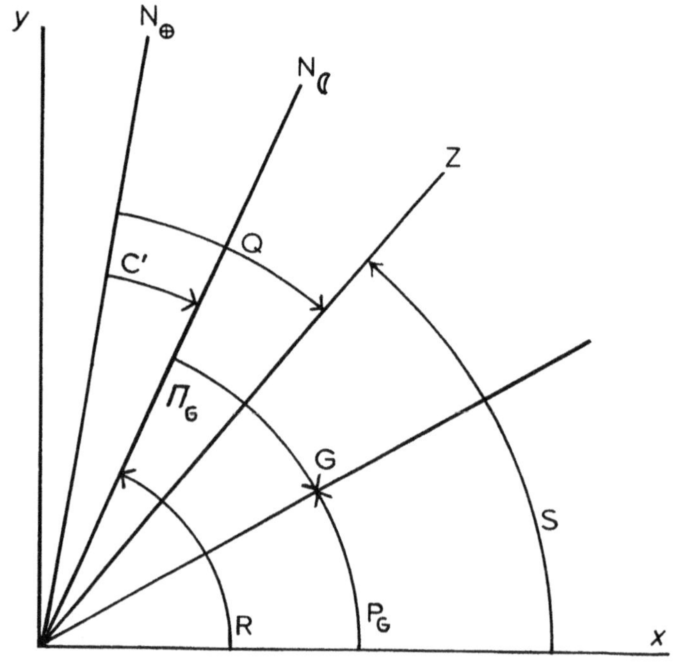

$$R = \Pi_G + P_G \qquad S = R + C' - Q$$

Fig. 4. Plate Orientation Geometry.

Q is the parallactic angle of the Moon between the meridian passing through the Moon's center and the great circle passing through the observer's zenith, Z, and the Moon.

Using selected orientation features after the technique described by WEIMER (1954) the angle, Π_G, is determined for each plate. Then from the plate angle P_G of orientation feature, G, the angle, R, is determined from $R=\Pi_G+P_G$ and the angle, S, from $S=R+C'-Q$. Table IV provides orientation angles for each plate.

TABLE IV

Plate Orientation Angles

Plate Nos.	Q	R	S
1-B1	− 5°.7	352°.4	335°.5
2-B2	− 12 .2	13 .6	11 .6
3-B3	0 .2	339 .6	316 .1
4-B4	− 6 .5	206 .4	236 .6
5-B5	− 5 .3	348 .3	343 .6
6-A1	− 13 .1	325 .2	317 .7
7-A2	− 11 .0	248 .7	248 .5
8-A3	− 16 .9	357 .1	356 .2
9-A4	− 16 .9	1 .4	355 .2
10-A5	− 12 .7	8 .2	35 .9
11-A6	− 11 .6	345 .0	20 .8

Differential Refraction

Differential refraction corrections were made for each measured lunar feature on each plate. However, conventional differential refraction corrections to apparent zenith distances were not made. Instead, differential refraction corrections were made directly to the direction cosines of each measured lunar feature and are unique to this evaluation.

Appendix C provides direction cosines of the 652 measured lunar features on all 11 photographic plates used in this evaluation. The v_1 and v_2 columns of feature direction cosines are corrected for differential refraction, the magnitude of these corrections being tabulated in the last two columns as δv_1 and δv_2. The v_3 column of direction cosines is not corrected for differential refraction as the magnitude of this correction is small enough to be neglected.

Each feature on each plate was evaluated on a scale of good, fair, and poor by each investigator for weighting purposes. Most features were rated fair to poor; less than one percent of the 1304 measures used were rated good by either investigator. An attempt was made to select features with small apparent diameters for measurement. Their apparent diameters ranged from 3 to 19 kilometers and of the 77 features used 55 had apparent diameters equal to or less than 10 kilometers.

Concluding Remarks

Resulting direction cosines of measured features corrected for differential refraction can be derived from an older collection of lunar photography and may be useful to selenodesy. The future use of these data in a selenodetic photogrammetry determination of the lunar physical libration constants will help decide the degree of their usefulness to selenodesy.

One cannot generalize on the utility to selenodesy of other older collections of lunar photographs based on a sample evaluation of one old collection. However, problem areas encountered in this evaluation may be common in some degree to other older lunar photography collections. Two are mentioned. First the thick emulsions of the Paris photographs made it necessary to increase the illumination intensity of the available linear comparator. Even with this modification it remained difficult to identify and measure many lunar features with the aid of a microscope attachment. This may have contributed to increasing the magnitude of the RMS residuals of the measures. Finally, the paucity and questionableness of old observational records made it difficult to provide plate information.

Acknowledgments

I am indebted to the following without whose interest and cooperation this technical investigation could not have been completed. To Professor A. Danjon, Director of Paris Observatory for permission to use the lunar photography collection, observatory facilities, and equipment; to Dr. T. Weimer at Paris Observatory for the secletion of photographic plates and lunar features, and for providing plate and measured data under a sub-contractual agreement through contract AF61(052)-524; to Dr. D. H. Eckhardt, AFCRL for his contributions to generating the plate information and to the analysis and reduction of the measured data; and to the Office of Naval Research and Dr. William Markowitz, U.S. Naval Observatory, for permission to make necessary modifications to the linear comparator located at Paris Observatory.

References

Blagg, M. and Müller, K.: 1935, 'Named Lunar Formations', in Catalog. I.A.U., vol. I, London.

Weimer, T.: 1954, 'Recherches sélénographiques: allongement du sélénoïde; libration physique, profils lunaires', *Bull. Astron.* (Paris) **17**.

Appendix A: Measured Data (77 Features)

AFCRL No.	I.A.U. No.	I.A.U. Name	ξ	η	Apparent Diameters*
1	40	Prom. Agarum N	+ .854	+ .229	9
2	224	Taruntius G	+ .759	+ .033	12
3	227	Taruntius N	+ .804	+ .042	7
4	362	Hook D	+ .627	+ .653	19
5	458	Hercules G	+ .435	+ .724	16
6	491	Luther	+ .342	+ .547	9
7	498	Bond K	+ .458	+ .581	7
8	534c	Ross F	+ .403	+ .189	5
9	536a	Arago B	+ .355	+ .060	7
10	623	Deseilligny	+ .328	+ .360	7
11	635	Linné F	+ .203	+ .534	5
12	645a	Plana D	+ .329	+ .666	7
13	724	Aristoteles M	+ .272	+ .803	7
14	727	Eudoxus A	+ .239	+ .717	14
15	798	Manilius D	+ .119	+ .229	7
17	819	Silberschlag A	+ .227	+ .121	7
18	855	Blagg	+ .026	+ .021	5
19	896	Aratus A	+ .081	+ .374	9
20	921	Aristillus B	− .027	+ .570	7
21	967	Egede C	+ .144	+ .767	5
22	987	W.C. Bond B	+ .055	+ .906	16
23	1073	Plato H	− .020	+ .820	14
25	1145	Archimedes A	− .098	+ .470	12
26	1214	Bode A	− .020	+ .156	12
27	1283a	Eratosthenes B	− .143	+ .320	5
28	1298d	Timocharis F	− .218	+ .519	7
29	1304c	Leverrier B	− .170	+ .644	5
30	1315	Laplace A	− .327	+ .691	10
31	1328b	Fontenelle G	− .159	+ .861	3
32	1368c	Condamine D	− .305	+ .803	10
34	1412	Draper C	− .350	+ .293	7
36	1419	Tobias Mayer D	− .440	+ .211	7
37	1500	Gambart C	− .204	+ .058	12
38	1522	Hortensius C	− .447	+ .103	7
39	1529	Milichius	− .495	+ .174	12
40	1555	Kepler A	− .584	+ .124	10
41	1559a	Kepler F	− .623	+ .145	7
42	1581	Brayley D	− .509	+ .342	5
43	1591	Diophantus B	− .469	+ .486	7
44	1604	Caroline Herschel C	− .429	+ .604	7
45	1737	Angström	− .576	+ .498	10
46	1814	Marius A	− .702	+ .218	14
47	1817	Marius D	− .693	+ .198	9
48	2383	Puiseux D	− .531	− .434	7
49	2424	Gassendi Y	− .580	− .356	7
50	2425	Herigonius	− .543	− .231	16
51	2445	Flamsteed B	− .687	− .103	9

* to nearest kilometer.

Appendix A (cont'd)

AFCRL No.	I.A.U. No.	I.A.U. Name	ξ	η	Apparent Diameters*
52	2447	Flamsteed D	− .704	− .055	7
53	2589	Hainzel A	− .365	− .685	16
54	2806	Kies D	− .287	− .421	5
55	2854	Guericke A	− .291	− .193	5
56	2898	Fra Mauro A	− .355	− .095	10
57	2933	Mösting A	− .090	− .056	12
58	2963	Ptolemaeus A	− .014	− .148	9
59	3024	Lassell A	− .116	− .287	7
60	3237	Clavius J	− .163	− .848	10
61	3377	Jacobi B	+ .141	− .814	14
62	3394a	Lilius S	+ .062	− .796	14
63	3434	Stöfler F	+ .064	− .677	17
64	3550	Airy A	+ .128	− .293	12
65	3616	Hipparchus N	+ .087	− .084	5
66	3683	Alfraganus C	+ .309	− .106	12
68	3763	Tacitus C	+ .330	− .237	9
69	3780	Abenezra B	+ .164	− .356	12
70	3815	Pontanus F	+ .178	− .467	10
71	4004	Nicolai A	+ .296	− .675	16
72	4083f	Piccolomini P	+ .506	− .508	9
73	4148	Rosse C	+ .536	− .318	3
74	4157a	Beaumont G	+ .428	− .348	7
75	4226	Torricelli B	+ .487	− .045	7
76	4286	Isidorus D	+ .559	− .074	16
77	4320	Gaudibert J	+ .620	− .193	10
78	4374a	Cook G	+ .711	− .324	10
Feature		*Source*			
79	I	*USAF Lunar Atlas*	+ .602	+ .248	6
80	II	*USAF Lunar Atlas*	+ .804	− .101	5
81	III	*USAF Lunar Atlas*	+ .780	+ .381	6
82	IV	*USAF Lunar Atlas*	− .220	− .704	5

Appendix B: Adjusted Plate Co-ordinates

(Units in Microns, μ)

Feature Number	x	y	δx	δy	R_m
PLATE NO. 1-B1					
1	108483.	151561.	6.	4.	8.
2	89569.	144579.	− 15.	11.	18.
3	91384.	148489.	− 2.	− 8.	8.
4	141369.	128542.	− 7.	− 9.	11.

Appendix B (cont'd)

Feature Number	x	y	δx	δy	R_m
5	124635.	134745.	− 19.	− 11.	22.
6	126328.	105171.	19.	− 1.	19.
7	131284.	113918.	2.	20.	20.
8	96206.	112197.	33.	− 8.	34.
9	84717.	109294.	14.	20.	24.
10	109668.	104315.	22.	33.	40.
11	123340.	92630.	7.	24.	25.
12	136988.	102443.	13.	34.	36.
13	149238.	97231.	− 12.	9.	15.
17	88099.	97997.	28.	− 2.	28.
18	77616.	82312.	− 27.	18.	32.
19	107882.	83581.	1.	22.	22.
20	124015.	73281.	− 12.	− 1.	12.
21	143988.	86426.	7.	− 27.	27.
22	156787.	78980.	4.	11.	12.
25	114499.	68132.	− 15.	− 7.	17.
26	88383.	77207.	− 4.	16.	17.
27	101217.	65684.	20.	− 14.	24.
28	117952.	58135.	6.	− 14.	15.
29	129597.	61245.	− 1.	− 6.	6.
30	132939.	48500.	− 19.	− 11.	22.
32	143857.	49933.	− 4.	10.	11.
34	97505.	49254.	6.	− 8.	10.
36	90161.	42766.	− 15.	− 1.	15.
37	78710.	63131.	− 33.	1.	33.
38	81097.	43271.	− 20.	− 14.	24.
39	86791.	38771.	1.	− 4.	5.
40	82459.	32286.	15.	− 10.	18.
41	84077.	29106.	− 1.	40.	40.
42	101113.	36229.	2.	− 11.	12.
43	113776.	38332.	7.	− 4.	8.
44	124666.	40889.	14.	− 1.	14.
45	114829.	30072.	− 12.	− 4.	12.
46	86572.	26265.	1.	12.	12.
48	38268.	42972.	− 1.	− 11.	11.
50	53191.	37728.	7.	− 7.	9.
51	63871.	26700.	0.	2.	2.
55	57816.	58925.	9.	− 2.	9.
56	65274.	52599.	− 12.	7.	13.
57	70270.	73608.	− 22.	− 23.	32.
58	63518.	80814.	5.	− 12.	13.
59	51567.	74418.	− 14.	37.	39.
60	10654.	78829.	21.	15.	26.
63	23978.	94306.	− 1.	− 17.	17.
64	53479.	94317.	− 12.	− 7.	14.
66	70564.	107314.	− 19.	− 42.	46.
68	60418.	110464.	1.	− 6.	6.
69	49139.	98121.	− 2.	− 21.	21.
71	27024.	113615.	− 39.	− 20.	44.
73	57362.	128998.	− 38.	− 11.	40.

Appendix B (cont'd)

Feature Number	x	y	δx	δy	R_m
75	78075.	121680.	13.	− 5.	14.
76	76946.	128164.	− 12.	8.	14.
77	68502.	134764.	15.	12.	20.
78	60294.	144408.	10.	− 10.	14.
79	104436.	128832.	27.	6.	28.
80	79748.	150069.	22.	0.	22.
81	119998.	143507.	27.	− 8.	28.
82	19736.	71677.	5.	8.	9.
PLATE NO. 2-B2					
3	75358.	156548.	6.	− 7.	9.
4	128823.	152069.	19.	4.	19.
5	112174.	154787.	19.	32.	37.
6	127638.	130005.	− 10.	43.	45.
8	97357.	129448.	41.	27.	49.
9	87708.	123288.	− 22.	12.	25.
10	112944.	126289.	− 15.	16.	22.
11	129720.	118861.	35.	5.	35.
12	137304.	130251.	− 11.	1.	11.
13	149127.	126686.	2.	3.	4.
15	106784.	107305.	− 18.	− 6.	19.
18	91486.	95933.	− 12.	10.	15.
19	119401.	106598.	− 15.	− 21.	25.
20	137368.	100295.	6.	15.	16.
21	149262.	116372.	24.	20.	31.
22	161526.	109576.	12.	− 12.	18.
25	130820.	93102.	36.	20.	41.
26	103634.	94660.	0.	− 13.	13.
27	119574.	87106.	4.	5.	6.
28	137017.	83666.	21.	2.	21.
29	146066.	89192.	− 11.	− 8.	14.
30	152538.	76085.	11.	− 25.	27.
32	160734.	78725.	− 11.	− 35.	37.
36	116185.	59679.	− 13.	44.	46.
37	99116.	77265.	17.	43.	46.
38	107366.	57116.	8.	2.	9.
39	114047.	54104.	− 12.	20.	23.
40	111477.	45243.	− 16.	− 8.	18.
41	113727.	42066.	− 27.	− 2.	27.
42	128033.	55573.	25.	− 41.	48.
43	138879.	61155.	29.	− 8.	30.
44	147554.	66202.	0.	14.	14.
45	141583.	51726.	− 25.	21.	33.
46	119218.	36536.	− 23.	0.	23.
47	116709.	39556.	6.	− 2.	6.
48	63513.	37964.	2.	9.	10.
49	71011.	35419.	− 11.	− 18.	21.
50	80783.	39575.	0.	− 34.	34.
51	94195.	31357.	4.	− 5.	7.
52	98430.	30811.	0.	− 10.	10.

Appendix B (cont'd)

Feature Number	x	y	δx	δy	R_m
53	38782.	45813.	12.	22.	25.
54	60323	59741.	9.	8.	12.
56	89197.	61234.	− 4.	23.	23.
57	87422.	84779.	− 3.	− 14.	15.
58	78148.	89185.	28.	− 21.	35.
59	68306.	77722.	− 9.	25.	27.
60	20505.	58016.	− 41.	− 3.	41.
63	31196.	82486.	− 34.	0.	34.
64	62979.	97664.	− 10.	− 26.	28.
66	74827.	116360.	− 9.	− 4.	9.
69	56919.	99135.	− 15.	− 13.	20.
71	26421.	101266.	14.	− 14.	19.
72	36169.	122237.	7.	− 13.	14.
73	51841.	129317.	17.	− 9.	20.
74	51754.	120328.	6.	− 10.	12.
75	75950.	131522.	23.	− 23.	33.
77	60386.	138401.	− 28.	0.	28.
78	46945.	141974.	− 19.	12.	22.
79	63291.	153661.	− 11.	− 28.	30.
82	34391.	57908.	− 7.	− 27.	28.
PLATE NO. 3-B3					
1	130536.	142527.	19.	18.	27.
2	113152.	139278.	1.	26.	26.
3	114913.	142952.	− 13.	− 4.	13.
4	157417.	111693.	9.	29.	31.
6	142416.	89502.	− 6.	13.	14.
7	148088.	98440.	− 3.	− 20.	20.
8	116381.	104208.	14.	3.	14.
9	104730.	103918.	− 5.	− 4.	6.
10	127650.	93209.	4.	15.	16.
11	137497.	78369.	− 6.	− 10.	11.
12	150954.	85608.	11.	4.	12.
13	159334.	77905.	− 14.	10.	17.
17	105927.	91611.	7.	1.	7.
18	91995.	78199.	− 13.	23.	27.
20	133304.	58729.	− 21.	− 1.	21.
21	153153.	68014.	2.	21.	21.
22	160303.	58303.	− 8.	20.	22.
25	123501.	55650.	− 24.	− 7.	25.
26	101388.	70498.	− 2.	− 7.	8.
27	110475.	56100.	− 6.	0.	6.
28	123460.	44991.	− 12.	16.	20.
29	134495.	45660.	− 20.	− 4.	20.
30	132786.	32436.	− 3.	− 24.	24.
32	141672.	31841.	− 3.	− 3.	5.
34	101704.	40635.	21.	− 7.	22.
36	92322.	35957.	− 2.	2.	3.
37	87779.	58747.	− 7.	− 7.	10.
38	83498.	38540.	− 4.	− 9.	10.

Appendix B (cont'd)

Feature Number	x	y	δx	δy	R_m
39	87472.	32826.	11.	4.	12.
40	80502.	27518.	− 9.	− 4.	10.
41	80734.	24056.	13.	− 8.	15.
42	100170.	27104.	2.	− 10.	10.
44	122936.	26669.	2.	− 3.	3.
45	109795.	18207.	5.	− 9.	10.
46	81870.	20761.	4.	− 4.	6.
47	82418.	17424.	− 16.	8.	18.
48	36880.	49028.	4.	− 45.	45.
49	41639.	42747.	5.	− 10.	11.
50	51887.	40019.	16.	8.	17.
51	58590.	26682.	− 18.	1.	18.
52	61878.	24002.	1.	3.	3.
53	21266.	70580.	− 14.	− 3.	15.
54	46095.	67239.	12.	− 7.	14.
55	64755.	59617.	7.	9.	11.
56	70591.	51529.	5.	− 1.	6.
57	82234.	71195.	12.	− 5.	13.
58	77124.	80123.	10.	− 17.	20.
60	13611.	92596.	13.	− 1.	13.
61	25951.	115497.	14.	15.	20.
62	25160.	108399.	0.	6.	6.
63	35497.	104063.	− 11.	21.	23.
64	69565.	96173.	− 21.	− 3.	21.
65	85336.	86242.	4.	− 1.	4.
66	90058.	105178.	− 2.	− 18.	19.
68	80058.	110922.	22.	28.	36.
69	65512.	101112.	4.	7.	8.
70	56600.	106013.	− 14.	20.	25.
71	42433.	123029.	4.	2.	4.
72	62420.	134763.	− 2.	− 7.	8.
73	79049.	130841.	4.	8.	9.
74	73634.	122747.	12.	− 19.	23.
75	99945.	118126.	− 2.	8.	8.
76	99549.	125111.	− 6.	− 25.	25.
77	91440.	133995.	− 32.	− 7.	33.
78	82915.	146187.	25.	− 15.	29.
79	126327.	119457.	− 8.	13.	15.
80	103369.	147356.	21.	17.	27.
82	24365.	82518.	1.	− 8.	8.
PLATE NO. 4-B4					
1	138313.	127723.	− 24.	9.	26.
2	119057.	130116.	7.	− 21.	22.
3	121949.	132797.	12.	− 17.	21.
4	159112.	91140.	− 23.	− 5.	24.
5	147235.	104924.	− 23.	29.	37.
6	137371.	75872.	− 15.	5.	15.
7	145474.	82496.	− 19.	8.	21.
8	113455.	97214.	2.	− 32.	32.

Appendix B (cont'd)

Feature Number	x	y	δx	δy	R_m
10	122605.	83722.	− 10.	25.	27.
11	129852.	66328.	8.	− 8.	11.
12	145763.	69101.	33.	5.	33.
13	153575.	58223.	20.	1.	20.
15	103017.	74532.	7.	4.	8.
18	83266.	77406.	2.	− 3.	3.
19	112051.	64890.	8.	10.	13.
20	121777.	47631.	− 2.	− 10.	11.
21	144764.	50660.	18.	5.	19.
22	151449.	37276.	− 19.	14.	24.
25	110964.	47126.	0.	− 3.	3.
26	91168.	67695.	− 10.	− 7.	12.
27	97721.	50930.	2.	7.	8.
28	109184.	35970.	− 8.	18.	20.
30	117287.	19565.	− 10.	− 6.	12.
31	138020.	23682.	− 14.	1.	14.
34	86261.	36772.	− 13.	− 23.	26.
36	76133.	33825.	− 7.	− 11.	12.
37	75386.	58655.	− 1.	2.	3.
38	67756.	38403.	− 4.	− 4.	5.
39	70802.	31449.	− 11.	1.	11.
42	82788.	22507.	14.	− 8.	16.
44	106167.	16114.	11.	10.	15.
54	36538.	74320.	− 2.	− 8.	8.
55	52963.	63756.	− 17.	25.	30.
56	57198.	54415.	− 25.	− 19.	31.
57	72189.	72377.	− 10.	8.	13.
58	68950.	82224.	11.	2.	11.
59	54101.	81230.	44.	1.	44.
60	11967.	102241.	4.	− 13.	14.
61	28327.	123257.	− 5.	3.	6.
62	25856.	116742.	− 5.	1.	5.
63	34278.	111770.	15.	18.	24.
65	78339.	86774.	31.	− 12.	34.
66	87096.	104078.	− 1.	− 9.	9.
68	78489.	111541.	2.	28.	28.
69	62081.	104772.	− 6.	− 7.	9.
70	54618.	110890.	31.	17.	35.
73	82509.	130054.	5.	− 13.	14.
74	75101.	123735.	− 8.	− 16.	18.
75	100081.	114159.	6.	− 12.	13.
77	95518.	130498.	− 1.	17.	17.
79	127361.	108861.	2.	− 6.	6.
80	111407.	139406.	− 16.	− 14.	21.
81	146023.	115083.	9.	8.	12.
82	19343.	91980.	4.	7.	8.
PLATE NO. 5-B5					
4	146571.	131618.	− 12.	− 13.	18.
5	130903.	140801.	− 17.	3.	17.

Appendix B (cont'd)

Feature Number	x	y	δx	δy	R_m
6	130637.	113489.	− 31.	− 6.	32.
7	136048.	121412.	− 11.	9.	14.
8	101775.	125277.	0.	2.	2.
9	90255.	123781.	− 12.	10.	16.
11	126813.	103066.	1.	11.	11.
12	140721.	109686.	− 1.	10.	10.
17	92707.	112871.	− 4.	10.	11.
21	146149.	92745.	16.	4.	17.
25	116127.	80290.	11.	− 24.	27.
26	91264.	91593.	1.	18.	18.
28	118504.	69452.	14.	2.	15.
29	130065.	70589.	− 7.	− 11.	14.
30	131964.	56272.	− 5.	− 4.	7.
32	142448.	55071.	1.	− 15.	16.
34	97594.	62912.	11.	11.	16.
36	89657.	56789.	27.	− 14.	30.
37	80314.	79158.	− 2.	− 30.	30.
38	80687.	58121.	− 14.	− 5.	15.
39	85836.	52757.	− 12.	− 30.	32.
40	80799.	45913.	5.	3.	6.
41	82024.	42170.	1.	− 18.	18.
42	99682.	48377.	− 3.	2.	3.
43	112333.	48866.	− 1.	− 8.	8.
44	123103.	49683.	18.	− 3.	19.
45	112270.	39338.	− 4.	0.	4.
46	84148.	38699.	− 3.	8.	8.
47	85803.	34992.	18.	− 24.	30.
48	37454.	58300.	9.	5.	10.
49	42806.	53175.	− 5.	0.	5.
50	52040.	53045.	16.	9.	18.
51	61489.	40468.	− 12.	5.	13.
52	65135.	38261.	− 18.	12.	22.
53	20972.	74722.	− 7.	− 9.	12.
54	41441.	78932.	− 32.	26.	42.
55	59047.	75837.	23.	23.	33.
56	65888.	68981.	18.	− 14.	23.
57	72941.	90359.	12.	− 3.	12.
58	68838.	97869.	− 9.	49.	50.
60	12363.	92284.	− 32.	− 13.	35.
64	57896.	111211.	27.	− 2.	27.
65	73648.	104997.	3.	47.	48.
66	75968.	122878.	− 6.	− 2.	6.
68	66080.	126239.	7.	− 44.	44.
69	53738.	114844.	16.	3.	16.
70	45332.	117230.	1.	21.	21.
71	32377.	128205.	− 16.	− 35.	38.
72	49118.	142672.	17.	− 7.	18.
74	59475.	135205.	− 5.	28.	28.
75	84460.	135476.	12.	− 4.	13.
76	83742.	141347.	− 1.	5.	5.

Appendix B (cont'd)

Feature Number	x	y	δx	δy	R_m
78	67876.	155562.	− 3.	− 6.	7.
79	110962.	139145.	6.	− 2.	6.
82	21382.	87174.	− 6.	8.	10.
PLATE NO. 6-A1					
6	146011.	98503.	2.	− 13.	13.
8	119259.	115160.	30.	1.	30.
9	107189.	115519.	− 6.	− 12.	13.
10	130461.	103399.	6.	− 4.	7.
11	140314.	87644.	− 20.	13.	24.
12	155115.	93686.	− 5.	− 31.	31.
13	164229.	84797.	3.	17.	17.
15	112828.	90732.	24.	− 23.	33.
17	107784.	103141.	4.	− 3.	5.
19	123261.	82914.	− 41.	1.	41.
20	135439.	67846.	7.	− 8.	11.
23	155751.	60615.	− 1.	11.	11.
25	124989.	65347.	− 2.	− 9.	10.
26	102341.	81972.	4.	6.	7.
27	111456.	66719.	2.	6.	7.
28	124974.	54242.	2.	33.	33.
30	135113.	40147.	13.	3.	13.
32	144875.	38458.	− 5.	6.	8.
34	102321.	51055.	13.	− 6.	15.
36	92682.	46571.	1.	− 17.	17.
37	88220.	70497.	− 1.	7.	7.
38	83819.	47759.	1.	− 12.	12.
39	87742.	43408.	− 12.	− 13.	18.
40	80775.	38070.	− 7.	20.	22.
42	100913.	36752.	9.	5.	10.
44	124724.	34768.	− 23.	− 9.	24.
45	111300.	26560.	− 8.	16.	18.
46	82418.	30713.	35.	9.	37.
47	83122.	27075.	− 11.	10.	15.
48	37696.	61239.	− 1.	8.	8.
49	42234.	54690.	− 9.	3.	10.
50	52211.	51817.	− 12.	9.	15.
51	59057.	37569.	− 20.	13.	24.
52	62362.	34655.	23.	6.	24.
53	23104.	83367.	1.	9.	9.
54	46861.	80165.	37.	7.	38.
55	65107.	72051.	1.	− 12.	12.
56	70786.	63499.	− 31.	0.	31.
58	78025.	92693.	23.	− 3.	23.
59	63363.	89376.	48.	− 3.	48.
60	16851.	105177.	− 12.	12.	17.
61	29739.	128356.	13.	− 2.	13.
62	28509.	121393.	− 11.	7.	13.
63	38058.	117257.	− 6.	− 4.	7.
64	71140.	109069.	− 15.	− 9.	18.

Appendix B (cont'd)

Feature Number	x	y	δx	δy	R_m
65	86551.	98687.	35.	13.	38.
66	92246.	117435.	4.	−15.	15.
69	67326.	114125.	−25.	23.	34.
70	58753.	119153.	−27.	−12.	30.
71	46095.	135997.	−12.	−25.	28.
74	76845.	135355.	−4.	−1.	4.
75	103122.	129828.	−29.	−15.	33.
82	26393.	95544.	17.	−12.	21.
PLATE NO. 7-A2					
2	73598.	152004.	19.	−18.	26.
6	118042.	125348.	4.	−14.	14.
7	120485.	134519.	1.	1.	1.
8	87372.	126792.	11.	15.	19.
10	102113.	122659.	8.	40.	41.
11	117948.	114432.	−28.	37.	46.
12	128703.	125188.	5.	1.	6.
13	141762.	121432.	−18.	14.	23.
14	133925.	118513.	−5.	6.	8.
15	92934.	104893.	10.	18.	20.
17	83083.	112324.	12.	14.	19.
18	77000.	95127.	−11.	16.	19.
19	105396.	103345.	9.	−8.	30.
20	123405.	96393.	29.	−8.	30.
21	139346.	111207.	−21.	−39.	44.
22	153574.	104574.	−14.	−14.	20.
23	145895.	98397.	1.	21.	21.
25	115568.	89702.	15.	−21.	26.
26	88407.	92965.	5.	21.	21.
27	103455.	84769.	−10.	3.	11.
28	121364.	80440.	−6.	−3.	7.
29	131763.	85294.	−40.	−26.	48.
30	138228.	72718.	10.	−16.	19.
31	151747.	87200.	13.	5.	14.
32	148391.	75065.	19.	17.	26.
34	104084.	67515.	7.	−1.	7.
37	82695.	76904.	−8.	−19.	21.
38	89987.	57436.	−26.	−26.	37.
39	96566.	54155.	−3.	−5.	6.
40	94089.	46100.	−33.	16.	37.
42	110879.	54639.	24.	−6.	25.
44	132194.	63648.	−7.	9.	12.
45	125557.	50410.	−1.	1.	2.
46	99609.	40547.	−4.	−14.	15.
47	102360.	37662.	−19.	−1.	19.
48	49741.	43348.	23.	5.	23.
49	56430.	40337.	−11.	10.	14.
50	65022.	43269.	6.	−18.	19.
51	77994.	34642.	−17.	−12.	21.
52	82129.	33817.	34.	−12.	36.

Appendix B (cont'd)

Feature Number	x	y	δx	δy	R_m
53	28980.	53119.	– 26.	18.	32.
54	46611.	63829.	– 17.	0.	17.
55	64013.	66690.	– 17.	– 19.	26.
56	72665.	62647.	– 4.	1.	4.
57	72149.	84823.	21.	– 2.	21.
58	63965.	89852.	– 1.	46.	46.
59	54304.	79849.	– 1.	3.	3.
60	15122.	66642.	11.	14.	18.
61	13821.	91762.	– 5.	– 23.	23.
62	15677.	85911.	– 29.	– 4.	29.
63	23750.	88183.	42.	– 5.	42.
65	67929.	98760.	3.	38.	38.
66	64226.	116032.	– 17.	11.	20.
68	53914.	115926.	– 3.	– 27.	27.
69	46227.	101221.	– 1.	– 27.	27.
70	37669.	100668.	7.	– 29.	30.
71	21952.	106568.	11.	– 2.	11.
72	32738.	125561.	7.	– 18.	19.
74	44836.	122014.	9.	0.	9.
75	67976.	130549.	4.	26.	27.
76	65354.	135712.	– 5.	– 11.	12.
77	55890.	138753.	– 1.	– 11.	11.
78	45950.	143703.	3.	– 5.	6.
PLATE NO. 8-A3					
6	135965.	112437.	– 2.	– 14.	14.
7	138832.	123070.	– 3.	– 10.	10.
9	95626.	113114.	– 7.	– 13.	14.
10	120643.	110630.	23.	23.	33.
11	134701.	100276.	27.	7.	28.
12	145513.	112030.	– 14.	– 17.	22.
14	149432.	104540.	– 15.	8.	17.
15	109252.	92323.	20.	– 21.	29.
17	100479.	101874.	9.	28.	30.
19	121166.	89272.	– 12.	– 43.	45.
20	136852.	80775.	– 9.	39.	40.
22	163591.	91113.	2.	6.	7.
23	156783.	83338.	24.	– 13.	27.
26	102717.	80681.	23.	– 38.	44.
27	115871.	70475.	5.	41.	42.
28	131893.	64849.	15.	– 16.	22.
30	145148.	57276.	10.	1.	10.
32	154070.	60280.	– 13.	11.	17.
34	112637.	53725.	22.	12.	25.
38	96146.	46101.	0.	– 12.	12.
39	101791.	42227.	– 14.	9.	17.
40	97035.	35419.	– 36.	5.	37.
42	115693.	41253.	– 14.	– 1.	14.
44	137611.	48600.	– 7.	9.	12.
45	127948.	36841.	– 6.	15.	16.

Appendix B (cont'd)

Feature Number	x	y	δx	δy	R_m
46	100668.	30038.	17.	− 3.	17.
47	102340.	27355.	− 35.	27.	45.
48	49691.	42428.	− 17.	14.	22.
49	56133.	37980.	12.	9.	15.
50	66406.	38275.	22.	19.	28.
51	77071.	28392.	− 13.	3.	13.
52	80990.	27011.	5.	2.	5.
54	52263.	61363.	9.	4.	10.
55	71923.	59520.	14.	− 38.	40.
56	79920.	53879.	− 28.	7.	29.
57	84483.	75412.	− 1.	− 4.	4.
58	76937.	82117.	1.	− 16.	16.
60	15002.	77074.	− 11.	37.	39.
63	31265.	93019.	− 9.	− 31.	33.
64	65002.	94919.	7.	− 5.	9.
66	81416.	109658.	22.	0.	22.
68	70358.	112100.	5.	− 2.	5.
69	59740.	98401.	15.	− 19.	24.
70	50049.	100566.	− 8.	31.	32.
71	32037.	113155.	− 10.	− 15.	19.
72	47069.	130097.	− 10.	− 13.	16.
73	63635.	131070.	6.	− 26.	27.
74	60920.	121558.	− 12.	− 16.	20.
75	86896.	125216.	− 20.	− 12.	23.
76	84494.	131870.	− 11.	1.	11.
79	111618.	135332.	10.	27.	29.
82	27662.	69997.	14.	3.	14.
PLATE NO. 9-A4					
2	92737.	145125.	2.	− 20.	20.
3	92976.	149388.	0.	9.	9.
4	142577.	137325.	− 20.	− 13.	24.
5	127588.	140741.	4.	8.	8.
6	137268.	110341.	− 33.	− 11.	34.
7	139133.	120862.	3.	10.	10.
8	108519.	113578.	17.	36.	40.
9	98093.	108970.	− 7.	2.	7.
10	122678.	107875.	40.	14.	42.
11	136868.	98291.	− 11.	16.	20
12	146311.	110351.	0.	− 11.	11.
13	156503.	107134.	12.	22.	25.
15	112804.	89213.	− 10.	9.	14.
17	103635.	98098.	− 4.	18.	18.
18	95868.	80713.	− 15.	− 11.	19.
19	124581.	86834.	− 33.	23.	40.
20	140105.	79115.	13.	− 11.	17.
22	164350.	90156.	− 3.	− 33.	33.
23	158671.	82429.	3.	1.	3.
25	132372.	72398.	8.	32.	33.
26	107075.	77283.	− 9.	1.	9.

Appendix B (cont'd)

Feature Number	x	y	δx	δy	R_m
27	120419.	67828.	18.	– 19.	26.
28	136137.	63013.	– 4.	– 16.	16.
30	148924.	55915.	– 5.	– 10.	11.
38	102246.	42400.	– 15.	– 4.	16.
39	107859.	38825.	– 16.	10.	19.
40	103388.	31678.	23.	– 23.	33.
42	121387.	38579.	8.	27.	28.
44	142101.	46982.	– 6.	– 6.	9.
45	133173.	34643.	– 19.	9.	21.
46	107008.	26433.	31.	8.	32.
47	108638.	23783.	– 2.	9.	10.
48	56019.	35638.	8.	10.	13.
49	62591.	31666.	– 16.	3.	16.
50	72943.	32654.	3.	24.	24.
51	83741.	23364.	– 3.	– 7.	8.
52	87662.	22194.	– 10.	9.	13.
53	34053.	49825.	27.	19.	32.
54	57969.	54811.	– 25.	– 22.	33.
55	77722.	54345.	– 6.	9.	11.
56	85876.	49215.	– 10.	10.	14.
57	89493.	70917.	27.	14.	30.
58	81643.	77135.	11.	8.	13.
59	69681.	68793.	6.	38.	39.
61	22043.	92373.	7.	5.	9.
62	24003.	85498.	21.	9.	23.
63	35041.	84740.	2.	– 40.	40.
64	69023.	88963.	– 26.	1.	26.
65	86860.	85703.	– 19.	– 2.	19.
66	84376.	104637.	14.	– 46.	48.
68	73168.	106353.	20.	– 19.	28.
70	53599.	93620.	– 19.	– 26.	32.
71	34468.	104650.	7.	3.	8.
72	48406.	122408.	13.	6.	14.
73	65010.	124534.	– 5.	– 35.	36.
74	63041.	115058.	– 2.	– 13.	13.
75	88574.	120273.	– 14.	0.	14.
76	85669.	126620.	– 27.	– 12.	29.
77	75072.	131975.	8.	– 10.	13.
78	62804.	140517.	– 1.	8.	8.
79	111932.	131535.	2.	10.	11.
80	79852.	149243.	23.	– 10.	25.
81	120458.	148851.	15.	– 24.	28.
PLATE NO. 10-A5					
1	87529.	152228.	31.	– 12.	33.
2	71629.	139218.	– 3.	– 41.	41.
3	72300.	144419.	4.	8.	9.
4	125190.	139800.	– 13.	– 25.	28.
5	107613.	140863.	– 10.	– 11.	15.
6	118086.	112549.	9.	– 7.	11.

Appendix B (cont'd)

Feature Number	x	y	δx	δy	R_m
7	119897.	123139.	− 3.	4.	5.
8	87311.	111499.	− 18.	15.	23.
9	77298.	105584.	11.	− 2.	12.
10	102329.	107970.	14.	− 30.	33.
11	118558.	100695.	12.	5.	13.
12	128603.	113894.	7.	23.	24.
13	141545.	112381.	16.	10.	19.
14	134092.	107510.	− 16.	1.	16.
15	93995.	88527.	5.	33.	34.
17	83723.	95835.	2.	33.	33.
18	78301.	77987.	− 21.	12.	25.
19	106508.	87778.	3.	− 18.	18.
20	124715.	82655.	− 12.	− 6.	13.
22	153808.	97268.	− 15.	− 45.	47.
23	146780.	88743.	0.	− 7.	7.
25	117190.	75045.	3.	3.	4.
26	89942.	76164.	− 9.	− 2.	10.
27	105332.	68931.	9.	− 7.	11.
28	123276.	66561.	4.	− 12.	12.
29	133318.	72834.	21.	34.	40.
30	140119.	61674.	15.	− 6.	17.
31	152636.	79141.	2.	17.	17.
32	149944.	66165.	− 6.	27.	28.
34	106471.	52287.	18.	1.	18.
37	84755.	60045.	− 43.	8.	44.
38	92560.	41950.	17.	− 13.	21.
39	99231.	39354.	− 11.	− 11.	15.
40	96886.	31913.	13.	− 29.	32.
42	113487.	40999.	− 10.	− 1.	10.
44	134367.	52081.	− 33.	6.	34.
46	102524.	27413.	0.	9.	9.
47	105294.	25163.	16.	− 29.	33.
48	51954.	29489.	13.	19.	23.
49	58882.	26529.	22.	5.	23.
50	67663.	28811.	− 6.	36.	36.
51	80841.	21401.	− 11.	− 12.	16.
52	84961.	20803.	− 11.	13.	17.
53	30318.	40360.	6.	20.	21.
54	48416.	48083.	9.	− 17.	19.
55	66158.	50173.	− 4.	3.	5.
56	75011.	46321.	17.	− 12.	21.
57	73756.	67609.	− 4.	3.	5.
58	65315.	72611.	− 28.	5.	28.
59	55817.	63000.	35.	26.	43.
60	15388.	55622.	− 25.	− 21.	32.
61	13334.	79964.	− 39.	− 14.	41.
65	68980.	81602.	− 10.	28.	29.
66	64409.	99548.	− 25.	− 1.	25.
69	46692.	84699.	4.	− 10.	11.
70	37909.	84938.	26.	− 6.	27.

Appendix B (cont'd)

Feature Number	x	y	δx	δy	R_m
71	21278.	93280.	−19.	7.	20.
72	31514.	112257.	13.	46.	47.
73	45656.	116321.	−30.	−35.	46.
75	67410.	115341.	6.	7.	9.
76	64475.	121137.	−17.	−22.	28.
77	54622.	124961.	−7.	−44.	45.
78	43972.	131765.	−3.	38.	38.
80	60754.	142608.	12.	2.	12.
81	100704.	147844.	28.	−5.	29.
82	26231.	51361.	23.	−11.	26.
PLATE NO. 11-A6					
2	107908.	151466.	11.	11.	16.
3	109016.	154492.	8.	−15.	17.
6	142021.	107724.	6.	−3.	7.
7	146247.	115893.	−13.	−13.	18.
8	115694.	120693.	11.	22.	24.
9	104669.	119613.	−3.	2.	3.
10	127730.	110968.	9.	12.	15.
11	138768.	97067.	−16.	12.	20.
13	159020.	95741.	10.	−28.	30.
14	152620.	95430.	8.	−1.	8.
15	113841.	97218.	−9.	11.	14.
17	107347.	108248.	10.	21.	23.
18	95583.	94201.	−34.	−26.	43.
19	124421.	91045.	−14.	−20.	25.
21	154496.	86495.	−26.	1.	26.
22	161979.	75686.	1.	13.	13.
23	154965.	72001.	15.	−23.	28.
25	128130.	74310.	2.	−4.	5.
26	105472.	87689.	13.	−10.	16.
27	115712.	74114.	21.	4.	22.
28	129202.	63448.	5.	−1.	5.
29	139211.	65031.	1.	−15.	15.
30	139151.	50247.	−17.	−40.	44.
32	147428.	49168.	−4.	46.	46.
34	108967.	57659.	−4.	−24.	24.
37	93778.	74692.	−4.	26.	27.
39	96105.	48267.	4.	11.	12.
40	89921.	41829.	36.	20.	41.
42	108838.	43374.	27.	−3.	27.
44	130425.	43936.	−21.	1.	21.
45	118697.	34191.	2.	35.	35.
46	91832.	34707.	1.	24.	24.
47	92642.	31050.	15.	−16.	22.
48	45361.	57227.	1.	38.	38.
49	50654.	51713.	−8.	27.	28.
50	60942.	50786.	43.	26.	50.
51	68767.	37860.	−5.	26.	26.
52	72165.	35383.	−18.	−38.	42.

Appendix B (cont'd)

Feature Number	x	y	δx	δy	R_m
53	27653.	75320.	− 16.	− 6.	17.
54	52443.	77278.	− 10.	− 5.	11.
55	71436.	72673.	− 21.	− 9.	23.
56	77952.	65431.	6.	11.	13.
57	87043.	86397.	− 4.	47.	47.
58	81079.	94431.	− 11.	26.	29.
61	26775.	118503.	− 30.	− 26.	40.
64	71889.	108589.	13.	− 21.	24.
66	90473.	119449.	− 10.	22.	25.
68	80113.	123647.	− 6.	− 29.	30.
70	58051.	115864.	31.	− 25.	40.
71	42043.	128758.	− 4.	18.	19.
72	59888.	142317.	− 2.	− 13.	14.
74	72438.	133480.	23.	0.	23.
75	98267.	132113.	2.	− 26.	26.
76	96944.	138306.	− 5.	− 26.	27.
78	78063.	154617.	15.	− 26.	30.
80	97344.	157412.	−35.	−21.	41.

Appendix C: Direction Cosines

Feature Number	ν_1	ν_2	ν_3	$10^{-5}\,\delta\nu_1$	$10^{-5}\,\delta\nu_2$
PLATE NO. 1-B1					
1	0.00121678	0.00358399	0.99999283	0.023	0.110
2	0.00017817	0.00320084	0.99999485	− 0.033	0.110
3	0.00027779	0.00341547	0.99999412	− 0.030	0.117
4	0.00302297	0.00231995	0.99999274	0.146	0.041
5	0.00210392	0.00266065	0.99999424	0.088	0.065
6	0.00219712	0.00103688	0.99999704	0.115	0.007
7	0.00246924	0.00151710	0.99999580	0.125	0.020
8	0.00054290	0.00142288	0.99999884	0.013	0.043
9	− 0.00008803	0.00126357	0.99999919	− 0.022	0.046
10	0.00128223	0.00099000	0.99999868	0.062	0.018
11	0.00203316	0.00034837	0.99999787	0.115	− 0.015
12	0.00278254	0.00088704	0.99999573	0.152	− 0.006
13	0.00345534	0.00060077	0.99999385	0.195	− 0.025
17	0.00009779	0.00064331	0.99999978	− 0.003	0.021
18	− 0.00047779	− 0.00021778	0.99999986	− 0.025	− 0.001
19	0.00118433	− 0.00014836	0.99999928	0.072	− 0.021
20	0.00207035	− 0.00071396	0.99999759	0.131	− 0.053
21	0.00316709	0.00000760	0.99999498	0.186	− 0.042
22	0.00387002	− 0.00040134	0.99999242	0.233	− 0.066
25	0.00154779	− 0.00099658	0.99999830	0.104	− 0.056

Appendix C (cont'd)

Feature Number	ν_1	ν_2	ν_3	$10^{-5}\,\delta\nu_1$	$10^{-5}\,\delta\nu_2$
26	0.00011353	− 0.00049816	0.99999987	0.013	− 0.019
27	0.00081843	− 0.00113092	0.99999902	0.063	− 0.051
28	0.00173751	− 0.00154550	0.99999729	0.123	− 0.078
29	0.00237701	− 0.00137485	0.99999622	0.158	− 0.081
30	0.00256059	− 0.00207463	0.99999456	0.178	− 0.108
32	0.00316016	− 0.00199601	0.99999301	0.212	− 0.113
34	0.00061469	− 0.00203296	0.99999774	0.064	− 0.080
36	0.00021144	− 0.00238913	0.99999712	0.045	− 0.087
37	− 0.00041757	− 0.00127094	0.99999910	− 0.007	− 0.039
38	− 0.00028635	− 0.00236134	0.99999716	0.015	− 0.079
39	0.00002640	− 0.00260842	0.99999659	0.037	− 0.092
40	− 0.00021149	− 0.00296446	0.99999557	0.028	− 0.102
41	− 0.00012261	− 0.00313905	0.99999506	0.035	− 0.109
42	0.00081292	− 0.00274812	0.99999589	0.085	− 0.108
43	0.00150833	− 0.00263274	0.99999539	0.124	− 0.113
44	0.00210635	− 0.00249245	0.99999467	0.157	− 0.116
45	0.00156619	− 0.00308624	0.99999400	0.134	− 0.130
46	0.00001442	− 0.00329506	0.99999456	0.045	− 0.116
48	− 0.00263836	− 0.00237744	0.99999369	− 0.123	− 0.048
50	− 0.00181883	− 0.00266543	0.99999478	− 0.071	− 0.069
51	− 0.00123223	− 0.00327099	0.99999388	− 0.028	− 0.098
55	− 0.00156496	− 0.00150170	0.99999765	− 0.072	− 0.032
56	− 0.00115538	− 0.00184909	0.99999762	− 0.043	− 0.049
57	− 0.00088116	− 0.00069560	0.99999937	− 0.042	− 0.013
58	− 0.00125203	− 0.00029993	0.99999917	− 0.070	0.006
59	− 0.00190829	− 0.00065104	0.99999797	− 0.104	0.003
60	− 0.00415513	− 0.00040850	0.99999128	− 0.240	0.042
63	− 0.00342355	0.00044110	0.99999403	− 0.208	0.062
64	− 0.00180345	0.00044149	0.99999827	− 0.112	0.040
66	− 0.00086528	0.00115496	0.99999895	− 0.067	0.052
68	− 0.00142249	0.00132799	0.99999810	− 0.102	0.066
69	− 0.00204179	0.00065037	0.99999770	− 0.129	0.051
71	− 0.00325641	0.00150124	0.99999356	− 0.213	0.097
73	− 0.00159044	0.00234558	0.99999598	− 0.126	0.104
75	− 0.00045291	0.00194366	0.99999800	− 0.053	0.075
76	− 0.00051493	0.00229968	0.99999722	− 0.062	0.088
77	− 0.00097869	0.00266208	0.99999597	− 0.094	0.107
78	− 0.00142952	0.00319163	0.99999388	− 0.128	0.132
79	0.00099475	0.00233612	0.99999677	0.027	0.069
80	− 0.00036123	0.00350231	0.99999379	− 0.069	0.128
81	0.00184920	0.00314174	0.99999335	0.066	0.086
82	− 0.00365628	− 0.00080125	0.99999299	− 0.205	0.022
PLATE NO. 2-B2					
3	− 0.00053043	0.00402467	0.99999175	− 0.013	0.119
4	0.00240506	0.00377878	0.99998996	0.099	0.117
5	0.00149095	0.00392799	0.99999117	0.064	0.120
6	0.00234002	0.00256750	0.99999397	0.094	0.081
8	0.00067740	0.00253690	0.99999655	0.031	0.077
9	0.00014763	0.00219868	0.99999757	0.010	0.066

Appendix C (cont'd)

Feature Number	ν_1	ν_2	ν_3	$10^{-5}\,\delta\nu_1$	$10^{-5}\,\delta\nu_2$
10	0.00153321	0.00236347	0.99999603	0.063	0.073
11	0.00245429	0.00195565	0.99999507	0.097	0.063
12	0.00287074	0.00258097	0.99999254	0.115	0.082
13	0.00351986	0.00238529	0.99999095	0.139	0.077
15	0.00119499	0.00132122	0.99999841	0.048	0.041
18	0.00035503	0.00069685	0.99999969	0.015	0.021
19	0.00188771	0.00128243	0.99999739	0.075	0.042
20	0.00287420	0.00093641	0.99999543	0.112	0.033
21	0.00352727	0.00181903	0.99999212	0.138	0.061
22	0.00420059	0.00144595	0.99999013	0.163	0.051
25	0.00251466	0.00054148	0.99999668	0.097	0.021
26	0.00102201	0.00062701	0.99999928	0.040	0.021
27	0.00189722	0.00021227	0.99999817	0.073	0.010
28	0.00285495	0.00002344	0.99999592	0.109	0.006
29	0.00335174	0.00032682	0.99999432	0.129	0.016
30	0.00370708	− 0.00039277	0.99999305	0.141	− 0.005
32	0.00415708	− 0.00024777	0.99999132	0.159	0.000
36	0.00171113	− 0.00129349	0.99999770	0.063	− 0.035
37	0.00077393	− 0.00032800	0.99999964	0.029	− 0.008
38	0.00122689	− 0.00143422	0.99999821	0.044	− 0.040
39	0.00159372	− 0.00159956	0.99999744	0.058	− 0.045
40	0.00145259	− 0.00208603	0.99999676	0.052	− 0.059
41	0.00157615	− 0.00226046	0.99999620	0.056	− 0.064
42	0.00236164	− 0.00151888	0.99999605	0.088	− 0.041
43	0.00295716	− 0.00121244	0.99999489	0.111	− 0.031
44	0.00343345	− 0.00093534	0.99999366	0.130	− 0.022
45	0.00310556	− 0.00173010	0.99999367	0.116	− 0.046
46	0.00187764	− 0.00256406	0.99999494	0.067	− 0.073
47	0.00173987	− 0.00239824	0.99999560	0.062	− 0.068
48	− 0.00118089	− 0.00248570	0.99999621	− 0.050	− 0.076
49	− 0.00076925	− 0.00262542	0.99999625	− 0.034	− 0.080
50	− 0.00023268	− 0.00239725	0.99999709	− 0.013	− 0.072
51	0.00050369	− 0.00284843	0.99999581	0.014	− 0.084
52	0.00073622	− 0.00287839	0.99999558	0.023	− 0.084
53	− 0.00253876	− 0.00205481	0.99999466	− 0.101	− 0.066
54	− 0.00135605	− 0.00129015	0.99999824	− 0.054	− 0.041
56	0.00022933	− 0.00120817	0.99999924	0.007	− 0.036
57	0.00013187	0.00008450	0.99999999	0.005	0.003
58	− 0.00037730	0.00032640	0.99999987	− 0.014	0.009
59	− 0.00091772	− 0.00030299	0.99999952	− 0.036	− 0.011
60	− 0.00354227	− 0.00138486	0.99999276	− 0.138	− 0.048
63	− 0.00295525	− 0.00004146	0.99999563	− 0.113	− 0.007
64	− 0.00121015	0.00079187	0.99999895	− 0.045	0.021
66	− 0.00055963	0.00181830	0.99999818	− 0.018	0.053
69	− 0.00154292	0.00087261	0.99999842	− 0.058	0.023
71	− 0.00321738	0.00098958	0.99999433	− 0.122	0.024
72	− 0.00268217	0.00214093	0.99999411	− 0.099	0.059
73	− 0.00182164	0.00252961	0.99999513	− 0.065	0.072
74	− 0.00182646	0.00203613	0.99999625	− 0.066	0.057
75	− 0.00049793	0.00265069	0.99999636	− 0.014	0.078

Appendix C (cont'd)

Feature Number	ν_1	ν_2	ν_3	$10^{-5}\,\delta\nu_1$	$10^{-5}\,\delta\nu_2$
77	− 0.00135250	0.00302835	0.99999449	− 0.046	0.088
78	− 0.00209045	0.00322449	0.99999261	− 0.074	0.092
79	− 0.00119295	0.00386614	0.99999180	− 0.039	0.113
82	− 0.00277986	− 0.00139077	0.99999516	− 0.109	− 0.046
PLATE NO. 3-B3					
1	0.00256337	0.00320039	0.99999159	0.070	0.098
2	0.00160893	0.00302210	0.99999414	0.035	0.099
3	0.00170559	0.00322381	0.99999335	0.037	0.106
4	0.00403937	0.00150733	0.99999070	0.140	0.022
6	0.00321588	0.00028901	0.99999478	0.119	− 0.016
7	0.00352724	0.00077973	0.99999347	0.127	− 0.001
8	0.00178636	0.00109655	0.99999779	0.058	0.026
9	0.00114668	0.00108070	0.99999876	0.034	0.031
10	0.00240515	0.00049264	0.99999698	0.087	− 0.002
11	0.00294585	− 0.00032222	0.99999560	0.114	− 0.037
12	0.00368470	0.00007517	0.99999321	0.139	− 0.028
13	0.00414480	− 0.00034777	0.99999134	0.160	− 0.048
17	0.00121247	0.00040500	0.99999917	0.043	0.005
18	0.00044758	− 0.00033132	0.99999984	0.020	− 0.016
20	0.00271574	− 0.00140050	0.99999533	0.115	− 0.075
21	0.00380550	− 0.00089080	0.99999236	0.152	− 0.065
22	0.00419811	− 0.00142403	0.99999017	0.171	− 0.088
25	0.00217751	− 0.00156951	0.99999639	0.096	− 0.077
26	0.00096334	− 0.00075422	0.99999925	0.043	− 0.036
27	0.00146231	− 0.00154476	0.99999773	0.068	− 0.070
28	0.00217532	− 0.00215477	0.99999531	0.101	− 0.098
29	0.00278118	− 0.00211808	0.99999388	0.123	− 0.102
30	0.00268742	− 0.00284414	0.99999234	0.126	− 0.128
32	0.00317527	− 0.00287685	0.99999081	0.144	− 0.134
34	0.00098080	− 0.00239380	0.99999665	0.057	− 0.097
36	0.00046571	− 0.00265064	0.99999637	0.040	− 0.103
37	0.00021615	− 0.00139936	0.99999899	0.020	− 0.054
38	− 0.00001880	− 0.00250879	0.99999685	0.020	− 0.093
39	0.00019946	− 0.00282250	0.99999599	0.031	− 0.107
40	− 0.00018325	− 0.00311390	0.99999513	0.019	− 0.114
41	− 0.00017050	− 0.00330399	0.99999452	0.021	− 0.121
42	0.00089664	− 0.00313674	0.99999467	0.060	− 0.124
44	0.00214663	− 0.00316069	0.99999270	0.108	− 0.136
45	0.00142512	− 0.00362529	0.99999241	0.084	− 0.147
46	− 0.00010809	− 0.00348491	0.99999391	0.025	− 0.129
47	− 0.00007800	− 0.00366814	0.99999326	0.028	− 0.136
48	− 0.00257842	− 0.00193269	0.99999480	− 0.082	− 0.050
49	− 0.00231706	− 0.00227759	0.99999472	− 0.069	− 0.065
50	− 0.00175440	− 0.00242740	0.99999551	− 0.046	− 0.076
51	− 0.00138632	− 0.00315969	0.99999405	− 0.026	− 0.106
52	− 0.00120575	0.00330689	0.99999380	− 0.018	− 0.113
53	− 0.00343578	− 0.00074934	0.99999381	− 0.124	0.001
54	− 0.00207256	− 0.00093290	0.99999741	− 0.071	− 0.017
55	− 0.00104795	− 0.00135148	0.99999853	− 0.028	− 0.041

Appendix C (cont'd)

Feature Number	ν_1	ν_2	ν_3	$10^{-5}\ \delta\nu_1$	$10^{-5}\ \delta\nu_2$
56	− 0.00072750	− 0.00179554	0.99999812	− 0.012	− 0.061
57	− 0.00008835	− 0.00071586	0.99999973	0.003	− 0.026
58	− 0.00036896	− 0.00022561	0.99999990	− 0.012	− 0.005
60	− 0.00385617	0.00045952	0.99999245	− 0.150	0.050
61	− 0.00317877	0.00171680	0.99999347	− 0.135	0.091
62	− 0.00322217	0.00132710	0.99999392	− 0.134	0.077
63	− 0.00265463	0.00108898	0.99999588	− 0.110	0.063
64	− 0.00078408	0.00065561	0.99999947	− 0.035	0.031
65	0.00008190	0.00011032	0.99999999	0.002	0.003
66	0.00034108	0.00114997	0.99999928	0.003	0.040
68	− 0.00020801	0.00146538	0.99999890	− 0.020	0.056
69	− 0.00100661	0.00092685	0.99999906	− 0.046	0.043
70	− 0.00149594	0.00119595	0.99999816	− 0.067	0.057
71	− 0.00227386	0.00213030	0.99999514	− 0.014	0.099
72	− 0.00117654	0.00277443	0.99999546	− 0.068	0.113
73	− 0.00026350	0.00255903	0.99999668	− 0.032	0.098
74	− 0.00056077	0.00211465	0.99999760	− 0.039	0.084
75	0.00088387	0.00186082	0.99999788	0.018	0.062
76	0.00086211	0.00224432	0.99999710	0.014	0.076
77	0.00041684	0.00273213	0.99999618	− 0.007	0.098
78	− 0.00005127	0.00340160	0.99999420	− 0.031	0.127
79	0.00233239	0.00193375	0.99999540	0.072	0.052
80	0.00107171	0.00346569	0.99999341	0.011	0.120
82	− 0.00326568	− 0.00009388	0.99999466	− 0.123	0.024
PLATE NO. 4-B4					
1	0.00302590	0.00300124	0.99999092	0.144	0.173
2	0.00196870	0.00313252	0.99999315	0.107	0.166
3	0.00212747	0.00327975	0.99999236	0.115	0.175
4	0.00416758	0.00099266	0.99999081	0.161	0.094
5	0.00351562	0.00174942	0.99999229	0.147	0.121
6	0.00297387	0.00015412	0.99999556	0.109	0.041
7	0.00341876	0.00051792	0.99999402	0.129	0.063
8	0.00166088	0.00132588	0.99999774	0.075	0.080
10	0.00216319	0.00058507	0.99999748	0.085	0.052
11	0.00256097	− 0.00036996	0.99999665	0.088	0.013
12	0.00343454	− 0.00021761	0.99999407	0.121	0.030
13	0.00386340	− 0.00081487	0.99999220	0.130	0.007
15	0.00108765	0.00008034	0.99999940	0.040	0.016
18	0.00000328	0.00023805	0.99999996	0.003	0.011
19	0.00158361	− 0.00044904	0.99999864	0.052	− 0.002
20	0.00211749	− 0.00139665	0.99999678	0.060	− 0.039
21	0.00337959	− 0.00123021	0.99999353	0.107	− 0.017
22	0.00374651	− 0.00196506	0.99999104	0.112	− 0.047
25	0.00152381	− 0.00142447	0.99999782	0.038	− 0.048
26	0.00043705	− 0.00029516	0.99999986	0.012	− 0.008
27	0.00079673	− 0.00121564	0.99999893	0.015	− 0.046
28	0.00142602	− 0.00203704	0.99999691	0.028	− 0.077
30	0.00187079	− 0.00293776	0.99999393	0.033	− 0.113
31	0.00300914	− 0.00271160	0.99999179	0.077	− 0.089

Appendix C (cont'd)

Feature Number	v_1	v_2	v_3	$10^{-5}\,\delta v_1$	$10^{-5}\,\delta v_2$
34	0.00016746	− 0.00199317	0.99999800	− 0.017	− 0.089
36	− 0.00038863	− 0.00215502	0.99999759	− 0.039	− 0.103
37	− 0.00042953	− 0.00079164	0.99999959	− 0.025	− 0.041
38	− 0.00084855	− 0.00190371	0.99999782	− 0.052	− 0.097
39	− 0.00068134	− 0.00228554	0.99999715	− 0.051	− 0.112
42	− 0.00002332	− 0.00277646	0.99999614	− 0.033	− 0.127
44	0.00126022	− 0.00312735	0.99999431	0.009	− 0.128
54	− 0.00256233	0.00006829	0.99999671	− 0.091	− 0.026
55	− 0.00166057	− 0.00051169	0.99999848	− 0.065	− 0.042
56	− 0.00142815	− 0.00102455	0.99999845	− 0.063	− 0.063
57	− 0.00060495	− 0.00003819	0.99999981	− 0.022	− 0.009
58	− 0.00078270	0.00050251	0.99999956	− 0.022	0.014
59	− 0.00159799	0.00044780	0.99999862	− 0.052	0.002
60	− 0.00391117	0.00160124	0.99999106	− 0.122	0.028
61	− 0.00301279	0.00275532	0.99999166	− 0.076	0.092
62	− 0.00314850	0.00239757	0.99999216	− 0.085	0.074
63	− 0.00268614	0.00212462	0.99999413	− 0.072	0.066
65	− 0.00026717	0.00075241	0.99999968	− 0.001	0.031
66	0.00021374	0.00170259	0.99999852	0.027	0.080
68	− 0.00025879	0.00211235	0.99999774	0.015	0.094
69	− 0.00115971	0.00174054	0.99999781	− 0.022	0.066
70	− 0.00156941	0.00207647	0.99999661	− 0.032	0.077
73	− 0.00003794	0.00312892	0.99999510	0.035	0.143
74	− 0.00044472	0.00278188	0.99999603	0.016	0.122
75	0.00092672	0.00225624	0.99999702	0.059	0.114
77	0.00067627	0.00315335	0.99999479	0.061	0.152
79	0.00242448	0.00196550	0.99999513	0.110	0.118
80	0.00154871	0.00364261	0.99999216	0.098	0.185
81	0.00344909	0.00230726	0.99999139	0.151	0.146
82	− 0.00350627	0.00103790	0.99999331	− 0.114	0.007
PLATE NO. 5-B5					
4	0.00328147	0.00242781	0.99999166	0.126	0.063
5	0.00242121	0.00293201	0.99999277	0.089	0.081
6	0.00240662	0.00143256	0.99999607	0.094	0.036
7	0.00270369	0.00186752	0.99999459	0.104	0.048
8	0.00082190	0.00207983	0.99999750	0.027	0.061
9	0.00018936	0.00199768	0.99999798	0.001	0.060
11	0.00219667	0.00086032	0.99999721	0.087	0.019
12	0.00296034	0.00122375	0.99999486	0.117	0.028
17	0.00032402	0.00139871	0.99999896	0.009	0.042
21	0.00325838	0.00029365	0.99999464	0.132	− 0.002
25	0.00161002	− 0.00039007	0.99999862	0.067	− 0.017
26	0.00024485	0.00023053	0.99999993	0.009	0.006
28	0.00174052	− 0.00098510	0.99999800	0.075	− 0.036
29	0.00237529	− 0.00092271	0.99999674	0.100	− 0.036
30	0.00247962	− 0.00170871	0.99999546	0.107	− 0.060
32	0.00305522	− 0.00177469	0.99999376	0.131	− 0.064
34	0.00059246	− 0.00134414	0.99999891	0.029	− 0.043
36	0.00015662	− 0.00168028	0.99999857	0.012	− 0.052

Appendix C (cont'd)

Feature Number	ν_1	ν_2	ν_3	$10^{-5}\,\delta\nu_1$	$10^{-5}\,\delta\nu_2$
37	− 0.00035638	− 0.00045217	0.99999983	− 0.013	− 0.013
38	− 0.00033587	− 0.00160714	0.99999864	− 0.008	− 0.048
39	− 0.00005315	− 0.00190163	0.99999818	0.004	− 0.058
40	− 0.00032972	− 0.00227738	0.99999735	− 0.006	− 0.068
41	− 0.00026240	− 0.00248285	0.99999688	− 0.002	− 0.075
42	0.00070714	− 0.00214210	0.99999745	0.036	− 0.068
43	0.00140172	− 0.00211529	0.99999677	0.064	− 0.069
44	0.00199307	− 0.00207043	0.99999586	0.088	− 0.070
45	0.00139831	− 0.00263838	0.99999554	0.066	− 0.085
46	− 0 00014578	− 0.00267341	0.99999641	0.003	− 0.081
47	− 0.00005492	− 0.00287695	0.99999586	0.007	− 0.088
48	− 0.00270963	− 0.00159723	0.99999505	− 0.106	− 0.040
49	− 0.00241577	− 0.00187858	0.99999531	− 0.093	− 0.049
50	− 0.00190879	− 0.00188576	0.99999639	− 0.072	− 0.051
51	− 0.00138994	− 0.00257628	0.99999571	− 0.048	− 0.074
52	− 0.00118975	− 0.00269743	0.99999565	− 0.040	− 0.078
53	− 0.00361468	− 0.00069560	0.99999322	− 0.146	− 0.009
54	− 0.00249078	− 0.00046450	0.99999679	− 0.101	− 0.006
55	− 0.00152411	− 0.00063443	0.99999863	− 0.060	− 0.014
56	− 0.00114847	− 0.00101087	0.99999882	− 0.044	− 0.027
57	− 0.00076123	0.00016279	0.99999969	− 0.032	0.008
58	− 0.00098656	0.00057514	0.99999934	− 0.042	0.021
60	− 0.00408737	0.00026860	0.99999160	− 0.169	0.022
64	− 0.00158733	0.00130765	0.99999788	− 0.069	0.045
65	− 0 00072247	0 00096647	0 99999926	− 0 033	0 032
66	− 0 00059510	0 00194814	0 99999792	− 0 031	0 062
68	− 0 00113799	0.00213268	0.99999707	− 0.054	0.069
69	− 0.00181565	0.00150709	0.99999721	− 0.080	0.052
70	− 0.00227724	0.00163810	0.99999606	− 0.099	0.058
71	− 0.00298855	0.00224069	0.99999302	− 0.130	0.079
72	− 0.00206940	0.00303488	0.99999325	− 0.095	0.100
74	− 0.00150068	0.00262494	0.99999543	− 0.070	0.085
75	− 0.00012886	0.00263978	0.99999650	− 0.014	0.081
76	− 0.00016830	0.00296211	0.99999560	− 0.017	0.091
78	− 0.00103944	0.00374255	0.99999245	− 0.055	0.118
79	0.00132629	0.00284115	0.99999508	0.045	0.082
82	− 0.00359217	− 0.00001195	0.99999354	− 0.147	0.012
PLATE NO. 6-A1					
6	0.00304667	0.00042053	0.99999527	0.107	− 0.007
8	0.00157780	0.00133517	0.99999785	0.047	0.035
9	0.00091508	0.00135489	0.99999866	0.023	0.040
10	0.00219290	0.00068939	0.99999736	0.074	0.008
11	0.00273391	− 0.00017569	0.99999624	0.100	− 0.026
12	0.00354651	0.00015602	0.99999369	0.127	− 0.020
13	0.00404692	− 0.00033209	0.99999175	0.148	− 0.041
15	0.00122481	− 0.00000602	0.99999924	0 044	− 0 009
17	0 00094781	0 00067532	0 99999931	0 029	0 017
19	0 00179764	− 0.00043528	0.99999829	0.068	− 0.028
20	0.00246636	− 0.00126262	0.99999616	0.098	− 0.062

Appendix C (cont'd)

Feature Number	ν_1	ν_2	ν_3	$10^{-5}\ \delta\nu_1$	$10^{-5}\ \delta\nu_2$
23	0.00358154	− 0.00165973	0.99999221	0.141	− 0.083
25	0.00189260	− 0.00139981	0.99999723	0.078	− 0.062
26	0.00064906	− 0.00048693	0.99999966	0.027	− 0.022
27	0.00114955	− 0.00132443	0.99999846	0.051	− 0.054
28	0.00189182	− 0.00200947	0.99999619	0.083	− 0.083
30	0.00244852	− 0.00278342	0.99999312	0.108	− 0.114
32	0.00298450	− 0.00287616	0.99999140	0.128	− 0.121
34	0.00064810	− 0.00218440	0.99999740	0.039	− 0.080
36	0.00011886	− 0.00243052	0.99999703	0.022	− 0.085
37	− 0.00012622	− 0.00111688	0.99999936	0.004	− 0.038
38	− 0.00036776	− 0.00236528	0.99999713	0.004	− 0.079
40	− 0.00053485	− 0.00289723	0.99999566	0.002	− 0.096
42	0.00057081	− 0.00296964	0.99999543	0.042	− 0.107
44	0.00187819	− 0.00307866	0.99999349	0.090	− 0.120
45	0.00114115	− 0.00352926	0.99999312	0.067	− 0.130
46	− 0.00044458	− 0.00330117	0.99999445	0.008	− 0.111
47	− 0.00040595	− 0.00350089	0.99999379	0.011	− 0.118
48	− 0.00290014	− 0.00162497	0.99999447	− 0.093	− 0.035
49	− 0.00265094	− 0.00198455	0.99999451	− 0.081	− 0.050
50	− 0.00210317	− 0.00214232	0.99999549	− 0.060	− 0.059
51	− 0.00172724	− 0.00292466	0.99999423	− 0.041	− 0.089
52	− 0.00154579	− 0.00308464	0.99999404	− 0.033	− 0.096
53	− 0.00370137	− 0.00041001	0.99999306	− 0.131	0.013
54	− 0.00239701	− 0.00058594	0.99999695	− 0.082	− 0.003
55	− 0.00139519	− 0.00103147	0.99999849	− 0.043	− 0.026
56	− 0.00108340	− 0.00150104	0.99999828	− 0.028	− 0.044
58	− 0.00068601	0.00010181	0.99999975	− 0.025	0.009
59	− 0.00149104	− 0.00008030	0.99999888	− 0.053	0.008
60	− 0.00404478	0.00078746	0.99999151	− 0.152	0.057
61	− 0.00333726	0.00206000	0.99999230	− 0.135	0.096
62	− 0.00340476	0.00167769	0.99999280	− 0.135	0.083
63	− 0.00288050	0.00145060	0.99999479	− 0.114	0.071
64	− 0.00106414	0.00100090	0.99999893	− 0.046	0.042
65	− 0.00021793	0.00043083	0.99999988	− 0.011	0.017
66	0.00009466	0.00146015	0.99999892	− 0.007	0.050
69	− 0.00127352	0.00127855	0.99999837	− 0.055	0.054
70	− 0.00174428	0.00155461	0.99999727	− 0.074	0.067
71	− 0.00243929	0.00247943	0.99999394	− 0.106	0.104
74	− 0.00075100	0.00244407	0.99999673	− 0.045	0.090
75	0.00069177	0.00214052	0.99999747	0.009	0.069
82	− 0.00352089	0.00025852	0.99999376	− 0.129	0.035
39	− 0.00015232	− 0.00260415	0.99999659	0.013	− 0.089
PLATE NO. 7-A2					
2	− 0.00070080	0.00358420	0.99999332	− 0.005	0.141
6	0.00173914	0.00212072	0.99999624	0.062	0.093
7	0.00187326	0.00262426	0.99999480	0.069	0.114
8	0.00005535	0.00219992	0.99999757	0.012	0.088
10	0.00086461	0.00197304	0.99999768	0.035	0.083
11	0.00173396	0.00152136	0.99999733	0.060	0.069

Appendix C (cont'd)

Feature Number	ν_1	ν_2	ν_3	$10^{-5}\,\delta\nu_1$	$10^{-5}\,\delta\nu_2$
12	0.00232446	0.00211200	0.99999506	0.080	0.095
13	0.00304138	0.00190577	0.99999356	0.101	0.090
14	0.00261115	0.00174548	0.99999506	0.087	0.082
15	0.00036065	0.00099757	0.99999943	0.015	0.042
17	− 0.00018019	0.00140554	0.99999899	0.001	0.055
18	− 0.00051417	0.00046130	0.99999975	− 0.014	0.016
19	0.00104484	0.00091261	0.99999903	0.036	0.041
20	0.00203354	0.00053092	0.99999779	0.064	0.030
21	0.00290874	0.00134433	0.99999486	0.095	0.067
22	0.00368982	0.00098020	0.99999271	0.117	0.056
23	0.00326826	0.00064104	0.99999445	0.102	0.040
25	0.00160326	0.00016351	0.99999870	0.049	0.014
26	0.00011205	0.00034261	0.99999993	0.005	0.014
27	0.00093821	− 0.00010738	0.99999955	0.028	0.000
28	0.00192143	− 0.00034502	0.99999809	0.057	− 0.005
29	0.00249234	− 0.00007845	0.99999689	0.075	0.008
30	0.00284725	− 0.00076895	0.99999564	0.083	− 0.018
31	0.00358946	0.00002626	0.99999355	0.109	0.017
32	0.00340523	− 0.00064004	0.99999399	0.101	− 0.010
34	0.00097271	− 0.00105470	0.99999896	0.025	− 0.038
37	− 0.00020156	− 0.00053923	0.99999983	− 0.009	− 0.022
38	0.00019873	− 0.00160814	0.99999868	− 0.001	− 0.063
39	0.00055993	− 0.00178831	0.99999823	0.009	− 0.069
40	0.00042389	− 0.00223055	0.99999741	0.003	− 0.087
42	0.00134572	− 0.00176166	0.99999753	0.033	− 0.064
44	0.00251594	− 0.00126695	0.99999603	0.071	− 0.039
45	0.00215155	− 0.00199386	0.99999569	0.057	− 0 070
46	0 00072694	− 0.00253543	0.99999652	0.011	− 0.098
47	0.00087795	− 0.00269386	0.99999598	0.015	− 0.104
48	− 0.00201088	− 0.00238174	0.99999513	− 0.072	− 0.104
49	− 0.00164361	− 0.00254706	0.99999540	− 0.061	− 0.109
50	− 0.00117192	− 0.00238606	0.99999646	− 0.046	− 0.101
51	− 0.00045979	− 0.00285973	0.99999580	− 0.027	− 0.116
52	− 0.00023273	− 0.00290501	0.99999575	− 0.020	− 0.117
53	− 0.00315064	− 0.00184532	0.99999332	− 0.104	− 0.088
54	− 0.00218264	− 0.00125724	0.99999682	− 0.072	− 0.060
55	− 0.00122728	− 0.00110013	0.99999864	− 0.042	− 0.049
56	− 0.00075226	− 0.00132209	0.99999884	− 0.029	− 0.056
57	− 0.00078052	− 0.00010446	0.99999969	− 0.024	− 0.008
58	− 0.00122986	0.00017163	0.99999923	− 0.037	0.001
59	− 0.00176024	− 0.00037763	0.99999838	− 0.055	− 0.023
60	− 0.00391144	− 0.00110285	0.99999174	− 0.124	− 0.061
61	− 0.00398280	0.00027641	0 99999202	− 0 120	− 0 007
62	− 0 00388093	− 0 00004486	0 99999246	− 0 118	− 0 019
63	− 0 00343767	0 00007987	0 99999408	− 0 104	− 0 012
65	− 0 00101216	0 00066077	0 99999926	− 0 028	0.022
66	− 0.00121546	0.00160912	0.99999796	− 0.030	0.059
68	− 0.00178156	0.00160327	0.99999712	− 0.047	0.056
69	− 0.00220362	0.00079585	0.99999725	− 0.063	0.022
70	− 0.00267347	0.00076542	0.99999613	− 0.078	0.019

Appendix C (cont'd)

Feature Number	ν_1	ν_2	ν_3	$10^{-5}\,\delta\nu_1$	$10^{-5}\,\delta\nu_2$
71	− 0.00353636	0.00108935	0.99999315	− 0.103	0.028
72	− 0.00294416	0.00213220	0.99999338	− 0.080	0.072
74	− 0.00227998	0.00193749	0.99999552	− 0.061	0.067
75	− 0.00100953	0.00240617	0.99999659	− 0.020	0.092
76	− 0.00115348	0.00268967	0.99999572	− 0.023	0.103
77	− 0.00167305	0.00285662	0.99999452	− 0.038	0.107
78	− 0.00221871	0.00312833	0.99999264	− 0.053	0.115
PLATE NO. 8-A3					
6	0.00282430	0.00104522	0.99999546	0.109	0.028
7	0.00298173	0.00162897	0.99999422	0.115	0.045
9	0.00060947	0.00108243	0.99999923	0.023	0.031
10	0.00198305	0.00094600	0.99999758	0.076	0.026
11	0.00275491	0.00037756	0.99999613	0.107	0.009
12	0.00334854	0.00102285	0.99999387	0.129	0.027
14	0.00356369	0.00061168	0.99999346	0.138	0.015
15	0.00135759	− 0.00005901	0.99999907	0.053	− 0.003
17	0.00087589	0.00046532	0.99999950	0.034	0.013
19	0.00201176	− 0.00022655	0.99999794	0.078	− 0.008
20	0.00287299	− 0.00069300	0.99999563	0.112	− 0.022
22	0.00434109	− 0.00012548	0.99999056	0.168	− 0.006
23	0.00396732	− 0.00055235	0.99999198	0.154	− 0.018
26	0.00099878	− 0.00069817	0.99999925	0.039	− 0.021
27	0.00172106	− 0.00125848	0.99999772	0.068	− 0.037
28	0.00260076	− 0.00156737	0.99999539	0.102	− 0.047
30	0.00332849	− 0.00198310	0.99999249	0.130	− 0.059
32	0.00381839	− 0.00181823	0.99999105	0.149	− 0.055
34	0.00154346	− 0.00217806	0.99999643	0.061	− 0.063
38	0.00063803	− 0.00259662	0.99999642	0.026	− 0.075
39	0.00094797	− 0.00280933	0.99999560	0.039	− 0.081
40	0.00068682	− 0.00318308	0.99999470	0.029	− 0.092
42	0.00171128	− 0.00286276	0.99999443	0.068	− 0.083
44	0.00291471	− 0.00245942	0.99999272	0.115	− 0.072
45	0.00238413	− 0.00310501	0.99999233	0.095	− 0.091
46	0.00088633	− 0.00347846	0.99999356	0.037	− 0.100
47	0.00097809	− 0.00362576	0.99999294	0.040	− 0.104
48	− 0.00191264	− 0.00279823	0.99999426	− 0.073	− 0.079
49	− 0.00155892	− 0.00304246	0.99999415	− 0.059	− 0.086
50	− 0.00099486	− 0.00302627	0.99999492	− 0.037	− 0.086
51	− 0.00040928	− 0.00356886	0.99999354	− 0.014	− 0.102
52	− 0.00019410	− 0.00364464	0.99999333	− 0.005	− 0.104
54	− 0.00177140	− 0.00175873	0.99999688	− 0.068	− 0.049
55	− 0.00069195	− 0.00185992	0.99999803	− 0.026	− 0.053
56	− 0.00025288	− 0.00216962	0.99999761	− 0.008	− 0.062
57	− 0.00000233	− 0.00098744	0.99999951	0.001	− 0.028
58	− 0.00041669	− 0.00061932	0.99999972	− 0.016	− 0.017
60	− 0.00381727	− 0.00089614	0.99999231	− 0.148	− 0.023
63	− 0.00292432	− 0.00002080	0.99999572	− 0.114	0.001
64	− 0.00107201	0.00008349	0.99999942	− 0.042	0.003
66	− 0.00017075	0.00089269	0.99999958	− 0.007	0.026
68	− 0.00077794	0.00102674	0.99999917	− 0.031	0.030

Appendix C (cont'd)

Feature Number	ν_1	ν_2	ν_3	$10^{-5}\,\delta\nu_1$	$10^{-5}\,\delta\nu_2$
69	− 0.00136088	0.00027465	0.99999903	− 0.053	0.009
70	− 0.00189301	0.00039357	0.99999813	− 0.074	0.013
71	− 0.00288195	0.00108467	0.99999525	− 0.113	0.033
72	− 0.00205663	0.00201482	0.99999585	− 0.081	0.059
73	− 0.00114707	0.00206821	0.99999720	− 0.046	0.060
74	− 0.00129611	0.00154599	0.99999796	− 0.051	0.045
75	0.00013011	0.00174681	0.99999846	0.004	0.050
76	− 0.00000179	0.00211214	0.99999776	− 0.002	0.060
79	0.00148748	0.00230219	0.99999624	0.056	0.065
82	− 0.00312215	− 0.00128472	0.99999429	− 0.121	− 0.035
PLATE NO. 9-A4					
2	0.00044566	0.00319474	0.99999479	0.016	0.090
3	0.00045877	0.00342879	0.99999401	0.017	0.096
4	0.00318248	0.00276647	0.99999110	0.156	0.072
5	0.00235942	0.00295402	0.99999285	0.114	0.079
6	0.00289102	0.00128506	0.99999499	0.144	0.031
7	0.00299337	0.00186265	0.99999378	0.148	0.047
8	0.00131234	0.00146278	0.99999806	0.064	0.039
9	0.00073983	0.00120981	0.99999899	0.035	0.033
10	0.00208984	0.00114967	0.99999715	0.104	0.029
11	0.00286906	0.00062350	0.99999569	0.144	0.012
12	0.00338756	0.00128561	0.99999343	0.169	0.030
13	0.00394722	0.00110897	0.99999159	0.197	0.024
15	0.00154766	0.00012516	0.99999879	0.078	0.001
17	0.00104415	0.00061294	0.99999926	0.052	0.015
18	0.00061768	− 0.00034151	0.99999975	0.032	− 0.011
19	0.00219438	− 0.00000550	0.99999759	0.111	− 0.004
20	0.00304679	− 0.00042929	0.99999526	0.155	− 0.018
22	0.00437810	0.00017688	0.99999040	0.221	− 0.003
23	0.00406632	− 0.00024737	0.99999169	0.206	− 0.015
25	0.00262220	− 0.00079802	0.99999624	0.134	− 0.028
26	0.00123309	− 0.00052980	0.99999909	0.063	− 0.017
27	0.00196583	− 0.00104889	0.99999751	0.101	− 0.033
28	0.00282895	− 0.00131327	0.99999513	0.146	− 0.042
30	0.00353112	− 0.00170295	0.99999231	0.182	− 0.055
38	0.00096791	− 0.00244489	0.99999654	0.054	− 0.071
39	0.00127616	− 0.00264116	0.99999569	0.070	− 0.077
40	0.00103062	− 0.00303354	0.99999486	0.058	− 0.088
42	0.00201902	− 0.00265465	0.99999443	0.107	− 0.079
44	0.00315645	− 0.00219336	0.99999261	0.164	− 0.068
45	0.00266622	− 0.00287076	0.99999232	0.140	− 0.086
46	0.00122946	− 0.00332147	0.99999373	0.068	− 0.096
47	0.00131895	− 0.00346696	0.99999312	0.073	− 0.101
48	− 0.00157051	− 0.00281608	0.99999479	− 0.074	− 0.077
49	− 0.00120961	− 0.00303416	0.99999466	− 0.056	− 0.084
50	− 0.00064118	− 0.00297991	0.99999535	− 0.027	− 0.083
51	− 0.00004823	− 0.00348996	0.99999391	0.004	− 0.099
52	0.00016712	− 0.00355419	0.99999367	0.015	− 0.101
53	− 0.00277673	− 0.00203719	0.99999406	− 0.137	− 0.052

Appendix C (cont'd)

Feature Number	v_1	v_2	v_3	$10^{-5}\,\delta v_1$	$10^{-5}\,\delta v_2$
54	− 0.00146341	− 0.00176349	0.99999737	− 0.071	− 0.047
55	− 0.00037874	− 0.00178910	0.99999832	− 0.016	− 0.050
56	0.00006901	− 0.00207074	0.99999785	0.007	− 0.059
57	0.00026758	− 0.00087929	0.99999958	0.015	− 0.025
58	− 0.00016348	− 0.00053792	0.99999984	− 0.007	− 0.015
59	− 0.00082033	− 0.00099588	0.99999916	− 0.040	− 0.027
61	− 0.00343623	0.00029874	0.99999405	− 0.175	0.015
62	− 0.00332864	− 0.00007870	0.99999445	− 0.169	0.004
63	− 0.00272251	− 0.00012035	0.99999628	− 0.138	0.002
64	− 0.00085647	0.00011144	0.99999962	− 0.044	0.005
65	0.00012298	− 0.00006754	0.99999999	0.006	− 0.002
66	− 0.00001342	0.00097194	0.99999952	− 0.003	0.028
68	− 0.00062888	0.00106620	0.99999923	− 0.034	0.031
70	− 0.00170343	0.00036715	0.99999847	− 0.087	0.014
71	− 0.00275401	0 00097273	0 99999573	− 0 142	0 033
72	− 0 00198865	0 00194762	0 99999612	− 0 105	0 059
73	− 0 00107685	0 00206430	0 99999729	− 0 059	0 060
74	− 0 00118498	0 00154407	0 99999810	− 0 063	0 046
75	0 00021710	0 00183038	0 99999829	0 008	0 051
76	0.00005758	0.00217884	0.99999762	− 0.001	0.062
77	− 0.00052435	0.00247282	0.99999680	− 0.031	0.071
78	− 0.00119800	0.00294178	0.99999495	− 0.066	0.086
79	0.00149971	0.00244861	0.99999587	0.071	0.066
80	− 0.00026186	0.00342083	0.99999411	− 0.020	0.097
81	0.00196789	0.00339925	0.99999228	0.093	0.093
PLATE NO. 10-A5					
1	0.00013542	0.00372731	0.99999304	0.037	0.137
2	− 0.00073763	0.00301292	0.99999518	− 0.006	0.104
3	− 0.00070078	0.00329848	0.99999431	− 0.002	0.114
4	0.00220323	0.00304510	0.99999293	0.118	0.130
5	0.00123816	0.00310339	0.99999441	0.078	0.124
6	0.00181305	0.00154888	0.99999715	0.089	0.072
7	0.00191253	0.00213034	0.99999589	0.098	0.094
8	0.00012329	0.00149114	0.99999887	0.018	0.055
9	− 0.00042652	0.00116631	0.99999923	− 0.008	0.039
10	0.00094787	0.00129743	0.99999870	0.051	0.055
11	0.00183895	0.00089809	0.99999790	0.085	0.048
12	0.00239052	0.00162278	0.99999582	0.114	0.079
13	0.00310114	0.00153979	0.99999400	0.143	0.082
14	0.00269189	0.00127229	0.99999556	0.124	0.069
15	0.00049020	0.00022989	0.99999984	0.023	0.013
17	− 0.00007379	0.00063105	0.99999979	0.002	0.022
18	− 0.00037161	− 0.00034891	0.99999987	− 0.019	− 0.016
19	0.00117722	0.00018880	0.99999928	0.051	0.017
20	0.00217691	− 0.00009238	0.99999762	0.091	0.015
22	0.00377441	0.00071004	0.99999262	0.165	0.058
23	0.00338848	0.00024199	0.99999423	0.144	0.037
25	0.00176368	− 0.00051027	0.99999831	0.070	− 0.004
26	0.00026756	− 0.00044893	0.99999986	0.007	− 0.014

Appendix C (cont'd)

Feature Number	ν_1	ν_2	ν_3	$10^{-5}\,\delta\nu_1$	$10^{-5}\,\delta\nu_2$
27	0.00111255	− 0.00084596	0.99999902	0.040	− 0.021
28	0.00209782	− 0.00097604	0.99999732	0.080	− 0.018
29	0.00264926	− 0.00063156	0.99999628	0.106	− 0.001
30	0.00302263	− 0.00124425	0.99999465	0.116	− 0.020
31	0.00370996	− 0.00028521	0.99999307	0.153	0.021
32	0.00356209	− 0.00099762	0.99999315	0.141	− 0.006
34	0.00117504	− 0.00175979	0.99999776	0.034	− 0.054
37	− 0.00001732	− 0.00133398	0.99999911	− 0.012	− 0.049
38	0.00041119	− 0.00232741	0.99999720	− 0.003	− 0.081
39	0.00077746	− 0.00246991	0.99999664	0.012	− 0.084
40	0.00064867	− 0.00287846	0.99999564	0.003	− 0.100
42	0.00156021	− 0.00237952	0.99999595	0.045	− 0.074
44	0.00270672	− 0.00177100	0.99999476	0.099	− 0.042
46	0.00095819	− 0.00312549	0.99999465	0.014	− 0.106
47	0.00111029	− 0.00324906	0.99999410	0.019	− 0.109
48	− 0.00181847	− 0.00301176	0.99999381	− 0.102	− 0.125
49	− 0.00143809	− 0.00317428	0.99999392	− 0.088	− 0.128
50	− 0.00095594	− 0.00304892	0.99999489	− 0.066	− 0.119
51	− 0.00023237	− 0.00345570	0.99999399	− 0.039	− 0.128
52	− 0.00000616	− 0.00348849	0.99999391	− 0.030	− 0.127
53	− 0.00300639	− 0.00241501	0.99999256	− 0.147	− 0.114
54	− 0.00201263	− 0.00199087	0.99999599	− 0.102	− 0.090
55	− 0.00103847	− 0.00187609	0.99999770	− 0.060	− 0.077
56	− 0.00055237	− 0.00208753	0.99999767	− 0.041	− 0.081
57	− 0.00062121	− 0.00091874	0.99999938	− 0.034	− 0.039
58	− 0.00108466	− 0.00064415	0.99999920	− 0.051	− 0.033
59	− 0.00160623	− 0.00117184	0.99999802	− 0.078	− 0.056
60	− 0.00382607	− 0.00157711	0.99999143	− 0.175	− 0.090
61	− 0.00393873	− 0.00024064	0.99999221	− 0.168	− 0.042
65	− 0.00088338	− 0.00015045	0.99999959	− 0.038	− 0.013
66	− 0.00113429	0.00083482	0.99999900	− 0.041	0.021
69	− 0.00210715	0.00001947	0.99999777	− 0.089	− 0.017
70	− 0.00258939	0.00003257	0.99999664	− 0.109	− 0.021
71	− 0.00350253	0.00049047	0.99999374	− 0.143	− 0.012
72	− 0.00294041	0.00153247	0.99999449	− 0.111	0.031
73	− 0.00216388	0.00175569	0.99999611	− 0.076	0.046
75	− 0.00096941	0.00170199	0.99999808	− 0.026	0.054
76	− 0.00113056	0.00202016	0.99999732	− 0.030	0.064
77	− 0.00167155	0.00223009	0.99999611	− 0.051	0.067
78	− 0.00225628	0.00260359	0.99999406	− 0.073	0.076
80	− 0.00133475	0.00319901	0.99999399	− 0.029	0.105
81	0.00085882	0.00348668	0.99999355	0.066	0.134
82	− 0.00323073	− 0.00181104	0.99999314	− 0.152	− 0.094
PLATE NO. 11-A6					
2	0.00131677	0.00353365	0.99999288	0.117	0.133
3	0.00137761	0.00369975	0.99999220	0.123	0.139
6	0.00318985	0.00113227	0.99999426	0.200	0.074
7	0.00342200	0.00158084	0.99999289	0.219	0.091
8	0.00174413	0.00184417	0.99999677	0.123	0.081
9	0.00113871	0.00178481	0.99999775	0.087	0.073

Appendix C (cont'd)

Feature Number	ν_1	ν_2	ν_3	$10^{-5}\ \delta\nu_1$	$10^{-5}\ \delta\nu_2$
10	0.00240505	0.00131032	0.99999624	0.156	0.071
11	0.00301116	0.00054717	0.99999531	0.183	0.052
13	0.00412330	0.00047450	0.99999138	0.248	0.062
14	0.00377184	0.00045736	0.99999277	0.227	0.058
15	0.00164223	0.00055529	0.99999849	0.103	0.037
17	0.00128569	0.00116083	0.99999849	0.089	0.053
18	0.00063957	0.00038954	0.99999972	0.042	0.020
19	0.00222324	0.00021646	0.99999750	0.133	0.032
21	0.00387482	− 0.00003314	0.99999248	0.228	0.042
22	0.00428564	− 0.0062654	0.99999061	0.245	0.027
23	0.00390044	− 0.00082894	0.99999204	0.220	0.016
25	0.00242681	− 0.00070231	0.99999680	0.135	0.004
26	0.00118257	0.00003209	0.99999929	0.070	0.014
27	0.00174487	− 0.00071315	0.99999822	0.095	− 0.004
28	0.00248563	− 0.00129865	0.99999606	0.132	− 0.016
29	0.00303529	− 0.00121167	0.99999465	0.165	− 0.007
30	0.00303191	− 0.00202337	0.99999335	0.156	− 0.034
32	0.00348641	− 0.00208257	0.99999174	0.182	− 0.031
34	0.00137432	− 0.00161662	0.99999774	0.063	− 0.039
37	0.00054031	− 0.00068157	0.99999961	0.024	− 0.017
39	0.00066794	− 0.00213234	0.99999750	0.015	− 0.064
40	0.00032829	− 0.00248587	0.99999685	− 0.009	− 0.080
42	0.00136714	− 0.00240094	0.99999618	0.053	− 0.065
44	0.00255262	− 0.00236990	0.99999393	0.124	− 0.051
45	0.00190853	− 0.00290500	0.99999396	0.080	− 0.076
46	0.00043321	− 0.00287687	0.99999576	− 0.007	− 0.092
47	0.00047763	− 0.00307765	0.99999514	− 0.007	− 0.098
48	− 0.00211874	− 0.00164074	0.99999640	− 0.144	− 0.079
49	− 0.00182808	− 0.00194346	0.99999644	− 0.130	− 0.086
50	− 0.00126310	− 0.00199426	0.99999721	− 0.097	− 0.081
51	− 0.00083348	− 0.00270390	0.99999599	− 0.080	− 0.100
52	− 0.00064687	− 0.00283987	0.99999575	− 0.070	− 0.103
53	− 0.00309107	− 0.00064748	0.99999501	− 0.190	− 0.057
54	− 0.00172969	− 0.00053981	0.99999835	− 0.108	− 0.038
55	− 0.00068669	− 0.00079258	0.99999944	− 0.050	− 0.034
56	− 0.00032888	− 0.00119014	0.99999923	− 0.033	− 0.044
57	0.00017053	− 0.00003899	0.99999998	0.010	0.001
58	− 0.00015694	0.00040208	0.99999990	− 0.005	0.012
61	− 0.00313901	0.00172337	0.99999359	− 0.166	0.022
64	− 0.00066158	0.00117937	0.99999908	− 0.026	0.032
66	0.00035909	0.00177570	0.99999835	0.041	0.064
68	− 0.00020984	0.00200611	0.99999796	0.010	0.065
70	− 0.00142148	0.00157867	0.99999774	− 0.066	0.037
71	− 0.00230048	0.00228650	0.99999473	− 0.110	0.051
72	− 0.00132042	0.00303102	0.99999452	− 0.044	0.087
74	− 0.00063128	0.00254593	0.99999656	− 0.009	0.078
75	0.00078718	0.00247104	0.99999663	0.074	0.092
76	0.00071459	0.00281106	0.99999579	0.074	0.102
78	− 0.00032220	0.00370644	0.99999308	0.023	0.121
80	0.00073662	0.00385999	0.99999227	0.087	0.138

SELENODETIC EQUIPMENT FOR OBSERVATIONS OF THREE ANNULAR ECLIPSES*

LAURENCE G. STODDARD**

The Editors regret to announce that, while the present volume went through the press, Dr. Stoddard passed away on November 29, 1966, aged 54 years.

Abstract. Annular eclipses of the Sun offer a unique opportunity to determine the shape of the section of the Moon presented in silhouette against the circular solar disk. This paper describes the equipment used to observe the annular eclipses of 1962, 1963 and 1965. Photographic techniques only were used at the two earlier eclipses. Both photographic and photoelectric scanning techniques were employed at the 1965 eclipse. The design philosophy and problems of the instruments are discussed and the adopted solutions presented. Finally, suggested improvements are treated, based on the experiences of the three expeditions.

Introduction

Several studies have been made in the past to determine the Moon's shape, and most of these require ultimate reference to the lunar limb. Observations of the lunar limb, because of the phase defect, have been generally limited to half of the circumference of the lunar disk at a time. Annular eclipses of the Sun offer a rare opportunity to see a complete lunar disk without a phase defect. Furthermore, it is seen silhouetted against a body whose shape is much better known, the circular disk of the Sun.

Expeditions sponsored jointly by Lockheed and the U.S. Air Force observed the annular eclipses of 1962 July 31 from Kheur Soce, Senegal, and 1963 Jan. 25 from the Oudtshoorn region of South Africa. Contract AF19(628)-5714 supported a successful expedition to the annular eclipse of 1965 Nov. 23 that was observed from Chiengmai, Thailand.

The prime objectives of these expeditions was to acquire data that would precisely define the limbs of both the Sun and the Moon during the annular phase of each eclipse. For the first two eclipses only photographic methods were used, but for the third eclipse a new photoelectric scanner was also used. This paper describes the instruments used on these expeditions.

* Work supported in part by the Air Force Cambridge Research Laboratories under Contracts AF19(628)-4162 and AF19(628)-5714.

** Astronomical Sciences Laboratory, Lockheed California Company, Burbank, Calif., U.S.A.

Photographic System

The photographic system was designed to produce a series of high-resolution photographs of the Sun during an annular eclipse. As both the Sun and the Moon subtend an angle of approximately half a degree, a small field-of-view system could be used. Furthermore, the specific object of photography was the bright ring of the Sun surrounding the Moon. Therefore, it was feasible to use a system of long focal length giving a large-scale image while maintaining short exposure times. For these conditions a long-focus achromatic mirror system has many advantages. To avoid moving this long optical system to track the Sun, the system was rigidly fixed and fed with a plane mirror driven by clockwork such that the reflected rays would continually fall on the image-forming concave mirror. This, in turn, would reflect and form the image on the film. Under annular eclipse conditions, scattered light is no problem since the Moon's presence across most of the Sun's disk reduces the total sunlight (and, hence, scattered light) by a factor between 100 and 1000. Thus, no light shielding is necessary.

The major optical system (Figure 1) employed a 17-inch diameter optically plane mirror loaned through the courtesy of the Mt. Wilson Observatory. This was mounted

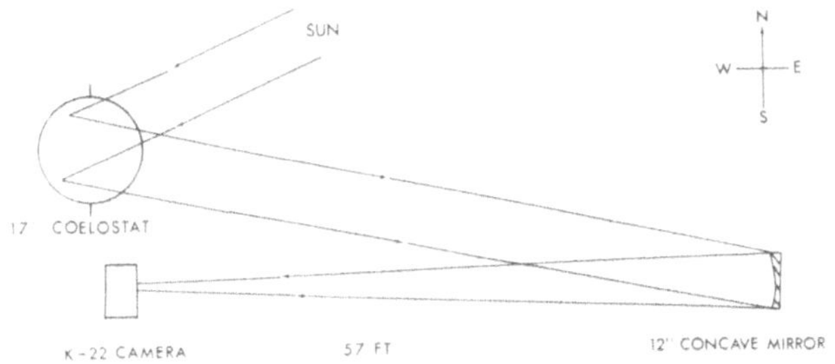

Fig. 1. Schematic Optical System.

as a clockdriven coelostat (Figure 2) that reflected the Sun's rays horizontally to a 12-inch diameter concave mirror approximately 60 feet away (Figure 3). The concave mirror of 57 feet focal length formed an image on a K-22 aerial camera body mounted just in front of the coelostat (Figure 4). This K-22 film transport system uses $9\frac{1}{2}$ inch wide film in 200 foot lengths and is capable of one exposure per 1.7 seconds. The camera body had been modified by the addition of a microswitch, so that the exposure times could be recorded. This system formed an image of the Sun (somewhat larger than 6 inches in diameter) at the focal plane of the K-22. For the two African eclipses, a smaller optical system of the same type was also used primarily as a back-up system. It consisted of a 12-inch coelostat and an 8-inch secondary mirror of 26 feet focal length. Its $3\frac{1}{2}$ inch solar image was recorded with a K-24 film transport system. This camera body, also modified with a microswitch, used $5\frac{1}{2}$ inch wide film and can take three frames per second.

Fig. 2. 17-inch Coelostat in Senegal.

Fig. 3. 12-inch Concave Mirror, South Africa.

Fig. 4. Coelostats and Cameras in Senegal. Uneclipsed Sun on Cameras.

Film systems were chosen in preference to plate systems. With modern Estar-based emulsions, dimensional stability is comparable to plates and the use of films offers many advantages. Reseau exposures were obtained at the South African site on the same roll as the eclipse films and were processed with them. Study of these reseau films indicate a negligible distortion on the original negatives. Reseau exposures were also obtained at the Thailand eclipse. The prime reason for the choice of films, however, was to obtain the maximum number of exposures during the brief annular phase. By combining the measurements of a larger number of frames, the random atmospheric distortion effects can be reduced.

Since an extremely bright source was to be photographed, the choice was made to use a very slow, fine grain emulsion in order to obtain maximum resolution and sharply defined limbs. This film also obviated the need to use filters and permitted the exposure times to remain within the capabilities of the camera shutters. The film supplied and processed by the Aeronautical Chart and Information Center (ACIC) of the U.S. Air Force was ideal for the purpose. It was the Eastman SO-105 emulsion, which has an ASA rating of 0.08; it is blue sensitive, with a resolution of the order of 1500 lines per mm.

Sky transparency can vary because of many factors. It is, therefore, necessary to determine exposure times immediately before an eclipse. For this purpose, ACIC provided a special exposure meter calibrated to the optical systems and film used. This meter, which could be inserted into any portion of the solar image forming rays just ahead of the film plane, could measure the intensity of light in any portion of the solar image. This versatility is required since the radiation is very intense at the center and diminishes to a small percentage at the limb which is the region to be photographed during the annular phase. At all three eclipses an exposure of $\frac{1}{350}$ second gave sharp outlines of the Sun and Moon with the larger system.

At the two African eclipses, the smaller system, being optically faster, was supplied with the even slower Eastman 649-0 emulsion. This proved somewhat too slow for the shutter capabilities of the K-24 and did not produce images as good as those of the larger system.

The Photoelectric Scanning System

For the third eclipse in Thailand, photoelectric scanning techniques were attempted in addition to the photographic program. The decision to do this was made less than four months prior to the eclipse thus leaving little time for construction and checking of the system. As a consequence, the scanner results were inferior to the photographs for selenodetic purposes. However, valuable experience was obtained with this system and its weak points revealed. Modifications will improve this system for any future similar applications.

The system consists of three major components: optics, scanner and recorder. The optics is a smaller version of that employed in the photographic system. It consists of an 8-inch coelostat, a 6-inch concave mirror of 31.5 feet focal length forming a $3\frac{1}{2}$ inch solar image on the scanner.

The scanner (Figures 5 to 10) consists of two mirrors at the front end. Both mirrors are mounted in a rotating disk that rotates at 6 R.P.M. One mirror is located

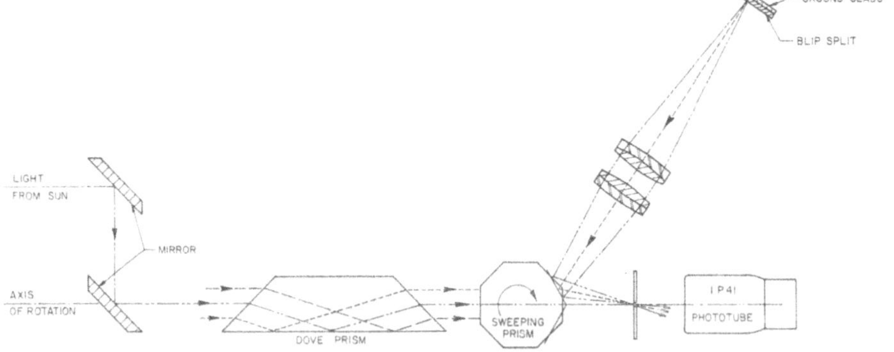

Fig. 5. Scanner Optical System.

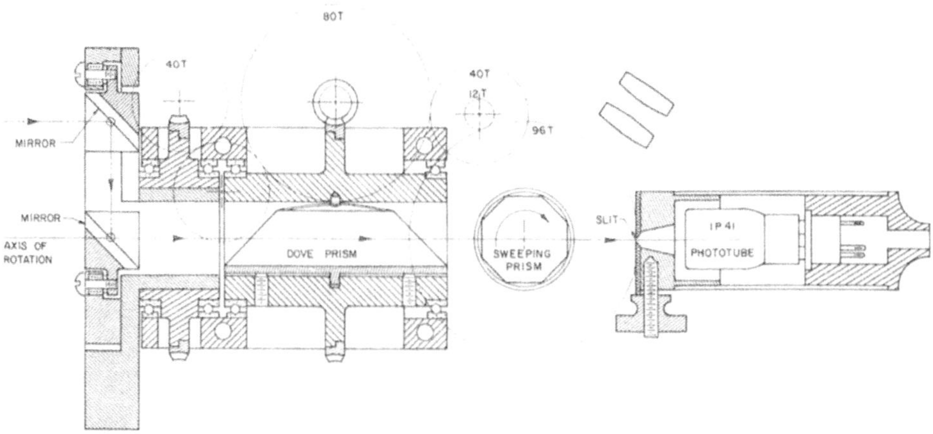

Fig. 6. Mechanical Details of the Scanner.

Fig. 7. Coelostats, Scanner and K-22 camera in Thailand.

near the outer edge of the disk and scans the annulus as the disk rotates. In proper alignment, the center of the image of the annulus lies very close to the axis of rotation of the scanning disk. The image of the annulus is somewhat out of focus at this point being actually focused on the slit in front of the phototube. The second mirror reflects the segment of the annulus along the axis of rotation and rotates it at the same frequency as the scanning disk rotates. This apparent rotation of the segment of the annulus is removed by passing the light through a Dove prism rotating in the same

Fig. 8. Front of the Scanner, showing Uneclipsed Sun.

Fig. 9. Rear View of the Scanner.

Fig. 10. Scanner with cover removed.

direction as the scanning head at half the angular velocity. By passing the light through a rotating octagonal prism, the light from the annulus is swept across the slit in front of the 1P41 phototube. During the eclipse, the slit opening was 0.004 inch. The gearing to the octagonal prism is such that the annulus is swept across the slit 72 times per revolution of the scanner, and thus makes a measurement of the annulus width at intervals of 5° in position angle. The time required for a section of the annulus to pass across the slit is a measure of its width. To help insure uniformity of motion, the entire scanning mechanism is driven by a synchronous motor.

In order to establish known references in the scanner data, an auxiliary signal is recorded when the octagonal prism is in known positions. This is accomplished by having a light beam sweep across the slit by reflection from one of the faces of the rotating prism. The orientation of the reference is such that its signal is inserted in the dark spaces between those of the annulus. In this manner the phototube sees in sequence annulus, dark space, reference, dark space, annulus, etc.

The output of the phototube was recorded ultimately on magnetic tape running at 15 inches per second. The direct output of the phototube was amplified by a Tectronix operational amplifier, then converted from voltage to frequency by a Vidar voltage to frequency converter. It was this frequency modulated signal that was actually recorded on tape, where the highest frequency corresponds to the most intense light signals. In this way the intensity profile of the limbs could be obtained. The maximum frequencies recorded were 25 kilocycles.

A 10 kilocycle reference frequency was recorded on the second channel to be used to evaluate irregularities of motion of scanner or tape recorder. This reference frequency signal was interrupted at known intervals such that Universal Time is also established on the magnetic tape record. The scanner head actuated a microswitch once per revolution and thus the correlation of the various scans with respect to location in the image could be established.

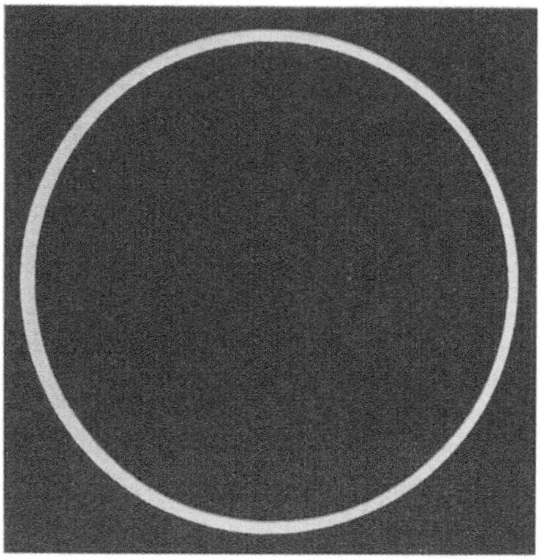

Fig. 11. Annular Eclipse of November 23, 1965

Miscellaneous Systems

Various types of auxiliary equipment are necessary or desirable to provide additional information needed to properly interpret the data.

It is important to know the Universal Time corresponding to each photograph or scan. This required the use of radios to obtain time signals. In remote parts of the world the reception of radio time signals is apt to be erratic. For this reason chronometers were used to carry U.T. and were considered to be the prime time standards. Their rates were well established from the time signals and during the eclipses their corrections were known to within a few hundreths of a second. Radios and chronometers were always taken in duplicate.

Two pen recorders were taken to each eclipse to record radio time signals, both chronometers and the camera shutter pulses. For the Thailand eclipse they also recorded impulses from a manually operated key that was simultaneously recorded on the scanner magnetic tape, and in this manner a U.T. time scale could be established on the scanner magnetic tape record.

For the two African eclipses two 500 watt, 110 volt, 60 cycle gasoline generators were taken. On the Thailand expedition, in addition, a 2500 watt portable power plant was also taken. A theodolite, useful for the alignment of the coelostats as well as position determination was a decided help on all the expeditions.

Conclusions and Recommendations

Experience with the large photographic system indicates that it is reasonably close to an optimum system for this purpose. Desirable improvements are in the area of refinements rather than major changes if the system is to remain practical for field work.

Telescopes with horizontal light paths are particularly sensitive to ground radiation and atmospheric turbulence that can deteriorate the image quality. Improvements could be made to the system to reduce this effect. However, this condition can be considerably reduced in choosing a site where ground radiation is minimized and uniform.

All the mirrors used were made of pyrex glass and no thermal problems were noticeable. However, the use of fused quartz mirrors would be desirable.

The choice of Estar based films in lieu of plates appears to have been correct. Measurements of reseau films taken at the South African eclipse shows film distortion to be negligible. Furthermore, more exposures of the eclipse were obtained with film than would have been possible with plates.

The photoelectric scanner was rather hurriedly designed and built. Time was not available to properly check the functioning and compatibility of the components. As a result, the data was inferior to the photographic results in Thailand. The optics and the scanner performed satisfactorily, but the recording system turned out to be the weak link. Excessive noise was present in the system and the tape recorder was marginal. As a consequence, the scanner slit width could not be reduced below 0.1 mm and limited the optical resolution to the order of two seconds of arc. Improvements in the recording system should produce a satisfactory scanning system for future applications of this sort.

References

Stoddard, L. G., Carson, D. G., and Fritzen, J. S.: 1966, 'Selenodetic Measurements and Error Analyses of Two Lockheed Annular Expeditions to Africa', *AFCRL-65-532.*

Stoddard, L. G.: 1966, 'Expedition to Thailand for the November 23, 1965 Annular Eclipse', *AFCRL-66-103.*

NOTE ON A PHOTOMETRIC DETERMINATION OF THE MOON'S FORM FROM ECLIPSE OBSERVATIONS

JEAN RÖSCH*

In reference to the foregoing communication presented by Dr. Eckhardt, I should like to propose another method for photometric measurement of the shape of the Moon's limb during annular solar eclipses. The method consists in rotating a system of two prisms A and B around the optical axis xx of the objective (see Figure 1). The prism A collects light from a sector of the annulus $\mathscr{A}$, the focal plane being located at the exit face of prism B. A slit S is located close to this face, rotating with the prism and oriented normally to the limbs of the Sun and the Moon. A field lens, following the slit, forms an image of the objective on the photocathode of a photomultiplier, and the photocurrent is recorded between the second and third contacts of the eclipse.

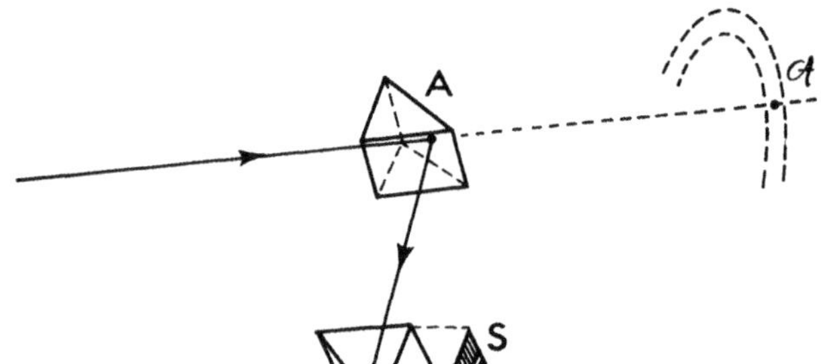

Fig. 1.

The change in the distance between the centres of the Sun and the Moon produces a sinusoidal variation of the intensity of the current with a period equal to the period of rotation of the prisms. The second harmonic would furnish the ellipticity of the Moon's limb, etc.; the mountains would reveal their presence by producing high-frequency oscillations akin to a noise which could be traced from one period of prism rotation to the next. The blurring of the image due to atmospheric anomalies would not affect the photocurrent; and neither would the motions of the image or of the instrument except for motions normal to the slit, which could change slightly the

* Observatoire du Pic-du-Midi, France.

amplitudes of the harmonics of very high orders. Some trouble could arise from bright faculae (or spots) near the limb. Such an effect could, however, be calibrated for before the first contact (or after the fourth one) of the eclipse, by scanning an artificial annular eclipse (of the same size) produced by occulting a part of the length of the above-mentioned slit. The first harmonic would reveal the off-center position of the Sun; the second harmonic would be due to differential refraction; and incidental photospheric phenomena on the limb would be reproduced as temporary irregularities during each period.

LUNAR PROFILES DETERMINED FROM THE ANNULAR SOLAR ECLIPSES OF 1962 AND 1963*

M. DAVIDSON,** C. L. GOUDAS, AND Z. KOPAL†

Abstract. The present communication contains a discussion of the techniques of measurements and the result of reductions of the photographs of the annular solar eclipses observed on July 31, 1962, in Senegal, West Africa; and January 25, 1963, in the Union of South Africa, by the Lockheed-Manchester Eclipse Expeditions; with the aim of improving our knowledge of the face of the Moon as seen in projection against the disc of the Sun. The lunar profile deduced from the 1962 eclipse plates was found to be insufficiently well-defined due to the effects of irradiation. The 1963 eclipse plates were, however, largely free from this defect, and their analysis led to satisfactory results.

The outcome of the reductions revealed that the deviations of the lunar profile from a circle were of the form $+0''.54\sin 2\beta - 0''.16\cos 2\beta + 0''.27\sin 3\beta - 0''.05\cos 3\beta - 0''.16\sin 4\beta + 0''.11\cos 4\beta$, where β denotes the lunar position angle measured from the eastern part of the equator in the counter-clockwise direction. These deviations were compared with the preceding lunar limb determination and found to be in satisfactory agreement; in particular, they agree within the limits of observational errors with the profiles obtained by POTTER (1962) and WATTS (1963). The elliptical component of the profile is found to have its major axis inclined by approximately 37° to the lunar axis of rotation – a value which is almost identical with that given by Watts.

1. Introduction

It is scarcely necessary to underline the need for improving our knowledge of the global shape of the Moon at the present time, on the eve of the efforts to land men on the surface of our satellite, when spacecraft launched from the Earth are to operate soon in close proximity of the lunar surface. Until these feats will have been accomplished, however, all information we possess on the shape of the Moon must be obtained by astronomical observations from the distance of the Earth, by visual or photographic methods; and is required to meet a number of purely terrestrial needs.

Thus, in astronometric observations of the Moon, the position of its center must be determined from observed transits (or altitudes) of its limb; and a transfer from the limb to the center (involved in any comparison of the observed positions of the Moon with the ephemeris) presupposes a knowledge of the limb's geometry. Secondly, when constructing the maps of the marginal zones of the Moon (HAYN, 1907; WEIMER, 1952; NEFEDIEV, 1957; WATTS, 1963) we have to refer the heights of the individual points to a mean limb defined for all sections of the limb and all angles of libration. Third, investigations of the axial rotation of the Moon and of its physical

* Work supported by Contract AF 61 (052)–524 and Grant EOAR 64–49 between the University of Manchester, and the Terrestrial Sciences Laboratory of the Air Force Cambridge Research Laboratories, U.S. Office of Aerospace Research, through their Brussels office.

** University of Manchester.

† Mathematics Research Laboratory, Boeing Scientific Research Laboratories (normally at the University of Manchester).

libration are based on the heliometric measurements of apparent motion of lunar points relative to the limb, and require again a knowledge of its general form. Last but not least, a knowledge of the exact figure of the Moon is very important for theories of the origin and evolution of our satellite.

A complete survey of the methods by which such a knowledge can be obtained is wholly beyond the scope of the present communication. Our aim will, more modestly, be to present certain new contributions to the problem of the exact form of the Moon's limb, as obtained from the observations of the annular solar eclipses of 1962 and 1963, and to compare them with the outcome of previous work.

A photographic determination of the form of the Moon's limb at any particular libration is, in principle, possible from plates of sufficiently high quality at any phase by measuring with sufficient accuracy the shape of its sunlit limb. However, on account of the phase effect, only one (i.e., the illuminated) half of the lunar circumference can be measured at any time – including the time of full moon, at which phase effect becomes minimum; but still sufficiently large to invalidate the measurements of more than one half of the entire circumference. Due to the geometry of the problem, no sunlit full moon is ever completely full; and the actual effect may fluctuate from month to month. Almost complete full moons of a phase approaching zero can occur only when the Moon enters the shadow cone of the Earth; and zero phase is attained at the mid-point of a central lunar eclipse when the low residual brightness of the lunar face (due to its illumination through the aureola of the terrestrial atmosphere) would necessitate relatively long exposures for photographic registration.

It is evident that the only times when the apparent lunar disc is free from a phase effect and the circumference of its limb can, therefore, be measured over arcs more than 180° in length occur when the centers of the Sun, Earth, and the Moon lie on the same line. Central eclipses of the Moon represent one class of such opportunities; the others are, of course, central eclipses of the Sun. If such eclipses are total, the dark limb of the Moon can be measured only on the background of the solar corona; the requisite exposure times are again relatively long; and (similarly as in the case of total eclipses of the Moon) determinations of the form of the limb would require precise angular measurements over distances up to half a degree.

If, however, the eclipse of the Sun becomes *annular*, a much more favorable situation obtains: for not only does the limb of the Moon continue to be free from any phase-effect, but the shape of the entire circumference of the Moon can be measured *differentially* with respect to the adjacent limb of the Sun, which should not deviate significantly from a circle. As a rotating mass of gas the Sun must, to be sure, be slightly flattened at the poles; but the actual difference between the equatorial and polar semi-axes due to axial rotation should be equal to only

$$\tfrac{3}{2}\left(\frac{\omega^2}{2\pi G\rho_m}\right)\Delta_2 = 2.10 \times 10^{-5}$$

in terms of the apparent solar radius (for $G = 6.668 \times 10^{-8}$ cm^3/g sec^2, $\rho_m = 1.41$ g/cm^3 $\omega = 2.85 \times 10^{-6}$ sec^{-1}, and $\Delta_2 = 1.02$ for the latest models of the interval structure of the Sun), which for the average value of $r_{\odot} = 900''$ corresponds to an angular difference of less than 0''02.

If quantities of this order can be regarded as ignorable (they are within the diffraction limits of a 300-inch aperture for visible light, and have not so for been measured), the limb of the Sun can be regarded as a conveniently situated reference circle, and all measurements carried out differentially with respect to it. This should constitute an inestimable advantage; for not only are the exposure times for photographs of the Moon on the bright background of the Sun thousand times shorter than those that would be required to record the image of the totally eclipsed Moon on the dark background of the sky, but all troublesome atmospheric as well as instrumental effects (such as air turbulence, off-axis field aberrations, emulsion shifts, and differential refraction, etc.) which might seriously impair the precision of astrometric measures extending over half a degree, enter only differentially (in as much as they affect the adjacent portions of the limbs of the Sun and the Moon by different amounts) and affect the results the less, the smaller the difference between the angular size of the Sun and the Moon at the time of the eclipse.

With these considerations which emphasize an overwhelming advantage of annular eclipses as a tool for the studies of lunar figure, one of us (Z. K.), then serving as consultant to the Lunar-Planetary Branch of the Basic Sciences Laboratory, Lockheed California Corp., suggested to Dr. L. G. Stoddard, then head of the Lunar-Planetary Branch (and now head of the Astronomical Sciences Laboratory) that suitably equipped expeditions be sent out to Africa to observe the annular eclipses of the Sun on July 31, 1962 and January 25, 1963. This project was approved by Dr. L. Larmore, then Chief Scientist of the Lockheed California Corp.,* and was granted logistic support by two branches of the U.S. Air Force – the Terrestrial Sciences Laboratory of the Air Force Office of Aerospace Research (Mr. Mahlon S. Hunt, Monitor) and the Aeronautical Chart and Information Center at St. Louis (Mr. R.W. Carder).

Both expeditions, organized and led by Dr. Stoddard ably assisted by Mr. D. Carson, met with a large measure of success. The results of the observations were then measured and reduced in the Department of Astronomy, University of Manchester. Preliminary results were obtained by Messrs. S. A. Brooks and M. Davidson in their unpublished M.Sc. theses written under the supervision of Professor Kopal and submitted to the Faculty of Sciences at Manchester in 1963, while the final results as presented in this report were obtained subsequently by a collaborative effort between Mr. Davidson, Dr. Goudas, and Professor Kopal at the University of Manchester and the Boeing Scientific Research Laboratories, Seattle, in 1965.

In the second section of this report, the circumstances of both eclipses will be briefly summarized; but for an account of the actual observations the reader must be referred elsewhere (cf. CARSON *et al.*, 1966 and pp. 128–137 of the present volume). The results obtained in the course of this study, their comparison with previous data, and conclusions arrived at, have already been summarized in the abstract, so that in what follows they will be substantiated in detail.

* Now Vice-President and Chief Scientist, Douglas Company.

2. Circumstances of the Eclipses

The annular eclipses of July 31, 1962 and January 25, 1963 which furnished the photographic data at the basis of the present study were characterized by the following circumstances:

July 31, 1962

Place of observation: Kheur Soce (Senegal)
Geographic coordinates of the observing site: $l = 16°4'12''$W, $b = 13°59'4''$N.
U.T. of the mid-eclipse: $11^h46^m17^s.2$
Duration of annular phase: $204^s.6$
Topocentric geometrical semi-diameter of the Moon: $917''.7$
Topocentric geometrical semi-diameter of the Sun: $945''.4$
Topocentric libration of the Moon: longitude: $+4^\circ.58$
latitude: $-0^\circ.17$
Position angle of the Moon's axis: $+16^\circ.32$
Position angle of second contact: $271^\circ.3$
Position angle of third contact: $110^\circ.9$
Position angle of line of motion of the Moon's center: 101°1
Zenith distance of the Sun: $20^\circ.0$
Position angle of zenith: $266^\circ.7$

January 25, 1963

Place of observation: Kruisrivier, South Africa
Geographic coordinates of the observing site: $l = 21°58'10''$E, $b = 33°36'34''$S.
U.T. of the mid-eclipse: $14^h48^m59^s.6$
Duration of annular phase: $37^s.7$
Topocentric geometrical semi-diameter of the Moon: $966''.2$
Topocentric geometrical semi-diameter of the Sun: $974''.7$
Topocentric libration of the Moon: longitude: $-5^\circ.36$
latitude: $+0^\circ.07$
Position angle of the Moon's axis: $-15^\circ.33$
Position angle of second contact: $257^\circ.6$
Position angle of third contact: $66^\circ.2$
Position angle of line of motion of the Moon's center: $71^\circ.9$
Zenith distance of the Sun :$55^\circ.9$
Position angle of zenith: $118^\circ.3$

For fuller details concerning the methods by which some of the foregoing quantities have been determined, cf. again CARSON *et al.* (1966).

3. Measurements of Data and Error Analysis

A. MEASUREMENTS

Typical examples of the photographic plates obtained from the 1962 and 1963 eclipses are shown in Figures 1 and 2, respectively, and our task now is to make an accurate mathematical presentation of the inner boundary of the bright annulus, i.e., the lunar profile. For reasons stated in Section 1 it is sufficient to assume that the outer boundary is practically a circle. The inner boundary can be described by meas-

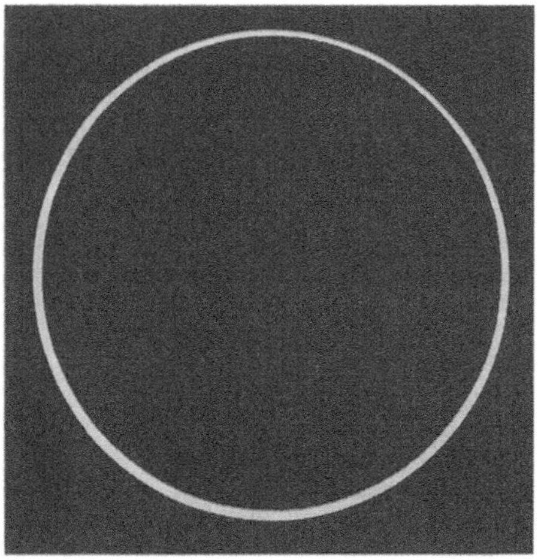

Fig. 1. A typical example of the photographs obtained during the 1962 annular solar eclipse.

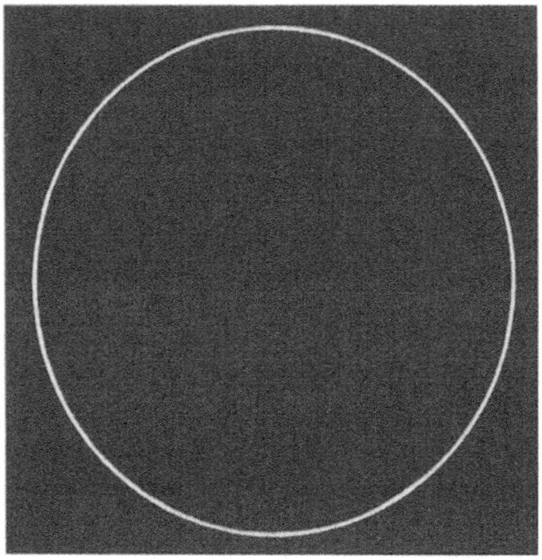

Fig. 2. A typical example of the photographs obtained during the 1963 annular solar eclipse.

urements of the coordinates of a series of points with respect to an arbitrary origin performed by means of a comparator. The same aim can essentially be accomplished by measuring the width or annulus of the bright belt versus the position angle using initially an arbitrary origin which later will be transferred to the center of the limb. The definition of this center will be discussed in the analysis of the data at a later stage.

The second method was found more practical as well as accurate in this case, and the width of the annulus between the solar and lunar limbs was measured by means of the Joyce-Loebl Mk IV twin beam microdensitometer of the Astronomy Department of the University of Manchester. The origin was located at the center of the solar disc, because the width was always measured along a direction perpendicular to the solar limb. In more detail, the work involved in the measurement was carried out as follows:

The original frames (about 180 from the 1962 eclipse and 25 from the 1963 one) were copied on two types of film: namely, the Kodak S.O.105 and S.O.278 which are high and low contrast, respectively. In the case of the 1962 eclipse plates it was found that only 15 of them, all on S.O.105 film, were suitable for measurement. Out of the

25 plates of the 1963 eclipse 20 were of good quality and were used for measurements.

Each negative was placed on a sheet of graph paper, the lines of which were visible through the transparent annulus. With a little adjustment it was easy to find the lines on the graph paper which intersected each other within a millimeter from the center of the solar disc. These lines were marked on the actual emulsion using a fine needle and with the aid of a ruler. A protractor was then used to mark 30° intervals round the annulus, measured from an arbitrary fiducial line coincident with one of the original markings, which was the same for all positives. The fiducial line was chosen to be parallel to the longest edge of the film. The microdensitometer table is transparent, runs on rails, and is connected by means of a lever arrangement to the pen recorder table so that a magnification of 50 is achieved. In addition, the table may be rotated with respect to the rails through an angle of $\pm 18°$. The film was placed on the table and centered as accurately as possible by observing the magnified image (which is five times larger) of the central cross previously marked on the emulsion on a screen in the microdensitometer as the table was rotated. When the centering was satisfactory, i.e., when the shift of the central cross was unobservable for a 12° rotation of the table and the fiducial line was parallel to the rails, the film was clamped with the aid of a glass slab and several clips.

The annulus at two points 180° apart was then traced on one piece of graph paper and then the prodecure was repeated. Two typical traces for position angles 180° apart are given in Figure 3. This last figure shows clearly that the boundary of the annulus is very sharp and the width of the trace at the half-height defines accurately a quantity linearly related to the angular width of the annulus at the given position angle.

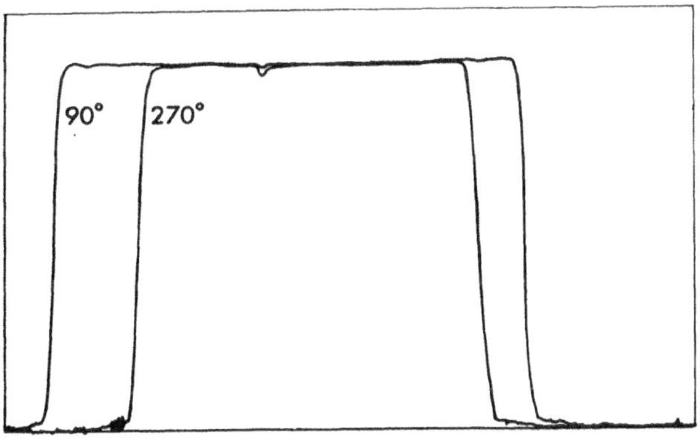

Fig. 3. Example of tracings obtained with the microdensitometer of Manchester University.

Because of the design of the apparatus, after five measurements had been made it was necessary to re-center the film, this time with the datum line at an angle of 30° to the rails. This whole procedure was repeated until the whole annulus had been measured – 60 measurements in all, per frame. The necessity of re-centering the film every 30° made the work time consuming but it also made centering errors less serious as these would not now extend round the entire annulus. The measurements were made by four different persons in order to diminish the probability of systematic errors arising from inaccurate centering.

After the measurements of each frame were completed, the width of the annulus was plotted against the position angle from the fiducial line, and a typical graph is shown in Figure 4 which corresponds to a 1962 plate. This graph immediately gives an idea of the extent to which the measurements can be affected by atmospheric turbulence and by the relief of the lunar limb. As we shall presently show, the curve on Figure 4 should be a cosine wave if the profile of the Moon is circular.

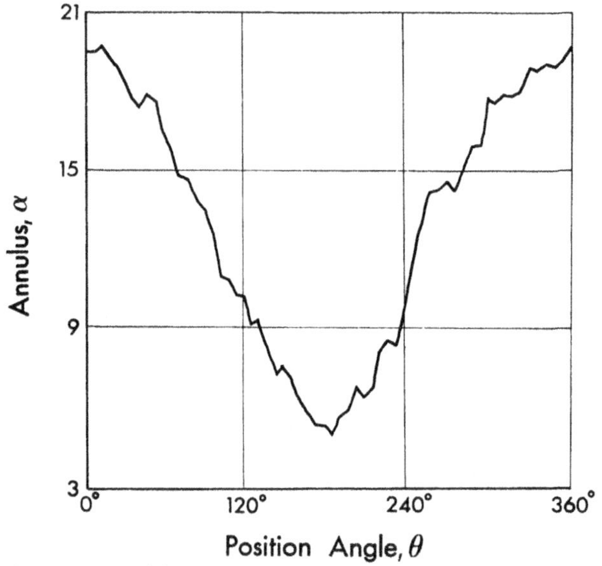

Fig. 4. Annulus versus position angle as given for one frame of the 1962 eclipse.

The marking of the center of the solar disc obviously cannot be exact. Therefore, we must set out to determine the errors in the width α of the annulus and in the position angle θ resulting from it. The case is illustrated in Figure 5, where S and M denote the centers of the Sun and Moon, respectively. The distance SM will be called Δ. We adopt the point S as origin and the direction SM as the reference line from which position angles θ are measured. Let R and r be the radial distances between the centers S and M and the limb B' and A', respectively. The latter points lie in the direction of the solar radius of position angle θ .The distance $A'B'$ varies with θ and, if both the solar and lunar profiles were circular, it would attain a minimum value, $AB=\alpha$, for $\theta=0$ (although in reality it is not necessarily a minimum).

If the line SB' is taken as the x-axis of a planar rectangular frame of reference, then the coordinates of point on the inner boundary of the annulus, i.e., the lunar profile satisfy the relation

$$r^2 = (x - \Delta \cos\theta)^2 + (y - \Delta \sin\theta)^2, \qquad (1)$$

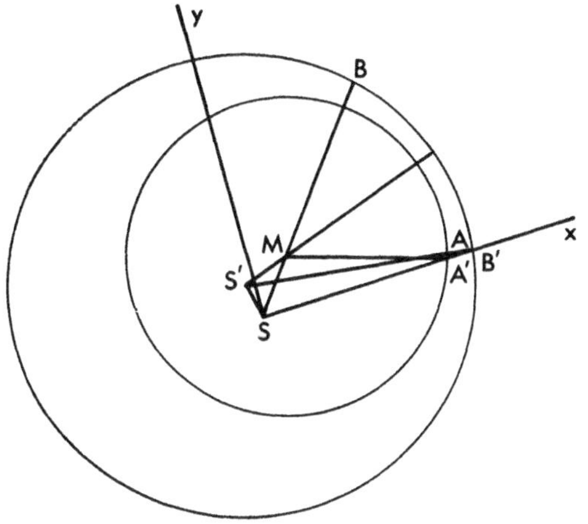

Fig. 5. Geometric representation of the annular phase.

where r depends on θ. Also the x-coordinate of the point A', if denoted by $x_{A'}$, is given from the expression,

$$x_{A'} = r + \Delta \cos\theta . \tag{2}$$

Therefore, the width α of the annulus is

$$\alpha = R - (r + \Delta \cos\theta) + \Delta x_0 . \tag{3}$$

The center M of the lunar profile will emerge as a by-product of our reduction; but the center S of the Sun may require the corrections $\delta\Delta$ and $\delta\theta$. We shall show that, in the present case, both corrections are smaller than the resolution of the plates. Let it be assumed that the coordinates of the true position S' of the center of the solar disc are δx_0, δy_0 instead of 0.0, The coordinates of the point B' (see Figure 5) are not $(R, 0)$ but $[(R^2 - \delta y_0^2)^{\frac{1}{2}} + \delta x_0, 0]$ and the correct annulus is the quantity $B'A''$ and not $B'A'$. Finally the correct position angle is the angle $B'S'M$ instead of $B'SM$. We shall first compute the error in the width of the annulus. To do so we determine the coordinates (X', Y'), (X'', Y'') of the points A' and A'', respectively. They are

$$\left.\begin{aligned} x' &= \Delta \cos\theta + r(1 - \Delta^2 \sin^2\theta/2r^2) \\ y' &= 0 \end{aligned}\right\} \tag{4}$$

$$\left.\begin{aligned} x'' &= \Delta \cos\theta + r(1 - \Delta^2 \sin^2\theta/2r^2 + y''\Delta \sin\theta/r^2) \\ y'' &= \delta y_0 (R - r + \delta x_0 - \Delta \cos\theta + \Delta^2 \sin^2\theta/2r)(R + \delta y_0 \Delta \sin\theta/4) . \end{aligned}\right\} \tag{5}$$

These coordinates suffice for the determination of both errors in the width of the annulus and its position angle. In particular, we find that

$$\delta\alpha = A - [(A - \Delta B \sin\theta/r)^2 + B^2]^{\frac{1}{2}} , \tag{6}$$

where

$$A = R + \delta x_0 - C ,$$

$$B = \delta y_0 \left(1 - \frac{C + \delta x_0}{R}\right),$$

and

$$C = \Delta \cos\theta + r(1 - \Delta^2 \sin^2\theta/2r^2).$$

On the other hand, the error $\delta\theta$ in the position angle is given by the expression

$$\delta\theta = \cot\theta - \frac{R(\Delta\cos\theta - \delta x_0) - \delta y_0(\Delta\sin\theta - \delta y_0)}{\sin\theta\,[R^2 + (\delta y_0)^2]^{\frac{1}{2}}\,[(\Delta\cos\theta - \delta x_0)^2 + (\Delta\sin\theta - \delta y_0)^2]^{\frac{1}{2}}}. \tag{7}$$

The following numerical values have been employed for the calculation of the errors $\delta\alpha$ and $\delta\theta$:

$$R = 8\text{ cm}, \quad r = 8\text{ cm}, \quad \Delta = 0.2\text{ cm}, \quad \delta x_0 = \delta y_0 = 0.02\text{ cm}.$$

The numerical computations showed that the errors in α never exceed 0.2μ; whereas the errors in θ never become large enough as to make it necessary to apply corrections to the measurements obtained.

The size of the image of the films make it possible to calculate its scale as well as the focal length of the telescope. Since the solar disc departs from a perfect circle by ignorable amounts, the scale of the photographs can be determined from the solar diameter only after the factors distorting it are properly treated. Since the focal length f of the telescope can be defined as the ratio of the linear diameter of the solar disc in the focal plane to the apparent (topocentric) angular diameter of the Sun at the time of observation, we obtain,

$$f = 17.67\ m. \tag{8}$$

The scale s of the photographs is, on the other hand, the inverse of the focal length and the numerical values obtained are

$$s = 11''.74 \text{ per millimeter, for the (1962) eclipse}$$
$$s = 11''.55 \text{ per millimeter, for the (1963) eclipse.} \tag{9}$$

The ratio of the radii of the two discs for the same eclipse is found to be 0.971 and 0.991, respectively. All the figures given here will have to be slightly modified after the analysis of the errors involved in their estimate are analyzed and estimated in the following section.

B. ERROR ANALYSIS

The departures of the lunar profile from a circle are rarely larger than 2''.0 geocentrically, and on the average are less than 1''.0. It is, therefore, essential in search for significant deformation of the lunar figure in its marginal zone that all sources of possible systematic errors capable of introducing effects larger than ±0''.2 be carefully considered. For this reason, a detailed study of errors was undertaken for the sets of photographs of both eclipses, concerning not only the systematic errors likely to be present, but also those which could possibly influence the final reductions by appreciable amounts. Our error analysis concentrated mainly on the following effects.

1. *Emulsion Shifts*

Such a shift after the exposure would be liable to distort appreciably the images of both the lunar and solar profile, and thus produce a measurable effect in the measurements. A uniform dilatation will, in general, change the scale of the photographs and leave the spectrum of the limb unaffected except for the zero harmonic, and for this reason need not be considered at all. On the other hand, non-uniform dilatation could be regarded as a single or multiple stretch and can be examined as follows: Since the two profiles recorded on the films, namely, that of the Sun and the Moon, are at a close proximity, any shift should deform both largely by the same amount. For this reason, measurements of the diameters of the solar disc at different directions for each frame were made. Analysis of these measurements showed that such an effect was not present in our case.

In the chemical development of the photographs the emulsion may be subjected to irregular effects causing differential displacements on the film. These displacements, if random, should cancel out when measurements are averaged over the 15 (in the first case) and 20 (in the second case) sequential photographs of each of the eclipses. If, on the other hand, substantial differential displacements are present, then their effect should appear on the solar profile and thus be detected. Such an effect has not been found.

2. *Refraction*

The altitude of the Sun during the 1962 eclipse was 60° and during the 1963 eclipse 56°. The maximum geocentric angular dimension of the annulus along the main direction of differential refraction was in both eclipses less than 15″, and this effect introduces a decrease in the annulus of at most 0.″2, which is again negligible and does not need to be taken into account.

However, the eclipse itself is expected to modify to an important extent the normal properties of a quiet atmosphere for the following reason: The sudden elimination of the energy flux inside the shadow cone will inevitably result in a fall of temperature and, hence, in local currents which are capable of causing anomalous refraction patterns on the recorded shape of the limbs.

This effect was discussed by Schuster in connection with the relativistic deflection of stellar images in the neighborhood of the solar limb. He concluded that the effect was unimportant, although doubted if this was generally the case.

3. *Turbulence*

Atmospheric turbulence can result in differential displacements of limb details during short exposure times. In the case of long exposures the instantaneous profiles will be smeared out around the true position of the limb and obliterate its details. For this reason, short exposures were taken in a sequential fashion every 1.5 seconds. The effects of refraction anomalies due to atmospheric turbulence can easily be seen by a comparison of any two successive photographs. These, however (being of random nature), can be eliminated by averaging the measurements on the entire sequence of photographs. This has been found to be a powerful technique in lunar photographic work in other cases as well, and is expected to increase the accuracy of selenodetic work in the marginal zone as well as on the remainder of the lunar surface. Reduction of the amount of work necessary for the measurement of many sequential photographs can be achieved by judicious choice of a small number of frames out of the entire series. The selection has to be based on the condition that the average of positions of features on the selected sample of photographs must be as close as possible to the average of the entire set. In order to do so, measurements of a certain number of identical points on all photographs are initially required. Their positions must then be harmonically analysed and the important frequencies of the atmospheric turbulence during the exposures be tedermined.

4. *Plate Tilt*

If the optical axis of the telescope is not perpendicular to the photographic plate, the width of the annulus resulting from our measurements will not be correct. Let ϕ (see Figure 6) be the angle between the optical axis of the telescope and the normal to

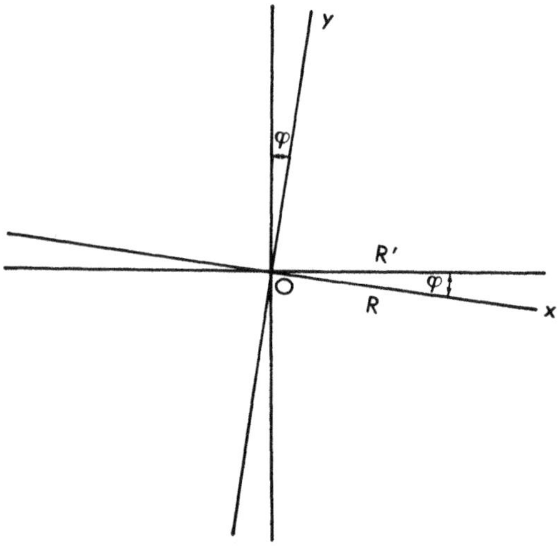

Fig. 6. Tilt Effect.

the plane of the film. On the assumption that their centers coincide, we find that

$$R - r = (R' - r')(1 - \phi^2/2), \tag{10}$$

where R' and r' are the radii of the solar and lunar images in the directions most affected by the plate tilt. If now the maximum error we can tolerate in R or r is about 0.″1 geocentric, then the angle ϕ must be less than 7°. But such a plate tilt cannot have been present for the instrumental arrangements used to observe both eclipses.

5. *Telescopic Aberrations*

If we assume that the mirror aberrations are linear in distance from the optical axis, these will simulate the stretching of the film, the effects of which we found before to be ignorable. Nonlinear (i.e., quadratic and higher) off-axis aberrations would, however, be reflected in the measured figure of the Moon. In our case – since the images of the solar annulus were formed slightly off-axis (due to the Herschelian nature of our system) – off-axis effects are bound to be present and could become serious if measurements were to be performed across the entire diameter of the annulus. Since, however, we are using the outer limb of the annulus as our reference circle, the form of the lunar profiles emerging from our measurements should be affected by off-axis effects only in so far as these change across the width of the annulus; and the latter amount only to about one-hundredth of its diameter. Such differential effects have proved to be theoretically negligible, and no trace of them was found in the observed results.

6. *Diffraction*

Due to the finite resolving power of the telescope, there may be some distorting effect when the annulus is narrow. This is the case with all the films available for both eclipses. Such distortion may, of course, be expected only when the width of the annulus is comparable in size with the diameters of the Airy disc imaging a light point; and this may be true only shortly after the second contact and before the third, when the two discs have one common point at internal tangency. The diffraction effect may appear in parts of the photographs where the geocentric width of the annulus is less than, or equal to, 0″.5 in blue light. The size of the error introduced by this effect can be computed in the following way: Let, in Figure 7, the x-axis be in the direction of the solar radius and the interval $(-a, +a)$ be the width of the annulus along which the light intensity I_0 is assumed to be distribution constant. The Fourier transform $f(k)$ of this annulus is

$$f(k) = \frac{1}{\sqrt{2\pi}} \int_{-\infty}^{\infty} f(x)\, e^{-ikx}\, \mathrm{d}x = \frac{1}{ik\sqrt{2\pi}} (e^{ika} - e^{-ika}). \tag{11}$$

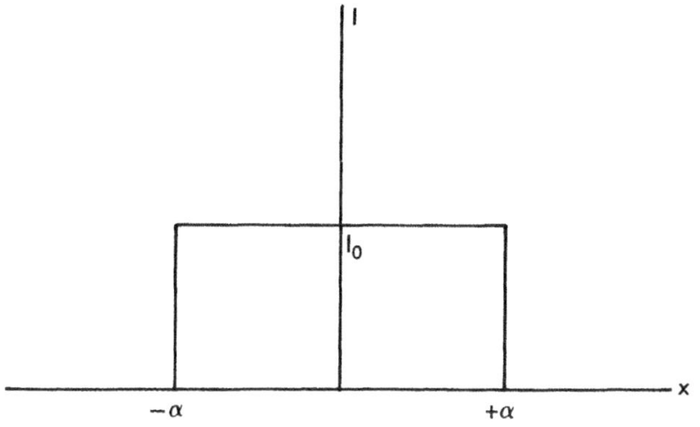

Fig. 7. Light intensity along a cross-section of the annulus.

Since the aperture of the telescope is finite the image in the photographic emulsion will be given by this Fourier transform multiplied by the aperture function. If it is assumed that the optics of the telescope is perfect and that the highest spatial frequency received by the telescope is q, then the intensity $f'(x)$ of the image in the emulsion will be given by the expression

$$f'(x) = \frac{1}{2\pi} \int_{-q}^{+q} \frac{e^{ika} - e^{-ika}}{ik} e^{ikx}\, dk, \tag{12}$$

which, in turn, can be written as

$$f'(x) = \frac{1}{2\pi} \int_{-q}^{q} \frac{\sin k(a + x) - \sin k(a - x)}{k} dk. \tag{13}$$

The substitution

$$k' = ak, \quad q' = aq, \quad x' = ax,$$

transforms expression (13) further into

$$f'(x'/a) = \frac{1}{\pi} \int_{0}^{q'} \frac{\sin k'(1 + x') + \sin k'(1 - x')}{k'} dk'. \tag{14}$$

We now define the aperture function $A(k)$ as follows:

$$A(k) = 1 \quad \text{for} \quad |k| \leq |q|,$$
$$A(k) = 0 \quad \text{for} \quad |k| > |q|.$$

Therefore,

$$A(x) = \frac{1}{\sqrt{2\pi}} \int_{-q}^{q} e^{kx}\, dk$$
$$= \frac{q}{\sqrt{2\pi}} \left(\frac{\sin qx}{qx} \right). \tag{15}$$

The aperture $A(x)$ given in this form vanishes for $qx = \pi$ and, as a result, if α is the Airy disc the integration in (14) can be terminated with $k' = \pi a/\alpha$.

The evaluation of this integral was performed numerically but prior advantage was taken of the following considerations. Since the measurements of the photographs with the microdensitometer can be carried out at different light intensities, a non-dimensional quantity z was introduced, defined as follows:

$$\text{light intensity during measurements} = zI_0. \tag{16}$$

What we actually seek is the function $x = x(a)$ for various values of the parameter z. In this way we shall have the relation between the width of the annulus recorded on the emulsion, and its correct value a in the absence of diffraction.

The results of the numerical integration performed with the "Atlas" Computer of

Manchester University are graphically shown on Figure 8 for four different values of the parameter z. The values x and a are plotted in terms of the Airy disc. In the absence of diffraction, the slope of these curves should be equal to 45°; but apparently this is not the case when the true size of the annulus is less than 0.7 Airy radii. When a is

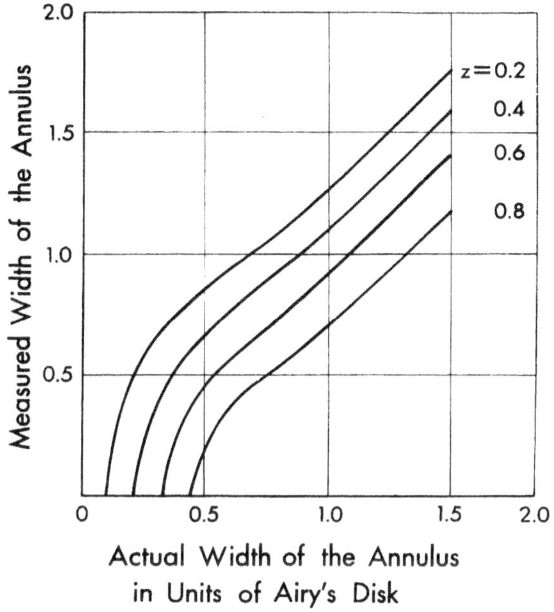

Fig. 8. Measured versus true width of the annulus for different values of the parameter z.

larger than one Airy radius the measured annulus x is a linear function of a with 45° slope, indicating the diffraction effects to be negligible. The annulus on the photographs of both eclipses was never less than 10 Airy radii – a fact which gave us confidence that no such effects would appear in our reductions.

7. *Error in the Position of the Lunar Center*

The path of the center of the lunar disc can be assumed to a very good approximation to be a straight line, such as $A'A$ in Figure 9. If now the adopted path of this center were different (say, $B'B$), then a systematic error in the reduction of the measurement is introduced which can be treated in the following way: Let P and Q be the adopted and true positions of the lunar center. The residuals of the points from the best fit circle with P as its center would then contain an additional term varying as $\cos\theta$, where θ denotes the angle between the position vector of the limb points and the direction PQ. In particular, if $PQ=\rho_0$ and the lunar limb were close to a circle, the radius $\rho(\theta)$ from the point P to the limb point of position angle θ could be written in the form

$$\rho = r + \rho_0 \cos\theta - \tfrac{1}{2}\frac{\rho_0^2}{r}\sin^2\theta + \tfrac{3}{8}\frac{\rho_0^4}{r^3}\sin^4\theta + \cdots .$$

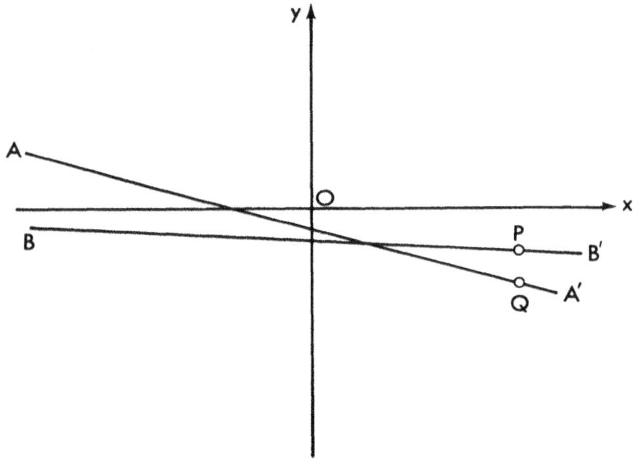

Fig. 9. Motion of the center of the Moon relative to the center of the Sun.

The term $\rho_0 \cos \theta$, would, of course, become immediately obvious if the constant ρ_0 were sufficiently large. In such a case a new determination of the path of the lunar center would be necessary. This was not the case with the photographs of the 1963 eclipse.

An examination of Figures 10 and 11 which correspond to the 1962 eclipse showed, however, that a systematic error was present in the positions of the lunar centers, because the separate treatment of the frames of serial numbers from 93 to 99 and from 138 to 147 did not give comparable values for the slope of the straight line interpolated by least squares. In addition, an inspection of the plot of the annulus versus position angle (see Figure 8) shows clearly an "attraction" of the lunar limb by the solar limb. This appeared to be the case with all individual frames. The two groups of sequential photographs mentioned above resulted in two profiles which together with their average proved to be unreliable. The two profiles are barely compatible.

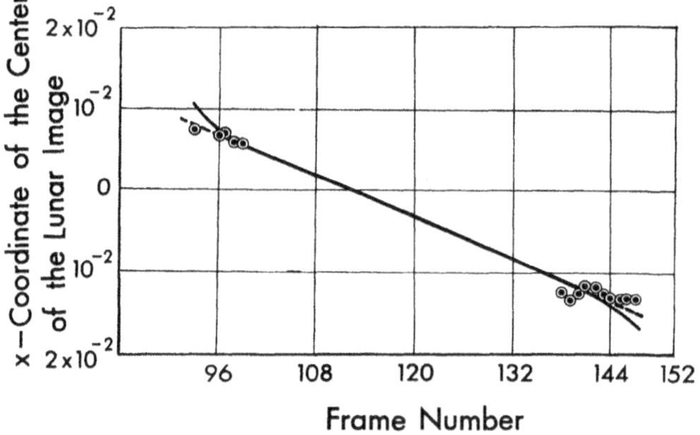

Fig. 10. Variation of the x-coordinate of the lunar center.

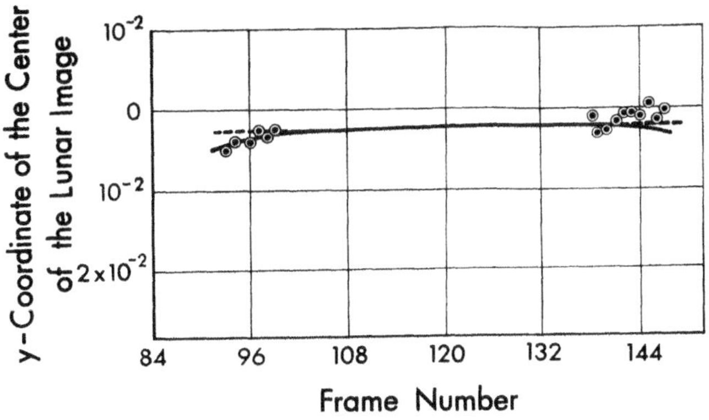

Fig. 11. Variation of the y-coordinate of the lunar center.

The discrepancy between the slopes of the paths of the two centers as well as between the profiles in the case of the 1962 eclipse is due to the effect of irradiation, which we shall discuss next.

8. *Irradiation in the Emulsion*

A careful study of all available photographs revealed that the 1962 eclipse data cannot produce good profiles of the Moon because of undisclosed systematic effects. In more precise terms, first – against expectation – the centers of the Moon in the series of photographs were not sufficiently well aligned. Secondly, the mean radius of the Moon was found to vary from frame to frame; and the value of r as borne out by measurements differed from the one listed in the Astronomical Ephemeris and Nautical Almanac by an amount greater than expected (although this could be explained in terms of the characteristic curve of the emulsion). Third, the profile of the Moon obtained from two sets of sequential photographs taken a few minutes apart in time are not in satisfactory agreement. In addition, none of the profiles determined from individual frames seemed to agree well enough with the rest. All these can be explained in terms of the effect of irradiation in the emulsion.

Various authors (see, e.g., SELWYN, 1950) have studied the effect of irradiation on the size of photographic stellar images and derived a number of relationships between the diameter of stellar images and exposure times. The extent of the blurring is represented by the law

$$I = I_0 \exp(-kx - \sigma x^2), \tag{17}$$

where I_0 and I are the intensities of the image at its geometrical center and at distance x from it, respectively; and k and σ are constants, the latter representing the effects of atmospheric turbulence. Therefore, the law of irradiation of the form

$$I = I_0 \exp(-kx), \tag{18}$$

can safely be adopted where for our photographs the exposure times ($\frac{1}{350}$ sec) do not allow the effects of atmospheric turbulence to be recorded on the emulsion.

Consider a section of the annulus of width x_0, along which the intensity I_0 is uniform. The effect of irradiation is to blur the image in such a manner that the intensity I at a point whose distance from a small element of the annulus is x will be given by the expression

$$I = 2I_0 \left[\int_0^\infty \exp\left(- k\sqrt{x^2 + z^2} \right) \mathrm{d}z \right] \mathrm{d}x\,, \tag{19}$$

where z is the width along the annulus which is assumed to be infinite. The right-hand side can be expressed in terms of Bessel functions, so that, to avoid complexity of numerical calculation, the z term will be neglected. This will not, in principle, alter the predicted effects qualitatively, but will make the result only approximate. It is thus assumed that

$$I = I_0 \exp(- k\,|x|)\, \mathrm{d}x\,. \tag{20}$$

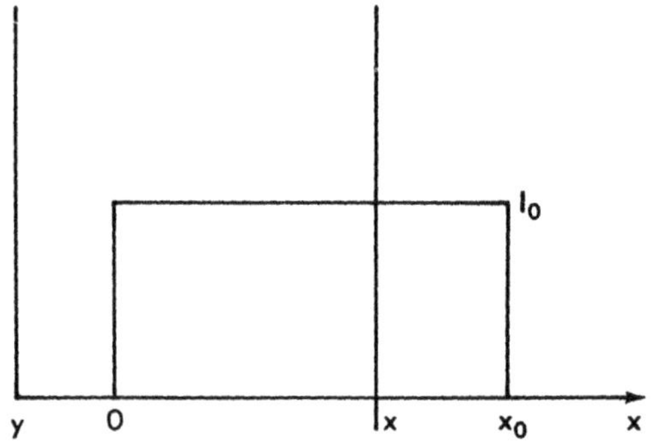

Fig. 12. Light intensity along the annulus.

Consider now Figure 12, in which one of the limbs of the annulus sets in for $x=0$; the other, for $x=x_0$. The intensity at the point x contributed by the entire annulus is

$$I_x = I_0 \left[\int_0^x e^{-kx}\, \mathrm{d}x + \int_0^{x_0 - x} e^{-kx}\, \mathrm{d}x \right]$$

$$= \frac{I_0}{k} \left[2 - e^{-kx} - e^{-k(x_0 - x)} \right]. \tag{21}$$

If the point y were on the other side of the annulus, then

$$I_y = \frac{I_0}{k} \left[e^{-ky} - e^{-k(y - x_0)} \right]; \tag{22}$$

whence, for $y<0$, we obtain

$$I_y = \frac{I_0}{k} \left[e^{-ky} - e^{-k(x_0 + y)} \right]. \tag{23}$$

The total of the energy E incident on the emulsion (on the assumption that we work on the linear part of the reciprocity curve) is given by the expression

$$E = n\frac{I_0}{k}\left\{\int_0^{x_0} [2 - e^{-kx} - e^{-k(x-x_0)}]\,\mathrm{d}x + \int_0^{\infty} [e^{-ky} - e^{-k(x_0-x)}]\,\mathrm{d}x + \int_0^{\infty} [e^{-k(y-x_0)} - e^{-ky}]\,\mathrm{d}x\right\},$$

which on evaluation of the integrals becomes

$$E = 2nx_0 I_0/k, \tag{24}$$

where the quantity n is a normalizing constant. If we now want to make the total energy equal to $x_0 I_0$, we assign to n the value $k/2$.

The effect of high contrast obtained in the copies selected for measurement had no effect on the results because the contrast of a copy merely provides a line of constant density in the original, thus making measurements easier. It is necessary to assume here that not only the originals but also the copies had identical exposures, so that the quantity does not vary from positive to positive.

Because of the relations (21) and (22), at the measured boundary of the annulus we have

$$2q = e^{-kx} - e^{-k(x_0+x)}$$

and, consequently,

$$e^{-k(x+x_0/2)} = 2q\exp(-kx_0/2)/[1 - \exp(-kx)].$$

If the measured width of the annulus is denoted by δ, then we can put

$$\delta = x_0 + 2x \tag{25}$$

and, as a consequence, we find that

$$k\delta - kx_0 = 2\log(1 - e^{-kx_0}) - 2\log 2q. \tag{26}$$

This formula relates the variables x and δ, the actual and measured widths, as long as the apparent width is greater than the actual one. If the opposite were true, the corresponding relation could be obtained by combining equations (20) and (22) leading to the relation

$$e^{-kx} = (1 - q) \pm \sqrt{(1 - q)^2 - e^{-kx_0}} \tag{27}$$

which can, in turn, be combined with equation (25) to give the desired relation. The two functions coincide in the line $\delta = x_0$.

The graphs δ versus x_0 are shown in Figure 13 for different values of q from 0.1 to 0.9. It will be seen that for x_0 approaching infinity, δ exceeds x_0 by the amounts

$$1/k\log 2(1 - q) \quad \text{for} \quad q < 0.5, \tag{28}$$

$$-2/k\log(2q) \quad \text{for} \quad q \geq 0.5. \tag{29}$$

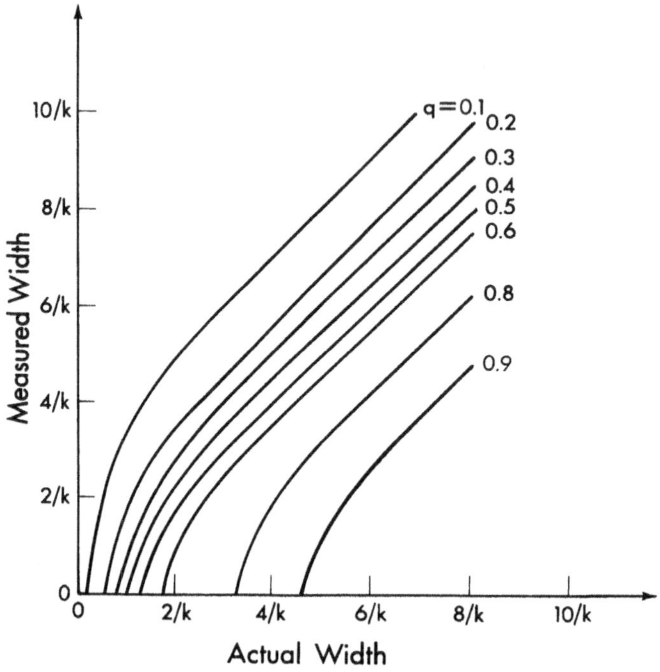

Fig. 13. Measured versus true width of the annulus for various values of the parameter q.

In any experimental curve that can be drawn these quantities may only roughly be approximated and it is more convenient to standardize the functions so that as x_0 approaches infinity, δ approaches x_0. The new functions are plotted in Figure 14. The maximum cut-off corresponds to $q=0.5$. These curves, as stated previously, cannot be expected to correspond exactly to the experimental curve, because of the initial assumptions, and also of the fact that solar limb darkening and the characteristics of the emulsion will affect the experimental curves.

A modified treatment which allows for solar limb darkening may be performed along the following lines. The intensity, I_θ, at a point on the solar disc where the heliocentric radius vector makes an angle θ with the line of sight can be represented empirically by a relation of the form

$$I_\theta = I_c(1 - u - v + u \cos\theta + v \cos^2\theta + \cdots) \tag{30}$$

where I_c is the intensity at the center of the solar disc, and u and v are parameters depending on the wavelength of the observation. This corresponds on our previous model to

$$I(x) = A + B\sqrt{x} + Cx\,, \tag{31}$$

where the solar limb has been assigned the coordinate x and A, B, C are constants; and leads to integrations involving the term $\sqrt{x}e^{-kx}$ which, again, has to be solved

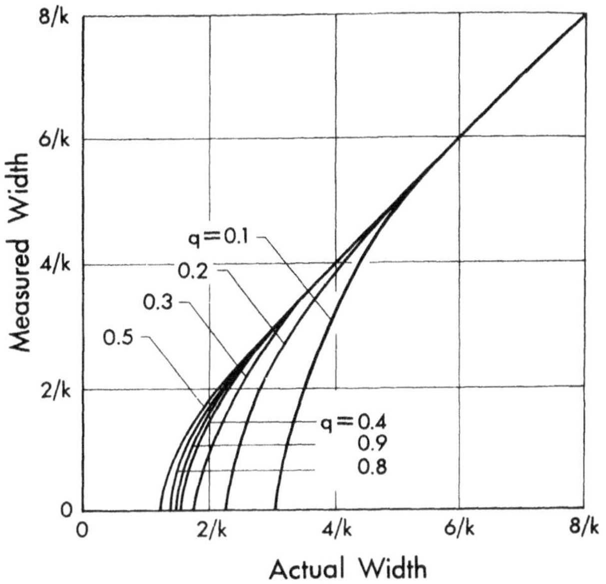

Fig. 14. Measured versus true annulus for five values of the parameter q.

numerically. Further, if for a zone close to the limb we approximate this law by the expression

$$I(x) = A' + C'x, \tag{32}$$

we can obtain a relationship between x_0 and δ which is rather complicated. Such an investigation would, however, be of limited interest in the present case, since it is not possible to determine the necessary parameters by least squares.

It is sufficient, at present, to investigate the effects of limb darkening qualitatively. The energy which reaches the emulsion does not then depend linearly on the width of the annulus, but increases more sharply as the annulus width increases. It is to be expected, therefore, that the experimental curves will be steeper than the theoretical ones for values of x_0 for which the distortion takes place.

It may be noted that in no frame among the SO-105 or the SO-278 positives is there any sign of limb darkening. If the copies are assumed to have been exposed correctly (which is likely since they were made in the laboratory) then this implies that the originals were considerably over-exposed – a fact which accounts for the importance of irradiation in the reductions.

The experimental curves of δ versus x_0 do, in fact, appear steeper than the predicted curves. The curves for Senegal and South Africa are shown in Figure 15. The methods of deriving the two curves are, however, different. In the case of the photographs of the South African eclipse there was a considerable number of high-contrast positives corresponding to partial phases. By measuring these, and extrapolating for the centers of the lunar images on the basis of the positions of the centers found for the central phases, we found it possible to plot the measured width against the width

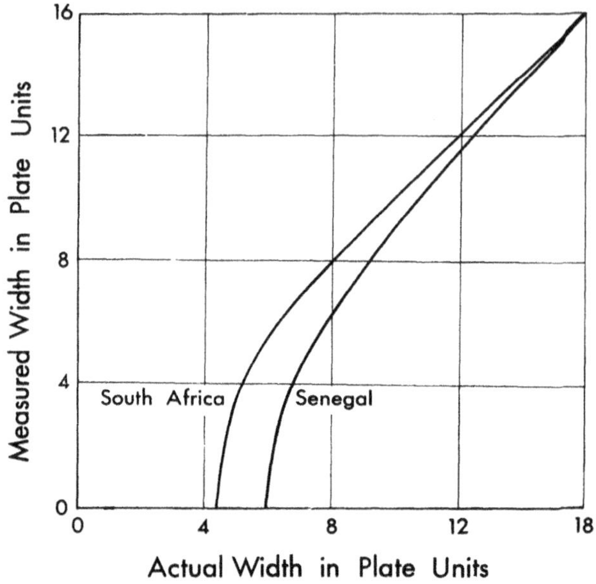

Fig. 15. Measured versus true width of the annulus for the two eclipses. The 1963 eclipse is apparently recorded satisfactorily as long as the width of the annulus is greater than 4.7 plate units.

found by adopting the extrapolated centers. For the Senegal eclipse, there were no high-contrast positives of the partial phases. An attempt to use the low contrast positives in a similar way failed because of a large dispersion on the tracings equal to that width found on the high-contrast positives of the same frame number. A histogram of these heights is shown on Figure 16. This being the case, it was clear that measurements of the partial phases would not produce a very reliable calibration curve and an iterative process was instead employed. This will be discussed later. A further prediction, on the grounds of irradiation in the emulsion, may be verified with reference to the low contrast positives. Equation (2) can be written in the form

$$I(x) = I_0 x_0 \left[2 - e^{-kx} - e^{-k(x_0 - x)}\right], \tag{33}$$

which attains the maximum value

$$I_{\max} = I_0 \left(1 - e^{-kx_0/2}\right) \tag{34}$$

when $x = x_0/2$. The quantity x_0 cannot be measured directly. It is estimated by plotting the maximum height of the trace against the measured width, bearing always in mind the effects of solar limb darkening, etc. Such a graph is given in Figure 17.

The continuous lines correspond to the theoretical curves (Equation (28)) for $k = 50\ \text{cm}^{-1}$ and $q = 0.2$, 0.3, and 0.4. It seems that the parameters have their best values when $k = 50\ \text{cm}^{-1}$ and $q = 0.3$. However, it was concluded that q was not independent of x_0.

Using these parameters, it is possible to recalculate the quantities calculated

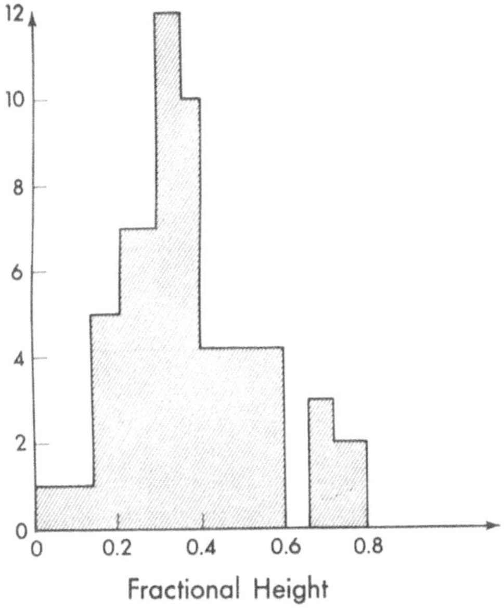

Fig. 16. Histogram of heights.

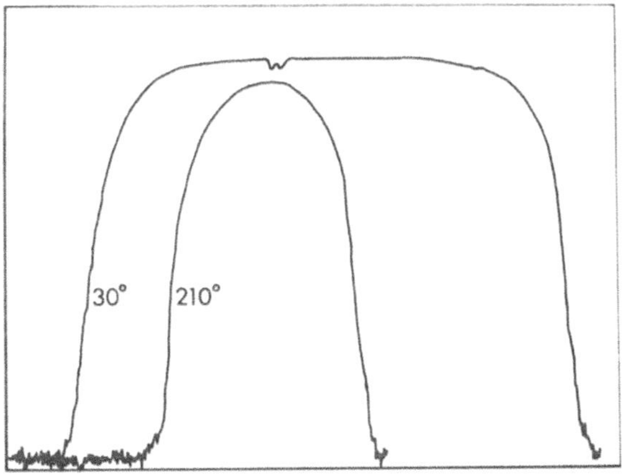

Fig. 17. Photometric tracing.

earlier. Namely, from Equation (28) we obtain

$$\Delta = \delta - x_0 = 1/k \log w(1 - q)$$
$$= 0.007 \text{ cm},$$

where $kx_0 \gg 1$.

4. Reduction of Measurements

The reduction of the measurements with the aid of a microdensitometer can be performed in different ways. Let (see Figure 18) the origin S of the Sxy rectangular

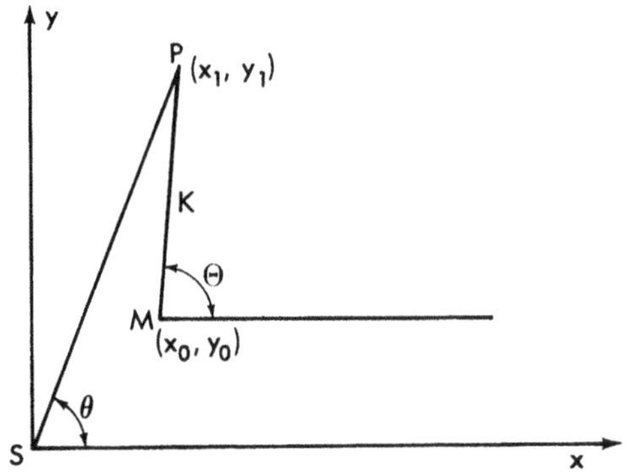

Fig. 18.

frame of reference (whose axes are arbitrarily orientated) be identical with the center of the solar disc and the point $M(x_0, y_0)$ be the center of the lunar disc. The radius of the Sun is denoted by R; and r_0 is the mean radius of the lunar disc. We can therefore express the two limbs by the equations

$$x^2 + y^2 = R^2, \tag{35}$$

$$(x - x_0)^2 + (y - y_0)^2 = (r_0 + \delta r)^2, \tag{36}$$

where δr is the departure of the particular point of the limb from the mean circle.

If now the point $P(x_1, y_1)$ on the lunar limb corresponds to a position angle θ, the straight line through it and the origin is represented by the equation $y = x \tan\theta$ and the coordinates of the point P are given by the expressions

$$x_1 = \{x_0 \cos\theta + y_0 \sin\theta + [(r_0 + \delta r)^2 - (x_0 \sin\theta - y_0 \cos\theta)^2]^{\frac{1}{2}}\} \cos\theta, \tag{37}$$

$$y_1 = \{x_0 \cos\theta + y_0 \sin\theta + [(r_0 + \delta r)^2 - (x_0 \sin\theta - y_0 \cos\theta)^2]^{\frac{1}{2}}\} \sin\theta. \tag{38}$$

Therefore,

$$x_1^2 + y_1^2 = x_0 \cos\theta + y_0 \sin\theta + [(r_0 + \delta r)^2 - (x_0 \sin\theta - y_0 \cos\theta)^2]^{\frac{1}{2}}$$

and for

$$\delta = R - (x_1^2 + y_1^2)^{\frac{1}{2}}$$

we obtain

$$\delta r = [(R - \delta)^2 - 2(R - \delta)(x_0 \cos\theta + y_0 \sin\theta) + x_0^2 + y_0^2]^{\frac{1}{2}} - r_0 .$$

It also follows that

$$\tan\Theta = \frac{y_1 - y_0}{x_1 - x_0},$$

and, hence,

$$\tan\Theta = \frac{\sin\theta\,[r_0^2 - (x_0 \sin\theta - y_0 \cos\theta)^2]^{\frac{1}{2}} + x_0 \sin\theta\cos\theta - y_0 \cos^2\theta}{\cos\theta\,[r_0^2 - (x_0 \sin\theta - y_0 \cos\theta)^2]^{\frac{1}{2}} + y_0 \sin\theta\cos\theta - x_0 \cos^2\theta}. \tag{39}$$

Assuming the Sun to appear as a circle, and the relative coordinates of the Sun and Moon, we can determine the deviations of the lunar profile from a circle. In theory, both the differences in celestial coordinates and the value of r_0 can be determined from the Astronomical Ephemeris. In this case, each negative can be dealt with separately, and the results from each negative combined. The methods to be presently described do not make any assumption concerning the knowledge of the relative coordinates of the two bodies.

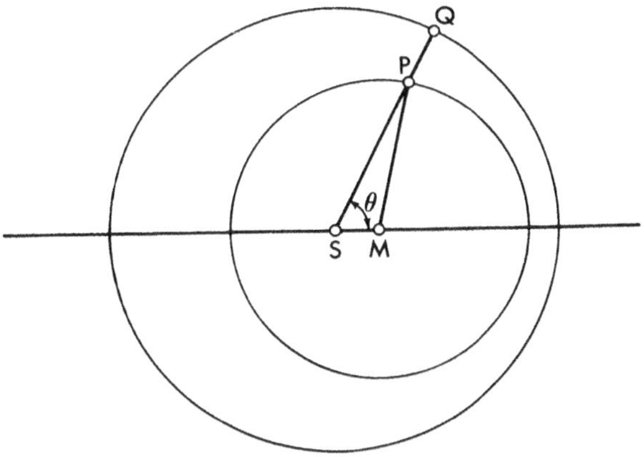

Fig. 19.

Let, in Figure 19, M be the position of the projected center of the Moon, and S, that of the Sun. If so, it evidently follows that

$$(MP)^2 = (SM)^2 + (SP)^2 - 2(MS)(SP)\cos\theta,$$

which can be rewritten as

$$(SP) = (MS)\cos\theta + [(MP)^2 - (MS)^2 \sin\theta]^{\frac{1}{2}}. \tag{40}$$

Remembering that $(SM) = \varDelta$, we can derive the expression of the width of the annulus PQ in the form

$$(PQ) = (R - r) - \varDelta\cos\theta + \frac{\varDelta^2}{2r}\sin^2\theta + \frac{3\varDelta^4}{8r^3}\sin^4\theta. \tag{41}$$

Thus, to a good approximation, as the Moon moves across the solar disc the width of the annulus changes according to a cosine law. As the distance between the centers changes the cosine will exhibit a phase change. We may generalize the formula to

$$\alpha = (PQ) = (R - r) - \varDelta\cos(\theta - \phi) + \frac{\varDelta^2}{2r}\sin^2(\theta - \phi) + \cdots. \tag{42}$$

Thus if, for each negative, we plot the width of the annulus against the position angle we will obtain graphs which resemble a cosine curve whose amplitude and phase varies from frame to frame. The angle ϕ, is, of course, equal to zero when the line joining the two centers is taken as the x-axis.

Equation (42) after elimination of the term containing Δ^2 can be used as one of condition for determination of the mean radius and of the coordinates of the center of the lunar disc. If X^j, Y^j and Z^j are defined by the expressions

$$\left.\begin{aligned} X^j &= \Delta^j \cos \phi^j, \\ Y^j &= \Delta^j \sin \phi^j, \\ Z^j &= -(R^j - r^j), \end{aligned}\right\} \tag{43}$$

where the superscript j indicates the frame number then the width α_i^j of the annulus corresponding to position angle θ_i^j will be related with these quantities by the expression

$$\alpha_i^j + a_i^j X^j + b_i^j Y^j + Z^j = 0, \tag{44}$$

where

$$a_i^j = \cos \theta_i^j, \quad b_i^j = \sin \theta_i^j.$$

By performing n measurements distributed along the entire annulus and applying the least squares technique, we can determine the central and mean radius corresponding to the jth plate.

The sequence of positions of the lunar center must practically define a straight line. In this way, we can determine corrections for the position of the lunar center for each frame and the new coordinates will be denoted by X_i^j and Y_i^j. The next step is to determine the radius r_i^j corresponding to position angle θ_i^j. This can be done by means of the relation

$$r_i^j = \alpha_i^j - R_i^j + a_i^j X^j + b_i^j Y^j. \tag{45}$$

By averaging over j we determine the radius r_i corresponding to the position angle θ_i. We must emphasize here that θ_i^j and θ_i^{j-k} are not even approximately the same due to the variation of the line through the solar and lunar disc and as a result, θ_i is by no means the average of the θ_i^j's, but a position angle measured from an arbitrary direction which is common to all frames.

An obvious alternative to the technique described here would consist of fitting a circle to the lunar limb for each frame, then to refer all to the same origin and arbitrary directions with respect to which the position angles are to be measured and, finally, to obtain the residuals for each position angle and their average over the entire sequence of frames. In order to ensure the good quality of the results obtained by the first technique we independently made use of the second one and the results were then compared. It was found that, for all position angles, the results from the two techniques were compatible within the marginal error 0".15 or 0.26 km, in elevation of limb points. This we consider as satisfactory agreement.

A spectrum analysis of these residuals seems highly useful because this will provide a comparison between the present reductions and results obtained independently and on the basis of different or similar photographic material or even different kinds of observation. Equation (42) shows that the values of the width of the annulus are already in a suitable form for Fourier analysis. Before presenting the results of such an analysis, it is necessary to determine constants defining the particular sections of the marginal zone recorded on the photographs of the two eclipses as well as constants of the eclipses themselves.

The Astronomical Ephemeris and Nautical Almanac give the predicted values of all variables defining both eclipses. Quantities of importance are the geographic coordinates of the site, the exact time of the four contacts (especially the internal ones) the apparent path of the Moon across the solar disc, etc. The determinations of the latitude and longitude of the site, and the measurement of time, are not very critical since, unlike the cases when a determination of the difference between ephemeris time and universal time is intended, the main objective of our expeditions was to procure photographs of the central phase for the measurement of lunar profile.

Precise prediction of the points of contact of the lunar and solar limbs, and a prediction of the apparent path of the Moon across the Sun are also not essential, but certainly desirable, in the reductions. These are required in the calculation of the orientation of the image on the plate, which may be calculated as is to be seen later by other means but it is useful to have an independent confirmation.

The geocentric right ascensions and declinations of the Moon and Sun are tabulated in the Ephemeris; it is necessary only to interpolate between the given values and determine them for the times of observations and then to compute the right ascensions and declinations as seen from the sites of observation in order to find the apparent path of the Moon across the Sun. In this calculation we shall have to allow for the oblateness of the Earth, which enters in the computation of the geographic latitude of the site of observation. Thus if ϕ and ϕ' are the geographic and geocentric latitudes of the site, their relationship is expressed by the formula

$$\tan\phi' = (1 - e^2)\tan\phi . \tag{46}$$

The determination of the topocentric coordinates of a point from its geocentric ones will be made as follows: Let α and δ be the geocentric right ascension and declination of the point in question, and α' and δ' their topocentric values for an observing site whose geocentric distance is ρ. Let also $Oxyz$ be a geocentric frame of reference with the axis Oz coinciding with the axis of ratation and the Ox-axis pointing toward the point of vernal equinox. The rectangular coordinates x, y, z, of the point of geocentric distance Δ will then be

$$\left.\begin{aligned} x &= \Delta\cos\delta\cos\alpha ,\\ y &= \Delta\cos\delta\sin\alpha ,\\ z &= \Delta\sin\delta . \end{aligned}\right\} \tag{47}$$

The coordinates x', y', z', of the same point with respect to a frame of axis parallel to those of the $Oxyz$ frame but with its origin at the observer's site, will be

$$\left.\begin{aligned} x' &= \Delta'\cos\delta'\cos\alpha' ,\\ y' &= \Delta'\cos\delta'\sin\alpha' ,\\ z' &= \Delta'\sin\delta' , \end{aligned}\right\} \tag{48}$$

where Δ' is the distance between the observer and the point. The transformation

$$\left.\begin{aligned} x' &= x - a, \\ y' &= y - b, \\ z' &= z - c, \end{aligned}\right\} \tag{49}$$

can be established, where

$$\left.\begin{aligned} a &= \rho \cos \phi' \cos t \\ b &= \rho \cos \phi' \sin t \\ c &= \rho \sin \phi' \end{aligned}\right\} \tag{50}$$

and t is the sidereal time at the observing station. From Equations (47) to (50) we can establish sufficient relations for the determination of the topocentric coordinates; the final expressions obtained are

$$\tan(\alpha - \alpha') = \frac{\rho \cos \phi' \sin(\alpha - t)}{\Delta \cos \delta - \rho \sin \phi' \cos(\alpha - t)}, \tag{51}$$

$$\tan(\delta' - \delta) = \frac{\rho \sin \gamma \sin(\delta - \gamma)}{\Delta \sin \gamma - \rho \sin \phi' \cos(\delta - \gamma)}, \tag{52}$$

$$\Delta' = \Delta \sin(\delta - \gamma)/\sin(\delta' - \gamma), \tag{53}$$

where

$$\tan \gamma = \frac{\tan \phi' \cos \frac{1}{2}(\alpha' - \alpha)}{\cos[\frac{1}{2}(\alpha' + \alpha) - t]}.$$

These equations make it possible to calculate the relative positions of the Sun and Moon, as seen in the sky, observed from a point on the surface of the Earth. The ephemeris gives values for the geocentric quantities used above in intervals of time of one hour or one day. It is necessary to interpolate between these and in order to obtain accurate interpolations the problem was solved in the computer.

The results of our reduction on the motion of the center of the lunar disc across that of the Sun are plotted in Figures 20 and 21. The first corresponds to the 1962 eclipse, and the second to that of 1963. The coordinates axes x and y are defined with respect to the photographic films and are common for all frames, as we already

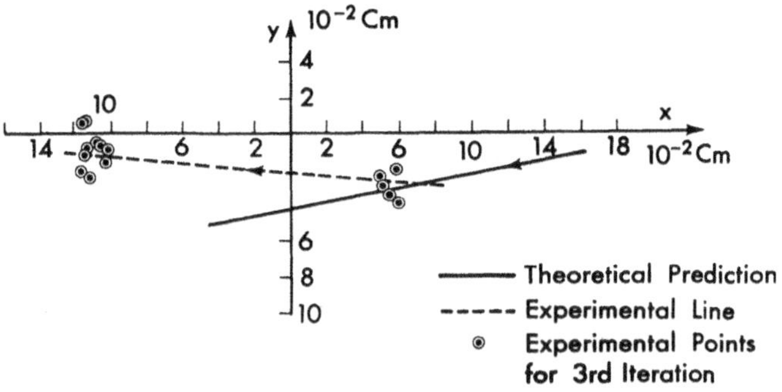

Fig. 20. Motion of the lunar center during the annular eclipse in 1962.

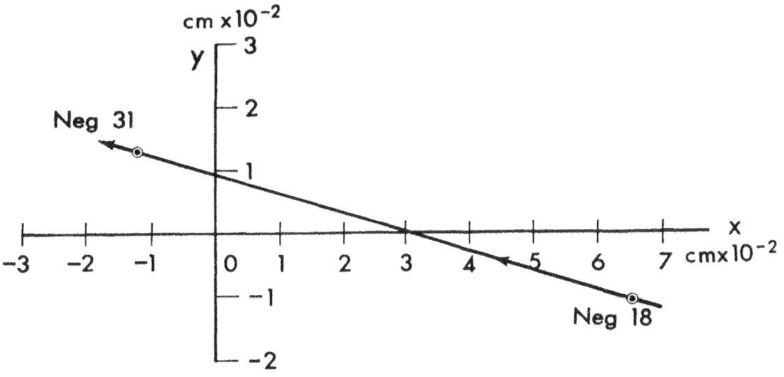

Fig. 21. Motion of the lunar center during the annular eclipse in 1963.

have mentioned. The straight line fitted by least squares to these points in either case is found to be identical within the permissible error of 0\".1 with the predicted lines which were determined from Equations (51) and (52).

The times of the second and third contact of each eclipse were computed as follows: The equation of the path of the lunar center across the solar disc is expressed in terms of the time parameter, t, by the equations

$$x = k_1 t + c_1$$
$$y = k_2 t + c_2$$

where x and y are the coordinates of the lunar center and the origin rests at the center of the Sun. If we denote by t_c the time at which any of the four contacts occurs, then the angle of contact of the limb of the Moon to that of the Sun is

$$\tan^{-1}\left(\frac{x_2 t_c + c_2}{x_1 t_c + c_1}\right) + \pi(2k + 1/2). \tag{54}$$

The U.T. of the second and third contact are found to be as follows: For the 1962 eclipse we had:

Time of second contact = 11h 44m 3.2sec
Time of third contact = 11h 48m 1.1sec

The angles of contact for the same eclipse were found to be 271° for the second contact and 111° for the third. In the case of the 1963 eclipse the equivalents were:

Time of second contact = 14h 48m 41.2sec
Time of third contact = 14h 49m 19.4sec

The angles of contact at the second and third contacts were 257° and 68°, respectively.

We shall consider now the problem of determining the orientation of the lunar image on the photographic film. The case is illustrated in Figure 22, where Z is the local zenith and z the angle of the normal to the film and the meridian plane. We shall make use of the following notation:

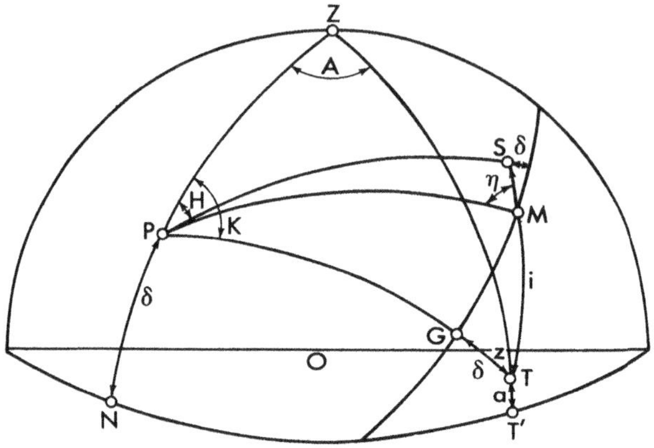

Fig. 22.

P = the North pole, δ = the declination of the Sun, H = the hour angle of the Sun, K = the hour angle of the telescope, A = the azimuth of the telescope, OT = the direction of the optical axis of the telescope, OM = the direction normal to the coelostat mirror, OS = the direction of the Sun, OT' = the direction the telescope would point if the altitude a of the telescope was equal to zero, and i = the angle of incidence of a ray of light from the Sun on the coelostat mirror.

Now we can establish the following relations:

$$-\sin\delta = \sin\phi \sin a + \cos\phi \cos a \cos A, \tag{55}$$

$$\frac{\sin z}{\sin(\pi/2 - \phi)} = \frac{\sin A}{\sin(\pi/2 + \delta)} = \frac{\sin K}{\sin(\pi/2 - a)}. \tag{56}$$

For $a=0$ we have

$$\cos A = -\sin\delta \sec\phi. \tag{57}$$

Therefore, Equation (56) yields

$$\begin{aligned} \sin z &= -\tan\delta \tan A \\ &= -(1 - \cos^2 A)^{\frac{1}{2}} \tan\delta/\cos A \\ &= \frac{(\cos^2\phi - \sin^2\delta)^{\frac{1}{2}}}{\cos\delta}. \end{aligned} \tag{58}$$

The last relation enables us to determine z. But it is now necessary to know the spherical angle SMP which we shall denote by η, and by which the image of the Moon is rotated inside the plane of film due to the fact that the normal to the coelostat mirror is not in the meridian plane. The determination of η can be done by means of the relations

$$\frac{\sin\eta}{\cos\delta} = \frac{\sin(K - H)}{\sin i},$$

$$\frac{\sin(\pi/2 - \eta)}{\sin\delta} = \frac{1}{\sin i}.$$

Therefore,

$$\tan\eta = \cot\delta \sin(K - H)/2. \tag{59}$$

This formula was used for the determination of η in all cases.

For the determination of the particular section of the marginal zone recorded on the eclipse photographs analyzed in the present investigation, we need to know in addition the geocentric zenith distance of the Moon ξ, the parallactic angle (the angle between the vertical and the declination circle through the Moon's center, in other words, the angle ZSP of Figure 22, Q, and the topocentric parallax π', which can be calculated from the geocentric right ascencion α, declination δ, and parallax π, of the Moon by the following formulae:

$$\cos\xi = \sin\phi \sin\delta + \cos\phi \cos\delta \cos H, \tag{60}$$

$$\sin Q = \sin H \cos\phi \operatorname{cosec}\xi, \tag{61}$$

$$\cos Q = \frac{\sin\phi - \cos z \sin\delta}{\sin z \cos\delta}, \tag{62}$$

$$\pi' = \pi(\sin\xi + 0.0084 \sin 2\xi). \tag{63}$$

If the geocentric librations of the Moon are l, b, C (longitude, latitude, and position angle) then the corrections to these, according to ATKINSON (1951), are

$$\Delta l = -\pi' \sin(Q - C) \sec b, \tag{64}$$

$$\Delta b = \pi' \cos(Q - C), \tag{65}$$

$$\Delta C = \sin(b + \delta b)\, \Delta l - \pi' \sin Q \tan\delta. \tag{66}$$

The numerical values of the topocentric libration angles are found to be,

$$\left.\begin{aligned} l' &= 4^\circ\!.58 \\ b' &= -0^\circ\!.17 \\ C' &= 16^\circ\!.32 \end{aligned}\right\} \tag{67}$$

for the 1962 eclipse, and

$$\left.\begin{aligned} l' &= -5^\circ\!.36 \\ b' &= 0^\circ\!.07 \\ C' &= -15^\circ\!.33 \end{aligned}\right\} \tag{68}$$

for the 1963 eclipse. The topocentric libration angles are necessary mainly when one intends to compare the lunar profiles obtained here with those derived by other surveys of the marginal zone, such as the ones of HAYN (1907) and WATTS (1963).

5. Analysis of the Profiles and Comparison with Preceding Results

The numerical values of the elevations along the profile of the 1962 eclipse were analyzed in the same way as those of the 1963 one – the latter case will be presently discussed in detail – but the answers obtained were not compatible with the corresponding data of HAYN (1907), WATTS (1963), YAKOVKIN (1962), DOMMANGET (1962), POTTER (1962), and others. This was anticipated by the extent of the distortion caused by unfavorable weather but mainly by the irradiation in the photographic emulsion. The magnitude of the effect is much larger than the deviations of the profile from a circle which we originally set out to determine. The effect was made serious by overexposure. The same exposure times were used for the South African eclipse photographs but, in the latter case, the annulus was much narrower than in Senegal, and hence the effect was negligible. The effect is difficult to correct because solar limb-darkening makes the construction of simple calibration curves impossible. Non-uniform irradiation impairs the quality of the statistics .The statistics are also deficient because clouds marred many of the photographs of the central phase of the 1962 eclipse.

The profile of the 1963 eclipse is given in Table I, where the standard selenographic coordinates (ξ, η, ζ) of the 62 points are listed. The standard deviation of these points referred to a 1738.0 km mean radius is 2.1 km, and the mean error is 0.44 km or 0".23.

TABLE I

Points of the Profile Obtained from the 1963 Annular Eclipse

ξ	η	ζ	ξ	η	ζ
−0.03490	0.99936	0.00076	0.03491	−0.99959	0.00727
−0.13910	0.98973	−0.00895	0.13928	−0.99105	0.01700
−0.24187	0.97009	−0.01853	0.24177	−0.96969	0.02655
−0.34183	0.93917	−0.02785	0.34173	−0.93888	0.03588
−0.43784	0.89770	−0.03681	0.43735	−0.89671	0.04480
−0.52933	0.84710	−0.04535	0.52882	−0.84629	0.05334
−0.61440	0.78639	−0.05331	0.61389	−0.78574	0.06129
−0.69291	0.71753	−0.06065	0.69247	−0.71707	0.06865
−0.76417	0.64121	−0.06732	0.76364	−0.64077	0.07532
−0.82651	0.55748	−0.07316	0.82605	−0.55718	0.08117
−0.87988	0.46784	−0.07817	0.87996	−0.46789	0.08623
−0.92393	0.37329	−0.08230	0.92341	−0.37308	0.09031
−0.95820	0.27476	−0.08551	0.95764	−0.27460	0.09352
−0.98172	0.17310	−0.08772	0.98167	−0.17309	0.09579
−0.99480	0.06956	−0.08895	0.99150	−0.06933	0.09670
−0.99692	−0.03481	−0.08915	0.99436	0.03472	0.09697
−0.98689	−0.13870	−0.08821	0.98575	0.13854	0.09617
−0.96510	−0.24063	−0.08617	0.96542	0.24071	0.09425
−0.93398	−0.33994	−0.08325	0.93606	0.34070	0.09150
−0.89561	−0.43682	−0.07964	0.89579	0.43690	0.08771
−0.84603	−0.52866	−0.07499	0.84566	0.52843	0.08301
−0.78748	−0.61525	−0.06949	0.78633	0.61435	0.07745
−0.71806	−0.69342	−0.06300	0.71865	0.69399	0.07111
−0.64103	−0.76395	−0.05580	0.64128	0.76425	0.06386
−0.55857	−0.82812	−0.04809	0.55734	0.82629	0.05600
−0.47005	−0.88403	−0.03981	0.46912	0.88228	0.04777
−0.37522	−0.92871	−0.03096	0.37458	0.92712	0.03894
−0.27547	−0.96068	−0.02167	0.27567	0.96137	0.02972
−0.17342	−0.98351	−0.01215	0.17366	0.98488	0.02020
−0.06973	−0.99724	−0.00249	0.06972	0.99711	0.01051

A harmonic analysis of these data was performed with the aim to separate harmonics of orders from zero to four, since the accuracy of the data would not allow the accurate determination of higher terms. For comparison, we used the profiles determined from the surveys of HAYN (1907) and WATTS (1963), which we list in Tables II and III. In order to check the results, the analysis was performed in two different ways. First, a separate Fourier analysis of the profiles presented by Hayn, Watts and the 1963 eclipses was made and the results were compared. Second, each profile was combined with the data of Schrutka-Rechtenstamm, the Army Map Service (AMS) (1964), and the Aeronautical Chart and Information Center (ACIC) (1965). An analysis into spherical harmonics was then applied in each case.

TABLE II

Points of the Marginal Zone (zero libration, $\zeta = 0$) obtained from Hayn's Maps

ξ	η	ξ	η
0.000000	.999571	0.000000	-1.000785
-.087155	.996194	.087155	-.996194
-.173666	.984913	.173890	-.986180
-.258846	.966029	.259124	-.967065
-.341983	.939591	.342020	-.939692
-.422482	.906016	.422573	-.906210
-.499892	.865839	.500107	-.866211
-.573453	.818976	.573760	-.819415
-.642718	.765962	.642925	-.766208
-.707334	.707334	.706879	-.706879
-.765798	.642580	.765633	-.642442
-.818976	.573453	.818888	-.573391
-.865468	.499678	.865932	-.499946
-.906016	.422482	.906307	-.422618
-.939491	.341946	.939793	-.342056
-.965925	.258819	.966547	-.258985
-.985124	.173704	.985441	-.173759
-.996621	.087193	.995981	-.087137
-1.000214	0.000000	.999035	0.000000
-.996515	-.087183	.995340	.087080
-.984702	-.173629	.984174	.173536
-.965407	-.258680	.965718	.258763
-.938483	-.341580	.939994	.342130
-.906016	-.422482	.906113	.422527
-.866118	-.500053	.866396	.500214
-.819591	-.573883	.819503	.573822
-.766701	-.643338	.766537	.643201
-.707182	-.707182	.707789	.707789
-.642511	-.765715	.642994	.766290
-.573268	-.818712	.573207	.818625
-.500160	-.866303	.499034	.864353
-.423071	-.907279	.422437	.905919
-.342130	-.939994	.342130	.939994
-.258819	-.965925	.258846	.966029
-.173573	-.984385	.173666	.984913
-.087071	-.995233	.087174	.996408

TABLE III

Points of the Marginal Zone (zero libration, $\zeta = 0$) taken from the maps by Watts

ξ	η	ξ	η
0.000000	.999785	0.000000	-1.000857
-.087146	.996087	.087118	-.995767
-.173611	.984596	.173927	-.986391
-.258763	.965718	.258791	-.965822
-.341910	.939390	.342093	-.939894
-.422618	.906307	.422165	-.905335
-.500000	.866025	.499946	-.865932
-.573268	.818712	.573760	-.819415
-.642925	.766208	.642649	-.765880
-.707334	.707334	.706955	-.706955
-.765798	.642590	.765387	-.642236
-.818888	.573391	.818185	-.572899
-.865653	.499785	.865746	-.499839
-.906016	.422482	.906502	-.422708
-.939894	.342093	.940297	-.342240
-.965615	.258735	.966547	-.258985
-.984913	.173666	.985863	-.173834
-.997369	.087258	.994806	-.087034
-1.000107	0.000000	.997104	0.000000
-.997476	-.087267	.994485	.087006
-.984913	-.173666	.984068	.173517
-.964579	-.258458	.964993	.258569
-.939390	-.341910	.939491	.341946
-.906307	-.422618	.906890	.422890
-.866861	-.500482	.865839	.499892
-.820820	-.574744	.819503	.573822
-.766783	-.643407	.766783	.643407
-.707485	-.707485	.708016	.708016
-.642580	-.765798	.642443	.765633
-.573453	-.818976	.573207	.818625
-.500750	-.867325	.499142	.864539
-.423071	-.907279	.422890	.906890
-.342790	-.941808	.341946	.939491
-.258930	-.966340	.258930	.966340
-.173387	-.983329	.173648	.984807
-.087081	-.995340	.087127	.995874

The results were then used in order to determine the values of the low-order Fourier terms of the limb. In principle, the latter terms must be independent of the front side control system with which the limb data are combined; but due to the finite number of points available for both areas, the values obtained for the same profile differed by amounts smaller than the mean error of the profile. For this reason, we shall present below the analytical form of the profile as obtained by means of the harmonic analysis of the entire figure. In such a case the radius r of the surface points whose selenographic coordinates will be denoted by λ, β will be given by the expression

$$r = r_0 + \sum_{i=0}^{\infty} \sum_{j=0}^{i} (J_{ij} \cos j\lambda + J'_{ij} \sin j\lambda)\, P_j^i(\mu) \tag{69}$$

where $r_0 = 1738.0$ km and $\mu \equiv \sin\beta$. The equation of the profile for zero librations is then

$$\begin{aligned} r = {} & r_0 + J_{00} + \tfrac{1}{4}J_{20} - \tfrac{3}{2}J_{22} + \tfrac{9}{64}J_{40} - \tfrac{45}{16}J_{42} + \tfrac{315}{8}J_{44} \\ & + (J_{10} + \tfrac{3}{8}J_{30} - \tfrac{15}{4}J_{32}) \sin\beta + (J'_{11} + \tfrac{3}{8}J'_{31} - \tfrac{45}{4}J'_{33}) \cos\beta \\ & + (\tfrac{3}{2}J'_{21} + \tfrac{5}{8}J'_{41} - \tfrac{105}{4}J'_{43}) \sin 2\beta \\ & + (-\tfrac{3}{4}J_{20} - \tfrac{3}{2}J_{22} - \tfrac{5}{16}J_{40} + \tfrac{15}{4}J_{42} + \tfrac{105}{2}J_{44}) \cos 2\beta \\ & + (-\tfrac{5}{8}J_{30} - \tfrac{15}{4}J_{32}) \sin 3\beta + (-\tfrac{15}{8}J'_{31} - \tfrac{15}{4}J'_{33}) \cos 3\beta \\ & + (-\tfrac{35}{16}J'_{41} - \tfrac{105}{8}J'_{43}) \sin 4\beta + (\tfrac{35}{64}J_{40} + \tfrac{105}{16}J_{42} + \tfrac{105}{8}J_{44}) \cos 4\beta \end{aligned} \tag{70}$$

where β is no longer the selenographic latitude but a position angle varying counter-clockwise between 0° and 360° from the positive direction of the 0ξ-axis of the standard frame of reference. By substituting in Equation (70) the numerical values of the coefficients determined from the profile of the 1963 eclipse and express the obtained coefficients in geometric seconds of arc, we find the following expression for the lunar profile:

$$\begin{aligned} r = r_0 & - 0''.21 \sin\beta - 0''.11 \cos\beta \\ & + 0''.54 \sin 2\beta - 0''.16 \cos 2\beta \\ & + 0''.27 \sin 3\beta - 0''.05 \cos 3\beta \\ & - 0''.16 \sin 4\beta + 0''.11 \cos 4\beta . \end{aligned} \tag{71}$$

The profile obtained by the analysis of the data obtained from HAYN's maps (1907), is expressed similarly by the equation

$$\begin{aligned} r = r_0 & - 0''.21 \sin\beta + 0''.05 \cos\beta \\ & + 0''.38 \sin 2\beta - 0''.38 \cos 2\beta \\ & + 0''.32 \sin 3\beta - 0''.11 \cos 3\beta \\ & - 0''.05 \sin 4\beta + 0''.16 \cos 4\beta . \end{aligned} \tag{72}$$

A third expression of the same profile is obtained by the analysis of the data obtained from the recent survey of the marginal zone by WATTS (1963). Its actual form is

$$\begin{aligned} r = r_0 & - 0''.32 \sin\beta - 0''.16 \cos\beta \\ & + 0''.59 \sin 2\beta - 0''.21 \cos 2\beta \\ & + 0''.27 \sin 3\beta - 0''.05 \cos 3\beta \\ & - 0''.32 \sin 4\beta + 0''.21 \cos 4\beta . \end{aligned} \tag{73}$$

YAKOVKIN (1962) has suggested that the lunar profile can be described by a model whose expansion in Fourier series is (GOUDAS, 1965)

$$\begin{aligned} r = {} & r_0 + \frac{3a}{16} - \frac{a}{4} \cos 2(\beta + \gamma) + \frac{a}{16} \cos 4(\beta + \gamma) \\ & - \frac{48a}{\pi} \sum_{j=1}^{\infty} \frac{\sin (2j+1)(\beta+\gamma)}{(2j-3)(2j-1)(2j+1)(2j+3)(2j+5)} . \end{aligned} \tag{74}$$

The numerical values of the constants involved have been determined by Yakovkin from a long series of observations of star occultations as well as measurements of photographs. In particular, he concludes that

$$a = 0''.96 + 0''.08b$$

for the eastern half of the limb and

$$a = 0''.96 + 0''.02b$$

for the western half. The phase constant is found to be 15°. Thus, an expression of the zero libration profile similar to these given by Equations (71), (72), and (73) is obtained:

$$\begin{aligned} r = r_0 &+ A'_1 \sin\beta + B'_1 \cos\beta \\ &+ 0''.17 \sin 2\beta - 0''.30 \cos 2\beta \\ &+ 0''.44 \sin 3\beta + 0''.44 \cos 3\beta \\ &- 0''.05 \sin 4\beta + 0''.08 \cos 4\beta . \end{aligned} \tag{75}$$

The coefficients A'_1 and B'_1 need not be discussed here.

From the figures quoted by POTTER (1962) we can present the limb in a Fourier form as follows:

$$\begin{aligned} r = r_0 &+ A''_1 \sin\beta + B''_1 \cos\beta \\ &+ 0''.36 \sin 2\beta - 0''.13 \cos 2\beta . \end{aligned} \tag{76}$$

This expansion corresponds to libration angles $l = 2^\circ.5$ and $b = -1^\circ.4$ and, therefore, to a profile different than all others given above. Nevertheless, there is little difference between it and the profile corresponding to zero libration. The constants A'' and B'', need not be considered here, also.

From Equations (71) to (76) we find the major axis of the elliptical component of the limb at zero libration and the axis of rotation of Moon make an angle between 15° and 37° measured clockwise from the North Pole. The lower limit given by YAKOVKIN (1962) is based on an assumption about the shape of the Moon and, of course, measurements. This assumption, which is undoubtedly unnecessary here, seems to have affected considerably the estimate of the above angle as well as the axes of the elliptical component of the limb. If we eliminate the estimates by Yakovkin and Hayn (23°), we find that the recent measurements are in almost perfect agreement. In particular, we have estimates as follows:

Potter (1962):	34°
Watts (1963):	35°
Present determination (1965):	37°

DOMMANGET (1962) gave also similar values, which correspond to libration angles different than zero.

References

Aeronautical Chart and Information Center: 1965, *Tech. Report No. 15*, U.S. Air Force, St. Louis, Mo.

Army Map Service: 1964, *Tech. Report 29*, Part Two. Corps of Engineers, U.S. Army, Washington, D.C.

Atkinson, R.: 1951, *Mon. Not. Roy. Astr. Soc.* **111,** 448.

Carson, D., Davidson, M. E., Goudas, C. L., Kopal, Z., and Stoddard, L. G.: 1966, *Icarus* **5,** 334.

Dommanget, J.: 1962, *Comm. Obs. Roy. Belg., No. 208.*

Goudas, C. L.: 1965, *Icarus* **4,** 218–220.

Hayn, F.: 1907, 'Selenographische Coordinaten, III', *Abhandl. d. K. Sächs. Ges. Wiss., Math.-Phys. Kl.* **30,** 1–105.

Nefediev, A.: 1957, *Bull. Engelhardt Obs. No. 30.*

Potter, H. I.: 1962, in: *The Moon* (Z. Kopal and Z. K. Mikhailov, eds.). Academic Press, London and New York, pp. 63–66.

Selwyn, E. W. H.: 1950, *Photography in Astronomy.* Eastman Kodak Co., p. 66.

Watts, C. B.: 1963, *The Marginal Zone of the Moon* (Astronomical Papers 17, Astronomical Ephemeris and Nautical Almanac). U.S. Govt. Printing Office, Washington, D.C.

Weimer, Th.: 1952, *Atlas de Profils Lunaires*, Obs. de Paris.

Yakovkin, A. A.: 1962, in: *The Moon, A Russian View* (A. V. Markov, ed.). Univ. of Chicago Press, Chicago, Ill., pp. 3–53.

THE USE OF THE 48-INCH SCHMIDT TELESCOPE FOR SELENODETIC OBSERVATIONS

ALBERT G. and DONNA S. WILSON*

Photographing the Moon against star-field backgrounds is recognized as a most useful technique for measuring the difference between universal and ephemeris times, determining various higher-order motions of the Earth and Moon, and investigating certain geodesic problems. Because simultaneous photographic observations of the Moon and stars might be effectively used with new reduction techniques developed by Kopal for determining lunar librations from star positions independent of reference to the lunar limb, Kopal recently requested that we investigate the feasibility of utilizing the 48-inch Palomar Schmidt as a moon-star camera. The Schmidt is rarely in use when the Moon is above the horizon and good supplementary use could be made of the instrument if it should prove adaptable for lunar work.

The problem of photographing the Moon against its star-field background is complicated by the motion of the Moon, the large difference in brightness between Moon and stars, and the brightness of the sky in the neighborhood of the Moon. It is also necessary that the Moon and stars be photographed so that the position of the telescope when the Moon is exposed be the same as the average position for the stars in order to minimize refraction and off-axis effects.

A highly successful, but sophisticated, camera for making simultaneous photographs of the Moon and a star field has been developed by MARKOWITZ (1954, 1962). In this camera the differential motion of the Moon is corrected by continuously changing the tilt of a plane parallel glass filter. An exposure of from about 10 to 25 seconds is made of the star field with the Moon being tracked by the changing tilt filter. The filter drive is set so that the filter is parallel to the photographic plate at exactly mid-exposure. At the instant the filter is parallel, the Moon image is exposed, being recorded in its correct position with respect to the stars.

In considering the Schmidt for this problem, it was felt that since the $f/2.5$ speed allows the recording of 12th-magnitude stars (under dark sky conditions) in about 5 seconds, that the Moon-motion problem could be surmounted by driving the telescope at lunar rate and making a very short exposure. The star images would be only slightly trailed being still quite useful as fiduciary marks. The Schmidt has no declination rate control but for the short exposure, this should introduce very little blurring of the Moon's image.

* Douglas Advanced Research Laboratory, Huntington Beach, Calif., U.S.A.

The Moon-brightness problem was to be solved by introducing a neutral filter having a dense central circle sufficiently large to mask the lunar image. Mr. George Kocher prepared a set of experimental filters by exposing 5 × 7 103a-O plates in contact with a template having a suitable central hole. The scale of the Schmidt is only 69″ arc/mm and marginal at best for recording lunar features. Only the sharpest of properly exposed images could be of use for libration measures.

The sky problem could be handled by photographing the star field through a suitable filter. However, since this extends the exposure time and the star images are trails, the usual gains from reducing sky brightness cannot be effected.

Time was secured on the nights of May 31 and June 1 through the courtesy of the Director of the Mt. Wilson-Palomar Observatories to make a feasibility study on full moon. The 5 × 7 plate holder with field flattening lens was employed. The chief limitation to the experiment was the Schmidt shutter which has an operating time of three seconds and a minimum effective exposure time of some 4 to 5 seconds. The mounting of an auxiliary shutter would constitute a major and expensive modification for which neither time nor money was available.

The results were highly unpromising. It was not possible to record the lunar surface and stars on the same plate. The intense brightness of the sky in the neighborhood of the Moon, although the transparency was good, vitiated the attempt.

The conclusions of the experiment are that the Schmidt is not useful as a Moon-camera without extensive and expensive modifications. While it may be possible to adapt the Schmidt for simultaneous moon-star field photography through diaphragms to change the *f* ratio, building a system of dual shutters, introducing suitable combinations of filters, and a declination rate drive, none of this seems worthwhile. Since telescopes built for one purpose are rarely adaptable to quite different purposes, we recommend development of instruments following the successful designs of Markowitz.

References

Markowitz, W.: 1954, *Astron. J.* **59**, 69–73.

Markowitz, W.: 1962, in *Telescopes* (ed. by G. P. Kuiper and B. M. Middlehurst). Univ. of Chicago Press, Chicago, pp. 107–114.

LASER MEASUREMENT OF EARTH-MOON DISTANCE AND SELENOGRAPHY*

A. ORSZAG**

1. Introduction

The possibility of obtaining radar echoes on the Moon is known to have been envisaged as early as 1947, and echoes were in fact observed in 1949. Since then, there has been practically no interruption in observation of the Moon by radar, which has brought us a great deal of important information, including, in particular, a new lunar parallax value (cf. YAPLEE *et al.*, 1965).

After the apparition of the laser, it therefore seemed natural to substitute a spatially coherent light source for the Hertzian wave source, also coherent, represented by the radar transmitter.

Several trials have already been made to this effect since the first successful attempt by SMULLIN and FIOCCO (1962), using a free-running laser, and Kokurin's much more recent 1965 experiment, this time using a Q switch laser (cf. KOKURIN *et al.*, 1966). It must be noted that the latter, which alone emits a pulse short enough to allow of precise distance measurement, disposes of only a low energy – a few joules maximum. Consequently, scarcely more than 1 to 2 echo photons per m^2 of receiving surface are picked up in return.

The detection of such a signal is of course very difficult, all the more so as the light received is situated at 0.7 micron†, a wavelength for which the photocathodes present a very limited yield. This means that in order to register the echo for certain at each shot, one would need a collecting surface of 20 to 30 m^2.

The construction of such a surface appears to be extremely expensive, even if one adopts tolerances far wider than those usual for telescope construction. The use of a solar furnace is likewise excluded, for the angular separating power of available furnaces is insufficient. We are thus brought to consider using a collector of lesser

* This project is supported by the following organizations: CNES, CNET, DGRST, DRME and the Committee for Laser Atmospheric and Space Applications, presided by Professor P. Grivet. The work has been undertaken at the Ecole Polytechnique and at the Pic-du-Midi Observatory, under the direction of Professor Rösch.

** Ecole Polytechnique, Paris, France.

† This wavelength is the most favorable. The light of the neodymium laser, emitting at 1.1 micron, is largely absorbed by the atmosphere. Moreover, the quantum yield at this wavelength is 20 to 30 times less than at 0.7 micron. It might later be interesting to double the frequency of the neodymium laser, which would give an appreciable power in the vinicity of 0.55 micron.

dimensions. It will then be necessary to correlate the results of several successive laser shots in order to be sure of picking up and identifying the echo. This leads us to treat the information by means of techniques similar to those sometimes used in radar detection, for example.

2. Aims of the Experiment

The chief characteristic of laser utilization is the possibility of measuring distance from a given point of the lunar ground. Consequently, besides measurements already taken by radar (measurement of the lunar radius, of distance to the center of gravity of the Moon and calculation of the lunar parallax; cf. YAPLEE *et al.*, 1965) we can also envisage taking simultaneous measurements from a single point on the Earth to different spots of the lunar surface in order to reach the physical libration of our satellite, or, inversely, measurement of distance from several points on the Earth to a given spot on the Moon. This would lead to geodetic applications (BENDER,1966).

Let us note in passing that both these categories of determination would be made far easier – and above all more precise – by the use of retroreflectors set on the surface of the Moon (ALLEY *et al.*, 1965).

Lastly, the multiplication of such localized measurements over numerous spots covering a fair portion of the lunar surface will enable us to obtain a polar representation of this surface, the pole being the Observatory. This representation would complete existing photographic or visual maps.

We may envisage adding to these various experiments, essentially devoted to the geometrical study of the Moon and the Earth-Moon system, such other determinations as, for example, the use of Earth and Moon together as a gravitation wave detector (cf. ALLEY *et al.*, 1965), the measurement of the density of charged particles in the space between Earth and Moon by simultaneous radar and laser measurements, etc. These experiments should also lead to improvements of the Hill-Brown ephemeris of the Moon, as envisaged already by Eckert (ECKERT *et al.*, 1966).

3. Description of the Equipment

The equipment described as follows is being installed at the Pic-du-Midi Observatory in the Pyrenees. It is hoped that the high altitude of this observatory will free us, as far as possible, from the absorption and diffusion due to the lower strata of the atmosphere.

TRANSMITTING APPARATUS

This equipment includes an optical system of which the final element will be the 105 cm telescope recently installed in the Observatory. The light of the laser, placed behind the mirror, will be reflected onto it by means of a Cassegrain fitting. The laser, triggered by a rotating mirror, will deliver pulses of an energy of 3 joules every 4 seconds; the laser beam divergency will be of about 1.5 minutes.

RECEIVING APPARATUS

This apparatus has to serve in the two following capacities: (1) to receive sufficient energy to permit detection, (2) to ensure the elimination of stray light, so that the signal/noise ratio permits echo identification.

We have thus been led to plan the construction of an optical system with a surface of 10 m^2 and an angular separating power of 10 seconds (ORSZAG, 1965).

Because of the difficulties involved in the construction of a single mirror with a diameter of over 3 m and the required separating power, we are trying to obtain instead a collector made up of small elements which can be made and adjusted independently (RÖSCH, 1966).

Then again, the stricter alignment tolerance imposed by the use of mirrors rather than lenses, as well as the need, in telescope-type assemblies, for a second set of mirrors to send the image back to the rear of the instrument, have brought us to prefer a dioptric system which will consist of a certain number of refracting telescopes juxtaposed on a common support.

For reasons of cost and production in series, a unit-diameter of 25 cm has been adopted. The optical assembly of each refracting telescope has been so designed as to reduce as far as possible the diffused light within the instrument. The aperture of each telescope will be of $f/10$, and the whole collector will thus be composed of 200 refracting telescopes, aperture 25 cm, juxtaposed on a support.

A prototype of one of the refracting telescopes has been assembled at the Pic-du-Midi Observatory. It has been used, in particular, to measure the diffused light from the crescent Moon.

The constituent elements of the collector are thus already well defined. But its financing is not yet ensured. That is why, at the first stage*, we contemplate using for reception the 105 cm telescope which also serves for transmission. Deflection between paths of transmission and reception will be effected by means of a mobile mirror. Such an installation will suffice for taking measurements, if a retro-reflector is used.

References

ALLEY, C. O., BENDER, P. L., DICKE, R. H., FALLER, J. E., FRANKEN, P. A., PLOTKIN, H. H., and WILKINSON, D. T.: 1965, *Journ. Geophys. Res.* **70**, 2267.

BENDER, P. L.: 1966, 'Geophysical Observations Obtainable from Lunar Distance Measurements', *Journ. Geophys. Res.* (in press).

ECKERT, W. J., WALKER, H. J., and ECKERT, D.: 1966, *Astron. J.* **71**, 314.

KOKURIN, Yu. L., KURBASOV, V. V., LOBANOV, V. F., MOZHZHERIN, V. M., SUKHANOVSKI, A. N., and CHERNYKH, N. S.: 1966, *Zhur. Exper. Theor. Fiziki* (Pisma) **3**, 219.

ORSZAG, A. G.: 1965, *Radio Science* **69**, 1681.

RÖSCH, J.: 1966, private communication.

SMULLIN, L. D. and FIOCCO, G.: 1962, *Nature* **194**, 4835.

YAPLEE, B. S., KNOWLES, S. H., SHAPIRO, A., CRAIG, K. J., and BROUWER, D.: 1965, *I.A.U. Sympos. No 21* (*Astron Bull.* **25**, 81).

* A term of two years being required for the construction of the 10 m^2 collector, this could be put into service in 1969.

INVESTIGATING THE MOON'S MOTION BY LASER RANGING*

RENNE S. JULIAN**

Introduction

Radar methods have for the past two decades been applied with considerable success to the study of the Moon [1] and planets. The resulting distance determinations have been especially interesting because radar measures distance (albeit in light-seconds rather than in meters) more directly than and independently of the classical astronomical triangulation methods. In addition to yielding precise distances, detailed examination of the reflected energy has revealed facts about the nature and motion of the reflecting objects not obtainable in any other way. This has been true even though the radar beams used have been too broad to appreciably resolve even the Moon. Success has resulted rather from the skillful application of pulse and Doppler data processing techniques. The possibility of extending radar astronomy to optical wavelengths, where angular resolution characteristic of optical astronomical telescopes might be expected, attracted attention almost as soon as the invention of the laser light source suggested such a possibility. This extension is now becoming a practical reality.

As soon as the first ruby laser was operated in 1960, and especially after Q-switching was described [2], it was clear that this new device would in most fundamental respects be an ideal transmitter for an optical radar. Besides the efficiency with which the light from a laser could be focussed because of its spatial coherence, and the ease with which it could be discerned through solar background light because of its time coherence (monochromaticity), the very short pulse produced by Q-switching would permit excellent range precision. The phenomenal spectral radiance resulting from these characteristics of a Q-switched laser as compared with other light sources may be cited as something like 10^{14} watts per square centimeter per steradian per degree Ångstrom, while the corresponding number for the surface of the Sun is less than unity. The potential advantages of the laser were soon demonstrated in practice, and by now many rangefinders operating over terrestrial distances of a few miles have been engineered for military and surveying purposes.

The possibility of ranging to the Moon was first demonstrated by SMULLIN and

* This report is based on a study sponsored by the Geodesy Branch of the Terrestrial Sciences Laboratory, Air Force Cambridge Research Laboratories through contract AF 19(628)-5172.

** Hughes Aircraft Company, Culver City, Calif., U.S.A.

FIOCCO [3] in 1962. They accomplished a somewhat marginal detection of laser light reflected from the Moon, but made no attempt to determine range because the poor precision obtainable with the long pulse they were using would have made the result uninteresting. Later a similar demonstration was made in the U.S.S.R. [4] with very similar results.

The signal-to-noise ratio of these early lunar experiments was low mainly because it was necessary to accept the background noise over an interval at least as long as the laser pulse, which was of the order of one millisecond. On the other hand, the pulse duration of a laser using Q-switching could be as short as 10^{-8} seconds, from which it was clear that considerably better signal-to-noise ratio, as well as interesting range accuracy, would be possible with this improvement. Several proposals to this effect have been published [5, 6, 7] and recently KOKURIN *et al.* have presented preliminary results of successful ranging to a crater near the center of the Moon's disc with a Q-switched ruby laser [8]. As expected, this result showed much improved signal-to-noise ratio and range precision over earlier experiments.

At the same time interest in accurately locating artificial earth satellites has led to the use of lasers for that purpose. Satellites carrying optical retroreflectors* have been detected both photoelectrically [9] and photographically [10] in laser illumination, and experiments with a combined laser ranging and camera system capable of giving precise position data have been reported [11]. Such a system, when fully developed, will provide a powerful tool for studying the motion of optically augmented satellites.

The effectiveness of an optical retroreflector in increasing the received signal has led to the suggestion that such reflectors be carried to the Moon by spacecraft, where they would remain as passive beacons available for studying the Moon and its motion by laser ranging from the Earth [12]. This proposal followed a previous suggestion [13] that radio beacons be placed on the Moon to facilitate observation of its motion, a suggestion that still has merit, although reflectors have the advantage of being passive. A distribution of such reflectors, once enplaced, would probably remain operational for a long time.

Things that might be accomplished with a high resolution optical radar powerful enough to do precision ranging to the lunar surface include improved measurements of the size of the Moon and its orbit, a determination of the gross figure of the Moon with more assurance than it is now known, and a measurement of the coordinates of earth based observing stations by methods independent of geodetic surveys.

A network of reflectors on the Moon would greatly facilitate and improve these measurements and would make several additional programs feasible. Besides increasing the number of reflected photons available for ranging, retroreflectors would be observed as point targets in angle as well as in range. Extended observations on the reflectors would result in their positions on the Moon and their motions in space becoming very precisely known. The network would then be available as a set of control points both for lunar mapping and for the navigation of spacecraft in the

* Devices such as cube-corner reflectors (three mutually perpendicular plane mirrors), which according to geometric optics reflect back the incident light in exactly the direction from which it came.

vicinity of the Moon. Continued range measurements would establish the motion of the Moon to a new degree of accuracy which would stimulate the development of orbit theory and might supply new information important to gravitational or cosmological physics.

Not only the orbital motion, but by differential ranging the librational motions also, would be sharply observable. The study* on which the present report is based has shown that by differential laser ranging on two properly located reflectors the current uncertainity in the physical libration parameters of the Moon could be reduced by about three orders of magnitude.

The present report considers the constraints within which any optical moon ranging system must operate and presents from an engineering viewpoint the applicable physical data. Where specific system characteristics are considered attention is given mostly to operation with retroreflectors on the Moon because this very promising method has not been previously examined extensively. Since the environmental constraints will not change as laser technology evolves, it may be hoped that this report will remain a useful introduction to the subject.

Laser Ranging

The setting in which a laser moon ranging system must operate, and an abbreviated block diagram of the essential parts of such a system, are illustrated in Figures 1 and 2. As a framework to support the later detailed discussion the operation of this system will be briefly described.

A pulse of light from a Q-switched laser is projected onto the Moon by an astronomical telescope. With present technology the light-packet incident on the Moon will resemble a very thin pancake moving broadside, perhaps three meters thick and a mile in diameter. A receiving telescope, which may be the same as the transmitter, collects a minute fraction of the light scattered from this packet either by the lunar surface or

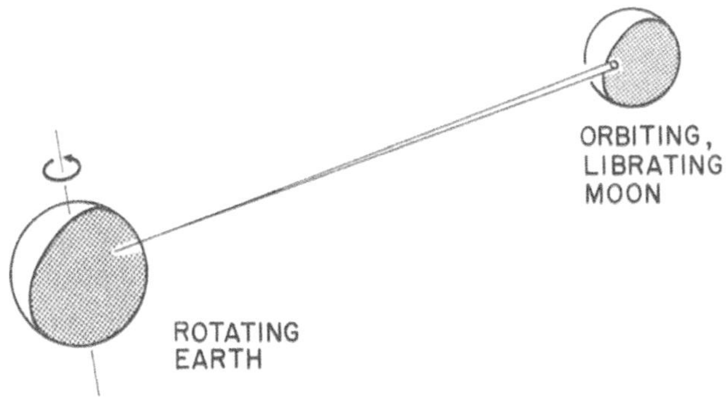

Fig. 1. Laser ranging to the Moon.

* R. S. JULIAN, 'Feasibility Study of Laser Ranging Systems'. Final Report, Hughes Aircraft Co., AFCRL-66-503, June 1966. The libration analysis was by P. Wong, M. Horstein, and J. McFerson.

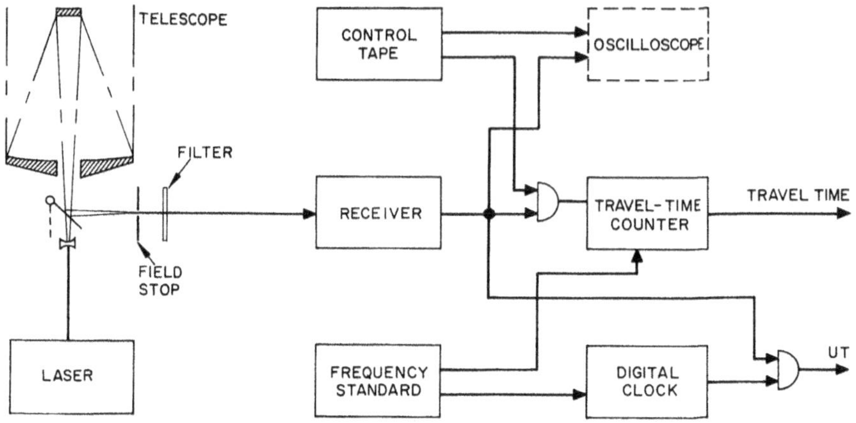

Fig. 2. Abbreviated system diagram.

by a retroreflector, and delivers it through a filter and field stop to a photomultiplier. Here the few photoelectrons released by the collected photons are amplified to an observable electrical pulse. The time interval between the initial flash and the return of the reflected echo, both of which flashes are detected by the same photomultiplier, is measured against a precision frequency standard. The number of seconds elapsed is apparently twice the distance in light-seconds between the observatory and the reflector at an instant near the center of the flight interval. The measured travel time and the instant of laser firing in UT, as determined by a precision electronic clock synchronized with standard radio time signals, are recorded digitally for later processing. A paper control tape punched with a precomputed travel time ephemeris controls an electronic gate which allows only those pulses near the expected instant of echo return to be counted.

The physical realization and practical operation of such a system is subject to many constraints imposed by basic physics, by limitations of technology, and by the photometry and motion of the Moon. Perhaps the most far reaching difficulty is that even under the best circumstances the detected echo will consist of only a few photoelectrons, which must be recognized as the signal in the inevitable presence of some background of random disturbances. In practice the noise generated within the electronic equipment can be made negligible compared with that caused by unwanted photons of scattered sunlight reaching the detector.

With reflectors on the Moon stray light will be a limiting factor only when the background surface is sunlit (or else very near new moon), but without reflectors the background of scattered sunlight not only complicates the equipment but severely restricts the operation of the system. Figure 3 illustrates the problem This figure shows how the background radiance at the center of the lunar disc varies with the phase of the Moon. Because of undesirable range spreading when the laser beam strikes the surface obliquely the region near the center of the disc is particularly important for laser studies. With existing (or anticipated) lasers and no reflectors on the Moon

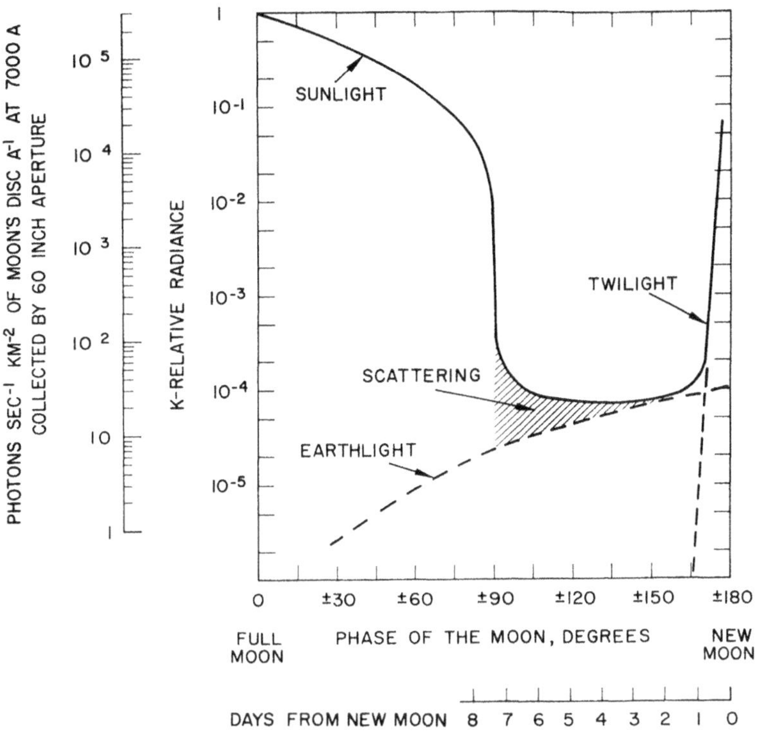

Fig. 3. Background radiance at the center of the Moon's disc.

operation on this central point is feasible only during first and third quarters but not too near new moon. As may be further appreciated by considering Figure 1, the center of the disc can be observed at small zenith angles only during the still more restricted periods just before first quarter and just after third quarter. With effort useful measurements of the Moon's motion will no doubt be obtained without the benefit of reflectors but the installation of reflectors will permit much more complete observation.

Radiometry

The magnitude of the background noise compared to the available signal will largely determine the practicality of the laser for studying the Moon, and predicting this comes down mainly to understanding the radiometry of the Moon as observed by telescope through the Earth's atmosphere. This section reviews lunar radiometry as it affects laser ranging.*

The optical properties of astronomical bodies are usually described in photometric terms lumens, candles, etc. –, but since lasers are highly monochromatic, radiometric quantities are much to be preferred, and will be used here. The symbols which will

* If desired this somewhat tutorial summary may be skimmed quickly and consulted later as specific questions arise.

be used to designate the basic radiation quantities are

- P for radiant power (watts),
- J for radiant intensity from a point source (watts sterad^{-1}),
- N for radiance of an extended source (watts sterad^{-1} meter^{-2}),
- H for irradiance incident on a surface (watts meter^{-2}).

The radiation involved may be the total radiation at all wavelengths, or that within some smaller specified range.

PLANETARY RADIOMETRY

The radiometry of the Moon and Earth are parts of the larger subject of planetary radiometry generally [14]. Familiarity with some of the concepts and data present in the literature of that subject is necessary to an understanding of any optical lunar ranging system.

The radiation scattered by an entire planet or satellite considered as a point is given by

$$J = Ha\frac{p}{\pi}\phi(\alpha), \tag{1}$$

where J is the radiant intensity of the planet at the angle α, H is the irradiance of the planet by the Sun or other source, a is the cross-sectional area of the planet's disc normal to the incident radiation, p is called the geometric albedo of the planet, α is the phase angle between the source and the observer measured at the planet, and $\phi(\alpha)$ is called the phase function.

This equation, with the additional stipulation that ϕ is unity at zero phase angle, simply defines the geometric albedo and the phase function. These two parameters, which may be functions of wavelength, completely describe the scattering at large distances from a spherically uniform planet. If the planet surface is not uniform, values of p and ϕ averaged over all planetary orientations may be useful. Since ϕ is unity at $\alpha=0$, p specifies the back-scattering, and ϕ describes the angular dependence of the scattering. The π factor in the denominator puts the formula in such form that if the scattering object under consideration were a flat disc oriented normal to the incident radiation, and if this disc scattered all the incident radiation into a Lambert law distribution, then p would be unity. This form is possibly the cause of occasional statements that the equation assumes Lambert's law to hold for the planet. Actually (1) makes no such assumption, but is a perfectly general description.

The total power P scattered by the planet in all directions is the integral of J over all solid angle, expressed as

$$P = \int J\,\mathrm{d}\Omega. \tag{2}$$

Putting J from (1), and $2\pi \sin\alpha\,\mathrm{d}\alpha$ for $\mathrm{d}\Omega$, into this equation gives

$$P = Hap\int_0^\pi 2\phi(\alpha)\sin\alpha\,\mathrm{d}\alpha. \tag{3}$$

The integral in (3), called the phase integral, is another parameter characteristic of the planet and is usually designated by q;

$$P = Hapq. \tag{4}$$

The word albedo in astronomy text-books usually refers to the spherical or Bond albedo A, which is the fraction of incident radiation scattered over all directions, and is seen from (4) to be given by

$$A = pq. \tag{5}$$

By itself A tells nothing about the scattering in any particular direction, and may be quite different from the parameter p that determines back-scattering. The existence of these two different quantities, both called albedo, is a source of occasional confusion.

Values of the foregoing parameters for the Earth and for the Moon, as well as for some idealized objects that are useful models of comparison, are displayed in Table I and Figures 4 and 5. The earth and moon data, which are for the visual region of the spectrum as given in standard references [15, 16, 20], are generally well established, even though more recent studies have revised the Moon's phase curve at small phase angles.

TABLE I

Photometric Parameters

Object	A	p	q	$\phi(\alpha)$
Lambert Disc	1	1	1	$\cos\alpha$
Lambert Sphere	1	2/3	1.5	$\left(1 - \frac{\alpha}{\pi}\right)\cos\alpha + \frac{1}{\pi}\sin\alpha$
Constant Radiance Sphere	1	1/2	2	$(1 + \cos\alpha)/2$
Polished Sphere	1	1/4	4	1
Moon	0.068	0.110	0.62	Figure 5
Earth	0.40	0.38	1.07	Figure 5

Light scattered by the Earth is important to laser ranging because of the background light it produces on the Moon, and in fact the observation of this earth-light on the Moon is how the earth-scattering is determined in the first place. There is considerable uncertainty in the Earth's phase curve because the measurements of earth-light must be corrected for various systematic disturbing factors, such as scattering by the atmosphere and by the telescope optics of light from the Moon's crescent into the field of view, and for the fact that data from a given observatory emphasizes the earth's hemisphere in which that observatory is located. There is also considerable random scatter to the earth-shine data, as might be expected from the fact that the Earth's albedo undoubtedly varies with the amount of cloud and snow cover.

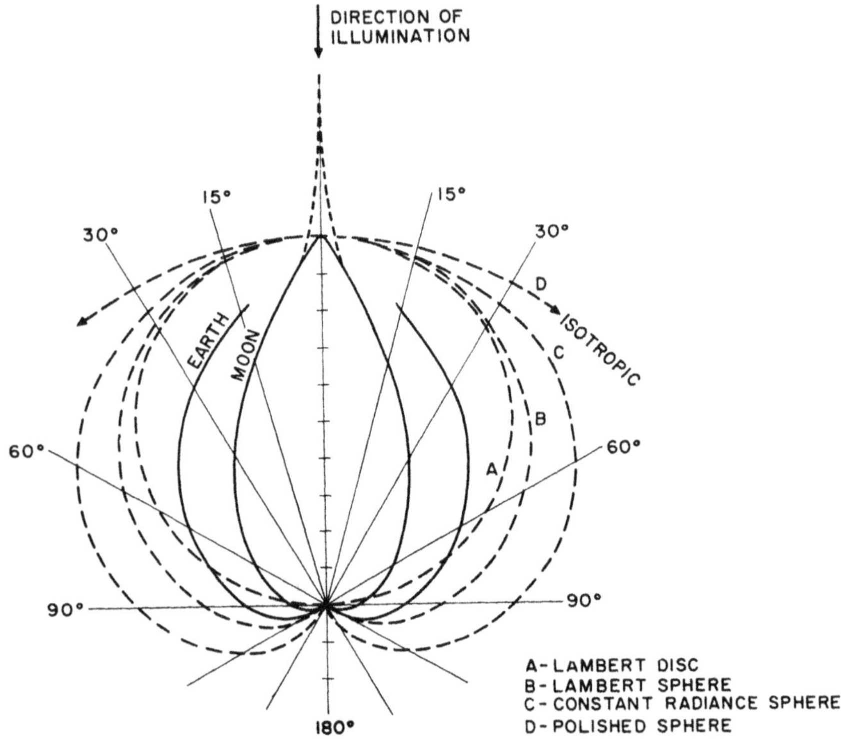

Fig. 4. Comparison of phase functions.

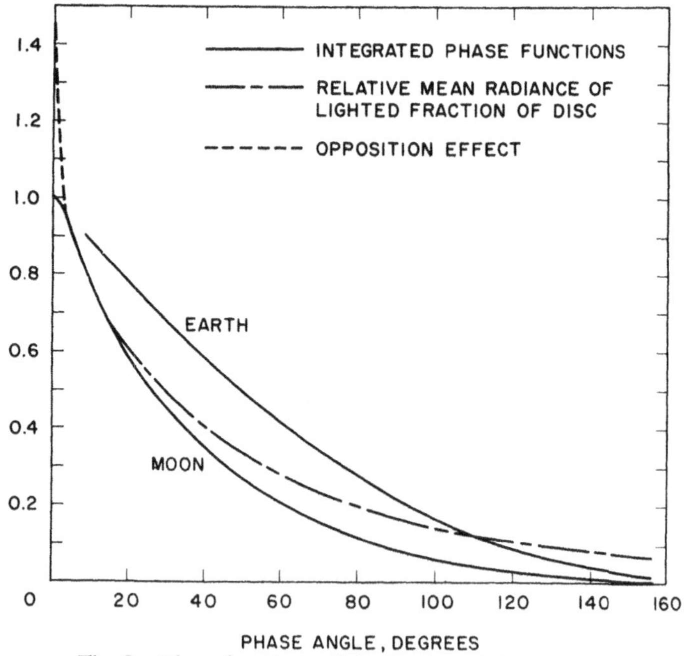

Fig. 5. Phase functions of the Earth and the Moon.

In most respects the data for the Moon are more certain. Since there is no weather on the Moon its reflecting properties presumably are constant.* The principal uncertainty is near zero phase, where the reduction of observations is complicated by the fact that below 1.5 degrees the Moon is partially or totally in the Earth's shadow. It has long been recognized that even at this small phase angle the radiance is still increasing appreciably with decreasing phase, but the extrapolation to zero phase is troublesome, and apparently the magnitude of the increase has been underestimated. Recent studies [17, 18] have produced a phase curve in which a spike at zero phase is about as indicated by the dashes in Figures 4 and 5. This change would give values of p and q of about 0.15 and 0.45 respectively, with the value of A not changed appreciably from that of Table I. If this strong "opposition effect" proves to be correct then the Moon is a more efficient reflector for laser ranging than the older published data indicate.

In principle the Moon's scattering at zero phase angle can be measured directly with a laser system by determining the fraction of transmitted energy that is returned from the Moon's surface. In practice, this method appears difficult because of the problem of accurately measuring the energies involved, the uncertainty of atmospheric effects, and the variation of the Moon's surface from place to place. Nevertheless, as lunar laser ranging data accumulate, a check on the extrapolation of earth-shine data should emerge.

LUNAR RADIOMETRY

The foregoing description of scattering by an entire planet is adequate when the planet may be considered to be a point object. For predicting the laser signal returned by the Moon, as well as for evaluating the background of scattered sunlight, the Moon must be described as an extended object.

The reflected radiance of any point on the Moon's surfaces produced by the incidence of sunlight, earth-light, or laser light, depends upon the incident irradiance, upon the intrinsic reflectivity, which varies from place to place possibly due to differences in composition of the surface material, and upon the angles between the surface normal, the irradiance direction, and the observer's direction. The radiance at any point and time is specified by a radiance factor ρ, defined to be the ratio of the observed radiance to the radiance of a perfectly reflecting Lambert screen on which sunlight is incident normally. In order to explicitly separate the intrinsic reflectivity from the angular dependence ρ is frequently written as a normal albedo ρ_0 times a photometric function f, which function is defined to be unity when the incidence and observation are both normal to the surface;

$$\rho = \rho_0 f(\text{angles}). \tag{6}$$

NORMAL ALBEDO AND BACK-SCATTERING

The normal albedo, which describes back-scattering at normal incidence, is some-

* Brightness differences of 10 to 20% between measurements made in different years have been ascribed to luminescence arising from bombardment by solar particles [17].

thing like the concept of geometric albedo except that it applies to a normal surface element instead of to an entire planet. Its use does not imply that the lunar surface follows Lambert's law. Strictly speaking this normal albedo can be observed from the Earth only for points near the center of the lunar disc. However, the photometric function for the lunar surface is found to have a peculiar form which permits the concept of normal albedo to be particularly useful even though we are not viewing the surface along a normal. Except for the crater rays which become prominent at full moon, the photometric function has essentially the same form for all types of lunar topography, as fact which has been interpreted as implying that the micro-structure of the surface is the same everywhere. The form of the function is such that all points simultaneously reach maximum radiance at full moon, at which time the Moon appears to be a flat disc. In other words, at full moon viewed from the Earth the dependence upon the direction of the surface normal disappears, which very likely means that we observe the normal albedo at all points even though they are not at the center of the disc. This behavior is quite different from that of a sphere whose surface material follows Lambert's law. For computing back-scattering from the Moon the normal albedo is sufficient.

Lunar reflecting properties have been studied in detail by VAN DIGGELEN [18] and others, and are well summarized by MINNAERT [20]. Minnaert's 1960 normal albedo values for various types of lunar topography are repeated here as the first column of Table II. These values, which are used throughout this report, are based upon many

TABLE II

Normal Albedo of Lunar Features

	Visual	7000 Å
Darkest spot (in Oceanus Procellarum)	0.051	0.08
Dark plains (maria)	0.065	0.10
Brighter plains (paludes)	0.091	0.14
Mountain regions (terrae)	0.105	0.16
Crater floors	0.112	0.17
Bright rays	0.131	0.20
Brightest spot (Aristarchus)	0.176	0.27
Average for whole Moon	0.105	0.16

observations, and when used with the other data and curves of this report will give reliable results except probably for phase angles less than 2 degrees. (The phase angle at "full moon" can be as large as 5 degrees.) For predicting true back-scattering the values are low in proportion to the increased opposition effect believed to exist, but no attempt at revision has been made here because the intensity of the opposition effect seems to vary considerably from one lunar feature to another and may be still somewhat uncertain. VAN DIGGELEN's revised determination [18 (1965)] of 39 crater floors gives an average ρ_0 of 0.125 as compared to 0.112 in Table II (12% increase), while his new value for Aristarchus is 0.257 (46% increase). In detailed calculation of the reflection of laser light from the lunar surface VAN DIGGELEN's paper [18 (1965)] should be consulted for the particular topographic feature involved.

It is interesting to note in passing that the brightness contrasts on the Moon due to variation in reflectivity have a range of only about 3.5 to 1. Of course, under oblique illumination by the Sun, the lunar background presents much greater contrasts caused by shadowing.

The values in the first column of the table are averages over the visible spectrum; at a particular wavelength the albedo may be significantly different. The albedos of all lunar features are found to vary with wavelength in about the same way, so a single curve giving the variation with wavelength for some average albedo will suffice for engineering purposes. Figure 6 is such a curve. It is based upon the data of Barabashev as summarized in [19], and upon the ultra-violet measurements of STAIR and JOHNSTON [21]. Both of these papers present their results as radiance relative to the radiance at some particular wavelength rather than as normal albedo, so to arrive at normal albedo the ordinate scale of Figure 6 has been chosen so that when weighted in accordance with the normal visibility curve the resulting albedo is about 0.105. The curve indicates that the average normal albedo at 7000 Å* is about 0.16 as compared to 0.105 for the visual. For convenience the second column of Table II gives values of normal albedo at 7000 Å, obtained by multiplying the values for the visible by 0.16/0.105.

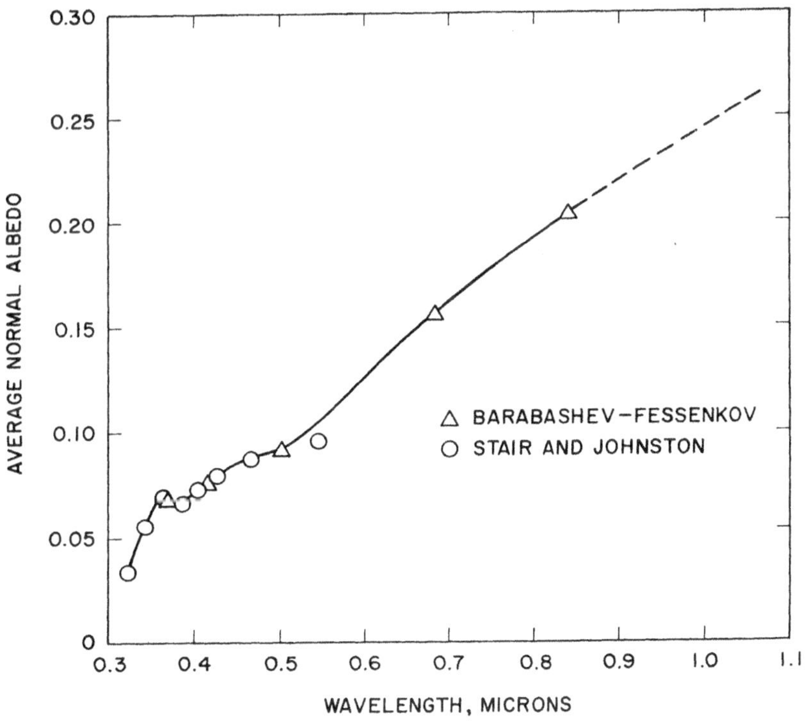

Fig. 6. Average normal albedo of the lunar surface

* The ruby laser operates at 6943 Å.

BACKGROUND

The laser light returned from the Moon must be discerned against the radiant background of the Moon's surface and the moonlit earth atmosphere. The background radiance of the lunar surface as illuminated by the Sun depends upon both the photometric function and the normal albedo; at other than full moon the radiance varies with position on the lunar disc in accordance with the photometric function. But here again, the Moon behaves in an unexpected way which may be summarized thus: reduced to a common normal albedo, the contours of constant radiance on the sunlit part of the lunar disc are approximately meridians through the ends of the terminator. One can therefore to this approximation display the radiance of any point on the bright part of the disc by a set of curves showing the radiance as a function of longitude* and phase angle for some specified normal albedo, as has been done by MINNAERT [20].

This peculiar behavior of the Moon's photometric function makes possible a fairly complete presentation of the optical background problem in the simple form of Figure 7. Intended as an engineering aid describing how the Moon appears at the surface of the Earth rather than as a study of the Moon, this figure incorporates Minnaert's curves for the sunlit region, and earth-light and atmospheric scattering for the dark region.

For an average normal albedo of 0.105, the background of scattered sunlight at any place not too far from the Moon's equator and at any phase of the Moon may be obtained with the help of this figure as

$$N_{\text{solar background}} = H_s T_a \frac{\rho}{\pi}, \tag{7}$$

where ρ for the proper meridian and phase angle is taken from the figure. Here H_s is the irradiance of sunlight and T_a is the transmittance of the Earth's atmosphere. At places or with wavelengths for which the normal albedo is different from 0.105 the value of N given by (7) should be multiplied by $\rho_0/0.105$.**

A description of the earth-light and atmospheric scattering data used in the construction of this figure follows.

EARTH-LIGHT

If there were no scattering, each curve of Figure 7 would fall at the terminator to the level of the earth-light at that phase, as indicated by the short arrows. The radiance of earth-light is constant across the disc to the same extent and for the same reason that the full moon is uniform. Of course the curves describe an imaginary smooth moon. The terminator on the actual Moon is a complicated pattern of highlights and shadows caused by the grazing sunlight on the lunar topography.

* The terms "photometric longitude", "photometric meridian", and "photometric equator" would seem to be useful additions to the language of lunar photometry. Their proposed meanings become obvious by thinking of the cusps as poles.

** But see further explanation under 'Scattering'.

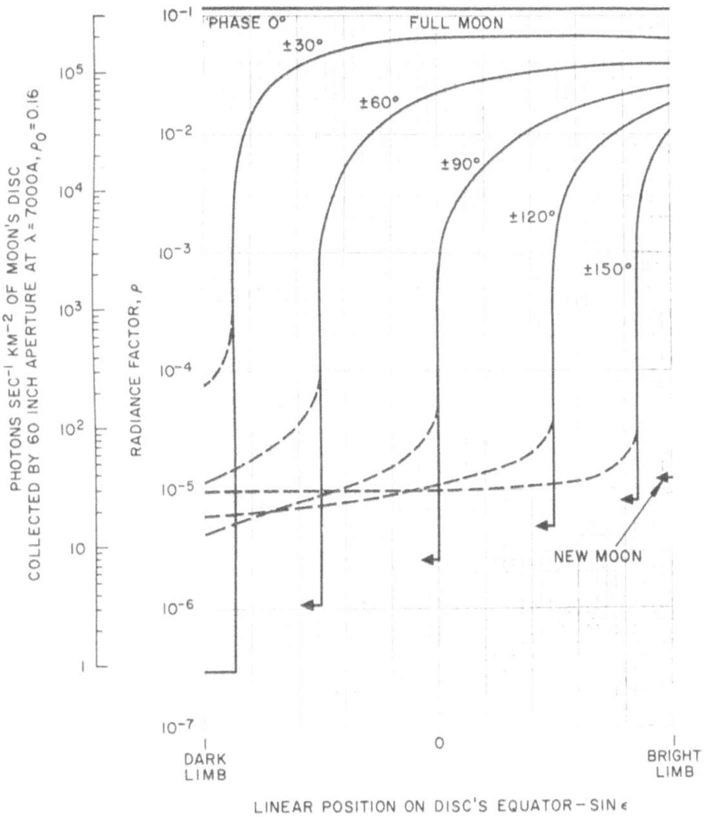

Fig. 7. Background radiance of the lunar surface as seen through the Earth's atmosphere

The irradiance H_E of the Moon by the sunlight scattered from the Earth is J_E/r_{EM}^2, where J_E is the radiant intensity of the Earth as a point source, and r_{EM} is the Earth-Moon distance. Taking J_E from (1) gives for the ratio of the irradiance of the Moon by the Earth to the irradiance by the Sun

$$\frac{H_E}{H_s} = \left(\frac{R_E}{r_{EM}}\right)^2 p_E \phi_E(\alpha_E), \tag{8}$$

where r_E, p_E ϕ_E, and α_E are the radius, geometric albedo, phase function and phase angle of the Earth as seen from the Moon. Putting in the numerical values

$$\frac{R_E}{r_{EM}} = 1/60,$$
$$p_E = 0.38,$$

and remembering that the phase of the Earth as seen from the Moon is opposite to the phase α of the Moon gives

$$\frac{H_E}{H_s} = 1.06 \times 10^{-4} \phi_E(180° - \alpha). \tag{9}$$

Evidently this is also the ratio of the radiance of any lunar feature due to earth-light to its radiance at full moon. The earthlit radiance values indicated on Figure 7 by the short arrows were obtained from (9) and the earth phase function of Figure 5. The value 0.38 for p_E is for the visible region of the spectrum. The variation of p_E with wavelength does not appear to be well established, but seems to be about constant out to one micron with some increase in the blue.

SCATTERING

Besides the lunar surface radiance due to direct sunlight and that due to the earth-light, Figure 7 includes an estimate of the radiance superimposed on the telescope image of the dark side of the Moon's disc by atmospheric scattering of light from the bright side, as indicated by the broken-curve appendages to the solid curves. The scattering shown has been calculated from the data given by VAN DE HULST [22] as typical of good astronomical sites. The scattering on good days at sites suitable for solar coronal observation is an order of magnitude less than that shown, while low altitude coastal sites may easily be an order of magnitude worse.

The scattering function $S(\theta)$ for a point source, which was integrated over the bright Moon to give the scattering in Figure 7, is shown in Figure 8*, where $S(\theta)$ has been defined so that the surface radiance caused by scattering at an angle θ from a star of radiant intensity J is $JS(\theta)$. The total background of scattered moonlight was obtained by a graphical summation, assuming the radiance of the sunlit part of the lunar

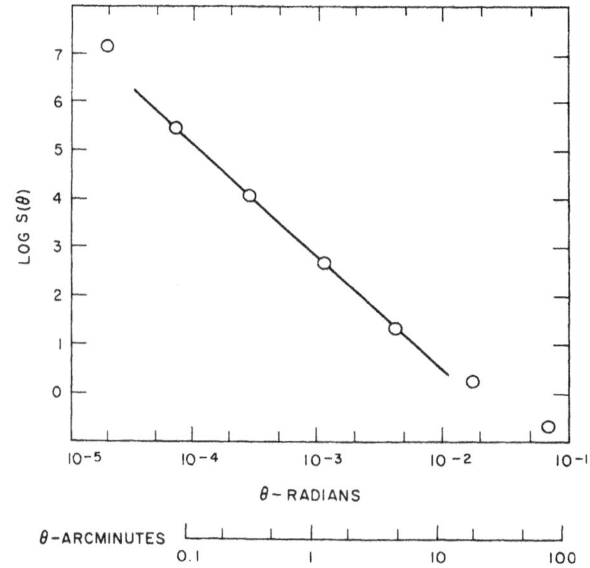

Fig. 8. Small angle scattering of visible light by the Earth's atmosphere.

* The plotted points of this figure are Van de Hulst's "adopted values" of Table 6 on p. 60 of [22], normalized to one steradian rather than to the Sun's solid angle.

disc to be constant along meridians, as previously described. The values plotted in Figure 7 are for points on the equator but since, as is intuitively evident, the contours of constant scattered radiance run about parallel to the terminator, the figure may be applied over much of the Moon.

In estimating atmospheric scattering from Figure 7 one should remember that all parts of the Moon, and not just the lunar feature being observed, contribute to the scattering, so this component should not be corrected for the local albedo as was the direct component. Since the scattering also varies with observing conditions, it has been included on Figure 7 more as a point of departure from which to estimate the atmospheric contribution under prevailing conditions than as an absolute fact.

A separate aspect of atmospheric scattering is the production of twilight [23]. For observations very near new moon sky brightness will become the limiting factor, as shown in Figure 3. Twilight is not included in Figure 7.

Besides scattering from the atmosphere, light may be scattered into the field of view by the components of the telescope. This can be controlled by careful design, but scattering from the optical surfaces themselves may be significant, particularly in the case of a reflecting telescope.

ATMOSPHERIC ATTENUATION

Besides the scattering of unwanted light into the field of view, a second undesirable effect of the atmosphere is the extinction caused to laser light as it passes twice through the air. Although the actual extinction will vary widely with atmospheric conditions, clear sky transmittance is a fairly definite function of wavelength. The curve given in Figure 9, which includes only continuous and not line absorption, is from astronomical experience [15] and should be applicable to moon ranging from good observing sites.

In addition to the above continuous absorption, there are scattered throughout the

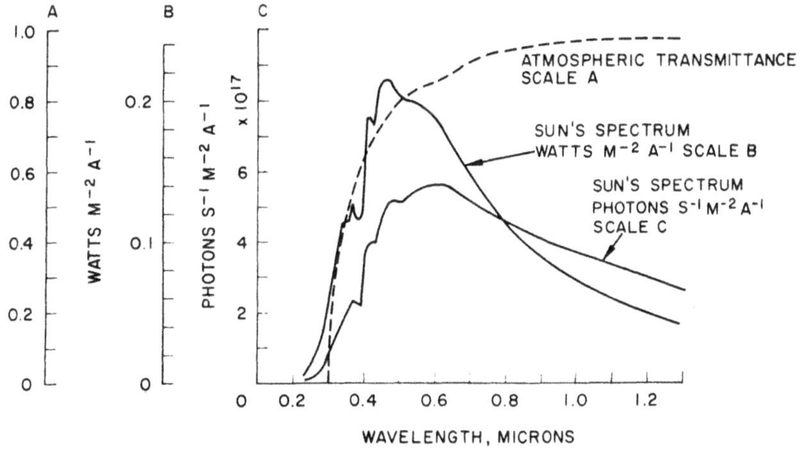

Fig. 9. Atmospheric transmittance T_a based on continuous absorption only. Solar irradiance H_s outside the Earth's atmosphere.

spectrum many line absorptions characteristic of the molecules making up the atmosphere. In some parts of the spectrum these lines are so intense and so closely spaced as to absorb a large fraction of the incident radiation. The effect of these lines is sometimes included in an atmospheric transmittance curve by averaging over some arbitrary bandwidth such as 100 Å, but this has not been done in Figure 9, because such averaging is not applicable to the very narrow spectral widths involved in laser technology; line absorptions must be considered separately and in detail.

Around the ruby wavelength of 6943 Å several water absorption lines exist [25]. Specifically at 6942.15 Å, 6942.37 Å and 6943.80 Å there are water lines capable of causing several decibels round trip attenuation at these wavelengths, but trouble from these absorptions can be avoided by close temperature control of the laser transmitter. A 1.4 Å band around 6943 Å is clear.* The vicinity of 1.06 μ, the operating wavelength of the neodymium laser, is less cluttered with troublesome absorption lines.

For calculating the background interference by equation (7) the values of H_s and T_a may be taken from Figure 9 for any of the wavelengths used in the present report. At some other wavelengths the effect of bands of absorption lines on T_a for use in (7) is not negligible [15, 24].

An example of the application of some of the data presented in this section to a specific laser system is the construction of the scale at the left in Figure 7. This scale allows one to read the background photon flux directly for a 1.5 m receiving telescope at the ruby wavelength. In associating this scale with the curves over their entire range the assumption is made that the Earth's albedo and the atmospheric scattering of moonlight into the telescope field are relatively the same at 7000 Å as for the visible, for which the curves were drawn. This appears to be a good approximation.

Received Signal Strength

Several of the most important things that must be considered in designing a laser ranging system are measured by the parameters appearing in the formula for received signal strength. Any such formula seems to be referred to as the radar (range) equation even though the form may vary.**

The energy W_M returned by the un-augmented lunar surface and collected by a receiving telescope on Earth from a laser transmission of W_L joules is

$$W_M = W_L\, T_t\, T_a \frac{\rho_0}{\pi} T_a \frac{(\pi/4)\, D^2}{r^2} T_r\,. \tag{10}$$

* The emitted wavelength from ruby is 6943 Å at 300°K with a temperature coefficient of + 0.068 Å per degree C.

** Radar cross section σ, which relates back-scattering to an isotropic point scatterer, is not a particularly natural way of expressing optical scattering either from the lunar surface or from a retroreflector. To permit comparison with radar results we note that for reflection from the bare lunar surface σ is $4\rho_0$ times the area of the spot of diameter S projected onto the Moon's disc by the laser, while for a retroreflector σ is $gA_F\, 4\pi/\Omega_F$. Because Ω_F will be very small compared to 4π, the radar cross section will be many times the aperture area of the reflector.

where each factor on the right-hand side of this equation has been written in the order in which that factor is encountered by the laser pulse in going to the Moon and back. In this equation T_t is the transmittance of the transmitting optics, T_a is the atmospheric transmittance, ρ_0 is the normal albedo of the irradiated spot on the lunar surface, D is the diameter of the receiving telescope, r is the range, and T_r is the transmittance of the receiving optics. In this form the truth of the equation is apparent from the system concept and from the definition of normal albedo.

The corresponding expression for the return W_F from a retroreflector is more complicated but may be set down in the same way,

$$W_F = W_L T_t T_a \frac{A_F}{(\pi/4) S^2} T_F g T_a \frac{(\pi/4) D^2}{r^2 \Omega_F} T_r. \tag{11}$$

Here, in addition to the previously defined symbols, A_F is the aperture area of the retroreflector, S is the diameter of the laser spot on the Moon's disc, T_F is the transmittance of the retroreflector, g is an efficiency factor required because of the aberration of light, and Ω_F is the solid angle of the reflected beam.

Rewritten in more compact form these formulas for the signal energy delivered to the detector become

$$\frac{W_M}{W_L} = \frac{\rho_0 D^2 T_t T_r T_a^2}{4r^2} \tag{12}$$

for scattering from the lunar surface, and

$$\frac{W_F}{W_L} = \frac{A_F D^2 g T_F T_t T_r T_a^2}{\Omega_F S^2 r^2} \tag{13}$$

for reflection from a retroreflector. The following discussion of some ways in which geometrical and physical principles restrict the factors in these formulae will begin to reveal the form of a practical moon ranging system.

DIFFRACTION

The spot diameter S, the reflected beam size Ω_F, and the abberration factor g require definition. The actual energy distributions measured by S and Ω_F have no sharp boundaries, but from the way in which these two parameters enter (13) it is apparent that each must refer to the equivalent flat-topped distribution with abrupt edges having the same on-axis intensity and same total power as the actual beam being considered. The relation of this definition to other frequently encountered measures of width is exhibited in Figure 10 for an idealized beam pattern.

The curve drawn in this figure, which will be useful in discussing the effects of aperture (transmitter, receiver, and reflector) on system performance, is the distant diffraction pattern produced by a uniform plane wave which has been limited by passing through a circular hole. It is also the familiar Airy disc star image formed in the focal plane of a perfect or diffraction limited telescope. The pattern has circular symmetry with this curve for its diametral section. From the analytic form of the curve

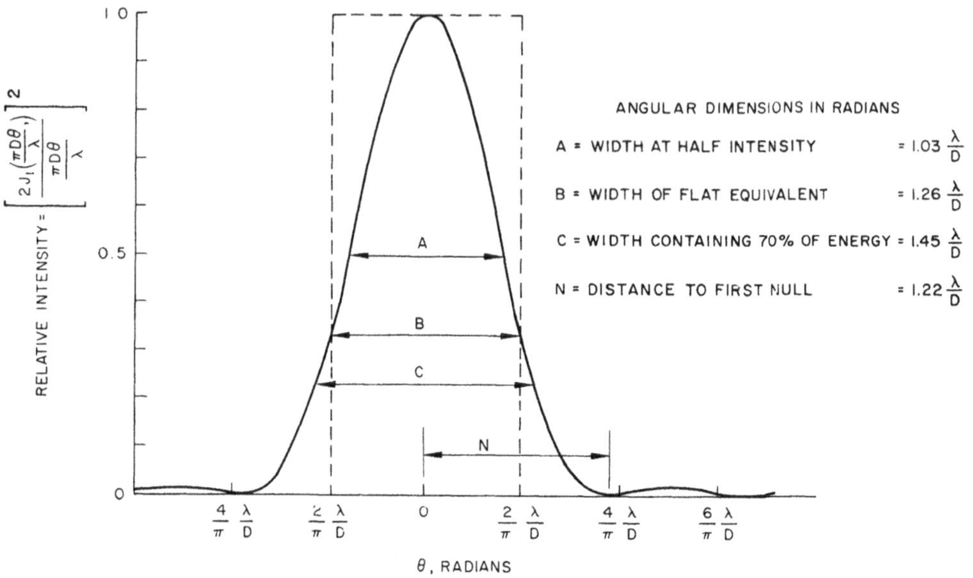

Fig. 10. Fraunhoffer diffraction pattern of a circular aperture of diameter D at wavelength λ. The angular distance from the axis of symmetry of the pattern is θ.

it follows that the dimension B, the width of the flat distribution of the same maximum intensity and same total power as the diffracted distribution, is $4\lambda/\pi D$. The solid angle of this equivalent beam is therefore

$$\Omega = \frac{\lambda^2}{\text{aperture area}}, \tag{14}$$

a relationship which holds not only for circular apertures but for any aperture transmitting a uniform plane wave.

RETROREFLECTORS AND THE ABERRATION OF LIGHT

The factor g, which decreases the effective area of the retroreflector, is needed because the transverse velocity of the reflector with respect to the observatory moves the station away from the center of the reflected beam by the time the light returns from the Moon. This aberration of light effect is depicted in Figure 11, where F is the reflector, S is the terrestrial station, v is the transverse velocity, and c is the speed of light.

In part A of this figure the observatory is shown as being stationary, while in B the reflector is stationary and the observatory moves. Since in electromagnetism only relative velocities matter, the magnitude of the aberration effect can be calculated by considering either the terrestrial station or the reflector to be at rest, and the answer will be the same either way. Use of the first reference frame would require that light rays be traced through a moving retroreflector with the attendant complication that the angles of incidence and reflection are not equal for a mirror moving with respect to an observer. On the other hand, to an observer on the retroreflector the situation is clear. As in Figure 11 B he sees light arriving from the direction of point 1,

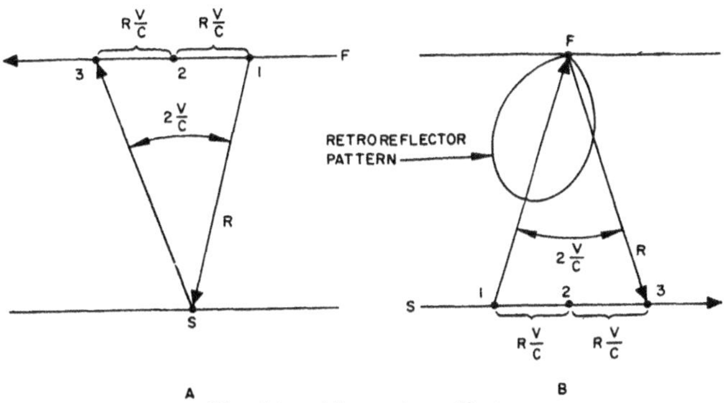

Fig. 11. Aberration effects.

when the terrestrial station has reached point 2. Since the reflector is an ordinary stationary retroreflector the light is sent back toward 1, arriving there when the station has reached point 3.

Of course, in actual fact the light does not return in a single direction only, but has a finite angular spread characteristic of the reflector as indicated by the pattern in the figure. Some reflected energy will return to the transmitting site at point 3, but this will be less than the return from a perfect reflector with no motion by the factor g. The apparent angular displacement of the reflected beam, which we will call the aberration angle a, is $2v/c$ independent of range. As will be shown in more detail later (Figure 14 and 15), under favorable circumstances this angle is about one arc-sec or one mile toward the east on the Earth's surface. In computing this, one must remember to measure the velocities in a non-rotating coordinate system; light retroreflected from a synchronous earth satellite would exhibit the effect.

Because of this velocity aberration effect a retroreflector must not be allowed to produce a beam that is too narrow if the receiver is to be located with the transmitter. For a diffraction limited reflector of diameter d and an aberration angle a the received signal strength, which is proportional to

$$\frac{d^4}{\lambda^2}\left[\frac{2J_1\left(\frac{\pi a d}{\lambda}\right)}{\frac{\pi a d}{\lambda}}\right]^2, \tag{15}$$

goes through alternate maxima and nulls as d is increased. The first maximum is at

$$d = 0.766\frac{\lambda}{a}, \tag{16}$$

with the first zero at

$$d = 1.22\frac{\lambda}{a}. \tag{17}$$

If there were no aberration, or if the receiver were placed so as to match the aberration displacement of the reflected beam, there would be no objection to a reflector of this type. Except for the practical difficulty of construction it could be made arbitrarily large, and since the reflected intensity would be proportional to d^4 while the mass would vary approximately as d^3 a single diffraction limited reflector would give the largest reflection for a given reflector weight. With aberration this type of reflector may not be best and in any case other means must be considered for increasing the received signal beyond that produced by reflector (16). A simple arrangement giving the necessary design freedom is a cluster of reflectors.

The angular distribution of the light reflected from a cluster will have a fine structure of lobes determined by the overall dimensions and population of the cluster, but this fine structure will have an envelope characteristic of the individual reflector size as given by Figure 10. The fine structure lobes, whose spacing becomes smaller as the overall dimension of the cluster is increased, will move through the fixed envelope as the viewing aspect varies. For a reflector fixed on the Moon the viewing aspect will be varied by libration. If the cluster has an overall diameter of one meter the fine structure lobes will be separated about 10^{-6} radian and the diurnal libration rate of about 10^{-6} rad s^{-1} will move the receiver from one lobe to another in about one half a second. Since the received signal is going to fluctuate widely from pulse to pulse anyway because of the effects of the Earth's atmosphere on the outgoing laser beam we may simply accept the statistical variation produced by the fine structure lobes and consider only the envelope, which will give the correct average received signal. Averaged over the fine structure the reflected energy at the aberration angle a from an array of n parallel diffraction limited circular reflectors each of diameter d is proportional to

$$\frac{nd^4}{\lambda^2}\left[\frac{2J_1\left(\frac{\pi a d}{\lambda}\right)}{\frac{\pi a d}{\lambda}}\right]^2 . \tag{18}$$

From this expression the value of d giving the greatest average received signal with various critical quantities held constant can be computed. The result is given in Table III.

TABLE III
Optimum Reflector Diameters

Quantity held constant	Optimum diameter d	d in cm*	value of g
n (number of reflectors)	$0.765\ \lambda/a$	10.7	0.186
nd^2 (total aperture area)	$0.586\ \lambda/a$	8.2	0.40
nd^3 (total mass)	$0.432\ \lambda/a$	6.0	0.62

* For $\lambda = 0.7$ micron and $a = 5 \times 10^{-6}$ radian.

An actual reflector in the lunar environment will suffer more or less from various imperfections of shape and optical index, but the above behavior of diffraction limited reflectors is a useful guide. The most likely type of retroreflector, an optical cube-corner, is discussed in more detail in the appendix.

The aberration effect we have been discussing is important only when the reflector produces a narrow beam. It is totally negligible for scattering from the lunar surface. But there is a second velocity aberration effect, evident from the terrestrial station viewpoint of Figure 11A, which comes in even for ranging on the bare lunar surface; the transmitting telescope should lead the receiving and guiding telescope by the angle $2v/c$. The importance of this second aspect of aberration depends upon the sizes of the transmitted beam and the receiver field stop, and at some future date when equipment operating at good observing sites becomes refined, the effect will need to be considered.

SPOT SIZE OF THE MOON

For various reasons a small spot diameter is desirable. With a retroreflector the strong effect of S on reflected intensity is seen in (13). For ranging on the unaugmented lunar surface the total returned energy is independent of S, but in this case S is the angular resolution of the system.

Even if high resolution were not needed to isolate particular lunar features a large spot causes range spreading of the reflected laser pulse which frustrates precise range measurement. This range spreading is large compared to the pulse length unless the beam strikes the surface very nearly normally. With presently feasible equipment the

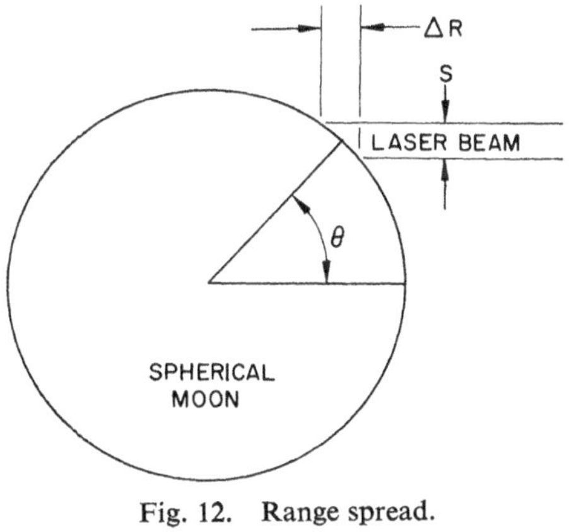

Fig. 12. Range spread.

most frequent detected signal from each laser flash against the unaugmented lunar surface will consist of a single photoelectron, so the range uncertainty from an individual flash will be set by the range spread ΔR, not by the transmitted pulse length. Table IV shows the magnitude of the spread for a featureless Moon.

BEAM DIVERGENCE

The effective size of the distribution of laser light projected on to the Moon will be set by the performance of the laser and its telescope as degraded by the effects of the Earth's atmosphere. The beam emerging from a laser oscillator is collimated light

TABLE IV

Range Spread

S – km	1	2	5	10
θ – Deg.	ΔR – meters			
0	0.07	0.29	1.8	7.2
1	17	35	87	175
2	35	70	175	350
5	87	175	440	880
10	175	350	880	1760

even though this light is generated throughout the bulk of the laser rod. This fact is part of the revolutionary difference between the laser and previous light sources. If the laser were ideal the beam spread at large distances would be given by Figure 10 with D equal to the diameter of the laser aperture. An actual laser will put out a more divergent beam, which may not resemble Figure 10 at all if more than one transverse mode of the laser cavity is active. Even if proper design has restricted the oscillation to a single mode the beam will diverge because of optical inhomogeneities in the laser rod. Ruby laser oscillators have been built which under low power operation are essentially diffraction limited, but with generally available materials present high power lasers produce a spread of several times the diffraction limit represented by Figure 10.

One would like to project onto the Moon a spot as small as allowed by the Earth's atmosphere, and so the laser divergence is de-magnified by projection through an astronomical telescope. The angular magnification of any telescopic optical system is equal to the ratio of entrance to exit pupil diameters – in this case the ratio of the laser aperture to the telescope aperture. Another way of saying this is that the divergence expressed as a multiple of the diffraction limit is invariant through the optical system.

As an illustrative numerical example consider what laser divergence would be permissible so that the spread when projected by the telescope would be the same order as the broadening caused by the atmosphere – perhaps one arc-second or 5×10^{-6} radian. For a 1.5 meter diameter telescope aperture and a 15 mm diameter laser rod the laser divergence comes out 0.5 milliradian. Present high power lasers are not this good, but careful laser oscillator cavity design, with selected ruby and with optical compensation for its inhomogeneities, could probably produce such a laser. This amount of divergence is about nine times the diffraction limit. The example shows that although transmitting apertures of only a few centimeters are usual on terrestrial laser rangefinders, for probing the Moon a much larger aperture is profitable. The poorer the laser and the better the observing site, the more is to be gained by a large transmitting aperture.

ATMOSPHERIC BEAM SPREAD

The beam as projected by the transmitting telescope will be further dispersed by the Earth's atmosphere. Several experimental studies of the spreading and scintillation caused by the passage of laser beams through the air have shown severe effects [26]. However, the propagation conditions of these investigations have been universally more severe than the conditions that will be encountered by a beam projected near the zenith from a good astronomical observing site. The projection of a laser spot on to the Moon by a telescope and the projection of the image of a star to the focus of the same telescope are closely related processes; astronomical experience should be the best guide to what to expect in moon ranging.

Atmospheric seeing usually blurs a star image to more than one arc-second in diameter, but under the best conditions images may be as small as 0.1 arc-second. Projection to the Moon, with allowance for the additional divergence due to the laser, should behave in the same way on the average. However, the Q-switched laser pulse is so short that no averaging takes place within a single shot, so an instantaneous description of atmospheric seeing is needed, and this is not so well understood. One would expect the instantaneous image to have a random fine structure, determined by the telescope aperture and the atmospheric turbulence scale, which would cause the signal returned from a retroreflector to vary widely from one laser shot to another depending on the instantaneous position of the reflector in this fine structure of the transmitted beam.

On the other hand the total energy reflected from the bare lunar surface will be constant from pulse to pulse because the entire transmitted beam is intercepted by the Moon. Although the total energy returned is constant in this case a second interference phenomenon, the sparkle effect observed with laser illumination of diffuse surfaces [27], should be considered. The spot on the Moon is itself an aperture of diameter S with some unknown amplitude and phase distribution, and since laser light has a coherence length of many meters this aperture will radiate with a random lobe structure of angular scale λ/S. This scale projected to the Earth will be $\lambda r_{\mathrm{EM}}/S$ or about 0.1 meter for a 2.5 km diameter spot. Since the receiving aperture will be large compared to 0.1 meter the signal fluctuation from this effect should be minor.

Motion

The motion significant to laser ranging is the relative motion of the point of reflection on the Moon with respect to the earth-based observatory. This motion may be separated into three parts:

(a) the orbital motion of the center of the Moon around the center of the Earth;
(b) the displacement of the reflector by the Moon's libration;
(c) the displacement of the observatory by the Earth's rotation.

Designating the vector displacement from the observatory S to the reflector F, it follows that

$$\mathbf{r}_{SF} = \mathbf{r}_{\mathrm{EM}} + \mathbf{r}_{\mathrm{M}F} - \mathbf{r}_{\mathrm{ES}}, \tag{19}$$

where $\mathbf{r}_{\mathrm{EM}}$ is the displacement of the moon center M from the earth center E, $\mathbf{r}_{\mathrm{M}F}$ is the displacement of the reflector from the moon center, and $\mathbf{r}_{\mathrm{ES}}$ is the displacement of the observatory from the earth center.

Equation (19) is useful for various purposes, and the coordinate system and mode most appropriate for describing the behavior of the component vectors will vary accordingly. For reducing the data obtained by laser ranging, the measured values of range (the magnitude of $\mathbf{r}_{SF}$) will be compared with values calculated from lunar theory, and for this purpose (19) will need to be evaluated as accurately as theory permits. This would be done by electronic computer making use of the ephemeris calculated from Brown's improved lunar theory which is now available on magnetic tape. For our present purpose of investigating the influences of motion on the design and performance of a ranging system a more conceptual approach, relying on either the American Ephemeris or approximate models for numerical values, will be followed.

VELOCITY ABERRATION

Aberration will be investigated by examining the vector $\mathbf{a}$ defined by

$$\begin{aligned}\mathbf{a} &= \frac{2}{c}\dot{\mathbf{r}}_{SF}\\ &= \frac{2}{c}(\dot{\mathbf{r}}_{\mathrm{EM}} + \dot{\mathbf{r}}_{\mathrm{MF}} - \dot{\mathbf{r}}_{\mathrm{ES}}),\end{aligned} \tag{20}$$

where c is the speed of light and the dot indicates differentiation with respect to time in an inertial coordinate system. Some of the terms in this formula are numerically more important than others. The orbital contribution $\dot{\mathbf{r}}_{\mathrm{EM}}$ is a vector of magnitude 950 to 1100 m sec^{-1} directed along the Moon's trajectory. The contribution $\dot{\mathbf{r}}_{\mathrm{MF}}$ from lunar (geocentric) libration may have any direction but is limited to 0.7 m sec^{-1}, while the station velocity $\dot{\mathbf{r}}_{\mathrm{ES}}$, directed toward the east point of the horizon, is given in magnitude very nearly by 465 $\cos\varphi$ meters per second at latitude φ. Aberration will therefore be described with sufficient accuracy by ignoring geocentric libration and considering the reflector to be at the center of the Moon; in which case

$$\mathbf{a} = \frac{2}{c}(\dot{\mathbf{r}}_{\mathrm{EM}} - \dot{\mathbf{r}}_{\mathrm{ES}}). \tag{21}$$

Formula (21) will be most useful if it is resolved into equatorial polar coordinates, which coordinates are related to the celestial sphere in Figure 13. In our present approximation O is the observatory and P is the reflector at the center of the Moon. Resolving $\mathbf{a}$ into components a_r, a_α, a_δ parallel to the unit vectors l_r, l_α, l_δ shown in the figure, (21) becomes

$$a_\alpha = \frac{2}{c}(v_\alpha - v_S \cos H) \tag{22}$$

$$a_\delta = \frac{2}{c}(v_\delta - v_S \sin\delta \sin H) \tag{23}$$

$$a_r = \frac{2}{c}(v_r + v_S \cos\delta \sin H). \tag{24}$$

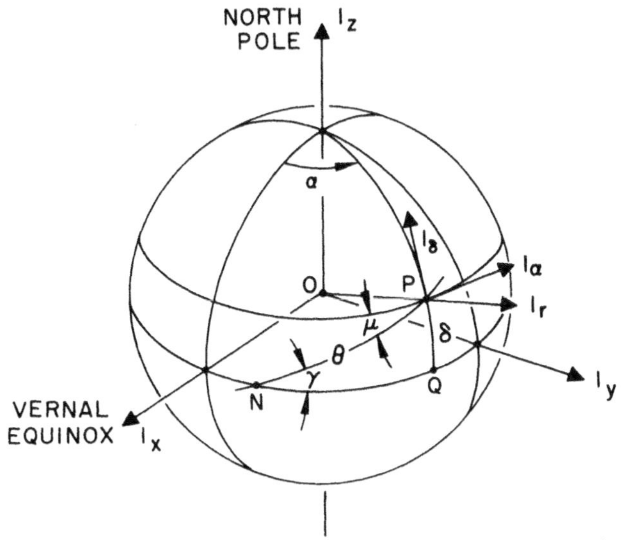

Fig. 13. Reference systems. The smaller central sphere is the Earth with the observing station at *S*. The large sphere is the celestial sphere with the projected positions of the Moon (at *M*) and a part of its orbit shown.

Here v_α, v_δ, v_r are the components of the Moon's orbital velocity, v_S is the eastward velocity of the station caused by the Earth's rotation, and δ and H are the declination and local hour angle of the Moon.

The quantities a_α and a_δ are angular aberration components, while a_r is the fractional Doppler shift of the wavelength of the reflected light.

Examination of a lunar ephemeris shows that the geocentric radial component of the Moon's orbital velocity lies in the range ± 75 m sec^{-1}, but topocentrically up to ± 20 m sec^{-1} must be added because of the effect of v_α when viewed from a position displaced by the Earth's radius. With values of ± 95 m sec^{-1} and 465 m sec^{-1} for v_r and v_S respectively (24) gives a maximum fractional doppler shift of about 3.7×10^{-6}. Unlike the angular aberration, which has consequences important to the operation of retroreflectors, this amount of Doppler shift is negligible with present techniques. For example, at 7000 Å the maximum shift is 0.026 Å or 1600 mc, which is small compared to the minimum filter bandwidth usable with a high intensity laser and is only comparable to the laser line width.

The behavior of the angular aberration components a_α and a_δ as functions of the Moon's condition, which behavior is exhibited in Figure 14, is calculated by putting into (22) and (23) some relationships derived from Figure 13. For this purpose we think of *P* as the position of the Moon on the celestial sphere and the arc *NP* as the projection of the Moon's orbit onto the sphere. Calling v_t the component of the Moon's orbital velocity transverse to the Earth-Moon line

$$v_\alpha = v_t \cos \mu \tag{25}$$

and

$$v_\delta = v_t \sin \mu , \tag{26}$$

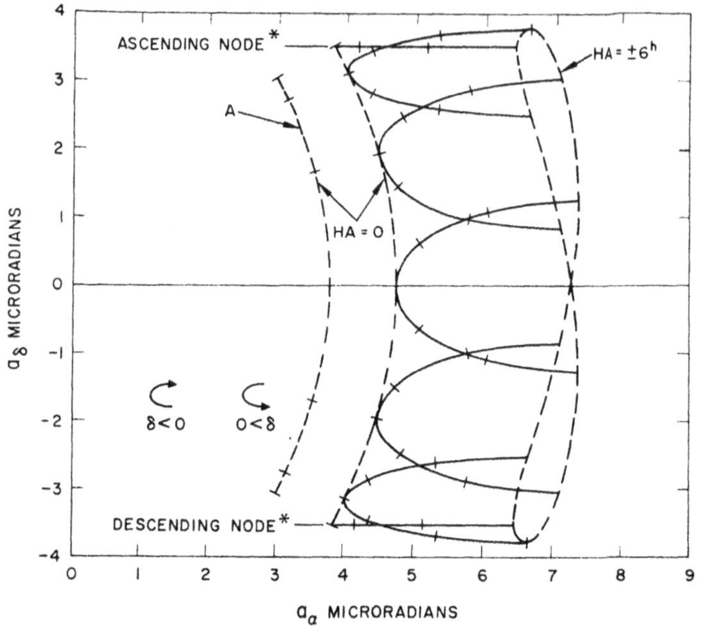

* Node of the Moon's orbit on the Earth's equator.

Fig. 14. Angular aberration displacements. Station latitude 32°. Orbit inclination 28°. The solid curves show the approximate variation over 12 hour intervals spaced every 30° of the Moon's longitude for an orbital velocity of 1100 m/s^{-1}. Depending upon the Moon's declination the curves are traversed in the directions indicated. The markers are at 2 hour intervals. For an orbital velocity of 950 m/s^{-1} the curves shrink so as to fit the zero hour angle curve *A*.

and from the spherical right triangle *NPQ*

$$\tan\mu = \tan\gamma\,\cos\theta\,, \tag{27}$$

where γ is the inclination of the Moon's orbit to the Earth's equator and θ is the Moon's longitude measured from the ascending node of its orbit on the Earth's equator. Also from triangle *NPQ*

$$\sin\delta = \sin\gamma\,\sin\theta\,, \tag{28}$$

so we can compute a_α and a_δ for given values of v_t, γ, and θ as a function of the hour angle H. Since the orbital speed does not change much in a twelve-hour interval the angular aberration displacement follows approximately one of the *C*-shaped curves of Figure 14 from moonrise to moonset each day.

If the light returned from the Moon by a retroreflector is to be received at the transmitter position the angular aberration displacement of the receiver from the axis of the reflected beam may be read from Figure 14. This angle (a) is given by the distance from the origin of the figure to the point on the curves corresponding to the condition of the Moon as

$$a = (a_\alpha^2 + a_\delta^2)^{1/2}.$$

The coordinates in Figure 14 may also be interpreted as the amounts by which the transmitted beam should lead the receiver field stop in right ascension and declination for perfect boresighting.

For some purposes the receiver and transmitter may be located at different sites. For example, for photographing a retroreflector in laser illumination the laser might be deliberately located to the west of some existing high-quality photographic telescope so that aberration would center the reflected beam on this telescope.

The angular offset of an arbitrary receiver position from the axis of a retroreflected beam may be obtained as follows.

The linear displacement of the transmitter from the axis of the reflected beam measured in a plane normal to the Earth–Moon line at range r from the reflector has components ra_α and ra_δ. The net displacement of the receiver from the axis of the reflected beam is obtained by subtracting the transmitter-receiver displacement from these aberration displacements. Specifically, calling the vector displacement of the receiver from the transmitter $\mathbf{D}$, and calling the components of the net displacement of the receiver from the axis of the reflected beam rb_α and rb_δ gives

$$b_\alpha = a_\alpha - \frac{D_\alpha}{r}, \tag{29}$$

$$b_\delta = a_\delta - \frac{D_\delta}{r}, \tag{30}$$

where D_α and D_δ are the components of $\mathbf{D}$ parallel to l_α and l_δ. If $\mathbf{D}$ is now defined by its components D_{E}, D_{N}, D_{Z}, the distances east, north, and up from the transmitter to the receiver, (29) and (30) become

$$b_\alpha = a_\alpha - \frac{D_{\mathrm{E}}}{r}\cos H + \frac{D_{\mathrm{N}}}{r}\sin\phi\sin H - \frac{D_{\mathrm{Z}}}{r}\cos\phi\cos H \tag{31}$$

and

$$b_\delta = a_\delta - \frac{D_{\mathrm{E}}}{r}\sin\delta\sin H - \frac{D_{\mathrm{N}}}{r}(\sin\phi\sin\delta\cos H + \cos\phi\cos\delta)$$
$$+ \frac{D_{\mathrm{Z}}}{r}(\cos\phi\sin\delta\cos H - \sin\phi\cos\delta). \tag{32}$$

These are the components of net angular offset. The formulas may be evaluated in conjunction with (22) and (23) as was done in producing Figure 14.

For deciding on the best location of a receiver relative to a transmitter a map showing where the reflected beam strikes the ground would be useful. An example of such a map is Figure 15, which shows the intersection of the reflected beam with level terrain. The values plotted here were obtained by setting b_α, b_δ, and D_{Z} all equal to zero and solving for D_{E} and D_{N} to obtain

$$D_{\mathrm{E}} = \frac{2r}{c}\left[\frac{v_\alpha(\cos\phi\cos\delta + \sin\phi\sin\delta\cos H) + v_\delta\sin\phi\sin H}{\cos\phi\cos\delta\cos H + \sin\phi\sin\delta} - v_{\mathrm{S}}\right] \tag{33}$$

$$D_{\mathrm{N}} = \frac{2r}{c}\left[\frac{-v_\alpha\sin\delta\sin H + v_\delta\cos H}{\cos\phi\cos\delta\cos H + \sin\phi\sin\delta}\right] \tag{34}$$

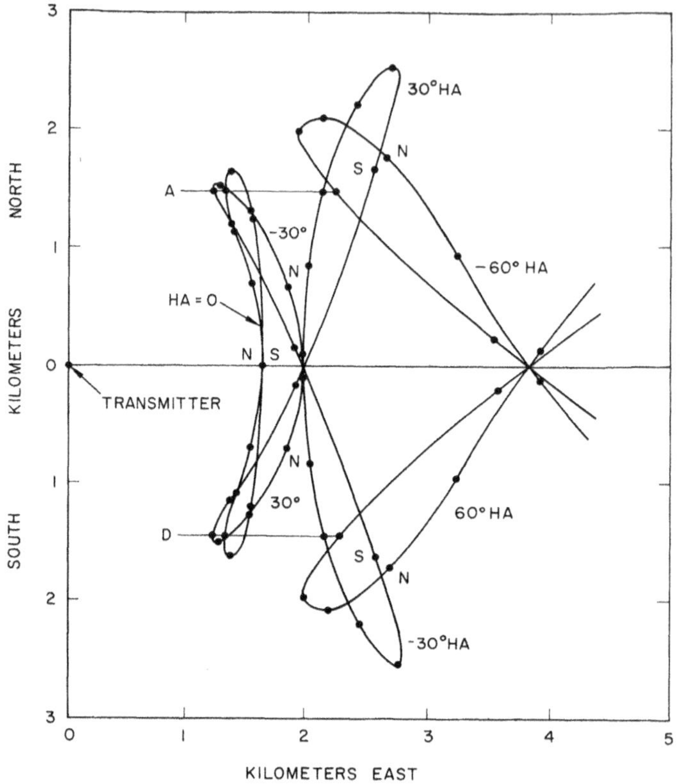

Fig. 15. Aberration displacements in ground coordinates. Station latitude 32°. Orbit inclination 28°. Orbital velocity 1100 m/s^{-1}. Each curve follows the motion for a month at fixed hour angle. The markers are at 30° intervals of the Moon's longitude. *N* and *S* mark the points corresponding to the maximum north and south declination, and *A* and *D* mark the points corresponding to the ascending and descending nodes of the Moon's orbit on the Earth's equator.

For the desirable observing condition of small zenith angle the Moon will be near its maximum north declination and not too far from the local meridian. Figure 15 shows that for this condition a receiver located about two kilometers east of the transmitter is in a good position.

RANGE AND RANGE RATE

For examining the range and its rate of change a different expansion of (19) is in order. Only the magnitude of r_{SF} of the station-to-reflector distance is needed rather than a resolution into components, but because of the sensitivity of a laser ranging system to range all three terms of (19) should be retained.

If r_{SF} is to be expressed entirely in terms of quantities given in the American Ephemeris at least eight parameters are involved. Besides the Earth-Moon distance r_{EM}, which is given in the Ephemeris as lunar parallax, the station latitude and the Moon's declination and local hour angle are needed to specify the center of the Moon with respect to the station. Two selenographic coordinates specify the point of reflection on the Moon, and two libration parameters are needed to define the topocentric libration condition. If all these are put analytically into (19), the resulting expression for the range is too complex to be easily comprehended.

The complexity may be divided by introducing first only two quantities (in addition to r_{EM}) which vary with time as the motion progresses, namely the two zenith angles Z_M and Z_S shown in Figure 16.* As may be seen from the figure all the librational effects are contained in Z_S, Z_M depending only upon the position of the center of the Moon and the observatory coordinates.

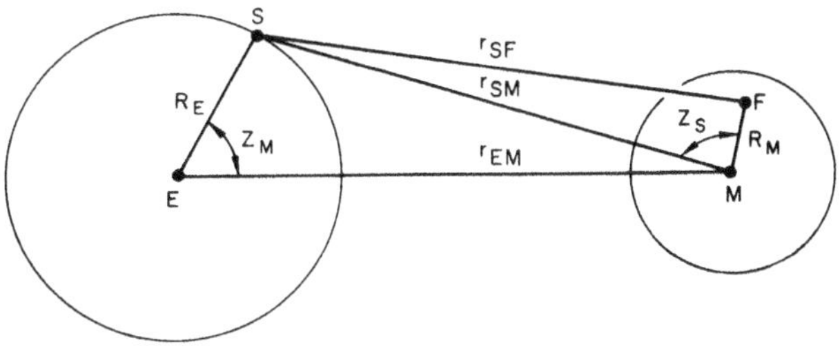

Fig. 16. Range geometry.

Applying the cosine law to triangles SMF and SEM

$$r_{SF} = [r_{SM}^2 + R_M^2 - 2r_{SM} R_M \cos Z_S]^{1/2}, \tag{35}$$

where

$$r_{SM} = [r_{EM}^2 + R_E^2 - 2r_{EM} R_E \cos Z_M]^{1/2}. \tag{36}$$

For exact numerical calculation these formulas would be evaluated as they are. For investigating the form of the motion we expand the square roots in powers of the parallaxes p and q, defined for this purpose to be R_E/r_{EM} and R_M/r_{EM} respectively. The result is the formula which has been written as Table V. Although p and q are variables they remain approximately 1/60 and 1/220.

The two numerical columns of the table give the maximum contribution of each term to the range change with time and to the range rate, with station and reflector positions chosen (separately for each entry) so as to maximize the entry. As may be seen by the range rate values, the first two terms alone describe the general form of the motion over a period short compared to a month. Although this fact was perhaps

* These variables are related to the larger set by

$$\cos Z_M = \cos \phi' \cos \delta \cos H + \sin \phi' \sin \delta$$

and

$$\cos Z_S = \cos \beta_F \cos \beta_S \cos (\lambda_F - \lambda_S) + \sin \beta_F \sin \beta_S,$$

where ϕ' is the geocentric latitude of the observatory, δ is the geocentric declination of the Moon, H is the local geocentric hour angle of the Moon, β_F and λ_F are the selenocentric latitude and longitude of the reflector on the Moon, and β_S and λ_S are the selenocentric latitude and longitude of the sub-observatory point on the Moon.

Through these two equations Z_M and Z_S can be computed from the observatory and reflector coordinates along with data given in the American Ephemeris. Standard methods are available for obtaining β_S and λ_S from the parameters listed in the ephemeris for the physical observation of the Moon [28].

TABLE V

Range Formula Expansion

Term	Max. Δr_{SF} km	Max. $\dot{r}_{SF}$ m s^{-1}	Classification
$r_{SF} = r_{EM}$	5×10^4	± 75	Orbit
$- R_E \left(1 - \frac{p^2}{2} - \frac{q^2}{2}\right) \cos Z_M$	6.4×10^3	± 450	Earth rotation
$+ R_E \frac{p}{2} \sin^2 Z_M$	57	± 3.8	
$- R_E \frac{p^2}{2} \cos^3 Z_M$	1	± 0.06	
$- R_M \left(1 - \frac{q^2}{2}\right) \cos Z_S$	650	± 2.8	Lunar libration (including diurnal libration)
$+ R_M \frac{q}{2} \sin^2 Z_S$	0.55	$\pm 10^{-2}$	
$- R_M \frac{q^2}{2} \cos^3 Z_S$	10^{-3}	$\pm 10^{-4}$	
$+ \frac{3}{2} R_E\, pq \cos^2 Z_M \cos Z_S$	0.72	± 0.05	Cross terms
$+ R_M\, pq \cos Z_M \cos^2 Z_S$	0.03	$\pm 10^{-2}$	
$+ O(Rp^3)$			Remainder

intuitively evident before the expansion was made, the expansion gives a valuable quantitative estimate of the error made in neglecting various aspects of the motion. Libration contributes a maximum of only 2.8 m s^{-1} to the range rate, and of this maximum three-fourths is the diurnal effect.

Considering only the first two terms of Table V,

$$\dot{r}_{SF} = \dot{r}_{EM} + R_E(\omega_E - \dot{\alpha}) \cos\phi \cos\delta \sin H - R_E\,\dot{\delta}(\cos\phi \sin\delta \cos H + \sin\phi \cos\delta). \tag{37}$$

This was obtained by differentiation after expressing $\cos Z_M$ in terms of the station latitude ϕ, and the Moon's declination δ and hour angle H. The earth rotation rate is ω_E, and the Moon's right ascension rate is $\dot{\alpha}$. Further numerical evaluation shows that the term of this expression containing $\dot{\delta}$ is only comparable to the neglected second harmonic $\sin^2 Z_M$ term in Table V, which term arises from the daily oscillation of the observatory transverse to the Earth-Moon line. Thus, to an accuracy of about ± 15 meters per second the first two terms of (37) may be taken to represent the range rate. These two terms, which are approximately sinusoids of period one month and one day respectively are plotted separately in Figure 17. Numerically these terms alone give

$$\dot{r}_{SF} = \begin{bmatrix} +75 \\ -75 \end{bmatrix} + \begin{bmatrix} 452 \\ 446 \end{bmatrix} \cos\phi \cos\delta \sin H \quad \text{m s}^{-1}, \tag{38}$$

where the brackets contain the limiting values.

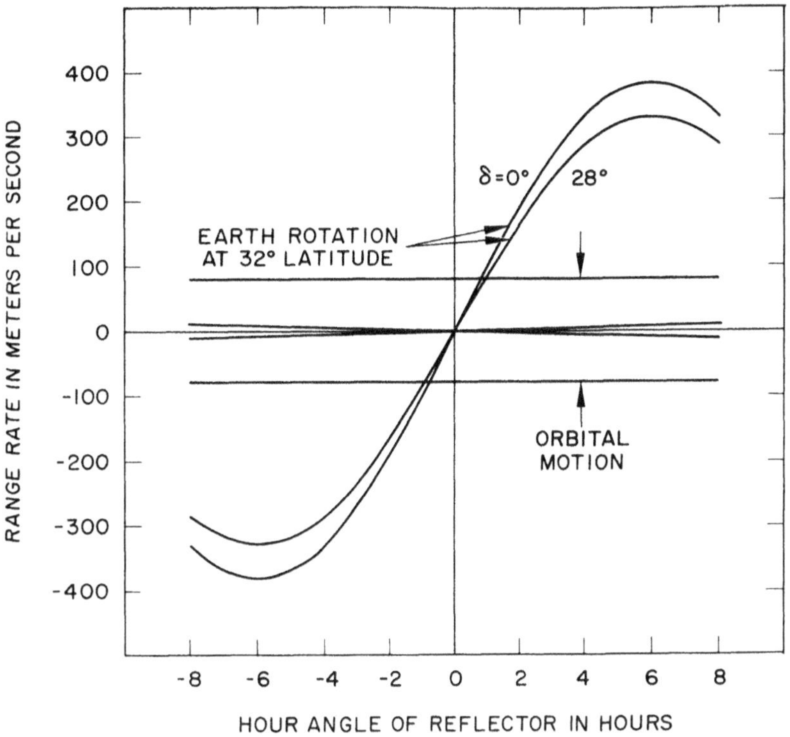

Fig. 17. The two principal components of the range rate.

It is interesting to note in passing that two body Keplerian motion is a particularly poor model of the Moon's actual motion from which to calculate $\dot{\mathbf{r}}_{EM}$. A Keplerian moon traveling in the real Moon's mean orbit would have equal maximum rates of approach and recession on each revolution around the Earth, and this maximum would always be 55.8 m s^{-1}. The actual maximum rates vary between about 40 m s^{-1} and 75 m s^{-1}.

As may be seen from Figure 17, the orbital contribution to the range rate varies little during any one daily passage of the Moon across the sky. The main effect of the orbital radial motion is to shift the moon position at which the total range rate passes through zero, the maximum shift away from the observer's meridian being about $\pm$ one hour. Figure 17 suggests that if the range be expanded as a power series in the time about the daily instant of minimum range the coefficients of the expansion should not vary greatly from one passage to another.

From this viewpoint the range rate is

$$\dot{r}_{SF} = a_1 h + a_2 h^2 + a_3 h^3 \ldots, \tag{39}$$

where h is the time (in hours) measured from the instant of minimum range. If, for some particular passage of the Moon, accurate values for the a's are wanted they would best be obtained on a computer by evaluating (19) and then taking derivatives numeri-

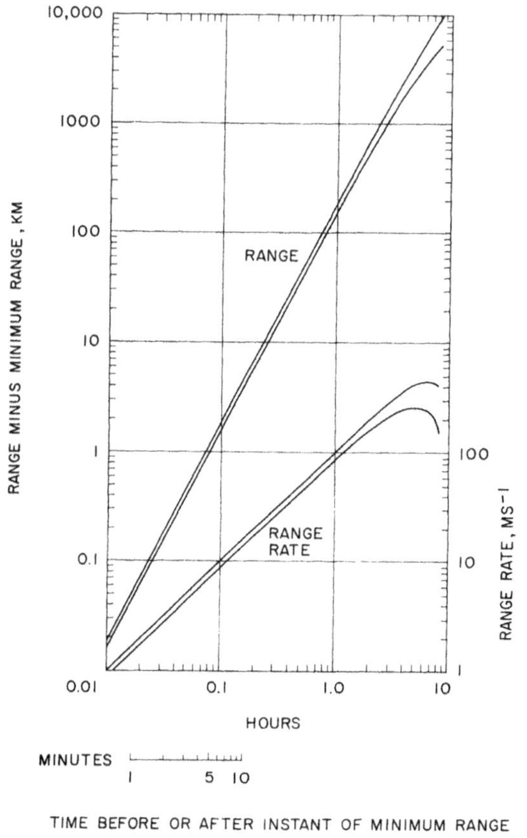

Fig. 18. Daily variation of range and range rate at a 32° latitude observatory.

cally. An examination of the possible values of the time derivatives of the terms in Table V has resulted in the curves of Figure 18, which figure is equivalent to a plot of (39) and its integral. The dual curves bracket the values of range and range rate possible from an observatory at 32° latitude. With a maximum error of about one m s^{-1} hr^{-1} the coefficient a_1 is given by

$$a_1 = \frac{2\pi}{24} R_E(\omega_E - \dot{\alpha}) \cos\phi \cos\delta \cos H_0 \quad \text{m s}^{-1}\ hr^{-1}, \tag{40}$$

where besides quantities already defined, H_0 is the hour angle of the reflector at $h=0$. Numerically a_1 lies between 85 and 101 m s^{-1} hr^{-1} for $\phi=32°$. The nearly straight parts of these curves illustrate that over a considerable interval surrounding the instant of minimum range the range rate varies almost linearly and the range almost quadratically with time.

OFFSET GUIDING

A third aspect of the Moon's motion which influences the design and operation of a ranging system is libration. The continually changing aspect presented to the observer by the Moon complicates offset guiding of the telescope. Either with or without reflectors on the Moon the ability to aim the telescope(s) at some desired point on the lunar surface by holding cross hairs to some other point would be convenient and may be essential. Although an eyepiece for providing a view through the receiving telescope of the spot being sensed by the detector is easily provided with essentially no sacrifice in efficiency by means of a dichroic beam splitting mirror, lunar surface details may not always be sufficiently clear for guiding on the earth-lit image. Some sort of automated guiding from the lunar limb might even be desirable.

The angular offset between any two points changes with time in accordance with the topocentric libration. The time rate of change of the offset angles in either right ascension or declination between two points a and b on the Moon is bounded approximately by

$$\dot{A} = 0.3 \times B, \tag{41}$$

where $\dot{A}$ is the rate of change of offset angle in arc-sec per hour, and B is the angle in arc-min subtended by the length ab when viewed broadside at the Moon's distance. The change is caused mostly by the diurnal motion of the observing station. Because the offsets vary slowly with time, completely mechanized programming of these quantities should be unnecessary. The motion could instead be followed sufficiently accurately by manually inputting the offsets and their time derivatives occasionally.

If the reflectors are to be photographed against a star background considerably higher angular offset rates are encountered. For this purpose an attachment like the plate holder of existing dual rate moon cameras might be used.

System Considerations

Besides staying within the comparatively fundamental boundaries implied by the foregoing discussions, a real system can be made only from existing technology. Laser technology itself is discussed extensively elsewhere [29, 30] and so is largely avoided in this report, but some points specific to moon ranging must be mentioned. A system block diagram is given in Figure 19. This diagram is an expansion of Figure 2, to show more details of the control and data equipment.

LASERS

At the present time only Q-switched lasers using ruby crystal or neodymium glass rods are powerful enough to be interesting for ranging to the Moon. Using these light sources with telescope apertures not unusual in astronomical work a few photons will be recovered from each laser shot (perhaps a few hundred photons with a retroreflector), but this small number is sufficient because the background count from all sources can be made small by comparison, and because the short pulse generated by Q switching gives a useful range determination from a single detected photon.

Since neither of the above lasers emits in a part of the spectrum (ruby 0.7μ, neodymium 1.06μ) where photocathodes operate best, frequency doubling the laser output

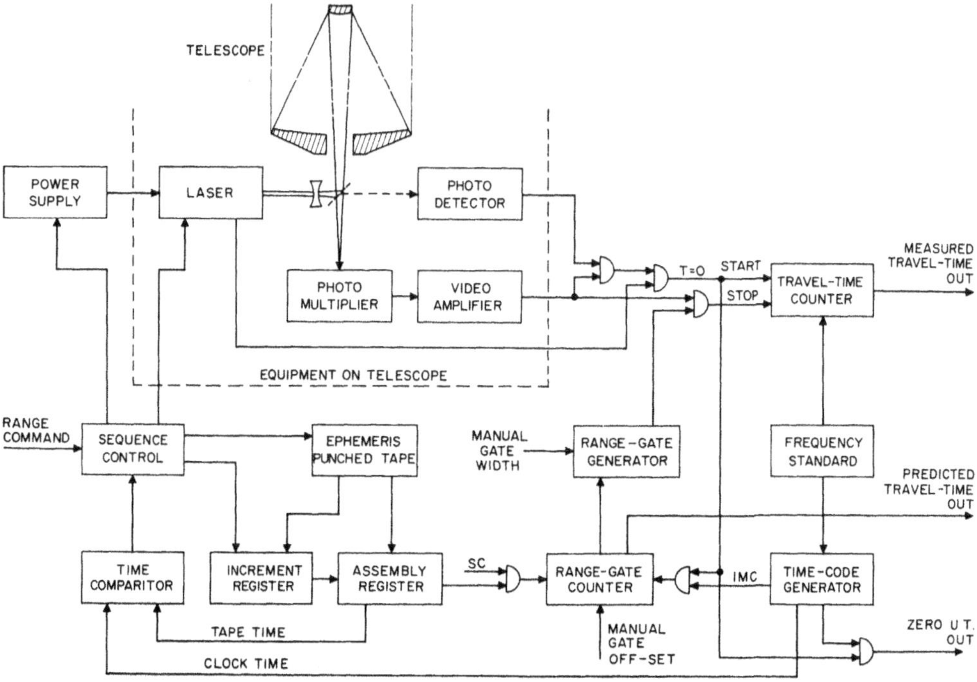

Fig. 19. System Diagram.

is worth considering.* Doubling is accomplished by passing the laser beam into a carefully arranged crystal of KDP, and although it complicates the transmitter the process is comparatively efficient. Thus four wavelengths are feasible: 0.7μ, 1.06μ, and half these values. Table VI compares calculated relative performances at these wavelengths. Although the one joule laser output at which the comparison is made is by no means the highest energy attainable, one joule is seen to be perhaps enough for ranging on retroreflectors. For ranging directly on the lunar surface one joule is very marginal.

The present state of laser technology is about 10 joules in 10 nanoseconds with a divergence of about a milliradian from a 15 mm diameter rod. Unless an arbitrarily large transmitting telescope is available, radiance (power per mm^2 per steradian) is the important output parameter and not simply output energy.

DETECTORS

Only photomultipliers have the necessary combination of fast response (10^{-8} seconds) and high sensitivity. By applying the well developed techniques used in astronomy and nuclear science for counting individual photoelectrons from refrigerated cathodes the

* Since the time of this study CO_2 gas lasers have produced average powers in the one kilowatt range at a wavelength of 10.6 microns. A system using this laser and a superhetrodyne receiver might be quite effective when operating with retroreflectors on the Moon. Velocities could be accurately measured from the Doppler shift.

TABLE VI

Wavelength Comparison

Comparison assumes one joule per shot from laser before frequency doubling. Transmitting and Receiving Telescopes are 1.5 m diameter. Transmitting and Receiving Optical Transmittances are 0.75 and 0.4.

Wavelength	3500	5300	7000	10600 Å
Efficiency of frequency doubling	0.3	0.2	–	–
Transmitted photons per shot	0.5	0.5	3.5	5.3×10^{18}
Atmospheric transmittance (one way)	0.47	0.81	0.91	0.96
Detector quantum efficiency	0.5	0.5	0.1 [a]	0.0015
Lunar surface ranging				
Average normal albedo	0.065	0.10	0.16	0.26
Photons at detector per shot	0.01	0.04	0.6	1.6
Photoelectrons per shot	0.005	0.02	0.06	0.002
Retroreflector ranging [b]				
Optimum diameter of single perfect reflector – cm [c]	5.4	8	11	16
Photoelectrons per shot with above reflector [c]	0.2	1.3	3.6	0.2
Photoelectrons per shot with a cluster of the same mass [d]	0.3	2.4	6.7	0.4
Photoelectrons per shot with 100 cm² reflector and no aberration [e]	18	22	22	0.3
Solar Background				
Solar constant – photons s^{-1} m^{-2} A^{-1}	2.0	5.2	5.1	3.5×10^{17}
Photoelectrons [f]	0.005	0.02	0.01	.0003

Notes:

[a] Assumes a photomultiplier assembly using light trapping in the glass substrate.

[b] Assumes a projected beam divergence of 2 arc-sec (diameter 3.7 km at the Moon). Reflector transmittance unity.

[c] With 5×10^{-6} radian velocity aberration. See Table III.

[d] With 5×10^{-6} radian velocity aberration. Cluster of six reflectors of diameters 3, 4.5, 6, 9 cm respectively at the different wavelengths

[e] Single perfect diffraction limited reflector 11.3 cm diameter.

[f] 3 arc-sec diameter field stop, 10 A filter, 30μs range gate, average albedo, dark moon conditions ($\rho = 10^{-5}$, Figure 7). Background at full moon is 10^4 times this.

ultimate goal of observing the individual arrival of every photon is frustrated only by the imperfect quantum efficiency of the photocathode. The 0.5 efficiency exhibited by the S 20 photosurface in the green drops to about 0.03 at 7000 Å and is much less still at 1.06μ.

The main reason for the poor efficiency at 7000 Å is that the photosurface is comparatively transparent to that wavelength. If the surface is made thick enough to absorb strongly, the photoelectrons do not escape. This has led to experiments [31] with trapping the incident light inside the glass plate which supports the photocathode. By means of a prism cemented to this glass the incident light is introduced at an angle so as to suffer repeated internal reflection at the two boundaries of the plate, thus encountering the photocathode a number of times. Photomultipliers thus modified have shown increases of as much as a factor of 5 in efficiency. There seems to be no reason why selected tubes to which this technique has been applied will not exhibit efficiencies as high as 0.1 at 7000 Å. No doubt the same technique could be applied at 1.06μ, but the resulting efficiency would still be poor.

RECEIVER OPTICAL PASS BAND

Besides the high spatial coherence of laser light, which enables a very intense beam to be focussed on the Moon, an equally important property is the time coherence which allows a very narrow band optical filter to be placed ahead of the detector. In practice the background of solar radiation can be attenuated by a factor of 1000 by an interference filter which transmits more than one half the received laser light.

One common type of filter (combination of multi-layer dielectric interference filter with a dye filter) has a band width of around 10A and a peak transmission of 0.5. Filters incorporating an additional interference element that narrows the passband to 1A are also available. These narrower filters are inconvenient because of their narrow acceptance angle and because of the difficulty of keeping the filter and the laser, both of whose operating wavelengths vary with temperature, tuned to the same wavelength. Furthermore for rejecting broadband background radiation the effect of low level transmission over all parts of the spectrum must be considered. It is doubtful if the very narrow filters have an equivalent bandwidth much less than 10A when transmission in the skirts of the transmittance curve is included. Typical 10A filters have about double their nominal bandwidths when measured in this way [32].

RECEIVER FIELD STOP

Background radiation at the detector is also controlled by excluding by means of a stop in the focal plane all light outside a small region. Since thermal electron emission can be reduced to a few per second by cooling the photocathode, it may be seen from Figure 7 that the optical background produces the only significant background detector count. This count is directly proportional to the solid field angle passed by the stop, which stop should therefore be just large enough to insure that most of the reflected laser light is passed. The stop size required for this will be determined by the image perfection as set by the telescopes and the atmosphere (and by the laser for surface ranging), plus an allowance for the fact that the receiving and transmitting optics may not be in perfect mutual alignment. If a common telescope is used for transmitting and receiving the alignment can be made so accurate that any effective misalignment will come from image wander, caused either by the atmosphere or by unsteadiness of the telescope mount. Image wander produces an effective misalignment because 2.5 seconds elapse between transmission and reception.

The fact that for all practical purposes the overall signal to noise ratio will be inversely proportional to the square of the angular diameter of the field stop must not be forgotten in the pursuit of a large receiving aperture. Because of the relatively bright lunar background large light-collectors having poor angular resolution and unsteady mounts are not so well suited to laser moon ranging as they are to the collection of light for stellar work.

TIME MEASUREMENTS

Any useful laser measurement of the distance between an earth station and a lunar station requires that two time quantities be determined: the interval elapsing while the light pulse travels to the Moon and back, and the instant on some standard time scale when the observation was made.

The basic scheme for determining the travel-time interval, as shown on the system diagram, involves starting a high speed counter when the laser light pulse leaves, and stopping the counter when the reflection returns, the counter being driven by at a precisely controlled rate. This simple method will be adequate for ranging on the comparatively strong reflection from a retroreflector against a dark moon background because the threshold level for stopping the counter may be set so that essentially perfect discrimination on the basis of amplitude is made between possible spurious pulses and the true signal.

Ranging on the reflection from the lunar surface requires some extension to this arrangement. Here the threshold must be set so low as to respond to the liberation of a single photoelectron at the cathode of the photomultiplier because the reflected laser flash is not likely to produce more than one photoelectron. In this case background solar photons will not be negligible, and the true signal can be identified only by correlating the results of several laser firings. A similar discrimination problem exists when ranging on a reflector against a sun-lighted background.

One way of integrating the results of several laser shots is to store the detected pulses in electronic time-bins as is frequently done in counting nuclear events. Each bin is of duration equal to the desired time resolution, and bins are provided in great enough number to cover the entire interval being observed. After a series of laser firings the number of responses recorded in each bin is read out. This method as applied to lunar ranging is complicated by the fact that the position of the bins must be constantly shifted to follow the changing range according to a pre-computed program. Even though the range rate is predictable accurately enough that observations spanning many minutes need not be degraded appreciably by this method a different scheme, which more clearly separates the recording and processing of data, may be preferable.

This alternate method is to record the delay in seconds of every receiver output pulse within the range gate along with the time at which the corresponding laser shot was fired, leaving all correlating to be done later on a computer. This more complete recording method allows the statistics of the detected photons to be examined in as much detail as may be desired.

This second system can be implemented by providing additional travel time counters to the system of Figure 19. The counters are all started in unison by the outgoing pulse and driven from a common frequency standard. They are fed through gates in such a way that the first detection of a photon after the opening of the range gate stops the first counter, the second detection stops the second counter, etc. Because the expected number of photons within the range gate is small the number of counters required is modest. On the other hand the number of range bins needed for the earlier system is large so the second more complete recording system may require the less equipment.

Accuracy of range measurement will therefore be considered in relation to Figure 19, where the single travel time counter shown may not be the only one provided. If the frequency f of the standard has a fractional error ε, and the value used for the speed of light has a fractional error u, then the error in a range measurement will be

$$\text{error in } R = (u - \varepsilon)\, r \pm 1.5 \text{ meters}, \tag{42}$$

where the ± 1.5 meters is the 10 nsec quantitizing error of the travel time counter. When two different range measurements are subtracted to give a measured range difference D,

$$\text{error in } D = (u - \varepsilon)\, D - r\delta\varepsilon \pm 3 \text{ meters}, \tag{43}$$

where $\delta\varepsilon$ is the change in ε between the two range measurements. There would also be a term $r\delta u$ except that we believe the speed of light is constant.

From this we see that even though the 10 nsec laser pulse has an equivalent range length of only 1.5 m, the uncertainty of range measurement is about ± 400 m because the fractional uncertainty u in the speed of light is about 10^{-6}. For most purposes this uncertainty in terms of meters is not important because light-seconds is an acceptable unit for expressing the size of the orbit, and because the 400 m uncertainty largely cancels out of difference measurements. Equation (43) shows that, provided D and $\delta\varepsilon$ are small enough, the accuracy of range difference measurements will be limited only by the quantitizing error, which has been chosen to match the pulse length. For example, with any reasonably good frequency checking scheme the first term in (43) will be dominated by uD, and for range differences less than 10^6 meters this term will be less than one meter. The error will then be determined by $r\delta\varepsilon$, and can be made less than one meter by making $\delta\varepsilon$ less than 2.5×10^{-9}. The stability of a quartz oscillator running continuously can be as good as a few parts in 10^{10} per day, and by checking the frequency of such an oscillator against either a low frequency standard radio station such as WWVL or against an atomic standard the ranging error caused by the inaccuracy of the time base can be made negligible.

As may be seen from Figure 18, the range to a point on the Moon from a 32° latitude earth station can vary as fast as ± 460 m per second, and will change at a rate less than 100 m per second for only about two hours each day. Hence, if the equivalent error in a range measurement caused by the uncertainty of the time to which the measurement refers is to be as small as one meter, the time will have to be recorded for each separate laser pulse with an accuracy of about 2×10^{-3} sec. Equipment which in practice will be accurate to one millisecond through synchronization with standard time broadcasts is commercially available.

In converting measured travel times to range a correction must be made for the additional delay caused by the contribution of the Earth's atmosphere to the optical path length. This delay is a function of the air mass along the path and so will vary with the zenith angle Z. It will also vary slightly with the barometric pressure. At normal pressure the total delay in meters produced by the atmosphere is given by 2.3 sec Z. For observations not too near the horizon the uncertainty as to the proper value for this correction will be small compared to one meter.

In summary, time base inaccuracy will introduce error into both the travel time data and the epoch data, and the atmosphere will add a slight perturbation, but the effects of all of these upon range can be reduced to less than one meter, leaving the quantitizing error dominant.

RANGE GATING

As previously inferred, one of the measures necessary to reduce the effect of stray background light is the use of an electronic gate which passes signals from the photomultiplier only during a short interval around the pre-computed instant at which the echo is expected. Although under favorable circumstances such as a good retroreflector on a dark moon background a range gate many kilometers long might be acceptable, the ability to control a gate of only a few microseconds duration with a tolerance of about a microsecond is desirable if the full potential of laser ranging is to be realized. To accomplish this when the Moon is far from the local meridian will require that the gate position be updated a few times per second.

A method of implementing such close control without storing an inordinate amount of range ephemeris data is to store range and its time derivatives. Between epochs for which the data are stored electronic adders update the range as often as necessary by adding differences. The required frequency of entries on the ephemeris tape depends upon the number of derivatives stored, and this frequency is predictable by examining the possible values of the coefficients in a power series expansion of the range about an arbitrary instant. If Δ is the spacing in seconds of the ephemeris data points and if T is the gate position tolerance in microseconds (the gate position may be in error by $\pm\ T\ \mu$s) the following approximate relations hold under extreme conditions at a 32° latitude station. If range only is stored

$$\Delta = 0.55T\,. \tag{44}$$

If range and its first derivative are stored

$$\Delta = 290T^{1/2}\,, \tag{45}$$

and if also the second derivative is used

$$\Delta = 2400T^{1/3}\,. \tag{46}$$

The second case appears to be operationally the most practical if the ephemeris is stored on punched paper tape. The entries could be spaced at five minute intervals, which would be economical of tape and punching time and still be often enough for convenience in synchronizing the tape to the clock. An implementation of this system is indicated on the system diagram.

PHOTOGRAPHY

Retroreflectors will be most useful on the Moon if they can be identified on photographs of the lunar surface. A knowledge of their location relative to topographic features is necessary if they are to be used as control points for lunar mapping. Photographing reflectors in ruby laser illumination appears to be feasible under good seeing conditions.

For one of the most sensitive films available Figure 20 shows exposure data expressed as photographic density produced by a given number of photons in an image of given size. Since a reflector on the Moon will photograph as a star, image density is not so directly significant as is the signal to noise value (S/N) associated with the image [33]. An S/N value of 5 is marginally satisfactory for deciding the existence of a star against a sky background, but for accurately locating a reflector a value between

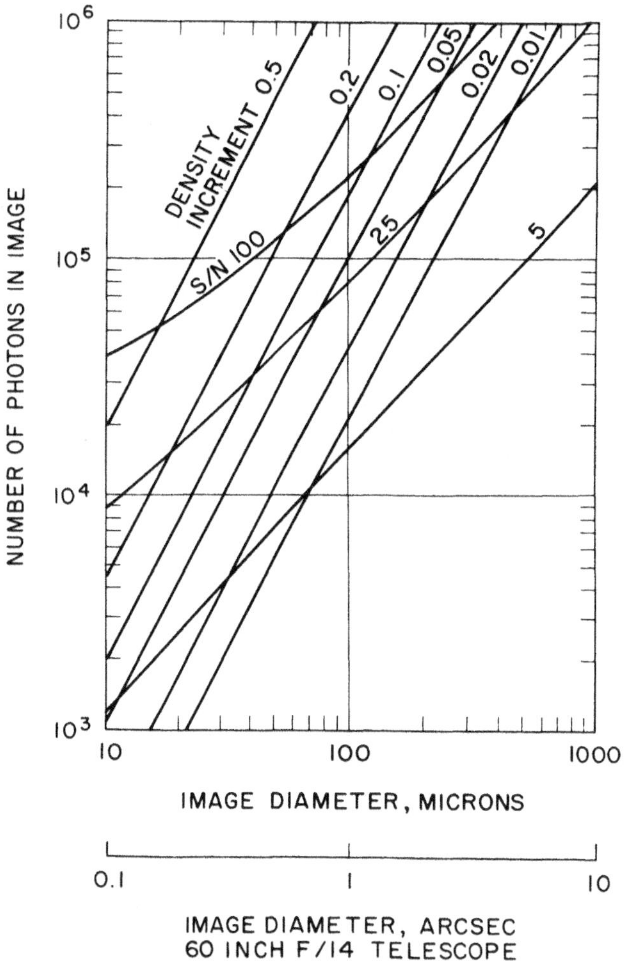

Fig. 20. Retroreflector photographic image data. Ruby laser light on Kodak 2475 film.

25 and 100 will probably be necessary even if the background exposure is controlled so that no background is discernable at the location of the reflector image. The background would be controlled by optical filtering and by using a mechanical camera shutter synchronized to the arrival of the reflections from the several laser shots necessary to produce a proper photographic image of the reflector.

If the reflector is to be recorded against a blank background the necessary topographic reference would be obtained by making the exposure at a phase of the Moon when the terminator appeared on the plate. Alternatively the background exposure could be controlled so that the topography surrounding the reflector was recorded in earth-light illumination, but this would probably require greater exposure by the reflector to produce a satisfactory image.

When illumating a reflector for photography Q-switching would not be used because a short pulse is no advantage and because a given laser will generate more energy per flash with the Q-switch removed. The 10 joule, 10 ns, Q-switched ruby laser cited earlier for example, would deliver about 40 joules in a 600 μs pulse without the Q-switch. The data of Table VI indicate that a single 100 cm^2 diffraction limited reflector should return about 11000 photons* from a 40 joule ruby laser pulse with the photographic telescope located at the center of the reflected beam. Under this condition Figure 20 shows that with a 100 μ diameter reflector image 7 laser shots would produce an S/N value of 25, and about 20 shots would produce an S/N value of 100 in the image. As would be expected Figure 20 shows the great value of good seeing conditions for photographing the reflectors.

Appendix: Cube-Corner Retroreflectors

This appendix describes some of the optical properties of cube-corner retroreflectors which are significant to their use on the Moon. It includes some calculated diffraction patterns of corners whose dihedral angles are in error.

Among the various forms of retroreflector available, the comparatively wide angular field over which cube-corner reflectors operate makes this type the best candidate for placement on the Moon. Topocentric libration has an extreme range of about $\pm$ 11 degress so even if a reflector is carefully oriented considerable tolerance to viewing aspect is needed. Figure 21 shows how the effective aperture area of an optical cube corner falls off as it is viewed away from its principal axis. This figure is for a corner filled with material of optical index n, which has been edged to circular outline.

A cube-corner retroreflector may be visualized as the corner cut from a hollow or transparent cube by a plane passed symmetrically through three edges. One body diagonal of the cube is normal to this plane. We call this body diagonal the principal axis and call the truncating plane the face of the reflector. In Figure 22, which is a view looking into the corner along the principal axis, the equilateral triangle is the boundary of the face, and the three cube edges meeting at the center of the view have

* The table shows 29 photoelectrons for one joule at 0.1 detector quantum efficiency; and $29 \times 10 \times 40 = 11600$.

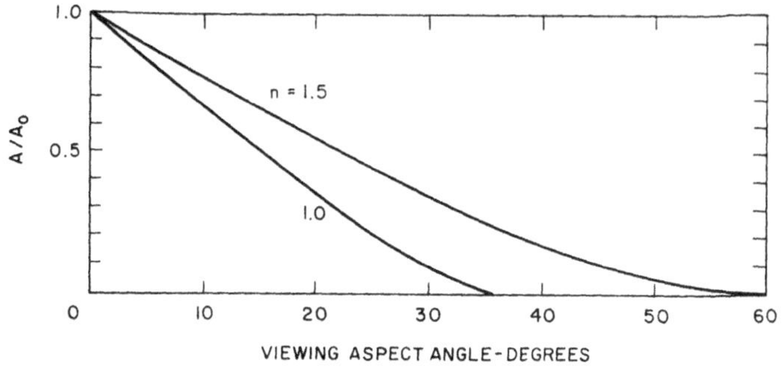

Fig. 21. Effective aperture area of a circularly edged cube-corner retroreflector. The viewing aspect angle is the angle between the principal axis and the line of sight. A/A_0 is the relative aperture area.

been designated as x, y, and z axes. If the internal faces of the original cube are mirrors then geometric optics shows that a ray incident on the cube corner normal to the plane of Figure 22 at the location of one of the round dots will, after reflection from each of the three mirrors, emerge parallel to itself from the other dot. The emergent and imergent rays are parallel and symmetrically disposed about the origin, and this relation remains true whether the principal axis of the reflector is parallel to the rays or not.

An effective way of picturing the passage of rays through a cube-corner is to think of the corner as one octant of a regular octahedron which is completed by the successive images of the corner in the three reflecting planes $x=0$, $y=0$, and $z=0$. The images of a ray make a straight line through these image spaces. From this picture we see that only those incident rays whose straight line continuations pass through the face of the octahedron which is opposite to the face of the cube corner will be retroreflected. Thus, for the axial aspect depicted in Figure 22 only the part of the aperture inside the regular hexagon is effective, and so in order to save space and weight reflectors are

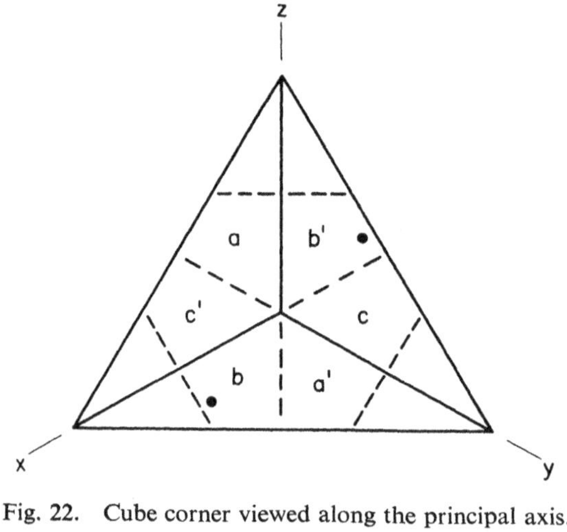

Fig. 22. Cube corner viewed along the principal axis.

frequently truncated to this outline even though the cut away parts might have contributed to the retroreflection of rays incident under some other aspect.*

A real cube corner will not be the perfect reflector just described. The surfaces will not be mathematically plane, the dihedral angles will not be exactly right angles, and a solid reflector will not be perfectly homogeneous to the propagation of light. In practice the dihedral angles are the feature most difficult to perfect.

According to geometric optics [34], small errors in the dihedral angles will cause an incident bundle of rays to emerge as six (in general) separate bundles (three centrally symmetric pairs) each slightly inclined to the retrodirection. In Figure 22 these bundles emerge from the six regions designated a, a', b, b', c, and c', which regions correspond to the six possible sequences in which a ray can be reflected from three planes. For incidence normal to the face of the reflector Figure 23 shows the angular displacements of bundles reflected from two particular imperfect reflectors. In these figures the ε's are the angular amounts by which the respective dihedral angles exceed 90°, and the letters inside the spots identify the regions of Figure 22 from which each beam emerged. The retro direction is at the center of symmetry of each figure. For solid corners the beam spacings shown should be multiplied by the index of refraction of the material because of refraction at the face.

The extension of the foregoing description from geometric optics to wave optics

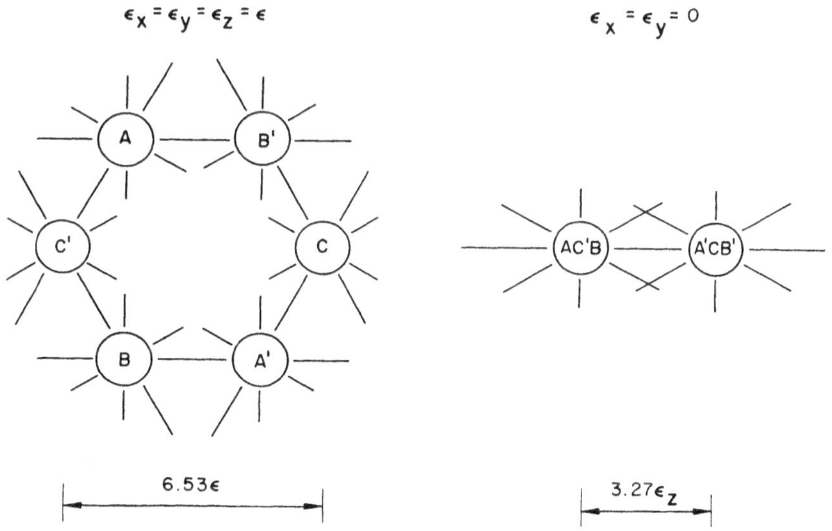

Fig. 23. Spot patterns from imperfect cube corners. For solid cube corners the constants 6.53 and 3.27 should be multiplied by the optical index.

* Optical cube corners are most often made of solid glass, the three reflections being internal. Instead of being cut to hexagonal outline they may be edged on a grinder to circular outline – the inscribed circle of the hexagon in Figure 22. If the diameter of this inscribed circle is D then the volume of the original cube corner is $0.306\ D^3$ ($0.306 = \sqrt{3}/4\sqrt{2}$), the volume of the hexagonally truncated corner is $0.255\ D^3$ ($0.255 = 5/6$ of 0.306), and the volume of the circularly edged corner is $0.249\ D^3$ ($0.249 = [\pi - \sqrt{3}]/4\sqrt{2}$).

is made by identifying the rays with wave normals. In this way the diffraction pattern of the reflected light can be calculated by adding up the waves that emerge from the six diamond shaped apertures a, a', b, b', c, and c'. If the dihedral errors are so large that the waves from these apertures are separated in angle by amounts large compared to the diffraction spread of each aperture the six beams may be considered independently, otherwise the fields must be added with due regard for phase.

A prominent feature of the diffraction pattern from an imperfect cube-corner reflector as seen with an autocollimator is the array of diffraction spikes indicated in Figure 23 by the short lines. These result from the fact that the boundaries of the diffracting apertures consist of straight line segments. Although very noticable to the eye these spikes account for little of the total reflected energy and are unimportant to laser ranging.

For the case of incidence parallel to the principal axis a formula for the far field diffraction pattern of the reflection of a plane wave by a cube corner with arbitrary small errors in the dihedral angles has been worked out.* Some examples of patterns computed from the formula are given in Figures 24, 25, and 26. The curves show the intensity along the sections through the geometric optics spot patterns as indicated by the small sketches at the upper right of each figure. The intensity is plotted relative to the on-axis intensity of a perfect cube corner. The numerical parameter on the

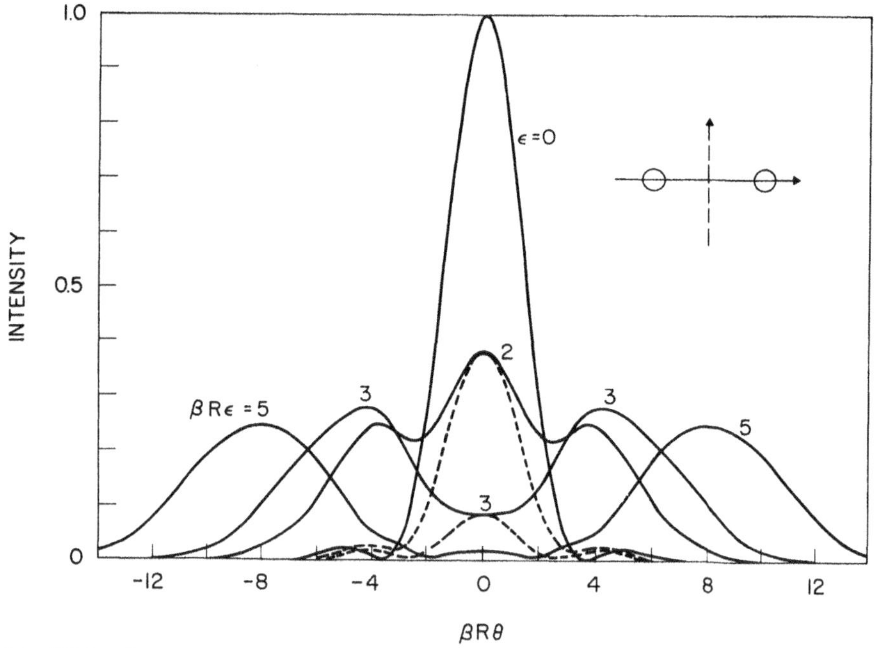

Fig. 24. Diffraction pattern of an imperfect cube corner. A single dihedral angle is in error by the amount ε.

* The calculations assume that the reflecting surfaces are plane boundaries on which the electric field vanishes. The slight modification of the polarization of incident light at the reflecting surfaces of actual optical reflectors hardly affects the patterns except to slightly fill in the nulls. The total effective aperture is the regular hexagon seen in Figure 22.

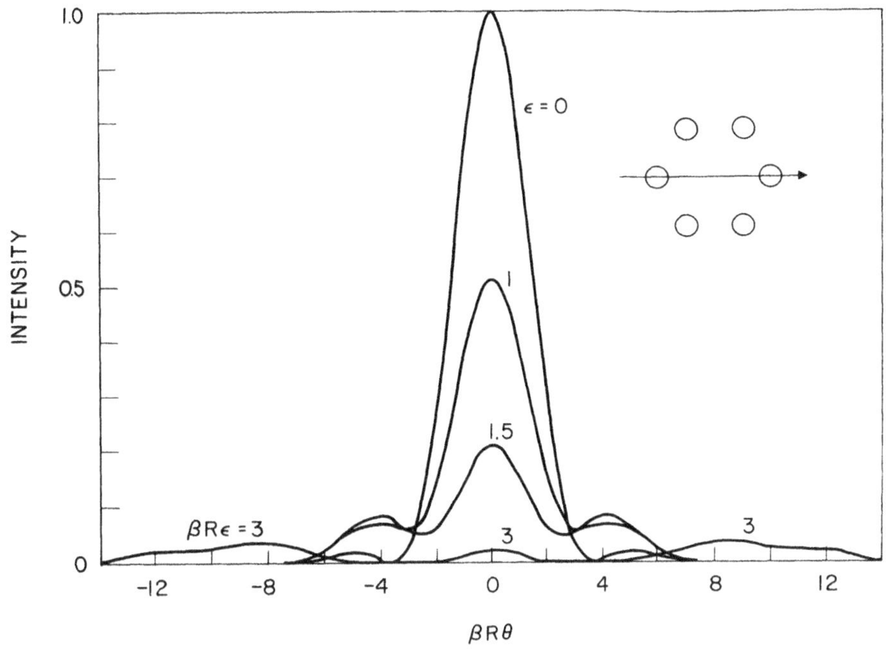

Fig. 25. Diffraction pattern of an imperfect cube corner. The three dihedral angles have equal errors ε.

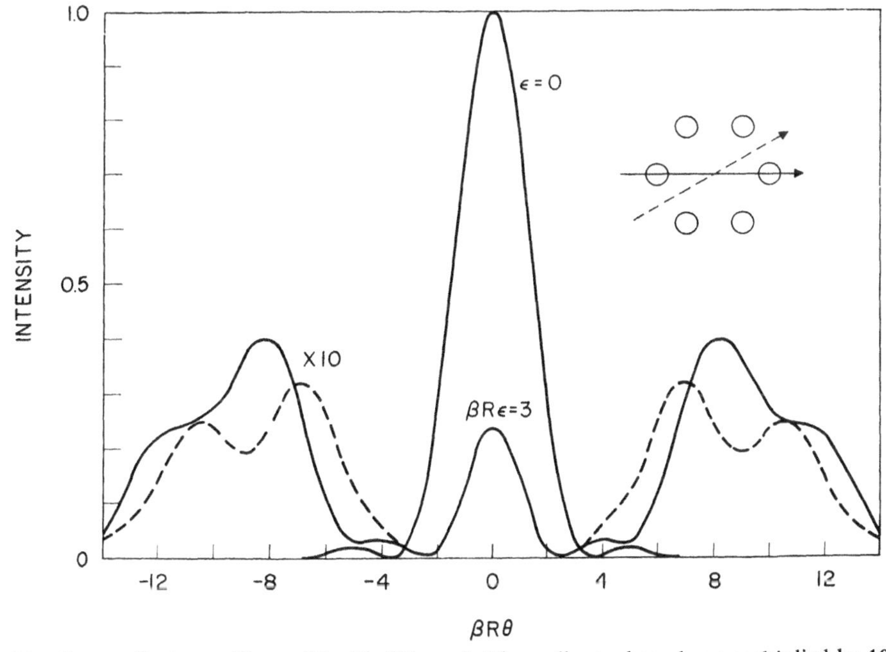

Fig. 26. Same reflector as Figure 25 with $\beta R\varepsilon = 3$. The ordinates have been multiplied by 10 and an additional section of the pattern plotted.

curves is $\beta R\varepsilon$, where β is $2\pi/\lambda$, R is the radius of the circle inscribable on the face of the reflector, and ε is the angular error in one or more of the dihedral angles as specified. The patterns apply to either hollow or solid corners and the wavelength λ is measured in the medium filling the corner.

As would be expected these patterns show that the on-axis reflected intensity is very sensitive to errors in the corner angles. (A $\beta R\varepsilon$ value of unity at one of the dihedrals corresponds to a tilt of one of the reflecting surfaces of about 0.4 wavelength across the extent of the surface.) If the angles could in practice be controlled precisely errors might be deliberately introduced as a way of broadening the reflected pattern as illustrated by Figures 25 and 26. Slightly altering all three angles by the same amount spreads the pattern about uniformly, and as the errors are increased the pattern passes through the form of a fairly uniform ring, as shown by Figure 26, before it breaks up into six separate beams.

As shown by Figure 24, error in only a single dihedral separates the reflection into two beams equally disposed about the retro direction. As the reflector size is increased with the angular corner error held constant these two beams become narrower but maintain constant displaced positions. The intensity of each reflected beam increases with the fourth power of the reflector diameter, and is equivalent to the reflection from a reflector of half the aperture area. Therefore, if a reflector can be placed on the Moon in a completely controlled orientation, as seems to be true for the likely methods of emplacement, then a single corner error could be built in to cancel the effect of aberration for one of the two reflected beams under some preferred observing situation. This would allow the reflector to be made arbitrarily large with the attendant large received signal per kilogram of reflector that is characteristic of large reflectors without aberration. However, in view of the difficulty of making a large corner which will have the required precision of shape and will maintain its performance in the lunar environment a cluster of small reflectors appears to be a more practical approach to building retroreflectors for the Moon.

References

1. *Radar Studies of the Moon.* Quarterly Progress Report, Lincoln Laboratory, M.I.T., 15 February 1966. A summary of the principal results to date in this field.
2. R. W. Hellwarth, 'Control of Fluorescent Pulsations', in *Advances in Quantum Electronics* (ed. by Jay R. Singer). Columbia University Press, New York, 1961.
3. L. D. Smullin and G. Fiocco, 'Project Luna See', *Proceedings of the IRE*, July 1962, 1703.
4. A. Z. Grasyk, V. S. Zuev, Yu. L. Kokurin, P. G. Kryukov, A. N. Sukhanovskii, V. V. Kurbasov, V. F. Lobanov, V. M. Mozhzherin, N. S. Chernykh, and K. K. Chuvacv, 'Optical Location of the Moon', *Soviet Physics-Doklady* **9**, No. 2, August 1964.
5. Robert L. Iliff and Michael S. Tavenner, *Feasability of a Lunar Optical Ranging Experiment.* Air Force Cambridge Research Laboratories, Research Report AFCRL-63-908, December 1963.
6. A. Orszag, 'Moon Distance Measurement by Laser', *Radio Science,* Special Issue on Planetary Atmospheres and Surfaces, **69D**, No. 12, Dec. 1965, pp. 1681 to 1689.
7. Robert L. Wildey, 'Measuring the Shape of the Moon', *Sky and Telescope* **31**, No. 3, March 1966.
8. Yu. L. Kokurin, V. V. Kurbasov, V. F. Lobanov, V. M. Mozhzherin, A. N. Sukhanovskii, and N. S. Chernykh, 'Measurement of the Distance to the Moon by Optical Radar', *JETP Letters,* 1 March 1966, 139–141.

9. H. H. PLOTKIN, T. S. JOHNSON, P. SPADIN, and J. MOYE, 'Reflection of Ruby Laser Radiation From Explorer XXII', *Proceedings of the IEEE*, March 1965.
10. Robert L. ILIFF, 'Photographing Satellite-Reflected Laser Pulses for Geodetic Stereo Triangulation', *Journal of Geophysical Research* **70** (1965), 3505–3508.
11. C. G. LEHR, L. A. MAESTRE, and P. H. ANDERSON, *Measurements of Satellite Range With a Ruby Laser*. Research in Space Science Special Report No. 211. Smithsonian Institution Astrophysical Observatory, May 13, 1966.
12. C. O. ALLEY, P. L. BENDER, R. H. DICKE, J. E. FALLER, P. A. FRANKEN, H. H. PLOTKIN, and D. T. WILKINSON, 'Optical Radar Using a Corner Reflector on the Moon', *Journal of Geophysical Research* **70** (1965), 2267–2269.
13. Mahlon S. HUNT, 'A Prototype Lunar Transponder', *Journal of Geophysical Research* **69** (1964), 2399–2401.
14. G. DE VAUCOULEURS, 'Geometric and Photometric Parameters of the Terrestrial Planets', *Icarus* **3** (1964), 187–235.
15. C. W. ALLEN, *Astrophysical Quantities*. The Athlone Press, 1963.
16. Andre DANJON, 'Albedo, Color, and Polarization of the Earth', in *The Earth as a Planet, The Solar System II* (ed. by G. P. Kuiper). The University of Chicago Press, Chicago, 1954.
17. T. GEHRELS, T. COFFEEN, and D. OWINGS, 'Wavelength Dependence of Polarization. III: The Lunar Surface', *The Astron. J.* **69**, No. 10 (December 1964).
18. J. VAN DIGGELEN, 'Photometric Properties of Lunar Crater Floors', *Recherches Astronomiques de l'Observatoire d'Utrecht* **14**, No. 2 (1960).
J. VAN DIGGELEN, 'The Radiance of Lunar Objects near Opposition', *Planetary and Space Science* **13** (1965), 271–279.
19. V. G. FESSENKOV, 'Photometry of the Moon', in *Physics and Astronomy of the Moon* (ed. by Zdeněk Kopal). Academic Press, New York–London, 1962.
20. M. MINNAERT, 'Photometry of the Moon', in *Planets and Satellites, The Solar System III* (ed. by G. P. Kuiper and B. M. Middlehurst). The University of Chicago Press, Chicago, 1961.
21. Ralph STAIR and Russel JOHNSTON, 'Ultraviolet Spectral Radiant Energy Reflected From the Moon', *Journal of Research of the National Bureau of Standards* **51** (1953), 81.
22. H. C. VAN DE HULST, 'Scattering in the Atmospheres of the Earth and the Planets', *The Atmospheres of the Earth and Planets* (ed. by G. P. Kuiper). The University of Chicago Press, Chicago, 1952.
23. M. J. KOOMEN, C. LOCK, D. M. PACKER, R. SCOLNIK, R. TOUSEY, and E. O. HULBURT, 'Measurements of the Brightness of the Twilight Sky', *Journal of the Optical Society of America* **42**, No. 5, May 1952.
24. P. R. GAST, A. S. JURSA, J. CASTELLI, S. BASU, and J. AARONS, 'Solar Electromagnetic Radiation', in *Handbook of Geophysics and Space Environments* (ed. by Shea L. Valley). Air Force Cambridge Research Laboratories, Office of Aerospace Research, U.S. Air Force, 1965.
25. R. K. LONG and C. H. BOEHNKER, *Measured Atmospheric Absorption at Ruby Optical Maser Wavelengths*. Report 1641–10 by The Ohio State University Research Foundation, Contract Number AF 33(657)–11195, Task Number 519102, 1 June 1965.
26. J. R. WHITTEN, G. F. PREHMUS, and K. TOMIYASU, 'Q-Switched Laser Beam Propagation Over a Ten-Mile Path', *Proceedings of the IEEE*, July 1965.
27. B. M. OLIVER, 'Sparkling Spots and Random Diffraction', *Proceedings of the IEEE*, January 1963.
28. R. d'E. ATKINSON, 'The Computation of Topocentric Librations', *Monthly Notices of the Royal Astronomical Society* **111** (1951), 448–454.
29. K. TOMIYASU, 'Laser Bibliography IV', *IEEE Journal of Quantum Electronics*, QE-2, No. 11 (November 1966).
30. *Proceedings of the IEEE* (October 1966). Special Joint Issue with *Applied Optics*, Laser Survey Papers.
31. J. L. GUMNICK and C. D. HOLLISH, *Multiple Reflection Effects Observed for S-1, S-11, and S-20 Photocathodes*. ITT Industrial Laboratories Research Memo No. 442, March 1966.
32. Irving H. BLIFFORD, Jr., 'Factors Affecting the Performance of Commercial Interference Filters', *Applied Optics* **5**, No. 1, January 1966.
33. J. C. MARCHANT and A. G. MILLIKAN, 'Photographic Detection of Faint Stellar Objects', *Journal of the Optical Society of America* **55** (1965), 907–910.
34. K. N. CHANDLER, 'On the Effects of Small Errors in the Angles of Corner-Cube Reflectors', *Journal of the Optical Society of America* **50** (1965), 203–206.

DISCUSSION

KOPAL: I should like to ask Dr. Weimer, or any colleague here present, a question that has worried me somewhat. A great deal of what has been done in recent years in selenodesy is based on the limb charts by Hayn and later by Watts. But, these charts were secured on the basis of photographic plates taken with telescopes of limited resolution. Between two-thirds and three-quarters of the plates which Watts used to obtain his profiles of the limb of the Moon were obtained with a Markowitz camera and the aperture of this telescope, and consequently the optical resolution, are very limited. The earlier charts by Hayn were extensively used by professor Koziel in his reductions of the libration observations. I think Dr. Weimer is the only astronomer who used for this purpose a telescope of aperture 24 inches, although even this has a theoretical optical resolution of one-quarter of a second under ideal conditions. I should, therefore, like to ask the colleagues here present if they think that, as a result of the optical limitations of the telescopes with which the underlying material has been secured, the charts now existing are adequate, or should the work be done over with more powerful optical means?

WEIMER: I will reply in French. (Professor Minnaert translates:) Dr. Weimer tells us that his measurements have been made on plates taken with an aperture of 65 cm and that the principal source of error was the plate grain. He considers the latter to be much more important source of error than the lack of resolution due to the finite diameter of the objective.

KOPAL: In view of the additional limitation pointed out by Dr. Weimer to be inherent in data we now have, should it not be desirable – in fact highly desirable – to take steps in the near future to re-do the maps of the limb of the Moon with more adequate optical means – not the least because this is one of the tasks which will not be materially advanced by photographs from spacecraft orbiting the Moon.

GOUDAS: It is desirable and necessary that the charts of the Marginal Zone be derived anew on the basis of sequential photography. I think Don Meyer has already computed absolute positions of 1500 points on the Marginal Zone. I think that the existing Marginal Zone work is rather outdated, and this can be shown by comparing the results. It is indeed a fact that the results by Hayn and Watts agree as far as the rough size of the second, third and fourth harmonics, good agreement is also found in a comparison between eclipse photographs and Watts' results. However, if you compare single measurements given by Hayn and Watts, you find that on occasions the difference exceeds one second of arc, which corresponds approximately to 2 kms. The charts of Watts are too detailed, I think, for their true accuracy, since they give contours differing by 0".2, which corresponds to about 390 metres, and I think one should not trust them to this extent.

KOPAL: I think what you said, Dr. Goudas, bears out the general story, but I do not know if you meant the words "out of date". The method is not, but the results may be.

GOUDAS: The photographs which were used are.

RÖSCH: Dr. Goudas mentioned work done by Mr. Meyer. I would like to ask whether the long series of photographs of the Moon which you get at the Pic could be used for such a purpose.

MEYER: We have not used Pic-du-Midi photography for the profile work mentioned by Dr. Goudas. We have thus far used long exposure plates taken at Flagstaff with the Navy 61" astrometric telescope, and measured 140 points per plate over eleven plates of different librations. We hope that Pic photography will be utilized successfully in this connection.

ARTHUR: Can I take up the point of resolution for a moment? As I understand, are you saying that there are sharper indentations on the profile than we can possibly see?

KOPAL: There may be; though these are not of primary interest for determination of the global form of the Moon. The principal trouble is the fact that we do not know the geometrical relation between the actual shape of the features on the Moon and the geometric diffraction pattern of these formations on the photographic plate.

ARTHUR: Yes, but I do not think that the modern high resolution photographs show up so very much more than the old photographs. We have been doing lunar profile work at JPL and I find that 400 to 450 points are quite sufficient to summarize the properties of the limb for 170° arc. Any more measurements would be superfluous. In other words, if I have 450 points, and I carefully place them at the bottom of indentations, and so on, then I can draw smooth curves through this and have everything correct to about 0.1 of arc.

KOPAL: For an analysis of the global form of the Moon this may be more than adequate. However, if you wish to use these limb data, say, for the reduction of occultation observations, or if you want to use them for the reduction of heliometric observations, then you may fall in between the available points.

ARTHUR: We need something better than we have already got.

ECKHARDT: I'd like to stress the importance of using occulations, together with these plate measurements; because otherwise your determination of the scale distance will be insufficient.

. . . .

ECKHARDT: Is it true, Mr. Orszag, that you will occasionally get two photons at one time?

ORSZAG: Yes, it should happen.

ECKHARDT: Therefore the separation between these will depend, say, on the local topography of the surface, and this will give you a measure of its roughness?

ORSZAG: Yes. In fact, that is one of the annoying things of having too large a spot on the Moon, and thus have relief effects. The average slope as far as we know might be of the order of one per cent or something of that order, and a four km spot in diameter, which is a reasonable estimate, gives you about 40 m curvature difference; and you will have to take this into account in your reduction even without having two photons emitted simply because you integrate your results over a series of firings and from one echo to another you might have a return current of two photons. We can, of course, get some information from that, but this should not encourage anyone to increase the detection probability by reducing the width of the channels, because below a certain width you will then have the signal returning into one or more adjacent channels and from that time on you gain nothing.

ECKHARDT: The noise level that you cited before – you did not give a unit of time. I presume that it was per second, it was 60 or 70?

ORSZAG: Oh, yes; it is 70 photons per square meter and per square kilometer of surface in one second, measured in a one Angstrom band.

RÖSCH: I want to mention two points. First, I am a little more optimistic than Mr. Orszag and Dr. Julian about the efficiency of the whole experiment when aiming, not at the reflector, but at the point on the surface – the reason being the existence of the well-known opposition effect, according to which the albedo of the lunar surface is somewhat larger for the phase angle equal to zero; and it happens that with the laser experiments, we shall have exactly zero phase. We can, therefore, expect an albedo slightly higher than the usual albedo taken for the surface of the Moon. Now I wish to refer to the light scattered in the instrument. You said that the earth-light would not be the main factor contributing to the optical background. Well, from your computations this is the case, and I think your estimate of the earth-light brightness based on Danjon's measurements must be correct. The fact is that one can observe or photograph the earth-light in a properly arranged instrument with a sufficiently high contrast on the background. This means that, if the instrument is properly built, you can have a background of scattered light lower than the earth-light, and if the earth-light is not very intense as your computations show, this means that the scattered light in the instrument would not be very intense either. So, I think one should try to have a good instrument, and you know as well as I do, that the composite collector that we plan to have is to be built in such a way that there will be a real focal plane. The primary image of the Moon will be formed in the focal plane and we shall isolate a small area by a small hole in the focal plane. Thus we shall eliminate the light of the crescent and we shall have subsequently a second selection to limit even further the spot to be exposed. If you use a Cassegrain system, then you have a mirror which receives the light from the crescent, and it is this mirror that can produce intense scattered light in the instrument.

ORSZAG: The preliminary results show that even with a very powerful laser, we need to develop a very special instrument and to take special precautions. A factor working in our favor is the fact that the colour temperature of the Moon's surface seems to vary with the albedo. Another factor to be considered is the problem of pointing at the right spot on the Moon, because, in fact, you will be firing to parts illuminated by the earth-light – that is, the very faint part where it is difficult to see details – in particular, details with sufficient angular resolution to point the telescope at (something like a fraction of a second of arc). Probably we can use an off-axis device by pointing on the crescent itself; but, of course, as the crescent must be pretty far away from the zone you are firing at, the distance you must travel in the focal plane of the instrument is pretty large; and that is an additional source of error.

JULIAN: And it varies with time, of course, as well with the lunar libration.

ORSZAG: Yes, of course. This should be taken into account every day.

JULIAN: Perhaps more frequently than that. I have some slides of the things that you were talking about, but I really have nothing new to add about scattered light. I think we have come to the same conclusions as yourself. But, I think that the intensity of scattered light in the atmosphere from the crescent will probably be larger than that of the earth-light, except possibly under very good high-altitude conditions. It depends, of course, on how close you are to the terminator. With regard to the lunar albedo, the pessimistic estimates by some people are due to several reasons. They assume the normal albedo of the Moon is 0.1; whereas, I think, it is definitely higher than average at 7000 Å – which is what the lasers' effective wavelength will be. It is 0.16 from data that professor Minnaert and other people have published.

ORSZAG: Yes, and I wonder whether the data experiments or the measurements at the sea level, in places where there is plenty of water vapour and water itself in the atmosphere, might give a pretty high diffusion.

. . . .

RÖSCH: I wish to put one question to Dr. Eckhardt. If I understand well, this scanner you have used during your recent expedition to Thailand does not scan along the annulus, but instead it scans along the radius vector at different position angles. Is this so?

ECKHARDT: Yes, it is; and, moreover, it is essential that you scan every five degrees.

RÖSCH: This is quite different from the scanning I had in mind, which I feel is by far superior and much to the point. May be I can give a short summary at the end of the conference (see pp. 138–139).

ECKHARDT: I have just a very short question about your scanner. Does it operate almost perpendicularly to the annulus?

RÖSCH: Yes, it has only these two first mirrors and rotates. If you like, you can measure the flux through this one slit while rotating all around; this avoids trouble with turbulence because of motion etc. This would give directly the harmonics, as Dr. Goudas mentioned. The first harmonic would give the distance between the centres, and the second harmonic would give the ellipticity of the Moon, etc.

KOZIEL: I wish to ask Dr. Goudas what he thinks of the Moon's figure based on the different sets of available data – namely: those of the eclipse, those obtained from Hayn's charts, and from the charts of Watts. Can he draw conclusions regarding the compatibility of these data?

GOUDAS: If you examine the details, you will see that the coordinates of individual points as given on the charts of Hayn and Watts are not identical, and are sometimes incompatible. This, of course, is true for all measurements of absolute positions, as well as the eclipse data which I presented earlier. But by fitting the harmonics, we find that the same shape is underlying as a background in the charts and the eclipse data for the Marginal Zone of the Moon. Comparisons of individual points of the Marginal Zone, as given by different observers, are liable to lead to false conclusions. The compatibility of the entire sets of data has to be examined by a technique of collective comparison of all data points, and this is what the analysis performed was intended to do. The conclusion of this analysis, as I mentioned in my communication, is that the three sets of data gave comparable coefficients for the Fourier expansion of almost all terms up to order four.

KOZIEL: Whose charts do you consider to be better: Hayn's or Watts's?

GOUDAS: I think the latter are better because I trust more the eclipse data, and Watts's charts are in closer agreement with them.

HOPMANN: I have seen the comparison of your estimates and wish to ask what you think about the differences.

GOUDAS: The differences are not very small. However, they are small enough to conclude that we have attained a level of significance.

KOPAL: They differ somewhat in absolute magnitudes, but agree remarkably well in the signs.

GOUDAS: They indeed agree in the signs as you saw. Their difference in magnitude can be as much as 0.″2 geocentrically. Such a difference for the elliptical component of the limb implies that the semi-major axis is unknown within half a kilometer. There is also a difference in the position angle of the semi-major axis of the elliptical component with respect to the axis of rotation; and we have found that, according to Hayn and Watts, this difference is about 12°.

HOPMANN: Can you repeat the angles of the directions of Hayn and Watts? Yours is about 37°.

GOUDAS: That is correct, though this latter value is not mine alone, since a group of five investigators have contributed to its determination, not counting those who have assisted in the picture-taking mission, such as Mrs. Stoddard.

HOPMANN: And Hayn's is about 23°. That is not a very large difference, and I think the results are very satisfactory. I have made calculations of the same quantities and found very similar answers, and I believe the mutual agreement is satisfactory.

KOZIEL: I have a second question, Mr. Chairman. What can you say, Dr. Goudas, from your results about the so-called Yakovkin effect?

GOUDAS: The Yakovkin effect is, I think, real and it is something you no doubt will find in all good measurements. As you saw, the first harmonic was positive always, and this means that the center of figure lies below the center of mass.

KOZIEL: Yes, but you have not answered my first question. I understand that, on the basis of only one libration, it is difficult to compare the charts by Hayn with those by Watts. But what can we say by comparison is: "Is Hayn better or Watts better?" That is my question.

GOUDAS: Well, I think I already answered it. Watts's charts are, on the whole, better. I made the assumption that the 25 sequential eclipse photographs allowed minimization – if not complete elimination – of the seeing effect. This is especially serious during an eclipse due to the fact that, as the shadow cone penetrates the atmosphere, you have air turbulence along the surface of the cone due to the sudden elimination of the solar radiation. In particular, it seems reasonable to expect hotter currents to come out from inside it, so along this you may have lots of air entering the colder space of the cone – especially along the direction of motion of the cone. This turbulence will be recorded on the photographs as lunar relief, and it is only when you have many sequential exposures that you can distinguish the time relief from the effects of the seeing. The 25 photographs of the 1963 eclipse have, for exactly these reasons, given us a very good lunar profile as well as a standard of comparisons. I took this as a standard of comparison in order to evaluate Watts' and Hayn's work, and concluded that Watts (for this particular profile, being in better agreement with the eclipse results) is better. I do not like the risk of extrapolating from this particular piece of evidence. However, if I were to use one or the other in my own work, I would employ Watts's.

KOZIEL: I would like to ask one question of Mr. Arthur. I noticed you did not consider at all the free libration, and I wonder why.

ARTHUR: I am not too sure about its existence.

KOZIEL: You will never be sure if you do not consider it.

GOUDAS: Mr. Arthur, in your paper I see that you mention Dr. Eckhardt as having computed the solutions of the physical librations by means of numerical integration.

ECKHARDT: And I would say that the coefficients are numerical.

GOUDAS: This is true and could not be otherwise. But you do not integrate numerically by means of the Runge-Kutta or any other conventional technique of numerical integration.

ARTHUR: Professor Koziel talked about that yesterday.

GOUDAS: Professors Koziel's work does not involve numerical integration but analytic approximations of the solution.

KOZIEL: This is true. I did not use numerical integration. I have another question for Mr. Arthur. I see in your paper that you have left out a term in the expression of the node that appears in the physical libration in longitude. In Hayn's table this term exists, and so does it exist in Brown's and Hansen's tables; yet in your paper it is missing. Its amplitude is 8″, and since you retain terms with coefficients of 0″.9, you have left out a much more essential term.

GOUDAS: Have you obtained thus far any values for the five libration constants using the analysis you present in this paper?

ARTHUR: This is the method we intend to apply in order to get these constants.

KOZIEL: I have one more comment regarding Mr. Arthur's technique. It is interesting to apply photographic methods to this problem, but because to date we have no estimates of libration constants from photography, we shall have to await your results. And I wish you good luck!

PART III

SHAPE AND GRAVITATIONAL FIELD OF THE MOON

HARMONIC ANALYSIS OF THE MOON'S SHAPE AND GRAVITATIONAL FIELD

C. L. GOUDAS*

Abstract. Detailed studies of the existing control systems and comparison of the results obtained from each one with the expected values for the second zonal harmonic show that all systems leave something to be desired in accuracy. Out of them, the ACIC (1965) system can be said with certainty to be closer to the required accuracy, whereas the quality of the new (1966) system has definitely dropped. The quality of the new system by SCHRUTKA-RECHTENSTAMM (1966) is low if compared with his earlier system (1958); this seems to be due to the unsuitable photography used.

The gravity zonal coefficients computed here can be wrong only if the data are not accurate enough or if the Moon is not sufficiently of uniform density – the former being a stronger reason than the latter. The tesseral and sectorial harmonics of both the figure and gravity field are affected by the assumption of symmetry, which can be true only to a small extent, and then not for all the harmonics. The preliminary Lunar Orbiter data have already indicated (MICHAEL *et al.*, 1966) that the symmetry assumption is reasonable for zonal harmonics. This is not entirely true for the others.

1. Introduction

A new control system of 140 points has recently been produced by SCHRUTKA-RECHTENSTAMM (1966). We shall call this system SR II and adopt the notation SR I for the system of FRANZ (1901) as reduced by SCHRUTKA-RECHTENSTAMM (1958). In an earlier publication (GOUDAS, 1963a) the SR I was studied, and various conclusions were drawn on the basis of a harmonic analysis that was performed. Later (GOUDAS, 1965) the same system was again analyzed after a number of points were eliminated. The points used in the last analysis had an inherent error in elevation less than 2.0 km, and it was found that the ellipsoidal component of the lunar figure, as determined from the physical libration and the assumption of homogeneity, was thus better approximated.

Because only a finite number of points are employed in the fit of orthogonal polynomials to the lunar surface, the coefficients do not remain invariant when more harmonics are added to the expansion. It is possible to apply a test in order to study the extent to which such a fit is orthogonal, and results from such a test have already been reported (GOUDAS, 1966). In an earlier report (GOUDAS, 1963) it was found that a least-squares fit of harmonics to the near lunar side only is not at all orthogonal: this was confirmed by ECKHARDT (1966). It was recently found useful to apply the orthogonality test indiscriminately in all cases, and use it as a criterion for eliminating

* Boeing Scientific Research Laboratories, Seattle, Wash., U.S.A.

control points with large mean errors in elevation, with the purpose of obtaining a set of invariant harmonic coefficients based on as few points as possible. As a result, this can allow us to eliminate many control points without introducing unacceptably large variances in the coefficients at the same time. For this reason the program was modified in such a way as to increase each time bij one the order of the harmonics included in the fit starting from the first harmonic. Thus we solve first 4 unknown coefficients, then for 9, 16, 25, etc., up to 81.

With these considerations we treated once more the system SR I and were able to eliminate from it all points with mean error 1.8 km in elevation. The same was done for the SR II system. These results will be discussed presently. Another important point in analyzing a control system are the additional necessary control points from the Marginal Zone. The reduction of eclipse photographs by CARSON *et al.* (1966) has resulted in very satisfactory expressions for the lunar profile, and comparisons (GOUDAS, 1965) have shown that these profiles are comparable to the ones given by HAYN (1914), YAKOVKIN (1962), POTTER (1962), and WATTS (1963). For this reason the Marginal Zone has been represented thus far in all harmonic expansions attempted by the above profile. It appears, however, that a certain adjustment is necessary for these points for the following reason. The profile in question is represented by 64 points which were chosen for measurement on the basis of their easy and certain indentification. They are, as a rule, rims of craters or peaks of mountains. Points on flat areas are impossible to identify and measure again on different plates of the sequence used for measurement. The need to use sequential photography makes it virtually impossible to apply a technique other than measuring the radius vectors of well-identifiable objects. As a result, the points measured do not truly represent the surface of the Moon at the Marginal Zone. In particular, out of the elevation of a rim of a crater one should subtract the height of the rim, etc. Such an adjustment has not been performed on the limb points because of the difficulties involved.

The new control system by Schrutka-Rechtenstamm has been discussed by HOPMANN (1966); the reader can also find there a detailed comparison of different systems. These comparisons are based on the degree of correlation of results of measurement of common points in these systems. The common points are indeed few, about 10% or so, and even for the common points it is very probable, if not certain, that the settings have not been made on exactly the same point of the crater. For this reason it is hard to believe that conclusions based on correlation coefficients always do justice to the systems. Nevertheless, for small craters and for all points which are found by different observers as having altitudes differing by more than one kilometer, criticism is entirely justified. We should point out, however, that the discrepancy cannot be interpreted as implying that both values are wrong. The recently gained experience on the effect of "seeing" speaks for the fact that systems derived on the basis of sequential photography should, in general, be better than the rest. However, because only a small number of sequential photographs are usually used, and the technique of selection is based on non-rigorous criteria, it is possible that the quality of a system based on sequential photography drops to the level of systems based on single photographs. Some comment on this will be made during the exposition of the system SR I immediately following.

The mean deviation of the elevations from a sphere of the SR II system is negative and one expects to find a negative zero order harmonic. We shall see later on that this is indeed the case. This system was analyzed together with both the non-adjusted and adjusted limb points.

2. A New Analysis of the System SR I

After subtracting from SR I all points with mean errors larger than 1.8 km in the elevation given by Schrutka-Rechtenstamm, the remaining points were treated, together with the limb points by Carson *et al.* (1966). The latter points were previously adjusted for a bias in the datum which was assumed to be −0.5 km. The modified program for the harmonic analysis solves, at stages for the harmonic coefficients increasing by one each time, the order of the harmonics included in the expansion and prints the obtained solution. In order to guarantee the orthogonality of the fit, the mirror images of all points, including those from the Marginal Zone with respect to the plane $\zeta=0$, are treated as points of the surface. This can lead to the conclusion that the coefficients $J_{n, 2m-1}$ and $J'_{n,2m}$ must be equal to zero, but this is not done in the program. It is instead asked to solve the normal equations for all coefficients and use the above coefficients, which for a perfectly symmetrical body would be equal to zero, as a criterion of correct functioning and accuracy of the program. The actual computations show that $J_{n, 2m-1}$ and $J'_{n, 2m}$ are, in general, more than five orders of magnitude smaller than the coefficients $J'_{n, 2m-1}$ and $J_{n, 2m}$, respectively. For example, we find that for the SR I the values of J_{11} and J'_{11} are 0.55×10^{-7} and $.10\times10^{-2}$, respectively. For the same set J'_{52} and J_{52} have the values 0.74×10^{-8} and -0.22×10^{-1}. Such a difference in magnitude can practically be interpreted as implying that the symmetry requirement is satisfactorily fulfilled.

The solution for the 81 unknown coefficients is given to two significant figures in Table I. Earlier estimates of mean errors in these coefficients on the basis of the uncertainties in the elevation and location of the control points have shown that the given mean errors are not reliable at present; for this reason no effort was made to compute mean errors in the coefficients. In Table II we list the solution of the coefficients of the zonal terms only as this was obtained at stages, whereas in Table III we give the coefficients of zonal, tesseral and sectorial terms up to order four. The equatorial and meridian sections of the Moon corresponding to SR I as expressed by the coefficients of Table I are given in Figures 1 and 2. According to the same coefficients, the diameters a, b, and c of the Moon along the axis ζ, ξ and η, respectively, are as follows:

$$a = 2r_0 + 2h(0, 0) = 2r_0 + 5.61\,,$$

$$b = 2r_0 + h(-\pi/2, 0) + h(\pi/2, 0) = 2r_0 - 0.46\,,$$

$$c = 2r_0 + h(0, \pi/2) + h(0, -\pi/2) = 2r_0 - 0.52\,.$$

TABLE I

Harmonic Coefficients derived from SCHRUTKA-RECHTENSTAMM's (1958) Control Points

$J_{00} = -0.58$		
$J_{10} = -0.34$	$J_{11} = 0$	$J'_{11} = 0.00$
$J_{20} = -0.37$	$J_{21} = 0$	$J'_{21} = 0.43$
	$J_{22} = 0.33$	$J'_{22} = 0$
$J_{30} = -0.34$	$J_{31} = 0$	$J'_{31} = 0.10$
	$J_{32} = -0.052$	$J'_{32} = 0$
	$J_{33} = 0$	$J'_{33} = 0.00005$
$J_{40} = 0.86$	$J_{41} = 0$	$J'_{41} = 0.31$
	$J_{42} = -0.038$	$J'_{42} = 0$
	$J_{43} = 0$	$J'_{43} = -0.0042$
	$J_{44} = 0.0068$	$J'_{44} = 0$
$J_{50} = -0.16$	$J_{51} = 0$	$J'_{51} = -0.19$
	$J_{52} = -0.022$	$J'_{52} = 0$
	$J_{53} = 0$	$J'_{53} = 0.00075$
	$J_{54} = 0.00038$	$J'_{54} = 0$
	$J_{55} = 0$	$J'_{55} = -0.00020$
$J_{60} = 0.47$	$J_{61} = 0$	$J'_{61} = -0.14$
	$J_{62} = 0.030$	$J'_{62} = 0$
	$J_{63} = 0$	$J'_{63} = 0.0015$
	$J_{64} = -0.00012$	$J'_{64} = 0$
	$J_{65} = 0$	$J'_{65} = -0.00025$
	$J_{66} = -0.000016$	$J'_{66} = 0$
$J_{70} = 0.83$	$J_{71} = 0$	$J'_{71} = 0.16$
	$J_{72} = -0.014$	$J'_{72} = 0$
	$J_{73} = 0$	$J'_{73} = 0.0012$
	$J_{74} = -0.00038$	$J'_{74} = 0$
	$J_{75} = 0$	$J'_{75} = 0.000007$
	$J_{76} = 0.0000075$	$J'_{76} = 0$
	$J_{77} = 0$	$J'_{77} = 0.0000021$
$J_{80} = -1.22$	$J_{81} = 0$	$J'_{81} = 0.038$
	$J_{82} = 0.0059$	$J'_{82} = 0$
	$J_{83} = 0$	$J'_{83} = -0.00035$
	$J_{84} = 0.00037$	$J'_{84} = 0$
	$J_{85} = 0$	$J'_{85} = 0.000012$
	$J_{86} = -0.0000023$	$J'_{86} = 0$
	$J_{87} = 0$	$J'_{87} = -0.00000020$
	$J_{88} = 0.00000016$	$J'_{88} = 0$

Thus the diameter pointing toward the Earth exceeds the mean of the other two diameters by 6.10 km.

The mean radius of the Moon, according to SR I is approximately 1737.45 km; this compares very well with the estimate based on the ACIC (1965) system which was found to be 1737.5 km. Subsequent analysis of the ACIC control system, with a few adjustments in the system as well as the limb points, has resulted in a similar value for the mean radius, as we shall see later on in this communication. The mean lunar radius corresponding to the AMS (Army Map Service) control system is (GOUDAS, 1965) about 1.4 km larger than the one corresponding to the ACIC system. It is not easy to conclude which of these two values is closer to the true mean radius

TABLE II

Orthogonality Test – Zonal Harmonics

Terms Included	J_{00}	J_{10}	J_{20}	J_{30}	J_{40}	J_{50}	J_{60}	J_{70}	J_{80}
4	– 0.14	– 0.77							
9	– 0.56	– 0.52	– 0.59						
16	– 0.57	– 0.43	– 0.59	0.24					
25	– 0.60	– 0.34	– 0.66	0.23	0.25				
36	– 0.59	– 0.34	– 0.68	0.32	0.26	– 0.15			
49	– 0.61	– 0.24	– 0.72	0.16	0.37	– 0.08	0.56		
64	– 0.58	– 0.35	– 0.70	– 0.08	0.44	– 0.11	0.46	0.05	
81	– 0.58	– 0.34	– 0.37	– 0.34	0.86	– 0.16	0.47	0.83	1.22
Totals	– 4.37	– 3.33	– 4.31	0.53	2.18	–.50	1.49	1.48	1.22
Means	– 0.55	– 0.42	– 0.62	0.09	0.44	.13	.50	.74	1.22
Mean Dev.	0.09	0.12	0.08	0.20	0.17	.03	.04	.09	0
Mean Dev. %	16	29	13	222	40	23	8	12	–

of the Moon, but from various arguments based on the derivation of the two systems, the most important of which is the sequential character of the photography employed, it appears that the true mean radius will be found to be very close to the value obtained from the ACIC system.

The first harmonic of the lunar figure when it is non-zero indicates that the origin selected for the *standard frame of reference* does not coincide with the center of figure. Thus if ξ_0, η_0, ζ_0 are the three coordinates of the center of figure, then the harmonic expansion will include the term

$$\eta_0 P_1^0(\sin\beta) + (\zeta_0 \cos\lambda + \xi_0 \sin\lambda)\ {}_1^1(\mathrm{s\ n}\,\beta) \tag{1}$$

where

$$P_1^0(\sin\beta) = \sin\beta\,, \quad P_1^1(\sin\beta) = \cos\beta\,.$$

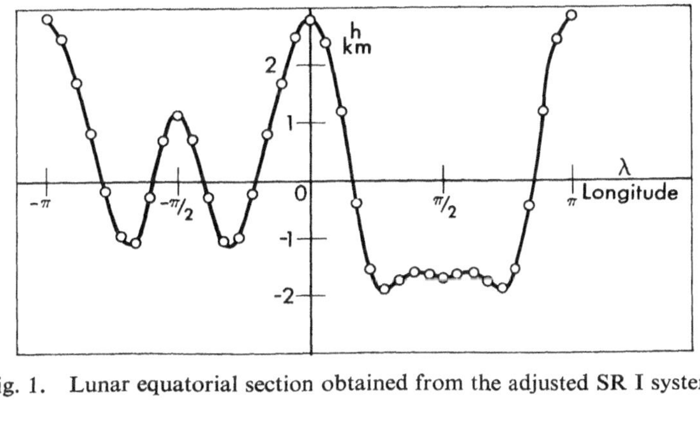

Fig. 1. Lunar equatorial section obtained from the adjusted SR I system.

TABLE III

Orthogonality Test – All Harmonics Up To Order Four

Terms Included	J_{00}	J_{10}	J'_{11}	J_{20}	J'_{21}	J_{22}	J_{30}	J'_{31}	J_{32}	J'_{33}	J_{40}	J'_{41}	J_{42}	J'_{43}	J_{44}
4	– 0.14	– 0.77	0.01												
9	– 0.60	– 0.52	– 0.13	– 0.59	0.36	0.41									
16	– 0.57	– 0.43	– 0.06	– 0.59	0.34	0.40	0.24	0.18	– 0.076	– 0.0020					
25	– 0.60	– 0.34	– 0.16	– 0.66	0.38	0.27	0.23	0.14	– 0.047	– 0.0038	0.25	0.32	– 0.069	– 0.0046	0.0066
36	– 0.59	– 0.34	– 0.03	– 0.68	0.36	0.28	0.32	0.15	– 0.028	– 0.0055	0.26	0.34	– 0.072	– 0.0053	0.0066
49	– 0.61	– 0.24	– 0.09	– 0.72	0.58	0.30	0.16	0.09	– 0.015	– 0.0034	0.37	0.31	– 0.077	– 0.0025	0.0071
64	– 0.58	– 0.35	0.03	– 0.70	0.61	0.29	– 0.08	0.03	– 0.033	– 0.0062	0.44	0.31	– 0.068	– 0.0005	0.0070
81	– 0.58	– 0.34	0.00	– 0.37	0.43	0.33	– 0.34	0.10	– 0.052	0.0000	0.86	0.31	– 0.038	– 0.0042	0.0008
Totals	– 4.37	– 3.33	– 0.43	– 4.31	3.06	2.31	0.53	0.69	– 0.252	– .0099	2.18	1.59	– .342	– .0171	.0341
Means	– 0.55	– 0.42	– 0.05	– 0.62	0.44	0.33	0.09	0.12	– 0.042	– .0017	.44	.32	– .068	– .0034	.0063
Mean Dev.	0.09	0.23	0.05	0.08	0.09	0.05	0.20	0.04	0.016	.0026	.17	.01	.09	– .0015	.0005
Mean Dev. %	16	29	100	13	20	15	222	33	38	153	39	3	13	45	82

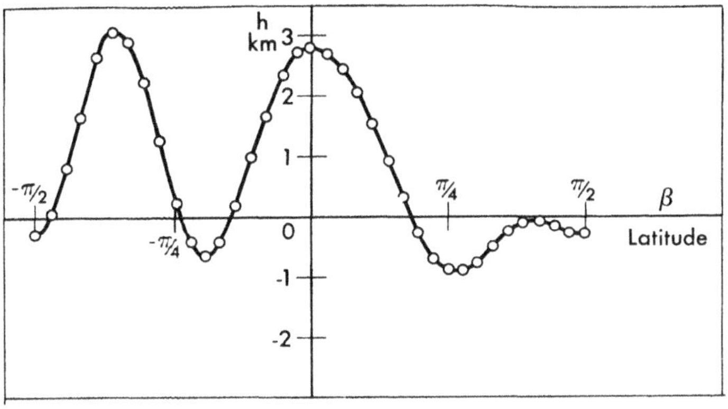

Fig. 2. Lunar meridian section obtained from the adjusted SR I system.

Our usual expansion of the function $h(\lambda, \beta)$ in terms of the coefficients J_{ij} and J'_{ij} includes the first harmonic component in the form

$$J_{10}P_1^0(\sin\beta) + (J_{11}\cos\lambda + J'_{11}\sin\lambda)\,P_1^1(\sin\beta), \tag{2}$$

and hence

$$\eta_0 = J_{10}, \zeta_0 = J_{11}, \xi_0 = J'_{11}. \tag{3}$$

The available data do not allow determination of ζ_0. For reasons explained in an earlier publication (GOUDAS, 1963) a non-zero value for ζ_0 will imply asymmetry between the two sides of the Moon. At the moment we can, in accordance with the assumed symmetry, adopt the value $\zeta_0 = 0$.

From Table III we find

$$\xi_0 = -0.05, \qquad \eta_0 = -0.42. \tag{4}$$

This means that the center of figure of the Moon, according to SR I, lies about half a kilometer south of the origin of this control system. Translation of the control system by this amount will eliminate the first harmonic with only trivial effect on the mean radius and the rest of the harmonics (see GOUDAS, 1966a). The corresponding values of the two coordinates ξ_0, η_0 of the center of figure obtained by analyzing the ACIC control system are (GOUDAS, 1965a, Eqs. (19) and (20))

$$\xi_0 = -0.40, \qquad \eta_0 = -0.39. \tag{5}$$

Only the η_0 coordinate of the SR I and the ACIC system are comparable, whereas the values of ξ_0 are somewhat different. The origin of both systems is based on limb measurements only (see SCHRUTKA-RECHTENSTAMM, 1958); and it is suggested by SPENCER JONES (1932), SCHRUTKA-RECHTENSTAMM (1958), YAKOVKIN (1952, 1962), and GOUDAS (1965a) that the geometrical center of the limb lies south of the center of mass of the Moon. (See also empirical correction for the Moon's center in the *Astronomical Ephemeris and Nautical Almanac*, e.g. 1967, p. 293.) The results presented by Equations (4) and (5) suggest that the center of figure corresponding to these two sets lies about one-and-one-half kilometers south of the center of mass. The assumption of symmetry has little or nothing to do with this estimate, especially with the estimate of η_0, which appears in our expansion as a coefficient of a zonal harmonic fulfilling the assumption of symmetry due to its character; hence the value of η_0 must be correct for both the symmetrical and asymmetrical cases provided the control system is accurate enough.

On the other hand, the analysis of the AMS control system produced the estimates

$$\xi_0 = 0.95\,, \qquad \eta_0 = 0.86 \tag{6}$$

which deviate substantially from the equivalent quantities obtained from the other systems. According to these values, the center of figure has the same η_0 coordinates as the center of mass, but the ξ_0 coordinates differ by one kilometer. In particular, the center of figure lies on the same side as Mare Crisium with respect to the center of mass. The quantities given by Equation (6) are rather large and so were other quantities obtained from the treatment of the AMS system, which again may imply that the ACIC estimate of the first harmonic is better. Regarding the validity of hypothesis that there indeed exists a departure between the center of mass and center of figure, it seems appropriate at the moment to defer discussion until sufficient information on absolute positions of the far side are available. As we already pointed out, the coefficient J_{10} or the coordinate η_0 of the center of figure is independent of the assumption of symmetry. Therefore, the differences in the values obtained from the different systems indicate the existence of errors in the systems themselves.

The present analysis of SR I shows that the second surface harmonic, $Y_2(\lambda, \beta)$, of the Moon is

$$Y_2(\lambda, \beta) = -\,0.31\,(3\sin^2\beta - 1) + 1.32\sin\beta\cos\beta\sin\lambda + 0.99\cos^2\beta\cos 2\lambda\,. \tag{7}$$

The equatorial and first meridian second harmonic waves $E_2(\lambda)$ and $M_2(\beta)$ are

$$E_2(\lambda) = Y_2(\lambda, 0) = 0.31 + 0.99\cos 2\lambda\,, \tag{8}$$

$$M_2(\beta) = Y_2(0, \beta) = 1.30 - 1.92\sin^2\beta\,. \tag{9}$$

If the Moon is homogeneous we should instead find the expressions

$$E_2(\lambda) = 0.30 + 0.21\cos 2\lambda\,, \tag{10}$$

$$M_2(\beta) = 0.51 - 1.11\sin^2\beta\,.^* \tag{11}$$

The expressions obtained from the ACIC system are

$$E_2(\lambda) = 0.23 + 0.87\cos 2\lambda\,, \tag{12}$$

$$M_2(\beta) = 1.10 - 1.56\sin^2\beta\,. \tag{13}$$

* In an earlier publication (GOUDAS, 1965b) the coefficient of $\sin\beta$ in this equation was incorrectly given as $-\,0.90$. The present value is the correct one.

The contribution of the second harmonic to the physical bulge of the Moon is equal to 1.86 km for the adjusted SR I. The analysis of the ACIC system gave the value 1.65 km, which compares better with the value of 0.76 km corresponding to the homogeneous Moon. A further discussion on the best available second order surface harmonic will be made after we give the result of the expansion corresponding to the ACIC (1965) adjusted system.

The third and fourth surface harmonics $Y_3(\lambda, \beta)$ and $Y_4(\lambda, \beta)$ corresponding to the adjusted SR I are found to be:

$$Y_3(\lambda, \beta) = 0.045(5 \sin^2 \beta - 3) \sin \beta + 0.18(5 \sin^2 \beta - 1) \cos \beta \sin \lambda - 0.63 \sin \beta \cos^2 \beta \cos 2\lambda - 0.026 \cos^3 \beta \sin 3\lambda, \quad (14)$$

$$Y_4(\lambda, \beta) = 0.055(35 \sin^4 \beta - 30 \sin^2 \beta + 3) + 0.80(7 \sin^2 \beta - 3) \sin \beta \cos \beta \sin \lambda - 0.51(7 \sin^2 \beta - 1) \cos^2 \beta \cos 2\lambda - 0.36 \sin \beta \cos^3 \beta \sin 3\lambda + 0.66 \cos^4 \beta \cos 4\lambda. \quad (15)$$

From Equations (14) and (15) we compute the equatorial and meridian intersections of the third and fourth harmonics which are found to be:

$$E_3(\lambda) = - 0.18 \sin \lambda - 0.026 \sin 3\lambda, \quad (16)$$

$$E_4(\lambda) = 0.17 + 0.51 \cos 2\lambda + 0.66 \cos 4\lambda, \quad (17)$$

$$M_3(\beta) = 0.86 \sin^3 \beta - 0.77 \sin \beta, \quad (18)$$

$$M_4(\beta) = 6.34 \sin^4 \beta - 7.33 \sin^2 \beta + 1.30. \quad (19)$$

For comparison we give here the same expressions obtained from the ACIC system:

$$E_3(\lambda) = - 0.13 \sin \lambda - 0.11 \sin 3\lambda, \quad (20)$$

$$E_4(\lambda) = 0.09 + 0.55 \cos 2\lambda + 0.87 \cos 4\lambda, \quad (21)$$

$$M_3(\beta) = 1.91 \sin^3 \beta - 1.51 \sin \beta, \quad (22)$$

$$M_4(\beta) = 5.77 \sin^4 \beta - 7.04 \sin^2 \beta + 1.51. \quad (23)$$

A comparison between Equations (16) and (20) shows that the difference is small or trivial; this is also true for Equations (17) and (21), and (19) and (23). We find substantial disagreement in the coefficients of Equations (18) and (22). Again it seems that the expression based on the ACIC system is closer to reality. However, we should stress the point that in all these determinations the signs obtained are in agreement.

The mean error of the fit decreases as more harmonics are added into the expansion; and for the present system after fitting 81 zonal, tesseral and sectorial terms, the mean error was found to be 1.11 km. That is less than the average of the mean error of the observations as estimated by Schrutka. The quality of the fit can be assessed by studying Appendix A, where we list the points of the system SR I together with the calculated heights as they are found by substituting the coordinates of the points in the expansion.

3. Analysis of the System SR II

This new control system is based on measurements on ten Lick plates made by SCHRUTKA-RECHTENSTAMM (1966), following the suggestion by WEIMER (1961) to establish a unique system of 140 points. As HOPMANN (1966) points out, the differences between SR I and SR II are large, and although comparison is possible for a few common points only, his conclusion is similar to the conclusion drawn here on the basis of the harmonic analysis of the two systems. The non-sequential character of the photography used makes it virtually certain that the two systems cannot be compatible, but certainly the mere "law of chance" can make one of the two systems considerably better than the other. In this case the system SR I is the better, although even

TABLE IV

Harmonic Coefficients derived from SCHRUTKA-RECHTENSTAMM'S (1966) Set of Absolute Heights

$J_{00} = -1.51$		
$J_{10} = -0.22$	$J_{11} = 0$	$J'_{11} = 0.26$
$J_{20} = 0.094$	$J_{21} = 0$	$J'_{21} = 0.48$
	$J_{22} = -0.17$	$J'_{22} = 0$
$J_{30} = 0.20$	$J_{31} = 0$	$J'_{31} = 0.041$
	$J_{32} = -0.0079$	$J'_{32} = 0$
	$J_{33} = 0$	$J'_{33} = 0.044$
$J_{40} = 0.22$	$J_{41} = 0$	$J'_{41} = -0.017$
	$J_{42} = -0.11$	$J'_{42} = 0$
	$J_{43} = 0$	$J'_{43} = -0.0053$
	$J_{44} = -0.00026$	$J'_{44} = 0$
$J_{50} = 0.29$	$J_{51} = 0$	$J'_{51} = -0.19$
	$J_{52} = 0.0092$	$J'_{52} = 0$
	$J_{53} = 0$	$J'_{53} = -0.00029$
	$J_{54} = 0.00078$	$J'_{54} = 0$
	$J_{55} = 0$	$J'_{55} = 0.00016$
$J_{60} = 0.15$	$J_{61} = 0$	$J'_{61} = 0.14$
	$J_{62} = 0.022$	$J'_{62} = 0$
	$J_{63} = 0$	$J'_{63} = -0.0029$
	$J_{64} = -0.00034$	$J'_{64} = 0$
	$J_{65} = 0$	$J'_{65} = -0.00026$
	$J_{66} = -0.000037$	$J'_{66} = 0$
$J_{70} = -0.20$	$J_{71} = 0$	$J'_{71} = 0.13$
	$J_{72} = 0.00000024$	$J'_{72} = 0$
	$J_{73} = 0$	$J'_{73} = 0.0018$
	$J_{74} = 0.00019$	$J'_{74} = 0$
	$J_{75} = 0$	$J'_{75} = 0.000019$
	$J_{76} = -0.0000032$	$J'_{76} = 0$
	$J_{77} = 0$	$J'_{77} = 0.00000035$
$J_{80} = 0.36$	$J_{81} = 0$	$J'_{81} = -0.069$
	$J_{82} = -0.018$	$J'_{82} = 0$
	$J_{83} = 0$	$J'_{83} = 0.0014$
	$J_{84} = 0.00020$	$J'_{84} = 0$
	$J_{85} = 0$	$J'_{85} = 0.000046$
	$J_{86} = 0.0000039$	$J'_{86} = 0$
	$J_{87} = 0$	$J'_{87} = 0.0000012$
	$J_{88} = 0.00000056$	$J'_{88} = 0$

this can be termed as "poor" in comparison to a system that takes advantage of the possibilities offered by sequential photography.

In Table IV we give the complete solution of the normal equations for the coefficients of all 81 harmonics, whereas in Tables V and VI we give the various solutions at stages, in the first for the coefficients of the zonal terms only and in the second for all terms, including zonal, tesseral and sectorial up to order four. The last two tables clearly show the fulfillment of the orthogonality requirement. In Table V we see that the zonal coefficients deviate from their mean value by less than 0.2 km on the average. In Figures 3 and 4 we give the equatorial and first meridian sections of the Moon as derived from all 81 surface harmonics on the basis of the coefficients presented in

TABLE V

Orthogonality Test – Zonal Harmonics

Terms Included	J_{00}	J_{10}	J_{20}	J_{30}	J_{40}	J_{50}	J_{60}	J_{70}	J_{80}
4	− 1.40	− 0.24							
9	− 1.24	− 0.23	− 0.43						
16	− 1.24	− 0.19	− 0.46	0.10					
25	− 1.22	− 0.19	− 0.23	0.07	0.65				
36	− 1.21	− 0.22	− 0.27	0.07	0.70	0.32			
49	− 1.14	− 0.22	− 0.33	0.02	0.79	0.34	− 0.08		
64	− 1.12	− 0.29	− 0.33	0.04	0.87	0.38	− 0.20	− 0.09	
81	− 1.51	− 0.22	0.09	0.20	0.22	0.29	0.15	− 0.20	0.36
Totals	−10.08	− 1.80	− 1.96	0.32	3.23	1.33	− 0.13	− 0.29	0.36
Means	− 1.26	− 0.23	− 0.28	0.06	0.65	0.33	− 0.04	− 0.15	0.36
Mean Dev.	0.10	0.023	0.12	0.04	0.17	0.03	0.13	0.05	0
Mean Dev. %	8	10	43	67	26	9	325	33	0

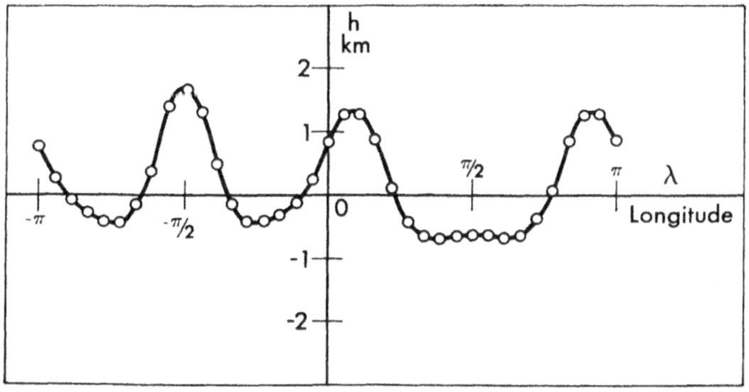

Fig. 3. Lunar equatorial section obtained from the adjusted SR II system.

TABLE VI

Orthogonality Test – All Harmonics Up To Order Four

Terms Included	J_{00}	J_{10}	J'_{11}	J_{20}	J'_{21}	J_{22}	J_{30}	J'_{31}	J_{32}	J'_{33}	J_{40}	J'_{41}	J_{42}	J'_{43}	J_{44}
4	– 1.40	– 0.24	0.56												
9	– 1.24	– 0.23	0.49	– 0.43	0.33	– 0.31									
16	– 1.24	– 0.19	0.24	– 0.46	0.32	– 0.32	0.10	0.071	– 0.022	0.049					
25	– 1.22	– 0.19	0.25	– 0.23	0.32	– 0.32	0.07	0.064	– 0.014	0.050	0.65	0.10	– 0.073	– 0.0017	– 0.00073
36	– 1.21	– 0.22	0.23	– 0.27	0.34	– 0.32	0.07	0.082	– 0.012	0.051	0.70	0.12	– 0.071	– 0.00083	– 0.0010
49	– 1.14	– 0.22	0.18	– 0.33	0.38	– 0.37	0.02	0.12	– 0.012	0.046	0.79	0.12	– 0.063	– 0.00086	– 0.0014
64	– 1.12	– 0.29	0.03	– 0.33	0.38	– 0.38	0.04	0.18	– 0.007	0.049	0.87	0.12	– 0.059	– 0.00076	– 0.0018
81	– 1.51	– 0.22	0.26	0.09	0.48	– 0.17	0.20	0.04	– 0.008	0.044	0.22	– 0.02	– 0.11	– 0.0053	– 0.0003
Totals	– 10.08	– 1.80	2.24	– 1.96	2.55	– 2.19	.50	2.51	– .0075	2.89	3.23	.44	– .376	– .00945	– .00523
Means	– 1.26	– .22	.28	– .28	.36	– .31	.08	.42	– .0013	.48	.65	– .09	– .073	– .00189	– .00105
Mean Dev.	0.10	.02	.25	.12	.04	.04	.04	.31	.0004	.04	.17	.04	.013	.00136	.00044
Mean Dev. %	8	91	89	43	11	13	50	74	31	8	26	44	18	72	42

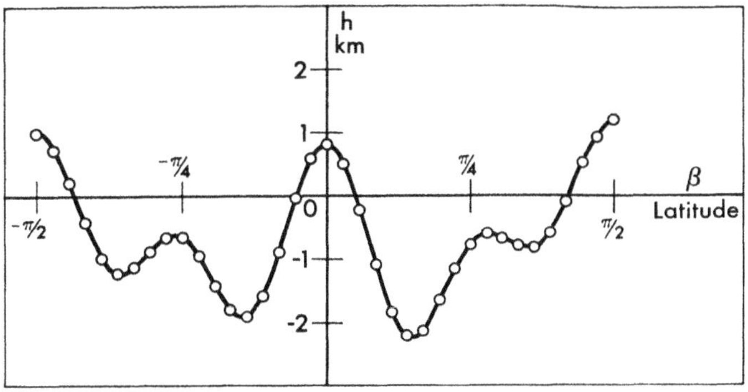

Fig. 4. First lunar meridian section obtained from the adjusted SR II system.

Table IV. The mean error of the fit of the 81 harmonics is 1.3 km and the errors in the individual points can be found in Appendix B.

The first major discrepancy between SR I and SR II appears in the mean radius which for the system SR I is found to be one kilometer larger than in the system SR II. As was mentioned earlier, this could be predicted from an inspection of the dispension of heights in the two systems. The orthogonality test indicates that the estimate of 1736.74 for the mean radius corresponding to SR II has a mean error of only 0.10 km; hence, the discrepancy is genuine and the reasons for the discrepancy should be searched for in the effects of the atmosphere, the accuracy of the measurements and reductions, and finally the large phase angles of eight out of the twelve photographs used for the determination of SR II.

For the center of figure we find from Table VI that

$$\xi_0 = 0.28\,, \qquad \eta_0 = -\,0.22\,, \qquad \zeta_0 = 0\,; \tag{24}$$

hence the center of figure of the near lunar side, including the Marginal Zone, virtually coincides, according to these figures, with the mean position of the center of the lunar disk. This again is not borne out completely by the analysis of SR I.

The second major discrepancy between the systems SR I and SR II and the main factor pointing to the existence of large systematic errors in the system SR II is the second harmonic. The coefficient of the zonal term has the expected sign although its magnitude is about 100% off the expected; but neither the sign nor the size of the coefficient of the sectorial term are correct. To deal with this case in some detail we write the second harmonic $Y_2(\lambda, \beta)$ in the form

$$Y_2(\lambda, \beta) = -\,0.14(3\sin^2\beta - 1) + 0.84\sin\beta\cos\beta\sin\lambda - 0.93\cos^2\beta\cos 2\lambda\,. \tag{25}$$

Thus, we obtain the expressions

$$E_2(\lambda) = 0.14 - 0.93\cos 2\lambda\,, \tag{26}$$

$$M_2(\beta) = -\,0.79 + 0.51\sin^2\beta\,, \tag{27}$$

from which it is deduced that the semi-diameters a, b and c of the ellipsoidal component are

$$a = r_0 + E_2(0) = r_0 - 0.79\,, \tag{28}$$

$$b = r_0 + E_2(\pi/2) = r_0 + 1.07\,, \tag{29}$$

$$c = r_0 + M_2(\pi/2) = r_0 - 0.28\,. \tag{30}$$

Therefore, according to SR II, the major semi-axis of the ellipsoidal component of the figure of the Moon coincides with the ξ-axis of the standard frame of reference and thus makes a 90° angle with the principal axis of least inertia. In addition, the diameter of the figure in the mean Earth–Moon direction is shown by Equations (28)–(30) to be the shortest of the three and smaller than the mean of the other two by roughly one kilometer. In view of this impossible outcome, the present author mildly concludes that SR II does represent a good approximation of control systems. We shall, therefore, not discuss further the remaining reductions of this system.

4. Adjustment and Analysis of the ACIC (1965) Control System

The present system has been analyzed and evaluated by the present author (1965b), and the conclusion was that it gives a good description of the lunar figure as this should be if the Moon is homogeneous. The announcement made during the Manchester meeting by Hopmann (see discussions at the end of Part I of this volume) that according to the evidence given by Luna 10 the Moon is very homogeneous, encouraged the further study of this system presented here.

The reasons suggesting the adjustment of the limb points obtained from the 1963 eclipse in South Africa have already been explained, and this is the first new factor rendering necessary a new reduction of this system. A check of the orthogonal character of the fit of harmonics to all data available to date has shown that even fewer data points can fulfil equally well this essential property. This allowed the elimination from the ACIC (1965) control system of a number of points with mean error in elevation exceeding one kilometer. As we shall see below, the effect of the elimination of these bad points led to an improvement of the harmonics obtained, if we assume that the Moon is homogeneous and the more reliable second zonal and sectorial terms of the lunar gravity field can be used as a criterion for evaluation of the figure analysis.

The fit of 81 harmonics to the adjusted set of control points resulted in the coefficients listed in Table VII, and the root of the mean squared error is 0.79 km. If we compare this error with the error obtained from the fit of all other control systems, we find that this is the smallest. This implies that the adjusted system represents a smooth surface where the unaccounted for higher harmonics will have small amplitudes in comparison to the already estimated harmonics. There is, of course, a limit to the number of harmonics one can obtain from a limited number of control points such as those of the systems under study, even when there is no error in the control points. For example, you cannot compute more harmonic coefficients than you have points available. In theory, the precise case would allow determination of as many

TABLE VII

Harmonic Coefficients derived from the Adjusted ACIC (1965) Control System

$J_{00} = 1.99$		
$J_{10} = -0.35$	$J_{11} = 0$	$J'_{11} = -0.43$
$J_{20} = -0.59$	$J_{21} = 0$	$J'_{21} = 0.59$
	$J_{22} = 0.36$	$J'_{22} = 0$
$J_{30} = 0.41$	$J_{31} = 0$	$J'_{31} = 0.08$
	$J_{32} = -0.059$	$J'_{32} = 0$
	$J_{33} = 0$	$J'_{33} = -0.0081$
$J_{40} = 0.11$	$J_{41} = 0$	$J'_{41} = 0.28$
	$J_{42} = -0.060$	$J'_{42} = 0$
	$J_{43} = 0$	$J'_{43} = 0.017$
	$J_{44} = 0.0068$	$J'_{44} = 0$
$J_{50} = 0.41$	$J_{51} = 0$	$J'_{51} = -0.25$
	$J_{52} = -0.0031$	$J'_{52} = 0$
	$J_{53} = 0$	$J'_{53} = -0.0031$
	$J_{54} = 0.00032$	$J'_{54} = 0$
	$J_{55} = 0$	$J'_{55} = 0.00029$
$J_{60} = 0.38$	$J_{61} = 0$	$J'_{61} = -0.079$
	$J_{62} = -0.0060$	$J'_{62} = 0$
	$J_{63} = 0$	$J'_{63} = -0.00027$
	$J_{64} = -0.00027$	$J'_{64} = 0$
	$J_{65} = 0$	$J'_{65} = -0.00014$
	$J_{66} = -0.000027$	$J'_{66} = 0$
$J_{70} = -0.39$	$J_{71} = -0$	$J'_{71} = 0.16$
	$J_{72} = 0.010$	$J'_{72} = 0$
	$J_{73} = 0$	$J'_{73} = 0.0007$
	$J_{74} = -0.000022$	$J'_{74} = 0$
	$J_{75} = 0$	$J'_{75} = 0.000009$
	$J_{76} = -0.000006$	$J'_{76} = 0$
	$J_{77} = 0$	$J'_{77} = 0.0000003$
$J_{80} = 0.085$	$J_{81} = 0$	$J'_{81} = -0.036$
	$J_{82} = -0.0025$	$J'_{82} =$
	$J_{83} = 0$	$J'_{83} = 0.0011$
	$J_{84} = 0.00013$	$J'_{84} = 0$
	$J_{85} = 0$	$J'_{85} = 0.000011$
	$J_{86} = -0.0000012$	$J'_{86} = 0$
	$J_{87} = 0$	$J'_{87} = -0.00000056$
	$J_{88} = 0.00000014$	$J'_{88} = 0$

harmonics as control points so long as the harmonics unaccounted for are very small in magnitude so as to be unable to impede the fulfillment of the orthogonality condition. This is strictly valid if the control points are free from errors, but a rigorous proof is beyond the requirements of this paper.

In Figures 5 and 6 we give the equatorial and first meridian sections corresponding to the coefficients listed in Table VII. Interpolated heights on the basis of the same coefficients are given in Appendix C.

By comparing the earlier values (GOUDAS, 1965b, p. 535, Table IV) obtained from an analysis of the original ACIC (1965) system with the values listed in Table VII we first notice a small – about 100 meters – change in the mean radius of the Moon, which, therefore, can be still considered as 1737.5 km with a small error in the first decimal.

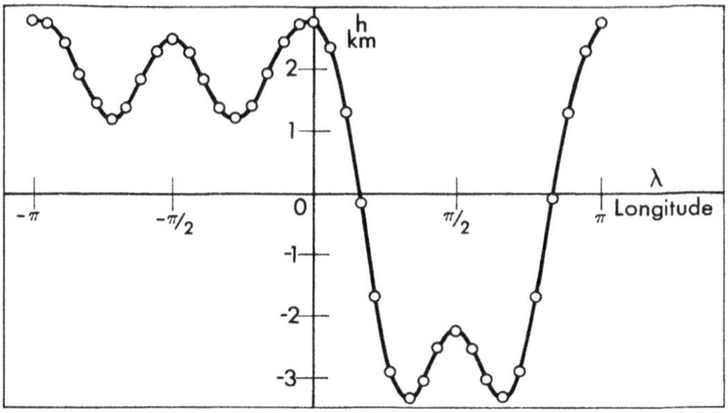

Fig. 5. Equatorial section obtained from the adjusted ACIC (965) system.

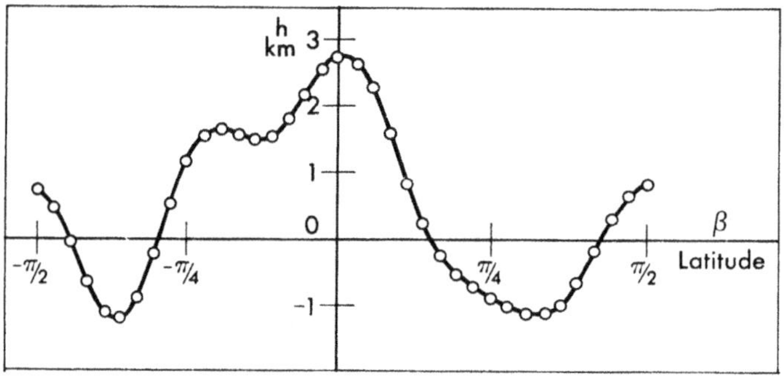

Fig. 6. First meridian section obtained from the adjusted ACIC (1965) system.

The adjustment of this control system resulted in a small change in the first harmonic, i.e., the position of the center of figure. It must be pointed out here that this change of the center of figure is entirely due to the elimination of the unsuitable points of the ACIC system and not to the adjustment of the eclipse measurements, since the latter was a uniform radial adjustment. In any case, this change is of the order of tens of meters and can be considered as unimportant.

A conspicuous change is found in the second zonal harmonic coefficient which jumped from -0.46 to -0.59. As shown earlier (Goudas, 1964, p. 381, Eq. (54)) the expected value of J_{20} in case the Moon is homogeneous, or nearly so, is -0.593; thus the above estimate is in more or less perfect agreement with the estimate obtained here. But the orthogonality test shows (see Table VIII) that this coefficient is subject to variations of about 6% of its value as more harmonics are added to the fit. Thus, the value we can give at the moment to this coefficient is -0.51 ± 0.06. The change of J_{20} from -0.46 (Goudas, 1965b, p. 535) to -0.51 represents an effect of the total adjustment both of the control system and the limb points, and justified the expectation of the author that improvement in the control systems can result in a reconciliation of the figure to the second order gravity harmonics. This being the case, one can perhaps

TABLE VIII

Orthogonality Test – Zonal Harmonics

Terms Included	J_{00}	J_{10}	J_{20}	J_{30}	J_{40}	J_{50}	J_{60}	J_{70}	J_{80}
4	2.46	– 0.61							
9	1.99	– 0.57	– 0.58						
16	1.98	– 0.41	– 0.57	0.10					
25	2.05	– 0.38	– 0.44	0.11	0.17				
36	2.05	– 0.40	– 0.43	0.23	0.13	0.44			
49	2.08	– 0.40	– 0.46	0.24	0.13	0.45	0.27		
64	2.09	– 0.40	– 0.49	0.42	0.10	0.45	0.29	– 0.39	
81	1.99	– 0.35	– 0.59	0.41	0.11	0.41	0.38	– 0.39	0.09
Totals	16.68	– 3.52	– 3.56	1.51	0.64	1.75	0.94	0.78	0.09
Means	2.09	– .44	– 0.51	0.25	0.13	0.44	0.31	0.39	0.09
Mean Dev.	0.10	0.08	0.06	0.11	0.02	0.01	0.04	0	0
Mean Dev. %	5	18	12	44	15	2	13	0	0

provide reasonable estimates of zonal harmonics of both the figure and gravity field, although the limitation posed by the observational error in the figure measurements will affect these estimates probably to a serious extent. As attempted in an earlier publication (GOUDAS, 1964), the connection between surface and gravity harmonics is invalid for inhomogeneous bodies, but for the case of the Moon the departure from homogeneity is, as far as it is known to date, less conspicuous than the effects of observational errors in the figure measurements.

The above statements are valid for the zonal harmonics only, since the assumption of symmetry is valid in this case. For the tesseral and sectorial terms we cannot make good predictions of the gravity harmonics, even when the assumption of homogeneity is valid, because we cannot guarantee the validity of the assumption of symmetry. This, of course, does not mean that the representation of the near side of the lunar surface is not accurate or adequate for preparing contour maps in case the errors of measurement were ignorable, or that we will improve the knowledge of the form of the near side if we have measurements of absolute positions on the far lunar side. It only means that estimates of the gravity harmonics in case of a nearly or entirely homogeneous Moon cannot be predicted from figure information of one of its sides. The assumption made thus far is that each term of the form

$$(J_{ij} \cos j\lambda + J'_{ij} \sin j\lambda)\, P_i^j (\sin\beta), \tag{31}$$

should be equal to

$$[J_{ij} \cos j(\pi - \lambda) + J'_{ij} \sin j(\pi - \lambda)]\, P_i^j (\sin\beta), \tag{32}$$

This false condition poses a serious limitation on values of coefficients J_{ij} and J'_{ij} for $j \neq 0$. Because of this, all coefficients of odd functions of λ are automatically set equal to zero and thus affect the values of the other non-zero coefficients. The effects of this assumption have been mentioned on various occasions in the past. We shall briefly discuss here two cases where undoubtedly the assumption of symmetry has produced erroneous results. The first is the case of the harmonic coefficient J'_{21} for which the various control systems analyzed thus far give a value significantly different than zero, whereas the true value is zero or very close to it. The reason why this should be so can be completely understood by considering the following argument. Let D', E' and F' be the products of inertia with respect to the standard frame of reference $0\xi\eta\zeta$.

By definition we have

$$D' = \int \eta\zeta \, \mathrm{d}M', \tag{33}$$

$$E' = \int \xi\zeta \, \mathrm{d}M', \tag{34}$$

$$F' = \int \xi\eta \, \mathrm{d}M', \tag{35}$$

where $\mathrm{d}M'$ is the mass-element of coordinates (ξ, η, ζ). The assumption of symmetry eliminates the integrals (33) and (34). Using now the formule representing the surface in spherical harmonics, and assuming that the density δ is *constant*, we can write integral (35) in the following successive forms:

$$F' = \delta \int_{-\pi/2}^{\pi/2} \int_{0}^{2\pi} \int_{0}^{r(\beta, l)} \rho^4 \sin\beta \cos^2\beta \sin\lambda \, \mathrm{d}\beta \, \mathrm{d}l \, \mathrm{d}\rho$$

$$= \frac{\delta}{5} r_0^5 \int_{-\pi/2}^{\pi/2} \int_{0}^{2\pi} [1 + \sum_{i=1}^{\infty} \sum_{j=0}^{i} (j_{ij} \cos j\lambda + j'_{ij} \sin j\lambda) P_i^j(\beta)]^5 \times$$

$$\times \sin\beta \cos^2\beta \sin\lambda \, \mathrm{d}\beta \, \mathrm{d}l, \tag{36}$$

where $j_{ij} = J_{ij}/r$, $j'_{ij} = J'_{ij}/r_0$ and r_0 is the mean lunar radius. By eliminating the second and higher order terms in the j_{ij}'s and j'_{ij}'s we obtain,

$$F' = \frac{\delta r_0^5}{5} \int_{-\pi/2}^{\pi/2} \int_{0}^{2\pi} [1 + 5 \sum_{i=1}^{\infty} \sum_{j=0}^{} (j_{ij} \cos j\lambda + j'_{ij} \sin j\lambda) P^j(\beta)] \times$$

$$\times \sin\beta \cos^2\beta \sin\lambda \, \mathrm{d}\lambda \, \mathrm{d}\beta$$

$$= \frac{\delta r_0^5}{5} \int_{-\pi/2}^{\pi/2} \int_{0}^{2\pi} \sin\beta \cos^2\beta \sin\lambda \, \mathrm{d}\beta \, \mathrm{d}\lambda$$

$$+ \delta r_0^5 \sum_{i=1}^{\infty} \sum_{j=0}^{i} j_{ij} \int_{-\pi/2}^{\pi/2} \int_{0}^{2\pi} P_i^j(\beta) \sin\beta \cos^2\beta \sin\lambda \cos j\lambda \, \mathrm{d}\lambda \, \mathrm{d}\beta$$

$$+ \delta r_0^5 \sum_{i=1}^{\infty} \sum_{j=0}^{i} j'_{ij} \int_{-\pi/2}^{\pi/2} \int_{0}^{2\pi} P_i^j(\beta) \sin\beta \cos^2\beta \sin\lambda \sin j\lambda \, \mathrm{d}\lambda \, \mathrm{d}\beta. \tag{37}$$

Now since

$$\int_{-\pi/2}^{\pi/2}\int_{0}^{2\pi} \sin\beta \cos^2\beta \sin\lambda \, d\beta \, d\lambda = 0,$$

Equation (37) becomes

$$F' = \delta r_0^5 \sum_{i=1}^{\infty}\sum_{j=0}^{i} j_{ij} \int_{-1}^{1} P_i^j(\mu)\,\mu(1-\mu^2)^{\frac{1}{2}}\,d\mu \int_0^{2\pi} \sin\lambda \cos j\lambda \, d\lambda$$
$$+ \delta r_0^5 \sum_{i=1}^{\infty}\sum_{j=0}^{i} j'_{ij} \int_{-1}^{1} P_i^j(\mu)\,\mu(1-\mu^2)^{\frac{1}{2}}\,d\mu \int_0^{2\pi} \sin\lambda \sin j\lambda \, d\lambda \tag{38}$$

where $\mu = \sin\beta$. We first notice that

$$\int_0^{2\pi} \sin\lambda \cos j\lambda \, d\lambda = 0, \qquad j = 0, \ldots, \tag{39}$$

and

$$\int_0^{2\pi} \sin\lambda \sin j\lambda \, d\lambda \begin{cases} = 0, & j = 0, 2, 3, \ldots, \\ = \pi, & j = 1. \end{cases} \tag{40}$$

But we also see that the function $P_i^j(\mu)\,\mu(1-\mu^2)^{\frac{1}{2}}$ is odd in the interval $(-1, 1)$, for $j=1$ and for odd values of i. For $i=4$, we obtain

$$P_4^1(\mu)\,\mu(1-\mu^2)^{\frac{1}{2}} = \tfrac{5}{2}(7\mu^4 - 3\mu^2)(1-\mu^2),$$

and hence,

$$\delta r_0^5 j'_{41} \int_{-1}^{1}\int_0^{2\pi} P_4^1(\mu)\,\mu(1-\mu^2)^{\frac{1}{2}} \sin^2\lambda \, d\mu \, d\lambda$$
$$= \frac{5\delta}{2} r_0^5 \left(\int_{-1}^{1} (7\mu^4 - 3\mu^2)(1-\mu^2)\,d\mu\right)\left(\int_0^{2\pi} \sin^2\lambda \, d\lambda\right) = 0. \tag{41}$$

The only non-vanishing term of F' is

$$\delta r_0^5 j'_{21} \int_{-1}^{1} P_2^1(\mu)\,\mu(1-\mu^2)^{\frac{1}{2}}\,d\mu \int_0^{2\pi} \sin^2\lambda \, d\lambda;$$

hence,

$$F' = (4\pi\delta r_0^5/5)\, j'_{21}. \tag{42}$$

Since F' must equal zero, we should expect for j'_{21} the value zero or a value very close to it. Instead we obtain very large values for the analysis of any system of absolute positions. Formula (42) can be written as follows:

$$F' = \frac{3Mr_0}{5} J'_{21}. \tag{43}$$

The value found for J'_{21} is 0.59; thus the value of F' is close enough to zero, indicating that the assumption of symmetry is not suitable for this case.

The second case of unacceptable value of a surface harmonic is the J_{22}. A first approximation of the moments of inertia of the Moon for constant density can be obtained from the evaluation of the integrals

$$A' = \delta \int_{-\pi/2}^{\pi/2} \int_{0}^{2\pi} \int_{0}^{r(\beta,\lambda)} \rho^4 \cos\beta (\cos^2\beta \sin^2\lambda + \sin^2\beta)\, d\rho\, d\lambda\, d\beta, \tag{44}$$

$$B' = \delta \int_{-\pi/2}^{\pi/2} \int_{0}^{2\pi} \int_{0}^{r(\beta,\lambda)} \rho^4 \cos\beta (\cos^2\beta \cos^2\lambda + \sin^2\beta)\, d\rho\, d\lambda\, d\beta, \tag{45}$$

$$C' = \delta \int_{-\pi/2}^{\pi/2} \int_{0}^{2\pi} \int_{0}^{r(\beta,\lambda)} \rho^4 \cos^3\beta\, d\rho\, d\lambda\, d\beta. \tag{46}$$

Integration with respect to ρ yields,

$$A' = \frac{\delta r_0^5}{5} \int_{-\pi/2}^{\pi/2} \int_{0}^{2\pi} [1 + \sum_{i=1}^{\infty} \sum_{j=0}^{i} (j_{ij} \cos j\lambda + j'_{ij} \sin j\lambda)\, P_i^j(\beta)]^5 \times \cos\beta (\cos^2\beta \sin^2\lambda + \sin^2\beta)\, d\lambda\, d\beta, \tag{47}$$

$$B' = \frac{\delta r_0^5}{5} \int_{-\pi/2}^{\pi/2} \int_{0}^{2\pi} [1 + \sum_{i=1}^{\infty} \sum_{j=0}^{i} (j_{ij} \cos j\lambda + j'_{ij} \sin j\lambda)\, P_i^j(\beta)]^5 \times \cos\beta (\cos^2\beta \cos^2\lambda + \sin^2\beta)\, d\lambda\, d\beta, \tag{48}$$

$$C' = \frac{\delta r_0^5}{5} \int_{-\pi/2}^{\pi/2} \int_{0}^{2\pi} [1 + \sum_{i=1}^{\infty} \sum_{j=0}^{i} (j_{ij} \cos j\lambda + j'_{ij} \sin j\lambda)\, P_i^j(\beta)]^5 \cos^3\beta\, d\lambda\, d\beta. \tag{49}$$

Finally, they can be linearized in terms of the j_{ij}'s and j'_{ij}'s as follows:

$$A_{,} = \frac{\delta r_0^5}{5} \int_{-\pi/2}^{\pi/2} \int_{0}^{2\pi} [1 + 5 \sum_{i=1}^{\infty} \sum_{j=0}^{i} (j_{ij} \cos j\lambda + j'_{ij} \sin j\lambda)\, P_i^j(\beta)] \times \cos\beta (\cos^2\beta \sin^2\lambda + \sin^2\beta)\, d\lambda\, d\beta, \tag{50}$$

$$B' = \frac{\delta r_0^5}{5} \int_{-\pi/2}^{\pi/2} \int_{0}^{2\pi} [1 + 5 \sum_{i=1}^{\infty} \sum_{j=0}^{i} (j_{ij} \cos j\lambda + j'_{ij} \sin j\lambda)\, P_i^j(\beta)] \times \cos\beta (\cos^2\beta \cos^2\lambda + \sin^2\beta)\, d\lambda\, d\beta, \tag{51}$$

$$C' = \frac{\delta r_0^5}{5} \int_{-\pi/2}^{\pi/2} \int_{0}^{2\pi} [1 + 5 \sum_{i=1}^{\infty} \sum_{j=0}^{i} (j_{ij} \cos j\lambda + j'_{ij} \sin j\lambda)\, P_i^j(\beta)] \cos^3\beta\, d\lambda\, d\beta. \tag{52}$$

After tedious integrations, we arrive at the expressions

$$A' = \frac{2M'r_0^2}{5}(1 + \tfrac{1}{2}j_{20} - 3j_{33}), \tag{53}$$

$$B' = \frac{2M'r_0^2}{5}(1 + \tfrac{1}{2}j_{20} + 3j_{33}), \tag{54}$$

$$C' = \frac{2M'r_0^2}{5}(1 - \tfrac{1}{2}j_{20}). \tag{55}$$

The moments of inertia of the Moon have been shown from observations to satisfy the following two conditions (JEFFREYS, 1959; KOZIEL, 1964)

$$\beta = (C' - A')/C' = 0.0006267 \pm 0.0000010 \text{ (s.e.)}. \tag{56}$$

$$f = \frac{B'(C' - B')}{A'(C' - A')} = 0.633 \qquad \pm 0.011 \text{ (s.e.)}. \tag{57}$$

This β should not be confused with the selenocentric latitude; and it should be remarked that its value was obtained by substituting in Jeffreys' relation $(1+0.00047f) = 0.0006280$ (cf. JEFFREYS, 1959, p. 165, Equation (8)) the latest value of f given by Koziel. In 1961 Jeffreys published the figure 0.0006279 ± 0.0000015 for β, but we will not gain accuracy by using this value. By substituting the expressions (53), (54), and (55) of A', B', and C', in the above two conditions, and by eliminating products of the j_{20} and j_{22}, we obtain the following two linear equations:

$$-1.4994j_{20} + 3j_{22} = 0.0006267, \tag{58}$$

$$0.1835j_{20} + 1.663j_{22} = 0. \tag{59}$$

The coefficients of the fourth order surface harmonic do not enter in these first order relations and hence they do not play an important part in the lunar differences in momenta. If we limit ourselves to three significant figures and remember that $J_{ij} = r_0 j_{ij}$, we obtain the solution,

$$J_{20} = -0.593, \tag{60}$$

$$J_{22} = 0.0666. \tag{61}$$

As has been remarked already, the value of J_{20} as given in Table VII is remarkably close to the value given by Equation (60). However, if we check the agreement between the expected value and actual value of J_{22}, we find a pronounced disagreement, the cause of which seems to be the assumption of symmetry.

The equatorial and meridian waves of second order given by the harmonic coefficients of Table VII are

$$E_2(\lambda) = 0.30 + 1.08 \cos 2\lambda, \tag{62}$$

$$M_2(\beta) = 1.38 - 1.98 \sin^2 \beta. \tag{63}$$

Although these are in agreement with the previous results (see Equations (17) and

(18); cf. GOUDAS, 1965b) as far as signs are concerned, the size of the coefficients is exaggerated. According to the values of the same table, the third and fourth order harmonic waves along the equator and first meridian are

$$E_3(\lambda) = -0.12 \sin\lambda - 0.12 \sin 3\lambda\,, \tag{64}$$

$$E_4(\lambda) = 0.04 + 0.45 \cos 2\lambda + 0.71 \cos 4\lambda\,, \tag{65}$$

$$M_3(\beta) = 1.89 \sin^3\beta - 1.49 \sin\beta\,, \tag{66}$$

$$M_4(\beta) = 1.20 - 5.43 \sin^2\beta + 4.34 \sin^4\beta\,. \tag{67}$$

In Figures 7 and 8 we give graphically the harmonic waves $E_2(\lambda)$, $E_3(\lambda)$, $E_4(\lambda)$ and $M_2(\beta)$, $M_3(\beta)$, and $M_4(\beta)$, whereas in Figures 5 and 6 we give the equatorial and meridian sections corresponding to 81 harmonics.

In Table VIII we see that the variation of all zonal harmonic coefficients over eight different solutions of the normal equations is in no case larger than 20% (except in the case of J_{30}). This implies that departures of these constants from their true values by more than this amount can be confidently attributed to errors in the control system.

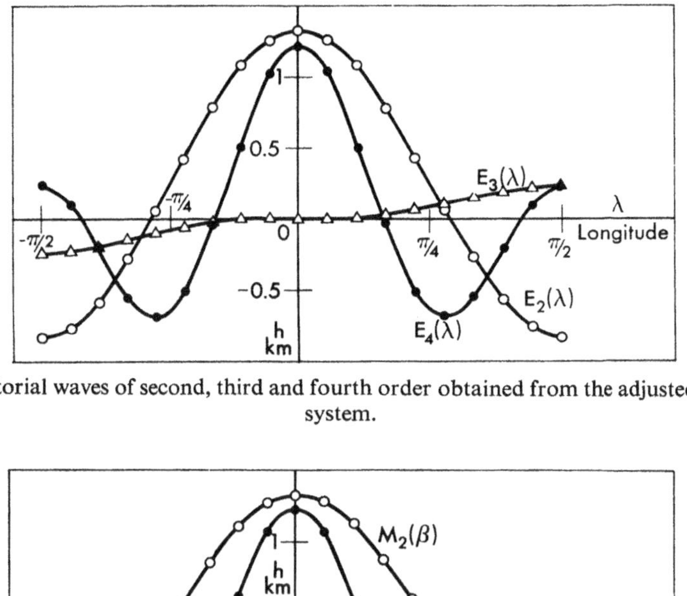

Fig. 7. Equatorial waves of second, third and fourth order obtained from the adjusted ACIC (1965) system.

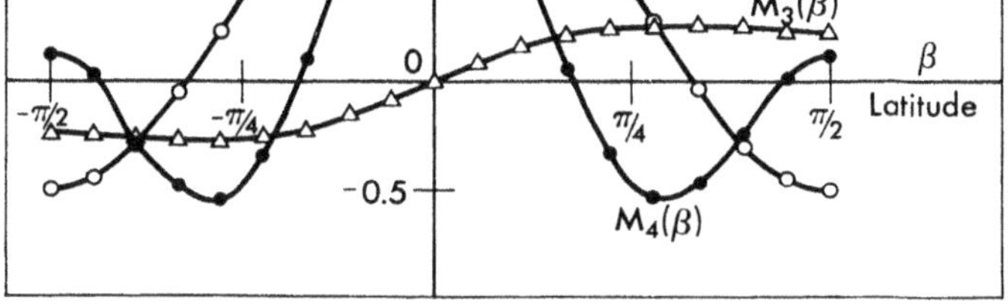

Fig. 8. First meridian waves of second, third and fourth order obtained from the adjusted ACIC (1965) system.

The assumption of symmetry has little or no effect at all on these coefficients. The gravity harmonics corresponding to these values can be computed from the formulae developed in an earlier publication (GOUDAS, 1964); and the actual values will be given below after warning the reader that their accuracy as predictions of the true lunar field will definitely decrease as we move to higher harmonics for two reasons. The first is that the data will, from a certain order of harmonic on, not contain sufficient information to give reliable figure determination. This in effect means that there one is searching for information after crossing the error limits of his observations. The second is that if there are any inhomogeneities in the Moon, the higher harmonics will be more affected by this than the lower ones. The zonal gravity harmonics corresponding to the last solution for the zonals as listed in Table VIII are:

$$\left.\begin{aligned}
K_{20} &= 3J_{20}/\;5r_0 = -\,2.037 \times 10^{-4}, &(68)\\
K_{30} &= 3J_{30}/\;7r_0 = 1.01 \times 10^{-4}, &(69)\\
K_{40} &= 3J_{40}/\;9r_0 = 0.21 \times 10^{-4}, &(70)\\
K_{50} &= 3J_{50}/11r_0 = 0.64 \times 10^{-4}, &(71)\\
K_{60} &= 3J_{60}/13r_0 = 0.50 \times 10^{-4}, &(72)\\
K_{70} &= 3J_{70}/15r_0 = -\,0.45 \times 10^{-4}, &(73)\\
K_{80} &= 3J_{80}/17r_0 = 0.09 \times 10^{-4}, &(74)
\end{aligned}\right\}\ \textit{Zonal Gravity Harmonics}$$

where r_0 is taken equal to 1737.5 km. As already mentioned, it is difficult to estimate the true accuracy of these figures, especially those from Equation (71) on, since the errors in the data are impossible to estimate because of the differential image wandering (usually called "seeing"). The effects of this "seeing" are difficult to estimate in the present data, although the present author feels confident that they can be eliminated by a sufficient number of carefully selected sequential photographs. In spite of the unknown uncertainties in the above gravity harmonics, it is appropriate to stress here that whatever these may be, their origin is not due to the method applied nor to the assumption of symmetry, but to a greater extent to errors in the control system and to a lesser extent to possible inhomogeneities.

The variation of the zonal, tesseral and sectorial coefficients up to order 4 due to incomplete fulfillment of the orthogonality condition is again, as Table IX reveals, never more than 20% of the individual value with the exception of J_{30} and J'_{31}, which vary by 44% and 25% respectively. This variation is due to the insufficient number of points, but it need not be considered as rendering the values obtained completely unreliable. It is possible that the true values are off by the amounts indicated above, or perhaps smaller.

In Appendix C we give a table of interpolated heights based on the present control system. If we again call a, b and c the three diameters of the Moon along the standard coordinates axes, we find that

$$a = 2r_0 + 2h(0, 0) = 2r_0 + 5.51 \tag{75}$$

$$b = 2r_0 + h(-\,\pi/2, 0) + h(\pi/2, 0) = 2r_0 + 0.29 \tag{76}$$

$$c = 2r_0 + h(0, \pi/2) + h(0, -\,\pi/2) = 2r_0 + 1.52, \tag{77}$$

TABLE IX

Orthogonality Test – All Harmonics Up To Order Four

Terms Included	J_{00}	J_{10}	J'_{11}	J_{20}	J'_{21}	J_{22}	J_{31}	J'_{31}	J_{32}	J'_{33}	J_{40}	J'_{41}	J_{42}	J'_{43}	J_{44}
4	2.46	− 0.61	− 0.47												
9	1.99	− 0.57	− 0.46	− 0.58	0.64	0.43									
16	1.98	− 0.41	− 0.39	− 0.57	0.63	0.43	0.09	0.16	− 0.060	0.0008					
25	2.05	− 0.38	− 0.41	− 0.44	0.55	0.31	0.11	0.18	− 0.050	0.004	0.17	0.21	− 0.066	0.011	0.0062
36	2.05	− 0.40	− 0.38	− 0.43	0.57	0.32	0.23	0.12	− 0.062	− 0.005	0.13	0.21	− 0.064	0.012	0.0066
49	2.08	− 0.40	− 0.41	− 0.46	0.65	0.32	0.23	0.12	− 0.058	− 0.006	0.13	0.21	− 0.067	0.014	0.0065
64	2.09	− 0.40	− 0.43	− 0.49	0.68	0.32	0.42	0.08	− 0.047	− 0.009	0.10	0.22	− 0.066	0.015	0.0066
81	1.99	− 0.35	− 0.43	− 0.59	0.59	0.36	0.41	0.08	− 0.059	− 0.008	0.11	0.38	− 0.60	0.017	0.0068
Totals	16.69	− 3.52	− 3.38	− 3.56	4.31	2.49	1.49	.74	− .336	− .023	.64	1.13	− .323	.069	.0327
Means	2.09	− 0.44	− 0.42	− 0.51	0.62	0.36	0.25	0.12	.056	− 0.004	0.13	0.23	− 0.065	0.014	0.0065
Mean Dev.	0.11	0.07	0.02	0.06	0.04	0.05	0.11	0.03	0.004	0.004	0.02	0.02	0.002	0.002	0.0002
Mean Dev. %	5	16	5	12	6	14	44	25	7	10	15	9	3	14	3

which imply that the radius pointing toward the Earth is longer by 2.3 km than the mean of the other two radii orthogonal to it. This bulging is not of an ellipsoidal nature. The contribution of the ellipsoidal component to this bulge can be computed from formulae (62) and (63).

5. The Enlarged ACIC System Contour Map

Shortly before the conference, ACIC developed an extended control system of more than 1200 points based on measurements of about 200 photographs, applying again the technique of averaging positions over photographs exposed sequentially during the same night. This new system was first treated as usual and the harmonics obtained were subsequently used in order to prepare the contour map presented in Figure 9. The Marginal Zone was represented in this analysis by points computed also at ACIC which are not published as yet. These latter points are determined by measuring

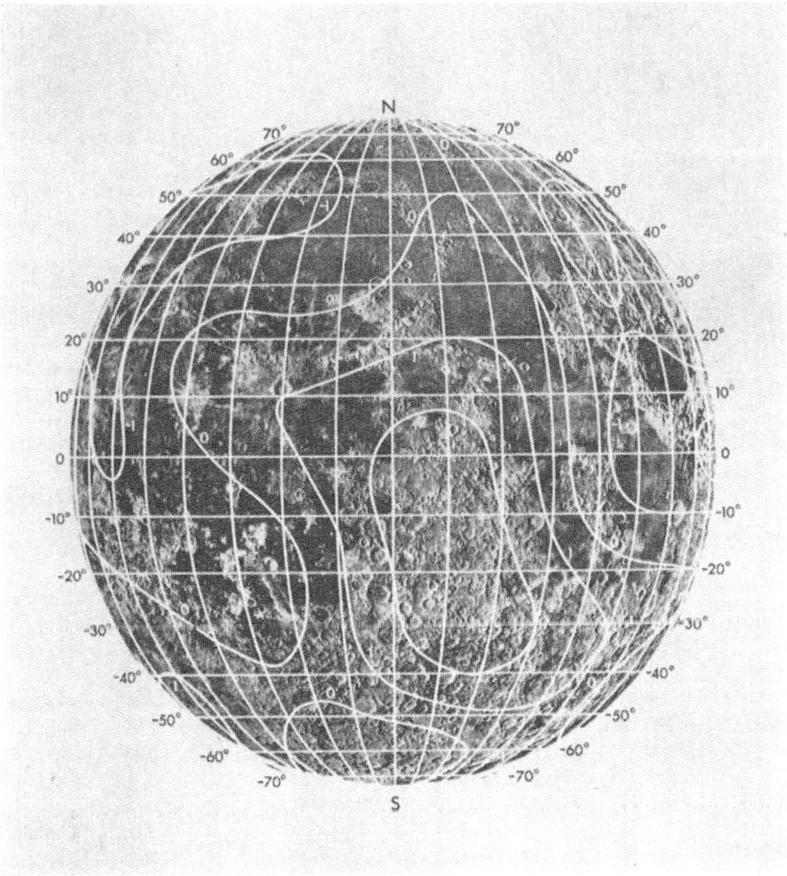

Fig. 9. Contour map of the Moon based on about 1200 points measured at ACIC.

long exposure photography obtained at the Lowell Observatory in Arizona. In this respect the combination is heterogeneous and has possibly given unreliable harmonics.

Then a subset of equal area distribution of about 750 points was selected mainly on the basis of two criteria – one that the crater is of regular circular shape, and the other that it is as small as possible. This homogeneous subset was then adjusted to the true lunar surface by subtracting from each elevation measure the height of the rim. This adjustment was only approximate. The craters were divided into three categories according to their diameters and the height was then collectively adjusted by 0.4 km for the large craters by 0.3 km for the medium size craters, and by 0.2 km for the small ones. The Marginal Zone was represented by the eclipse data.

The coefficients obtained from the analysis of this system are given in Table X. In

TABLE X

Harmonic Coefficients derived from a Homogeneous Subset of the Enlarged ACIC System

$J_{00} = 2.02$		
$J_{10} = -0.49$	$J_{11} = 0$	$J'_{11} = 0.10$
$J_{20} = -0.14$	$J_{21} = 0$	$J'_{21} = 0.08$
	$J_{22} = 0.35$	$J'_{22} = 0$
$J_{30} = 0.56$	$J_{31} = 0$	$J'_{31} = 0.13$
	$J_{32} = -0.031$	$J'_{32} = 0$
	$J_{33} = 0$	$J'_{33} = 0.031$
$J_{40} = -0.02$	$J_{41} = 0$	$J'_{41} = 0.12$
	$J_{42} = -0.039$	$J'_{42} = 0$
	$J_{43} = 0$	$J'_{43} = -0.01$
	$J_{44} = 0.0071$	$J'_{44} = 0$
$J_{50} = 0.42$	$J_{51} = 0$	$J'_{51} = -0.11$
	$J_{52} = 0.0090$	$J'_{52} = 0$
	$J_{53} = 0$	$J'_{53} = 0.0025$
	$J_{54} = 0.00079$	$J'_{54} = 0$
	$J_{55} = 0$	$J'_{55} = 0.00022$
$J_{60} = 0.11$	$J_{61} = 0$	$J'_{61} = -0.12$
	$J_{62} = -0.0099$	$J'_{62} = 0$
	$J_{63} = 0$	$J'_{63} = -0.00075$
	$J_{64} = -0.00033$	$J'_{64} = 0$
	$J_{65} = 0$	$J'_{65} = -0.000049$
	$J_{66} = -0.000019$	$J'_{66} = 0$
$J_{70} = -0.31$	$J_{71} = 0$	$J'_{71} = 0.17$
	$J_{72} = 0.0066$	$J'_{72} = 0$
	$J_{73} = 0$	$J'_{73} = 0.0011$
	$J_{74} = 0.00022$	$J'_{74} = 0$
	$J_{75} = 0$	$J'_{75} = 0.000034$
	$J_{76} = -0.000013$	$J'_{76} = 0$
	$J_{77} = 0$	$J'_{77} = 0.0000017$
$J_{80} = 0.22$	$J_{81} = 0$	$J'_{81} = 0.040$
	$J_{82} = -0.0054$	$J'_{82} = 0$
	$J_{83} = 0$	$J'_{83} = -0.00029$
	$J_{84} = -0.0000048$	$J'_{84} = 0$
	$J_{85} = 0$	$J'_{85} = 0.000017$
	$J_{86} = 0.00000076$	$J'_{86} = 0$
	$J_{87} = 0$	$J'_{87} = -0.00000040$
	$J_{88} = 0.00000014$	$J'_{88} = 0$

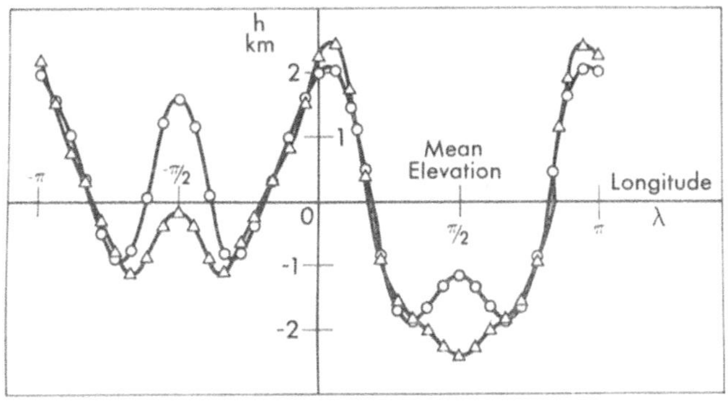

Fig. 10. Equatorial section obtained from the 1200-point new ACIC system with different data for the Marginal Zone.

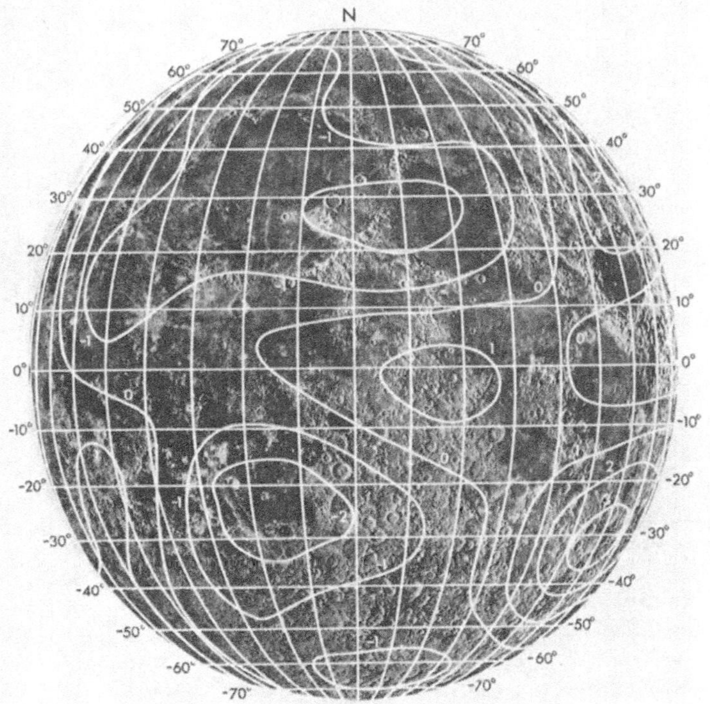

Fig. 11. Contour map of the Moon based on a homogeneous subset of the 1200-point set measured by ACIC.

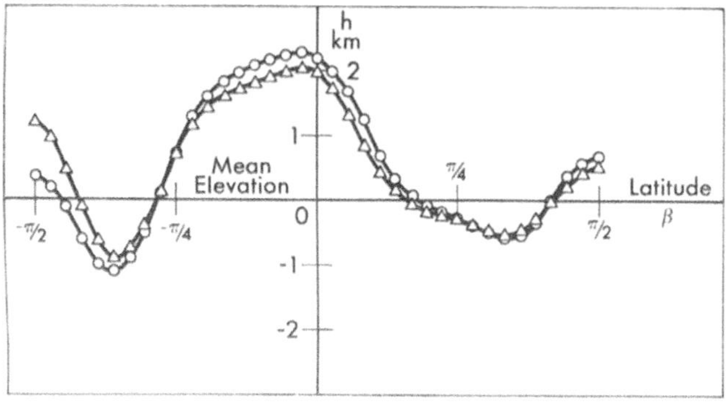

Fig. 12. First meridian section obtained from the 1200-point new ACIC system with different data for the Marginal Zone.

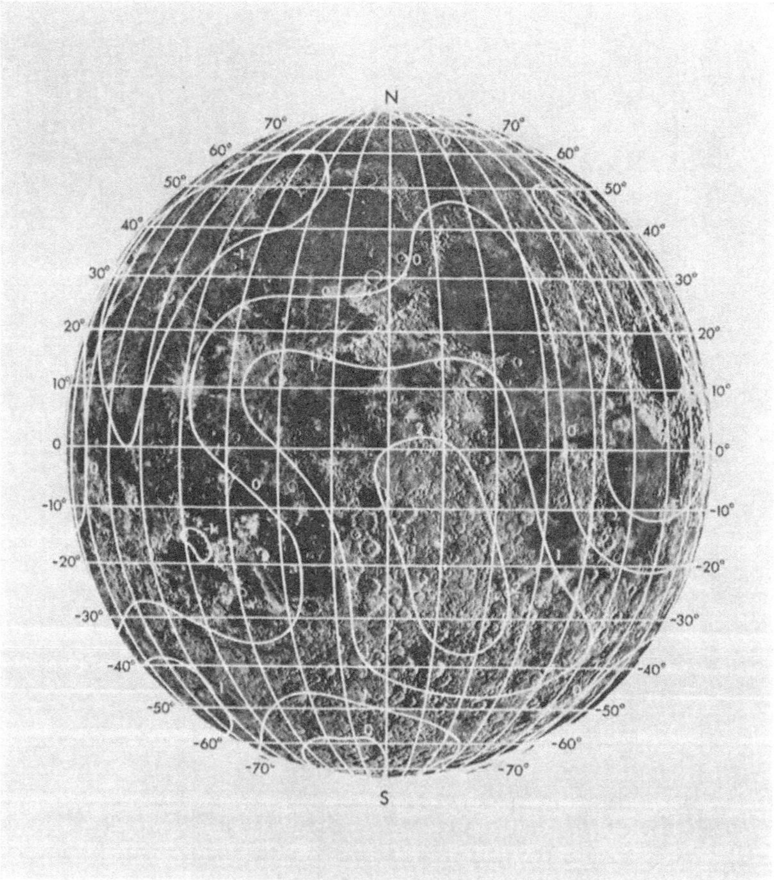

Fig. 13. Contour map derived from the SR II system.

Figures 10 and 12 we give the equatorial and meridian section of the Moon corresponding to the original and the homogeneous sets. It is easy to see that the difference is small except near the limb where different data were used. In Figure 11 we give a contour map of the Moon corresponding to the homogeneous subset. A comparison with the map in Figure 12 shows only small differences everywhere again except in the limb regions. The earlier contour map (BRAY and GOUDAS, 1966) based on the ACIC (1965) set is in agreement with the present ones.

In Figure 13 we give a contour map of the Moon corresponding to the system SR II. Only general agreement can be seen between the contour maps based on the ACIC measurements and the present one.

Appendix A: Table of Observed and Computed Lunar Heights of the System SR I

Name of position	λ (longitude)	β (latitude)	h (observed)	ε (error)	h (computed)
Hansen A.	74.88	13.38	− 2.0	0.8	− 1.5
Hahn A.	69.93	29.69	− 0.9	0.8	0.5
Tralles A.	47.02	27.43	0.5	1.1	0.7
Macrobius A.	40.35	19.55	0.1	0.5	− 0.5
Macrobius B.	40.84	20.93	− 0.3	1.0	− 0.3
Proclus	46.91	16.10	0.4	0.5	− 0.8
Proclus A.	42.29	13.36	− 2.0	1.1	− 1.5
Taruntius A.	49.94	7.28	− 2.4	0.9	− 2.1
Tarantius G.	49.53	1.87	− 2.9	1.2	− 2.3
Jansen B.	26.68	10.68	0.4	0.9	− 1.7
Vitruvius A.	33.81	17.76	− 0.2	1.1	− 1.2
Maury	39.66	37.08	1.1	1.1	0.0
Cepheus A.	46.50	41.02	− 0.3	0.8	− 0.2
Endymion G.	55.54	56.37	− 0.1	0.8	− 0.9
Thales	50.47	61.78	− 1.3	0.5	− 0.7
Hercules G.	39.23	46.42	− 1.9	0.8	− 0.9
Posidonius A	29.52	31.69	− 2.0	1.3	− 0.4
Dawes	26.35	17.22	− 1.3	0.6	− 1.3
Plinius β	23.57	15.32	0.1	1.4	− 1.3
Maclear	20.12	10.53	− 1.4	1.8	− 0.9
Manners	19.99	4.59	0.3	1.6	− 0.4
Dionysius	17.32	2.79	− 0.3	1.1	0.4
Cayley	15.11	3.96	0.1	1.2	0.6
Sosigenes	17.63	8.73	− 1.4	1.1	− 0.4
Sosigenes A.	18.46	7.77	0.8	1.8	− 0.4
Taquet	19.19	16.64	− 0.8	0.9	− 0.9
Taquet A.	20.29	14.35	− 2.5	1.3	− 1.0
Bessel	17.91	21.72	− 0.9	0.8	− 0.8
Bessel A.	21.03	24.77	− 2.6	1.5	− 0.8
Linné	11.80	27.73	− 1.1	1.2	− 0.6
Eudoxus A.	20.06	45.75	0.9	1.0	− 1.3
Silberschlag A.	13.23	6.95	− 0.7	1.0	0.6

Appendix A (cont'd)

Name of position	λ (longitude)	β (latitude)	h (observed)	ε (error)	h (computed)
Rhaeticus A.	5.19	1.75	1.9	1.4	1.9
Rhaeticus B.	6.82	1.63	3.5	1.4	2.2
Triesnecker	3.62	4.19	3.4	1.4	2.3
Aratus	4.53	23.58	2.2	0.8	0.3
Cassini C.	7.82	41.75	− 2.1	1.2	− 1.3
Egede A.	10.51	51.54	− 0.8	0.8	− 1.3
W. C. Bond B.	7.55	65.07	− 1.8	1.1	− 0.7
Pico β	− 8.19	43.24	− 1.8	0.9	− 1.5
Archimedes A.	− 6.39	28.04	0.3	0.8	0.4
Bode	− 2.42	6.73	1.9	1.0	2.5
Bode A.	− 1.14	9.00	2.8	1.4	2.3
Bode B.	− 3.06	8.75	2.0	1.5	2.4
Condamine A.	− 30.10	54.38	− 1.1	1.0	− 2.5
Maupertuis A.	− 24.72	50.68	− 2.7	1.3	− 2.4
Carlini	− 24.07	33.75	− 2.1	1.5	− 0.6
Lambert γ	− 18.40	26.45	1.7	1.5	0.7
Tobias Mayer A.	− 28.31	15.28	0.0	1.2	0.7
Gambart A.	− 18.70	0.99	1.0	0.8	1.0
Milichius	− 30.16	10.02	0.5	1.0	0.4
Kepler	− 37.93	8.14	− 0.2	1.0	− 0.4
Hipparchus C.	8.21	− 7.35	5.1	1.4	2.5
E. Pickering	6.99	− 2.84	3.4	1.6	2.4
Hipparchus G.	7.41	− 4.98	3.4	0.9	2.4
Moltke	24.22	− 0.56	− 3.2	1.3	− 0.3
Alfraganus	18.98	− 5.39	2.2	1.0	1.3
Alfraganus C.	18.09	− 6.09	2.5	1.3	1.6
Abulfeda A.	10.77	− 16.38	1.0	1.0	2.1
Abulfeda E.	10.16	− 16.75	− 1.4	1.3	2.0
Abulfeda F.	13.08	− 16.14	2.5	0.8	2.2
Abenezra B.	10.07	− 20.75	2.7	1.3	1.5
Sacrobosco C.	15.85	− 22.96	− 0.5	1.1	1.5
Busching E.	18.41	− 36.67	− 0.6	1.2	− 0.7
Nicolai A.	23.64	− 42.46	− 1.7	0.6	− 1.3
Piccolomini L.	33.75	− 26.09	0.0	1.0	0.9
Pons B.	20.75	− 28.70	1.6	1.1	0.7
Polybius B.	25.51	− 25.49	1.9	0.9	1.3
Rosse	34.87	− 17.82	2.1	1.6	1.4
Beaumont D.	26.14	− 17.01	2.9	1.0	2.0
Censorinus	32.70	− 0.39	− 2.5	1.8	− 1.4
Messier	47.66	− 1.87	− 1.7	0.7	− 2.0
W. H. Pickering	46.97	− 1.99	− 1.8	0.9	− 1.9
Messier G.	52.87	− 5.38	− 1.3	1.1	− 1.7
Isidorus D.	34.10	− 4.25	− 1.4	1.0	− 0.7
Gutenberg A.	39.91	− 8.99	0.7	0.7	− 0.2
Bellot	48.21	− 12.43	− 0.8	1.5	− 0.3
Janssen K.	42.34	− 46.15	− 3.2	0.4	− 3.1
Kepler A.	− 36.08	7.16	0.3	1.0	− 0.3
Brayley	− 36.79	20.89	0.1	1.4	0.0
Diophantus	− 34.21	27.61	− 0.4	1.0	− 0.2
Mairan E.	− 37.16	37.77	− 0.9	1.1	− 1.8

Appendix A (cont'd)

Name of position	λ (longitude)	β (latitude)	h (observed)	ε (error)	h (computed)
Sharp A.	− 42.70	47.65	− 3.0	0.4	− 2.8
Sharp B.	− 45.34	47.02	− 2.6	0.6	− 2.8
Foucault	− 39.83	50.46	− 2.7	1.4	− 2.8
Bouguer	− 35.84	52.35	− 3.1	0.5	− 2.7
Pythagoras	− 63.60	63.61	− 3.3	0.7	− 1.8
Marius A.	− 46.01	12.62	− 2.2	0.5	− 1.0
Reiner	− 54.92	6.93	− 1.2	0.8	− 1.7
Reiner A.	− 51.40	5.16	− 1.3	1.4	− 1.6
Galilei	− 62.76	10.51	− 2.1	1.3	− 1.6
Lohrmann A.	− 62.66	− 0.72	− 1.5	0.7	− 1.8
Damoiseau E.	− 58.27	− 5.17	− 0.8	1.3	− 1.6
Byrgius A.	− 63.68	− 24.49	2.0	0.4	− 1.5
Sirsalis F.	− 60.17	− 13.57	− 1.5	1.5	− 1.3
Billy	− 50.10	− 13.80	− 2.2	1.0	− 0.1
Mersenius C.	− 45.88	− 19.72	− 0.1	1.1	0.5
Mersenius S.	− 46.97	− 19.17	− 1.2	0.8	0.4
Palmieri A.	− 48.46	− 32.24	− 0.9	0.7	− 0.7
Drebbel	− 49.07	− 40.92	− 1.8	1.0	− 1.5
Vitello ξ	− 37.32	− 30.29	− 0.8	1.1	− 0.2
Gassendi ζ	− 42.80	− 16.40	− 0.1	0.9	0.6
Gassendi G.	− 44.51	− 16.69	0.9	1.6	0.5
Gassendi J.	− 36.94	− 21.54	0.6	1.2	0.7
Herigonius	− 33.84	− 13.29	1.5	1.1	0.5
Flamsteed	− 44.25	− 4.45	− 0.9	0.8	− 0.5
Euclid	− 29.47	− 7.36	− 0.5	1.6	0.2
Lansberg A.	− 31.05	0.21	2.1	1.5	0.0
Agatharchides A.	− 28.31	− 23.19	1.8	1.1	0.0
Campanus centr. mtn.	− 27.70	− 27.91	0.5	1.0	− 0.4
Darney	− 23.47	− 14.54	1.4	1.1	0.1
Guericke B.	− 15.24	− 14.54	− 0.8	0.6	0.1
Guericke C.	− 11.52	− 11.52	0.9	1.5	0.7
Parry A.	− 15.93	− 9.48	0.6	0.6	0.6
Lalande A.	− 9.76	− 6.60	2.3	1.4	1.4
Lalande C.	− 6.87	− 5.57	0.8	0.9	1.8
Turner	− 13.21	− 1.37	− 1.2	0.9	1.5
Mösting	− 5.82	− 0.66	3.2	0.7	2.3
Mösting A.	− 5.16	− 3.18	1.4	0.4	2.2
Herschel C.	− 3.16	− 4.97	3.3	1.4	2.2
Flammarion A.	− 2.47	− 1.93	2.4	1.7	2.5
Ptolemaeus A.	− 0.80	− 8.49	− 0.6	1.5	2.1
Alphonsus	− 2.69	− 13.29	3.0	1.4	1.4
Davy A.	− 7.71	− 12.19	− 0.7	1.0	1.0
Nicollet	− 12.44	− 21.90	− 1.1	0.8	− 0.6
Birt	− 8.54	− 22.32	− 0.6	1.3	− 0.5
Thebit A.	− 4.90	− 21.54	0.1	0.8	0.2
Tycho centr. mtn.	− 11.32	− 43.48	− 3.0	1.1	− 0.8
Maginus H.	− 10.00	− 52.33	2.4	1.8	1.0
Werner D.	3.25	− 27.02	1.4	1.3	− 0.3
Airy A.	7.68	− 16.99	1.1	1.3	1.8
Argelander D.	4.48	− 17.57	2.7	1.0	1.5

Appendix B: Table of Observed and Computed Heights of the System SR II

Name of position	λ (longitude)	β (latitude)	h (observed)	ε (error)	h (computed)
Hahn A.	69°.76	29°.72	− 0.1	1.0	0.6
Tralles B.	50°.61	27°.25	0.5	0.9	0.5
Macrobius A.	40°.36	19°.55	− 0.1	0.5	0.2
Macrobius B.	40°.84	20°.92	0.5	1.4	0.1
Proclus	46°.92	16°.10	0.2	0.5	1.1
Proclus A.	42°.26	13°.34	0.1	1.9	0.9
Taruntius A.	49°.89	7°.27	− 0.4	1.7	0.1
Taruntius E.	40°.18	5°.53	2.3	0.9	0.5
Taruntius G.	49°.62	1°.89	− 3.3	1.2	− 0.9
Taruntius H.	49°.80	0°.35	1.9	1.6	− 1.3
Sinas	31°.61	8°.85	1.6	1.0	− 0.3
Newcomb H.	42°.58	28°.97	− 0.2	1.1	− 0.8
Maury	39°.72	37°.10	0.0	1.2	− 2.1
Thales	50°.51	61°.80	− 1.4	0.4	− 1.3
Atlas A.	49°.57	45°.32	− 1.5	1.9	− 2.0
Posidonius P.	27°.70	33°.70	− 7.1	1.6	− 2.7
Dawes	26°.36	17°.22	− 1.5	1.9	− 1.8
Manners	20°.02	4°.60	− 1.0	1.8	− 0.8
Whewell	13°.76	4°.18	− 2.5	1.4	− 0.4
Cayley	15°.12	3°.97	− 1.0	0.9	− 0 5
Sosigenes	17°.61	8°.73	0.4	1.3	− 1.1
Sulp. Gallus	11°.71	19°.67	− 4.6	1.9	− 2.4
Bessel	17°.94	21°.73	− 1.4	1.1	− 2.7
Bürg A.	33°.09	46°.87	− 2.5	1.2	− 2.3
Bailey A.	31°.39	48°.69	− 2.3	1.3	− 2.1
Gärtner F.	30°.26	57°.48	− 1.9	1.9	− 1.6
Galle	22°.36	55°.87	− 2.4	1.3	− 1.5
Eudoxus A.	20°.13	45°.84	− 1.0	1.0	− 1.8
Lamech	13°.18	42°.77	0.1	0.9	− 1.8
Chladni	1°.15	4°.01	0.3	1.2	0.2
Conon	1°.97	21°.70	− 2.0	2.0	− 2.5
Aristillus B.	− 1°.90	34°.84	− 4.0	1.7	− 2.7
Theaetetus	6°.10	37°.03	− 2.1	1.3	− 2.3
Cassini C.	7°.81	41°.70	− 0.1	2.0	− 1.9
Egede A.	10°.55	51°.57	− 2.6	0.8	− 1.5
W. C. Bond B.	7°.57	65°.05	− 1.3	0.5	− 1.8
Plato H.	− 2°.00	55°.19	− 2.4	1.0	− 2.1
Piazzi Smyth	− 3°.21	41°.93	− 2.6	1.0	− 2.3
Kirch	− 5°.60	39°.29	− 2.4	1.6	− 2.6
Archimedes A.	− 6°.39	28°.05	− 1.2	1.0	− 3.2
Bode A.	− 1°.16	9°.02	0.3	1.3	− 0.6
Leverrier	− 20°.58	40°.35	− 2.7	1.4	− 3.1
Fontenelle A.	− 16°.18	67°.66	− 2.6	1.4	− 2.6
Condamine A.	− 30°.19	54°.48	− 3.3	1.0	− 3.2
Condamine B.	− 31°.58	58°.88	− 2.5	1.0	− 3.2
Carlini	− 24°.05	33°.74	− 1.7	1.1	− 3.2
Carlini D.	− 16°.06	33°.08	− 6.8	1.8	− 3.5

Appendix B (cont'd)

Name of position	λ (longitude)	β (latitude)	h (observed)	ε (error)	h (computed)
Draper	− 21°.77	17°.59	− 4.2	0.8	− 3.2
Draper C.	− 21°.51	17°.10	− 3.6	1.2	− 3.2
Tob. Mayer C.	− 26°.00	12°.23	− 2.5	0.5	− 2.6
Gambart A.	− 18°.76	0°.98	− 2.7	1.5	− 2.0
Gambart C.	− 11°.81	3°.34	− 1.0	1.5	− 1.2
Hortensius	− 27°.99	6°.48	− 2.1	1.2	− 2.1
Milichius	− 30°.22	10°.02	− 2.1	1.4	− 2.1
Encke C.	− 26°.41	0°.67	− 1.1	0.7	− 1.1
Kepler A.	− 26°.09	7°.15	− 0.9	1.0	− 1.5
Kepler C.	− 41°.88	10°.03	− 2.9	1.4	− 1.7
Bessarion A.	− 39°.84	17°.09	− 2.5	1.4	− 2.0
Diophantus	− 34°.24	27°.62	− 0.3	1.8	− 2.3
Car. Herschel	− 31°.25	34°.50	− 2.5	1.5	− 2.6
Sharp A.	− 42°.70	47°.65	− 2.9	0.4	− 2.4
Bouguer	− 35°.82	52°.36	− 3.0	0.5	− 3.0
Angström	− 41°.66	29°.91	− 2.9	1.3	− 1.9
Marius A.	− 45°.99	12°.62	− 1.6	0.5	− 2.3
Marius B.	− 47°.33	16°.34	− 2.4	1.7	− 2.6
Byrgius A.	− 63°.70	− 24°.51	1.8	0.4	0.0
Zupus D.	− 53°.43	− 19°.69	− 0.5	0.7	0.5
Palmieri A.	− 48°.46	− 32°.24	− 0.8	0.6	− 0.6
Drebbel	− 49°.03	− 40°.92	− 1.3	0.8	− 1.0
Puiseux D.	− 36°.14	− 25°.70	− 1.0	1.3	− 2.3
Gassendi ξ	− 48°.80	− 16°.40	0.1	0.9	− 0.3
Gassendi J.	− 37°.09	− 21°.62	− 3.7	0.5	− 1.8
Herigonius	− 33°.98	− 13°.34	− 2.4	1.0	− 1.7
Flamsteed A.	− 42°.91	− 7°.85	− 0.7	1.2	− 0.2
Euklid B.	− 30°.34	− 11° 76	− 1.9	0.8	− 2.3
Landsberg C.	− 29°.16	− 1°.45	0.1	1.7	− 1.8
Heinzel A.	− 30°.14	− 43°.25	− 2.8	1.0	− 2.5
Bayer A.	− 30°.44	− 51°.36	− 2.0	1.2	− 2.0
Wilhelm O.	− 17°.24	− 43°.18	− 2.3	0.8	− 2.5
Wurzelbauer	− 15°.44	− 35°.81	− 4.0	1.0	− 3.2
Hesiod B.	− 17°.50	− 27°.13	− 2.2	1.2	− 3.9
Darney	− 23° 54	− 14°.58	− 0.9	1.6	− 3.3
Cuericke B.	− 15°.30	− 14°.58	− 4.6	2.0	− 2.9
Fra Mauro A.	− 20°.97	− 5°.46	− 3.4	1.3	− 2.4
Turner	− 13°.22	− 1°.38	− 1.8	1.5	− 1.3
Mösting A.	− 5°.16	− 3°.18	1.4	0.4	− 0.1
Ptolemaeus A.	− 0°.81	− 8°.49	0.3	1.2	− 0.2
Alpetragius B	− 6°.87	− 15°.12	− 3.7	0.7	− 1.9
Nicollet	− 12°.48	− 21°.95	− 4.6	1.1	− 3 4
Thebit A	− 4°.88	− 21°.56	− 1.3	0.8	− 2.5
Orontius C.	− 4°.07	− 37°.94	− 0.9	0.7	− 2.4
Clavius J.	− 18°.08	− 58°.09	− 4.1	1.0	− 2.1
Clavius E.	− 12°.66	− 51°.43	0.2	1.4	− 2.0
Cysatus A.	− 0°.80	− 64°.27	− 1.0	0.7	− 2.4
Jacobi B.	13°.99	− 54°.55	− 3.0	0.5	− 2.3
Lilius S.	5°.86	− 52°.83	− 2.1	1.5	− 2.1
Stöfler F.	4°.95	− 42°.72	− 3.9	1.0	− 2.0

Appendix B (cont'd)

Name of position	λ (longitude)	β (latitude)	h (observed)	ε (error)	h (computed)
Walter A.	0°.78	− 32°.42	− 3.7	1.2	− 2.5
Airy A.	7°.67	− 17°.02	− 1.0	1.8	− 1.5
Hipparchus C.	8°.24	− 7°.38	− 1.2	1.8	− 0.2
Hipparchus L.	9°.03	− 6°.84	− 1.5	1.6	− 0 2
Alfraganus	18° 97	− 5°.39	2.7	1.1	− 0.8
Alfraganus C.	18°.13	− 6°.10	− 0.5	1.3	− 0.8
Descartes A.	15°.23	− 12°.06	− 0.1	1.7	− 1.2
Abenezra B.	10°.07	− 20° 80	− 0 2	1.2	− 2.0
Sacrobosco C.	15°.87	− 23°.01	− 3.4	0.4	− 2.1
Vlacq II	34°.87	− 47°.83	0.6	1.0	− 1.0
Nicolai A.	23°.65	− 42°.46	− 1.8	0.6	− 1.3
Rabbi Levi F.	20°.60	− 36.°15	− 2.1	1.6	− 1.6
Zagut A.	21°.70	− 32°.04	− 1.1	1.3	− 1.6
Piccolomini F.	35°.88	− 30°.46	2.2	1.4	− 0.1
Fracastorius C.	34°.57	− 24°.58	0.1	2.0	− 0.5
Rosse	35°.06	− 17°.92	− 3.3	1.7	− 0.8
Torricelli C.	26°.08	− 2°.70	− 4.7	1.9	− 0.7
W. Pickering	46°.98	− 1°.99	− 2.3	0.9	− 1.1
Gutenberg A.	39°.93	− 9°.00	− 0.1	0.7	− 0.8
Goclenius A.	50°.44	− 6°.90	− 1.6	1.4	− 2.4
Santbech F.	42°.02	− 25°.58	− 1.3	1.5	− 0.2
Janssen M.	35°.47	− 41°.83	0.9	1.5	− 0.4
Janssen K.	42°.34	− 46°.15	− 3.1	0.6	− 0.5

Appendix C: Interpolated Heights on the Basis of the Adjusted ACIC (1965) System

λ = − 1.570796		λ = − 1.483530		λ = − 1.396263	
β	h	β	h	β	h
− 1.5708	0.68	− 1.5708	0.68	− 1.5708	0.68
− 1.4835	0.28	− 1.4835	0.29	− 1.4835	0.29
− 1.3963	− 0.15	− 1.3963	− 0.15	− 1.3963	− 0.15
− 1.3090	− 0.47	− 1.3090	− 0.47	− 1.3090	− 0.48
− 1.2217	− 0.56	− 1.2217	− 0.56	− 1.2217	− 0.57
− 1.1345	− 0.38	− 1.1345	− 0.39	− 1.1345	− 0.40
− 1.0472	− 0.05	− 1.0472	− 0.06	− 1.0472	− 0.07
− 0.9599	0.24	− 0.9599	0.24	− 0.9599	0.23
− 0.8727	0.32	− 0 8727	0 32	− 0 8727	0.32
− 0.7854	0.09	− 0.7854	0.09	− 0.7854	0.10
− 0.6981	− 0.43	− 0.6981	− 0.43	− 0.6981	− 0.40
− 0.6109	− 1.08	− 0.6109	− 1.07	− 0.6109	− 1.03
− 0.5236	− 1.59	− 0.5236	− 1.58	− 0.5236	− 1.52
− 0.4363	− 1.71	− 0.4363	− 1.70	− 0.4363	− 1.65
− 0.3491	− 1.32	− 0.3491	− 1.32	− 0.3491	− 1.30
− 0.2618	− 0.45	− 0.2618	− 0.46	− 0.2618	− 0.48

Appendix C (cont'd)

$\lambda = -1.570796$		$\lambda = -1.483530$		$\lambda = -1.396263$	
β	h	β	h	β	h
− 0.1745	0.67	− 0.1745	0.64	− 0.1745	0.57
− 0.0873	1.76	− 0.0873	1.72	− 0.0373	1.59
0.0000	2.50	0.0000	2.45	0.0000	2.29
0.0873	2.70	0.0873	2.64	0.0873	2.47
0.1745	2.34	0.1745	2.28	0.1745	2.12
0.2618	1.54	0.2618	1.50	0.2618	1.37
0.3491	0.56	0.3491	0.53	0.3491	0.45
0.4363	− 0.36	0.4363	− 0.37	0.4363	− 0.41
0.5236	− 1.05	0.5236	− 1.06	0.5236	− 1.08
0.6109	− 1.48	0.6109	− 1.48	0.6109	− 1.49
0.6981	− 1.69	0.6981	− 1.70	0.6981	− 1.72
0.7854	− 1.82	0.7854	− 1.83	0.7854	− 1.87
0.8727	− 1.94	0.8727	− 1.96	0.8727	− 2.02
0.9599	− 2.10	0.9599	− 2.12	0.9599	− 2.19
1.0472	− 2.23	1.0472	− 2.25	1.0472	− 2.31
1.1345	− 2.21	1.1345	− 2.23	1.1345	− 2.27
1.2217	− 1.95	1.2217	− 1.96	1.2217	− 1.98
1.3090	− 1.41	1.3090	− 1.42	1.3090	− 1.42
1.3963	− 0.67	1.3963	− 0.67	1.3963	− 0.66
1.4835	0.13	1.4835	0.13	1.4835	0.14
1.5708	0.83	1.5708	0.83	1.5708	0.83

$\lambda = -1.308997$		$\lambda = -1.221730$		$\lambda = -1.134464$	
β	h	β	h	β	h
− 1.5708	0.68	− 1.5708	0.68	− 1.5708	0.68
− 1.4835	0.29	− 1.4835	0.29	− 1.4835	0.30
− 1.3963	− 0.15	− 1.3963	− 0.15	− 1.3963	− 0.15
− 1.3090	− 0.48	− 1.3090	− 0.49	− 1.3090	− 0.50
− 1.2217	− 0.58	− 1.2217	− 0.59	− 1.2217	− 0.60
− 1.1345	− 0.41	− 1.1345	− 0.43	− 1.1345	− 0.44
− 1.0472	− 0.08	− 1.0472	− 0.10	− 1.0472	− 0.10
− 0.9599	0.22	− 0.9599	0.22	− 0.9599	0.24
− 0.8727	0.33	− 0.8727	0.36	− 0.8727	0.43
− 0.7854	0.14	− 0.7854	0.22	− 0.7854	0.36
− 0.6981	− 0.32	− 0.6981	− 0.18	− 0.6981	0.04
− 0.6109	− 0.92	− 0.6109	− 0.71	− 0.6109	− 0.39
− 0.5236	− 1.39	− 0.5236	− 1.14	− 0.5236	− 0.75
− 0.4363	− 1.53	− 0.4363	− 1.27	− 0.4363	− 0.86
− 0.3491	− 1.21	− 0.3491	− 0.99	− 0.3491	− 0.63
− 0.2618	− 0.46	− 0.2618	− 0.35	− 0.2618	− 0.10
− 0.1745	0.50	− 0.1745	0.49	− 0.1745	0.57
− 0.0873	1.43	− 0.0873	1.28	− 0.0873	1.18
0.0000	2.05	0.0000	1.78	0.0000	1.52
0.0873	2.20	0.0873	1.85	0.0873	1.49
0.1745	1.85	0.1745	1.50	0.1745	1.09
0.2618	1.15	0.2618	0.83	0.2618	0.45
0.3491	0.29	0.3491	0.05	0.3491	− 0.25

Appendix C (cont'd)

$\lambda = -1.308\,997$		$\lambda = -1.221\,730$		$\lambda = -1.134\,464$	
β	h	β	h	β	h
0.4363	− 0.51	0.4363	− 0.67	0.4363	− 0.89
0.5236	− 1.13	0.5236	− 1.22	0.5236	− 1.37
0.6109	− 1.52	0.6109	− 1.59	0.6109	− 1.70
0.6981	− 1.76	0.6981	− 1.84	0.6981	− 1.94
0.7854	− 1.94	0.7854	− 2.03	0.7854	− 2.15
0.8727	− 2.11	0.8727	− 2.23	0.8727	− 2.37
0.9599	− 2.29	0.9599	− 2.42	0.9599	− 2.56
1.0472	− 2.40	1.0472	− 2.51	1.0472	− 2.64
1.1345	− 2.33	1.1345	− 2.41	1.1345	− 2.50
1.2217	− 2.01	1.2217	− 2.05	1.2217	− 2.09
1.3090	− 1.42	1.3090	− 1.43	1.3090	− 1.43
1.3963	− 0.65	1.3963	− 0.64	1.3963	− 0.62
1.4835	0.15	1.4835	0.16	1.4835	0.17
1.5708	0.83	1.5708	0.83	1.5708	0.83

$\lambda = -1.047\,197$		$\lambda = -0.959\,931$		$\lambda = -0.872\,664$	
β	h	β	h	β	h
− 1.5708	0.68	− 1.5708	0.68	− 1.5708	0.68
− 1.4835	0.30	− 1.4835	0.31	− 1.4835	0.32
− 1.3963	− 0.15	− 1.3963	− 0.15	− 1.3963	− 0.15
− 1.3090	− 0.51	− 1.3090	− 0.52	− 1.3090	− 0.53
− 1.2217	− 0.62	− 1.2217	− 0.64	− 1.2217	− 0.66
− 1.1345	− 0.46	− 1.1345	− 0.48	− 1.1345	− 0.50
− 1.0472	− 0.11	− 1.0472	− 0.10	− 1.0472	− 0.10
− 0.9599	0.27	− 0.9599	0.33	− 0.9599	0.38
− 0.8727	0.53	− 0.8727	0.66	− 0.8727	0.80
− 0.7854	0.56	− 0.7854	0.79	− 0.7854	1.03
− 0.6981	0.35	− 0.6981	0.71	− 0.6981	1.07
− 0.6109	0.02	− 0.6109	0.50	− 0.6109	0.97
− 0.5236	− 0.24	− 0.5236	0.32	− 0.5236	0.87
− 0.4363	− 0.32	− 0.4363	0.28	− 0.4363	0.86
− 0.3491	− 0.13	− 0.3491	0.43	− 0.3491	0.99
− 0.2618	0.27	− 0.2618	0.74	− 0.2618	1.19
− 0.1745	0.77	− 0.1745	1.06	− 0.1745	1.37
− 0.0873	1.17	− 0.0873	1.25	− 0.0873	1.39
0.0000	1.32	0.0000	1.20	0.0000	1.18
0.0873	1.14	0.0873	0.88	0.0873	0.74
0.1745	0.68	0.1745	0.34	0.1745	0.14
0.2618	0.05	0.2618	− 0.28	0.2618	− 0.48
0.3491	− 0.58	0.3491	− 0.87	0.3491	− 1.03
0.4363	− 1.13	0.4363	− 1.34	0.4363	− 1.45
0.5236	− 1.53	0.5236	− 1.67	0.5236	− 1.73
0.6109	− 1.82	0.6109	− 1.92	0.6109	− 1.95
0.6981	− 2.04	0.6981	− 2.14	0.6981	− 2.18
0.7854	− 2.27	0.7854	− 2.38	0.7854	− 2.45
0.8727	− 2.51	0.8727	− 2.64	0.8727	− 2.73
0.9599	− 2.71	0.9599	− 2.84	0.9599	− 2.93

Appendix C (cont'd)

$\lambda = -1.047197$		$\lambda = -0.959931$		$\lambda = -0.872664$	
β	h	β	h	β	h
1.0472	− 2.76	1.0472	− 2.87	1.0472	− 2.95
1.1345	− 2.58	1.1345	− 2.65	1.1345	− 2.70
1.2217	− 2.13	1.2217	− 2.15	1.2217	− 2.15
1.3090	− 1.43	1.3090	− 1.41	1.3090	− 1.38
1.3963	− 0.60	1.3963	− 0.57	1.3963	− 0.53
1.4835	0.19	1.4835	0.22	1.4835	0.25
1.5708	0.83	1.5708	0.83	1.5708	0.83

$\lambda = -0.785398$		$\lambda = -0.698132$		$\lambda = -0.610865$	
β	h	β	h	β	h
− 1.5708	0.68	− 1.5708	0.68	− 1.5708	0.68
− 1.4835	0.33	− 1.4835	0.34	− 1.4835	0.35
− 1.3963	− 0.14	− 1.3963	− 0.14	− 1.3963	− 0.13
− 1.3090	− 0.54	− 1.3090	− 0.55	− 1.3090	− 0.57
− 1.2217	− 0.69	− 1.2217	− 0.72	− 1.2217	− 0.75
− 1.1345	− 0.52	− 1.1345	− 0.55	− 1.1345	− 0.59
− 1.0472	− 0.10	− 1.0472	− 0.10	− 1.0472	− 0.13
− 0.9599	0.43	− 0.9599	0.47	− 0.9599	0.48
− 0.8727	0.92	− 0.8727	1.01	− 0.8727	1.06
− 0.7854	1.24	− 0.7854	1.40	− 0.7854	1.48
− 0.6981	1.37	− 0.6981	1.58	− 0.6981	1.67
− 0.6109	1.36	− 0.6109	1.60	− 0.6109	1.68
− 0.5236	1.31	− 0.5236	1.56	− 0.5236	1.61
− 0.4363	1.31	− 0.4363	1.56	− 0.4363	1.57
− 0.3491	1.41	− 0.3491	1.63	− 0.3491	1.62
− 0.2618	1.55	− 0.2618	1.74	− 0.2618	1.74
− 0.1745	1.63	− 0.1745	1.80	− 0.1745	1.85
− 0.0873	1.56	− 0.0873	1.72	− 0.0873	1.84
0.0000	1.26	0.0000	1.43	0.0000	1.65
0.0873	0.76	0.0873	0.95	0.0873	1.27
0.1745	0.14	0.1745	0.36	0.1745	0.76
0.2618	− 0.47	0.2618	− 0.23	0.2618	0.20
0.3491	− 1.00	0.3491	− 0.74	0.3491	− 0.30
0.4363	− 1.39	0.4363	− 1.14	0.4363	− 0.73
0.5236	− 1.66	0.5236	− 1.44	0.5236	− 1.09
0.6109	− 1.89	0.6109	− 1.71	0.6109	− 1.43
0.6981	− 2.14	0.6981	− 2.02	0.6981	− 1.81
0.7854	− 2.45	0.7854	− 2.37	0.7854	− 2.23
0.8727	− 2.76	0.8727	− 2.72	0.8727	− 2.61
0.9599	− 2.97	0.9599	− 2.96	0.9599	− 2.87
1.0472	− 2.98	1.0472	− 2.97	1.0472	− 2.89
1.1345	− 2.71	1.1345	− 2.67	1.1345	− 2.59
1.2217	− 2.13	1.2217	− 2.08	1.2217	− 2.00
1.3090	− 1.34	1.3090	− 1.28	1.3090	− 1.20
1.3963	− 0.48	1.3963	− 0.43	1.3963	− 0.36
1.4835	0.28	1.4835	0.31	1.4835	0.35
1.5708	0.83	1.5708	0.83	1.5708	0.83

Appendix C (cont'd)

$\lambda = -0.523\,599$		$\lambda = -0.436\,332$		$\lambda = -0.349\,066$	
β	h	β	h	β	h
− 1.5708	0.68	− 1.5708	0.68	− 1.5708	0.68
− 1.4835	0.36	− 1.4835	0.38	− 1.4835	0.39
− 1.3963	− 0.13	− 1.3963	− 0.12	− 1.3963	− 0.11
− 1.3090	− 0.58	− 1.3090	− 0.60	− 1.3090	− 0.62
− 1.2217	− 0.79	− 1.2217	− 0.83	− 1.2217	− 0.88
− 1.1345	− 0.64	− 1.1345	− 0.70	− 1.1345	− 0.78
− 1.0472	− 0.17	− 1.0472	− 0.24	− 1.0472	− 0.33
− 0.9599	0.45	− 0.9599	0.40	− 0.9599	0.31
− 0.8727	1.06	− 0.8727	1.02	− 0.8727	0.95
− 0.7854	1.49	− 0.7854	1.44	− 0.7854	1.37
− 0.6981	1.66	− 0.6981	1.58	− 0.6981	1.49
− 0.6109	1.62	− 0.6109	1.48	− 0.6109	1.35
− 0.5236	1.49	− 0.5236	1.28	− 0.5236	1.10
− 0.4363	1.40	− 0.4363	1.14	− 0.4363	0.91
− 0.3491	1.43	− 0.3491	1.16	− 0.3491	0.93
− 0.2618	1.59	− 0.2618	1.37	− 0.2618	1.20
− 0.1745	1.79	− 0.1745	1.70	− 0.1745	1.63
− 0.0873	1.93	− 0.0873	2.00	− 0.0873	2.09
0.0000	1.91	0.0000	2.16	0.0000	2.40
0.0873	1.67	0.0873	2.09	0.0873	2.45
0.1745	1.27	0.1745	1.79	0.1745	2.23
0.2618	0.76	0.2618	1.33	0.2618	1.78
0.3491	0.24	0.3491	0.79	0.3491	1.20
0.4363	− 0.23	0.4363	0.24	0.4363	0.59
0.5236	− 0.67	0.5236	− 0.28	0.5236	0.00
0.6109	− 1.11	0.6109	− 0.79	0.6109	− 0.55
0.6981	− 1.56	0.6981	− 1.30	0.6981	− 1.08
0.7854	− 2.03	0.7854	− 1.80	0.7854	− 1.58
0.8727	− 2.45	0.8727	− 2.24	0.8727	− 2.01
0.9599	− 2.73	0.9599	− 2.53	0.9599	− 2.29
1.0472	− 2.76	1.0472	− 2.57	1.0472	− 2.34
1.1345	− 2.47	1.1345	− 2.30	1.1345	− 2.09
1.2217	− 1.88	1.2217	− 1.74	1.2217	− 1.56
1.3090	− 1.10	1.3090	− 0.98	1.3090	− 0.85
1.3963	− 0.29	1.3963	− 0.20	1.3963	− 0.11
1.4835	0.39	1.4835	0.43	1.4835	0.48
1.5708	0.83	1.5708	0.83	1.5708	0.83

$\lambda = -0.261\,799$		$\lambda = -0.174\,533$		$\lambda = -0.087\,266$	
β	h	β	h	β	h
− 1.5708	0.68	− 1.5708	0.68	− 1.5708	0.68
− 1.4835	0.41	− 1.4835	0.43	− 1.4835	0.45
− 1.3963	− 0.10	− 1.3963	− 0.08	− 1.3963	− 0.06
− 1.3090	− 0.63	− 1.3090	− 0.64	1.3090	− 0.65
− 1.2217	− 0.94	− 1.2217	− 1.00	− 1.2217	− 1.05
− 1.1345	− 0.87	− 1.1345	− 0.98	− 1.1345	− 1.09
− 1.0472	− 0.44	− 1.0472	− 0.57	− 1.0472	− 0.72
− 0.9599	0.21	− 0.9599	0.08	− 0.9599	− 0.06

Appendix C (cont'd)

$\lambda = -0.261799$		$\lambda = -0.174533$		$\lambda = -0.087266$	
β	h	β	h	β	h
− 0.8727	0.86	− 0.8727	0.76	− 0.8727	0.65
− 0.7854	1.31	− 0.7854	1.26	− 0.7854	1.22
− 0.6981	1.44	− 0.6981	1.44	− 0.6981	1.49
− 0.6109	1.29	− 0.6109	1.34	− 0.6109	1.48
− 0.5236	1.02	− 0.5236	1.09	− 0.5236	1.31
− 0.4363	0.82	− 0.4363	0.91	− 0.4363	1.17
− 0.3491	0.84	− 0.3491	0.95	− 0.3491	1.21
− 0.2618	1.15	− 0.2618	1.26	− 0.2618	1.51
− 0.1745	1.65	− 0.1745	1.78	− 0.1745	1.98
− 0.0873	2.19	− 0.0873	2.33	− 0.0873	2.47
0.0000	2.59	0.0000	2.73	0.0000	2.79
0.0873	2.71	0.0873	2.83	0.0873	2.80
0.1745	2.51	0.1745	2.58	0.1745	2.47
0.2618	2.04	0.2618	2.06	0.2618	1.89
0.3491	1.41	0.3491	1.39	0.3491	1.19
0.4363	0.75	0.4363	0.71	0.4363	0.51
0.5236	0.13	0.5236	0.10	0.5236	− 0.04
0.6109	− 0.41	0.6109	− 0.40	0.6109	− 0.45
0.6981	− 0.92	0.6981	− 0.82	0.6981	− 0.77
0.7854	− 1.37	0.7854	− 1.19	0.7854	− 1.03
0.8727	− 1.76	0.8727	− 1.51	0.8727	− 1.25
0.9599	− 2.02	0.9599	− 1.73	0.9599	− 1.42
1.0472	− 2.07	1.0472	− 1.77	1.0472	− 1.45
1.1345	− 1.84	1.1345	− 1.57	1.1345	− 1.29
1.2217	− 1.36	1.2217	− 1.14	1.2217	− 0.91
1.3090	− 0.70	1.3090	− 0.54	1.3090	− 0.37
1.3963	− 0.01	1.3963	0.08	1.3963	0.19
1.4835	0.53	1.4835	0.58	1.4835	0.63
1.5708	0.83	1.5708	0.83	1.5708	0.83

$\lambda = 0.000000$		$\lambda = 0.087267$		$\lambda = 0.174533$	
β	h	β	h	β	h
− 1.5708	0.68	− 1.5708	0.68	− 1.5708	0.68
− 1.4835	− 0.47	− 1.4835	0.50	− 1.4835	0.52
− 1.3963	− 0.03	− 1.3963	− 0.00	− 1.3963	0.02
− 1.3090	− 0.65	− 1.3090	− 0.65	− 1.3090	− 0.63
− 1.2217	1.11	− 1.2217	− 1.15	− 1.2217	− 1.19
− 1.1345	− 1.20	− 1.1345	− 1.32	− 1.1345	− 1.42
− 1.0472	− 0.88	− 1.0472	− 1.04	− 1.0472	− 1.22
− 0.9599	− 0.23	− 0.9599	− 0.42	− 0.9599	− 0.62
− 0.8727	0.53	− 0.8727	0.37	− 0.8727	0.18
− 0.7854	1.18	− 0.7854	1.10	− 0.7854	0.97
− 0.6981	1.57	− 0.6981	1.61	0.6981	1.57
− 0.6109	1.67	− 0.6109	1.83	− 0.6109	1.88
− 0.5236	1.59	− 0.5236	1.85	− 0.5236	1.96
− 0.4363	1.50	− 0.4363	1.80	− 0.4363	1.92
− 0.3491	1.55	− 0.3491	1.83	− 0.3491	1.91
− 0.2618	1.79	− 0.2618	2.00	− 0.2618	2.00

Appendix C (cont'd)

$\lambda = 0.000000$		$\lambda = 0.087267$		$\lambda = 0.174533$	
β	h	β	h	β	h
− 0.1745	2.18	− 0.1745	2.27	− 0.1745	2.16
− 0.0873	2.55	− 0.0873	2.52	− 0.0873	2.32
0.0000	2.75	0.0000	2.61	0.0000	2.33
0.0873	2.65	0.0873	2.42	0.0873	2.12
0.1745	2.24	0.1745	1.96	0.1745	1.68
0.2618	1.61	0.2618	1.31	0.2618	1.08
0.3491	0.89	0.3491	0.62	0.3491	0.47
0.4363	0.25	0.4363	0.04	0.4363	− 0.01
0.5236	− 0.22	0.5236	− 0.34	0.5236	− 0.32
0.6109	− 0.53	0.6109	− 0.54	0.6109	− 0.44
0.6981	− 0.71	0.6981	− 0.62	0.6981	− 0.44
0.7854	− 0.86	0.7854	− 0.66	0.7854	− 0.42
0.8727	− 0.99	0.8727	− 0.72	0.8727	− 0.44
0.9599	− 1.11	0.9599	− 0.79	0.9599	− 0.48
1.0472	− 1.13	1.0472	− 0.81	1.0472	− 0.50
1.1345	− 1.00	1.1345	− 0.71	1.1345	− 0.42
1.2217	− 0.67	1.2217	− 0.43	1.2217	− 0.20
1.3090	− 0.20	1.3090	− 0.02	1.3090	0.14
1.3963	0.30	1.3963	0.41	1.3963	0.52
1.4835	0.69	1.4835	0.74	1.4835	0.79
1.5708	0.83	1.5708	0.83	1.5708	0.83

$\lambda = 0.261800$		$\lambda = 0.349066$		$\lambda = 0.436333$	
β	h	β	h	β	h
− 1.5708	0.68	− 1.5708	0.68	− 1.5708	0.68
− 1.4835	0.55	− 1.4835	0.58	− 1.4835	0.61
− 1.3963	0.06	− 1.3963	0.11	− 1.3963	0.16
− 1.3090	− 0.60	− 1.3090	− 0.56	− 1.3090	− 0.51
− 1.2217	− 1.21	− 1.2217	− 1.21	− 1.2217	− 1.19
− 1.1345	− 1.51	− 1.1345	− 1.58	− 1.1345	− 1.62
− 1.0472	− 1.38	− 1.0472	− 1.53	− 1.0472	− 1.64
− 0.9699	− 0.83	− 0.9699	− 1.05	− 0.9599	− 1.24
− 0.8727	− 0.04	− 0.8727	− 0.29	− 0.8727	− 0.55
− 0.7854	0.77	− 0.7854	0.51	− 0.7854	0.20
− 0.6981	1.41	0.6981	1.15	− 0.6981	0.80
− 0.6109	1.77	− 0.6109	1.50	− 0.6109	1.10
− 0.5236	1.87	− 0.5236	1.56	− 0.5236	1.09
− 0.4363	1.90	− 0.4363	1.42	− 0.4363	0.85
− 0.3491	1.72	− 0.3491	1.24	− 0.3491	0.55
− 0.2618	1.71	− 0.2618	1.14	− 0.2618	0.34
− 0.1745	1.79	− 0.1745	1.15	− 0.1745	0.32
− 0.0873	1.89	− 0.0873	1.26	− 0.0873	0.45
0.0000	1.91	0.0000	1.36	0.0000	0.67
0.0873	1.77	0.0873	1.35	0.0873	0.87
0.1745	1.43	0.1745	1.21	0.1745	0.95
0.2618	0.97	0.2618	0.93	0.2618	0.91
0.3491	0.48	0.3491	0.61	0.3491	0.78

Appendix C (cont'd)

$\lambda = 0.261800$		$\lambda = 0.349066$		$\lambda = 0.436333$	
β	h	β	h	β	h
0.4363	0.09	0.4363	0.33	0.4363	0.62
0.5236	− 0.14	0.5236	0.16	0.5236	0.50
0.6109	− 0.21	0.6109	0.10	0.6109	0.45
0.6981	− 0.19	0.6981	0.11	0.6981	0.43
0.7854	− 0.15	0.7854	0.14	0.7854	0.42
0.8727	− 0.15	0.8727	0.12	0.8727	0 38
0.9599	− 0.19	0.9599	0.07	0.9599	0.31
1.0472	− 0.22	1.0472	0.03	1.0472	0.26
1.1345	− 0.16	1.1345	0.07	1.1345	0.28
1.2217	0.01	1.2217	0.22	1.2217	0.41
1.3090	0.31	1.3090	0.47	1.3090	0.62
1.3963	0.63	1.3963	0.74	1.3963	0.84
1.4835	0.85	1.4835	0.90	1.4835	0.95
1.5708	0.83	1.5708	0.83	1.5708	0.83

$\lambda = 0.523599$		$\lambda = 0.610866$		$\lambda = 0.698132$	
β	h	β	h	β	h
− 1.5708	0.68	− 1.5708	0.68	− 1.5708	0.68
− 1.4835	0.64	− 1.4835	0.67	− 1.4835	0.70
− 1.3963	0.22	− 1.3963	0.28	− 1.3963	0.35
− 1.3090	− 0.44	− 1.3090	− 0.36	− 1.3090	− 0.27
− 1.2217	− 1.14	− 1.2217	− 1.07	− 1.2217	− 0.97
− 1.1345	− 1.62	− 1.1345	− 1.59	− 1.1345	− 1.51
− 1.0472	− 1.72	− 1.0472	− 1.74	− 1.0472	− 1.71
− 0.9599	− 1.39	− 0.9599	− 1.49	− 0.9599	− 1.52
− 0.8727	− 0.78	− 0.8727	− 0.95	− 0.8727	− 1.05
− 0.7854	− 0.08	− 0.7854	− 0.33	− 0.7854	− 0.49
− 0.6981	0.44	− 0.6981	0.12	− 0.6981	− 0.09
− 0.6109	0.66	− 0.6109	0.27	− 0.6109	− 0.00
− 0.5236	0.55	− 0.5236	0.05	− 0.5236	− 0.28
− 0.4363	0.19	− 0.4363	− 0.40	− 0.4363	− 0.83
− 0.3491	− 0.22	− 0.3491	− 0.94	− 0.3491	− 1.46
− 0.2618	− 0.53	− 0.2618	− 1.35	− 0.2618	− 1.98
− 0.1745	− 0.60	− 0.1745	− 1.49	− 0.1745	− 2.20
− 0.0873	− 0.43	− 0.0873	− 1.32	− 0.0873	− 2.09
0.0000	− 0.09	0.0000	− 0.90	0.0000	− 1.68
0.0873	0.29	0.0873	− 0.37	0.0873	− 1.10
0.1745	0.61	0.1745	0.13	0.1745	− 0.48
0.2618	0.80	0.2618	0.53	0.2618	0.05
0.3491	0.87	0.3491	0.77	0.3491	0.45
0.4363	0.84	0.4363	0.88	0.4363	0.70
0.5236	0.79	0.5236	0.91	0.5236	0.85
0.6109	0.74	0.6109	0.90	0.6109	0.91
0.6981	0.70	0.6981	0.87	0.6981	0.93
0.7854	0.65	0.7854	0.81	0.7854	0.90
0.8727	0.58	0.8727	0.74	0.8727	0.84
0.9599	0.50	0.9599	0.66	0.9599	0.77

Appendix C (cont'd)

$\lambda = 0.523599$		$\lambda = 0.610866$		$\lambda = 0.698132$	
β	h	β	h	β	h
1.0472	0.45	1.0472	0.60	1.0472	0.73
1.1345	0.46	1.1345	0.62	1.1345	0.75
1.2217	0.57	1.2217	0.72	1.2217	0.85
1.3090	0.76	1.3090	0.88	1.3090	0.99
1.3963	0.93	1.3963	1.02	1.3963	1.11
1.4835	0.99	1.4835	1.04	1.4835	1.08
1.5708	0.83	1.5708	0.83	1.5708	0.83

$\lambda = 0.785399$		$\lambda = 0.872665$		$\lambda = 0.959931$	
β	h	β	h	β	h
− 1.5708	0.68	− 1.5708	0.68	− 1.5708	0.68
− 1.4835	0.72	− 1.4835	0.75	− 1.4835	0.78
− 1.3963	0.41	− 1.3963	0.48	− 1.3963	0.55
− 1.3090	− 0.17	− 1.3090	− 0.06	− 1.3090	0.05
− 1.2217	− 0.85	− 1.2217	− 0.70	− 1.2217	− 0.55
− 1.1345	− 1.39	− 1.1345	− 1.24	− 1.1345	− 1.06
− 1.0472	− 1.62	− 1.0472	− 1.48	− 1.0472	− 1.30
− 0.9599	− 1.48	− 0 9599	− 1.37	− 0.9599	− 1.21
− 0.8727	− 1.06	− 0.8727	− 0.99	− 0.8727	− 0.85
− 0.7854	− 0.55	− 0.7854	− 0.50	− 0.7854	− 0.38
− 0.6981	− 0.18	− 0.6981	− 0.14	− 0.6981	− 0.01
− 0.6109	− 0.12	− 0.6109	− 0.08	− 0.6109	− 0.07
− 0.5236	− 0.42	− 0.5236	− 0.36	− 0.5236	− 0.16
− 0.4363	− 1.01	− 0.4363	− 0.96	− 0.4363	− 0.72
− 0.3491	− 1.72	− 0.3491	− 1.70	− 0.3491	− 1.46
− 0.2618	− 2.33	− 0.2618	− 2.39	− 0.2618	− 2.21
− 0 1745	− 2.67	− 0.1745	− 2.87	− 0.1745	− 2.82
− 0.0873	− 2.68	− 0.0873	− 3.04	− 0.0873	− 3.17
0.0000	− 2.36	0.0000	− 2.89	0.0000	− 3.22
0.0873	− 1.83	0.0873	− 2.48	0.0873	− 2.99
0.1745	− 1.20	0.1745	− 1.92	0.1745	− 2.54
0.2618	− 0.59	0.2618	− 1.30	0.2618	− 1.96
0.3491	− 0.07	0.3491	− 0.70	0.3491	− 1.30
0.4363	0.32	0.3463	− 0.17	0.4363	− 0.64
0.5236	0.60	0.5236	0.26	0.5236	− 0.04
0.6109	0.79	0.6109	0.60	0.6109	0.44
0.5981	0.90	0.6981	0.83	0.6981	0.79
0.7854	0.93	0.7854	0.94	0.7854	0.99
0.8727	0.91	0.8727	0.97	0.8727	1.05
0.9599	0.86	0.9599	0.94	0.9599	1.04
1.0472	0.83	1.0472	0.93	1.0472	1.02
1.1345	0.86	1.1345	0.96	1.1345	1.05
1.2217	0.96	1.2217	1.06	1.2217	1.14
1.3090	1.09	1.3090	1.18	1.3090	1.25
1.3963	1.18	1.3963	1.25	1.3963	1.31
1.4835	1.12	1.4835	1.16	1.4835	1.19
1.5708	0.83	1.5708	0.83	1.5708	0.83

Appendix C (cont'd)

$\lambda = 1.047\,198$		$\lambda = 1.134\,464$		$\lambda = 1.221\,731$	
β	h	β	h	β	h
– 1.5708	0.68	– 1.5708	0.68	– 1.5708	0.68
– 1.4835	0.80	– 1.4835	0.82	– 1.4835	0.84
– 1.3963	0.61	– 1.3963	0.67	– 1.3963	0.71
– 1.3090	0.16	– 1.3090	0.26	– 1.3090	0.35
– 1.2217	– 0.40	– 1.2217	– 0.24	– 1.2217	– 0.10
– 1.1345	– 0.87	– 1.1345	– 0.68	– 1.1345	– 0.50
– 1.0472	– 1.10	– 1.0472	– 0.89	– 1.0472	– 0.69
– 0.9599	– 1.01	– 0.9599	– 0.80	– 0.9599	– 0.60
– 0.8727	– 0.67	– 0.8727	– 0.47	– 0.8727	– 0.29
– 0.7854	– 0.21	– 0.7854	– 0.04	– 0.7854	0.11
– 0.6981	0.15	– 0.6981	0.33	– 0.6981	0.47
– 0.6109	0.29	– 0.6109	0.49	– 0.6109	0.65
– 0.5236	0.10	– 0.5236	0.36	– 0.5236	0.54
– 0.4363	– 0.39	– 0.4363	– 0.08	– 0.4363	0.16
– 0.3491	– 1.11	– 0.3491	– 0.75	– 0.3491	– 0.45
– 0.2618	– 1.89	– 0.2618	– 1.53	– 0.2618	– 1.20
– 0.1745	– 2.59	– 0.1745	– 2.28	– 0.1745	– 1.96
– 0.0873	– 3.10	– 0.0873	– 2.89	– 0.0873	– 2.61
0.0000	– 3.33	0.0000	– 3.26	0.0000	– 3.05
0.0373	– 3.28	0.0873	– 3.34	0.0873	– 3.20
0.1745	– 2.96	0.1745	– 3.12	0.1745	– 3.03
0.2618	– 2.42	0.2618	– 2.62	0.2618	– 2.54
0.3491	– 1.73	0.3491	– 1.90	0.3491	– 1.79
0.4363	– 0.97	0.4363	– 1.05	0.4363	– 0.89
0.5236	– 0.23	0.5236	– 0.21	0.5236	0.01
0.6109	0.39	0.6109	0.51	0.6109	0.78
0.6981	0.84	0.6981	1.02	0.6981	1.31
0.7854	1.10	0.7854	1.29	0.7854	1.56
0.8727	1.18	0.8727	1.36	0.8727	1.57
0.9599	1.15	0.9599	1.30	0.9599	1.45
1.0472	1.11	1.0472	1.22	1.0472	1.32
1.1345	1.13	1.1345	1.20	1.1345	1.27
1.2217	1.21	1.2217	1.27	1.2217	1.33
1.3090	1.32	1.3090	1.37	1.3090	1.42
1.3963	1.36	1.3963	1.40	1.3963	1.44
1.4835	1.22	1.4835	1.24	1.4835	1.26
1.5708	0.83	1.5708	0.83	1.5708	0.83

$\lambda = 1.396\,264$		$\lambda = 1.308\,997$		$\lambda = 1.483\,530$	
β	h	β	h	β	h
– 1.5708	0.68	– 1.5708	0.68	– 1.5708	0.68
– 1.4835	0.86	– 1.4835	0.85	– 1.4835	0.87
– 1.3963	0.79	– 1.3963	0.76	– 1.3963	0.80
– 1.3090	0.49	– 1.3090	0.43	– 1.3090	0.52
– 1.2217	0.09	– 1.2217	0.00	– 1.2217	0.15
– 1.1345	– 0.23	– 1.1345	– 0.35	– 1.1345	– 0.16
– 1.0472	– 0.38	– 1.0472	– 0.52	– 1.0472	– 0.30

Appendix C (cont'd)

$\lambda = 1.396264$		$\lambda = 1.308997$		$\lambda = 1.483530$	
β	h	β	h	β	h
− 0.9599	− 0.31	− 0.9599	− 0.44	− 0.9599	− 0.23
− 0.8727	− 0.04	− 0.8727	− 0.15	− 0.8727	0.02
− 0.7854	0.31	− 0.7854	0.23	− 0.7854	0.35
− 0.6981	0.61	− 0.6981	0.57	− 0.6981	0.63
− 0.6109	0.76	− 0.6109	0.73	− 0.6109	0.76
− 0.5236	0.68	− 0.5236	0.65	− 0.5236	0.67
− 0.4363	0.35	− 0.4363	0.30	− 0.4363	0.36
− 0.3491	− 0.16	− 0.3491	− 0.26	− 0.3491	− 0.12
− 0.2618	− 0.80	− 0.2618	− 0.95	− 0.2618	− 0.73
− 0.1745	− 1.48	− 0.1745	− 1.68	− 0.1745	− 1.36
− 0.0873	− 2.07	− 0.0873	− 2.32	− 0.0873	− 1.91
0.0000	− 2.49	0.0000	− 2.76	0.0000	− 2.29
0.0873	− 2.63	0.0873	− 2.93	0.0873	− 2.40
0.1745	− 2.44	0.1745	− 2.77	0.1745	− 2.19
0.2618	− 1.92	0.2618	− 2.27	0.2618	− 1.64
0.3491	− 1.11	0.3491	− 1.49	0.3491	− 0.81
0.4363	− 0.15	0.4363	− 0.55	0.4363	0.15
0.5236	0.78	0.5236	0.38	0.5236	1.09
0.6109	1.55	0.6109	1.16	0.6109	1.83
0.6981	2.01	0.6981	1.67	0.6981	2.26
0.7854	2.15	0.7854	1.87	0.7854	2.35
0.8727	2.02	0.8727	1.81	0.8727	2.16
0.9599	1.75	0.9599	1.62	0.9599	1.85
1.0472	1.51	1.0472	1.43	1.0472	1.57
1.1345	1.39	1.1345	1.34	1.1345	1.42
1.2217	1.40	1.2217	1.37	1.2217	1.42
1.3090	1.48	1.3090	1.45	1.3090	1.49
1.3963	1.48	1.3963	1.46	1.3963	1.49
1.4835	1.29	1.4835	1.28	1.4835	1.30
1.5708	0.83	1.5708	0.83	1.5708	0.83

$\lambda = 1.570797$					
β	h	β	h	β	h
− 1.5708	0.68	− 0.4363	0.36	0.6981	2.35
− 1.4835	0.87	− 0.3491	− 0.11	0.7854	2.42
− 1.3963	0.81	− 0.2618	− 0.70	0.8727	2.21
− 1.3090	0.53	− 0.1745	− 1.32	0.9599	1.88
− 1.2217	0.17	− 0.0873	− 1.85	1.0472	1.59
− 1.1345	− 0.13	0.0000	− 2.22	1.1345	1.43
− 1.0472	− 0.27	0.0873	− 2.32	1.2217	1.43
− 0.9599	− 0.20	0.1745	− 2.09	1.3090	1.50
− 0.8727	0.04	0.2618	− 1.53	1.3963	1.50
− 0.7854	0.36	0.3491	− 0.70	1.4835	1.30
− 0.6981	0.64	0.4363	0.27	1.5708	0.83
− 0.6109	0.76	0.5236	1.21		
− 0.5236	0.67	0.6109	1.94		

Acknowledgement

The author wishes to acknowledge the assistance of Miss Zdenka Kopal in the preparation of many parts of this communication.

References

ACIC: 1965, *Tech. Rept. No. 15*, by D. L. Meyer and B. W. Ruffin.
ACIC: 1966, Unpublished (private information).
BRAY, T. A. and GOUDAS, C. L., 1966, *Icarus* **5** (in press).
CARSON, D., DAVIDSON, M., GOUDAS, C. L., KOPAL, Z., and STODDARD, L. G.: 1966, *Icarus* **5**, 334.
ECKHARDT, D. H.: 1966, *Icarus* **5**, 313–316.
FRANZ, J.: 1901, *Mitt. Sternw. Breslau* **1**, 1–48.
GOUDAS, C. L.: 1963, Univ. Manchester Dept. Astronomy, *Techn. Rept. No. 1*, under Contract AF 61 (052)–524.
GOUDAS, C. L.: 1963a, *Icarus* **2**, 423–439.
GOUDAS, C. L.: 1964, *Icarus* **3**, 375–409.
GOUDAS, C. L.: 1965, *Icarus* **4**, 188–206.
GOUDAS, C. L.: 1965a, *Icarus* **4**, 218–221.
GOUDAS, C. L.: 1965b, *Icarus* **4**, 528–543.
GOUDAS, C. L.: 1966a, in *Advances in Astronomy and Astrophysics* (ed. by Z. Kopal). Academic Press, New York, vol. 4, p. 37.
HAYN, F.: 1914, *Abh. k. sächs. Ges. Wiss., Math.-phys., Kl.* **30**, 1–105.
HOPMANN, J.: 1967, in the present volume, pp. 282–294.
MICHAEL, W. H., TOLSON, R. H., and GAPCYNSKI, J. P.: 1966, *Science* **153**, No. 3740.
POTTER, H. I.: 1962, in *The Moon* (ed. by Z. Kopal and Z. K. Mikhailov). Academic Press, New York and London, pp. 63–66.
SCHRUTKA-RECHTENSTAMM, G. V.: 1958, *Österr. Akad. Wiss., Math.-Naturw. Kl., Sitzber. Abt. II* **167**, 71–106.
SCHRUTKA-RECHTENSTAMM, G. V.: 1966, private information.
SPENCER JONES, H.: 1932, *Ann. Cape. Obs.* **13**, 70.
WATTS, C. B.: 1963, *Astr. Paper No. 17, Am. Ephem. and Naut. Almanac*, U.S. Govt. Printing Office, Washington, D.C.
WEIMER, TH.: 1961, in *Problems of Lunar Topography* (ed. by Z. Kopal and E. B. Finlay) (Astr. Contr. Univ. Manchester, Ser. III, No. 90), p. 91.
YAKOVKIN, A. A.: 1952, *Trans. I.A.U.* **8**, 231.
YAKOVKIN, A. A.: 1958, *Trans. I.A.U.* **10**, 268.
YAKOVKIN, A. A.: 1962, in *The Moon, A Russian View* (ed. by A. V. Markov). University of Chicago Press, Chicago, pp. 3–53.

WHAT CAN WE SAY ABOUT THE SHAPE OF THE MOON?

J. HOPMANN*

Abstract. After a brief history, we propose to discuss the seven principal series of observations and their characteristics. From the subsequent comparison there will emerge, on one hand, systematic amounts by which to reduce the altitude data to a common level; and, on the other, values of the absolute accuracy which amounts to approximately ± 1.1 km for three series, and for the others to approximately ± 1.8 km. The causes of the various differences will be investigated.

Plans are disclosed to compile (similarly as was done for trigonometric parallaxes and radial velocities) a 'Preliminary General Catalogue of Absolute Heights on the Moon'. At the present stage of preliminary studies we can expect to be able to list approximately 410 certain and 600 less reliable heights. It is also planned to draw a contour map. Preliminary studies of the map of the Mare Cognitum and the LAC charts Nos. 59 and 78 of ACIC showed that local characteristics (as, for example, degrees of steepness at the shores of the maria) must be taken into account.

We shall discuss the various attempts to represent the shape of the Moon by a tri-axial ellipsoid or by a series of spherical harmonics. It seems to be more appropriate to regard the Moon as a sphere covered with extended "bulges" (mostly the continents) and "craters" (largely the maria). The departures of the heights of these formations from the mean Moon-level exceed but rarely ± 2 km.

1. History: Observational Material

In this investigation we regard as the absolute height of a small crater its distance from the surface of a sphere with a radius of 1738.0 km. This value was used as a basis for all investigations made by Schrutka-Rechtenstamm. We can approach the problem of the exact shape of the Moon only if we have a sufficient number of sufficiently accurate absolute heights. They must also be homogeneously distributed over the entire surface of the Moon.

Selenodesy was not sufficiently developed in the 19th century. Only around 1890, Franz observed in Königsberg Mösting A and eight additional craters with the Fraunhofer heliometer (constructed in 1830) and their selenographic coordinates λ and β. Soon afterwards Hayn (1904–1914) observed four additional craters in Leipzig by means of a 12-inch refraction. Both measurements were completely reduced by Schrutka-Rechtenstamm (1956, 1958); and after over 70 years they have remained the basis of selenodesy.

Around 1900, Franz – still in Königsberg – measured five plates obtained around 1890 with the large refractor of the Lick Observatory at different angles of libration. From them he derived – in an unsatisfactory manner – the absolute heights of sixty craters and established the first contour map of the Moon. It showed already very

* University Observatory, Wien, Austria.

Kopal and Goudas (eds.), Measure of the Moon.

clearly what was confirmed later on: namely that the "continents" of the SW quadrant lie higher than those in the NE, Oceanus Procellarum and other maria.

After moving to Breslau, Franz remeasured the same five Lick plates and derived the selenographic coordinates λ and β for 150 objects, but not the absolute heights (FRANZ, 1901). Since these measurements were published, HOPMANN (1954) could do this by means of an approximation process.

On the other hand, SAUNDER (1907) measured hundreds of objects on four plates obtained with the Equatoreal-Coudé in Paris and combined them with the 150 points measured by Franz. For 38 of them, close to the central meridian, he also derived absolute heights. These heights will, in what follows be referred to as "Sau".

SCHRUTKA-RECHTENSTAMM (1958a, b) made a complete reduction of the Breslau measurements in 1958 and obtained also the heights of the same 150 objects. Their list formed the basis of five additional catalogues. In what follows, it will be referred to as "I". Baldwin's book *The Measure of the Moon* appeared in 1963. In it he gave the results of measurements of five glass copies of the plates obtained with the 36-inch refractor of the Lick Observatory and added them to I. His catalogue with 696 objects hereafter is listed as "Ba".

In 1965 the U.S. Army Map Service (1964) published a list of 256 objects on the basis of the measurements of glass copies of twelve suitable Lick photographs, again added to I. This work will hereafter be referred to as "AMS".

In 1966 an additional catalogue of 140 points was established by SCHRUTKA-RECHTENSTAMM (1966), hereafter referred to as "II". Then glass copies of Lick plates were carefully added to I.

The same is true of a study made by SCHÖFFL (1966), who determined the coordinates λ and β for 320 points close to the edge on the same Lick plates (glass copies) and in close relation with I and II. Absolute heights were determined – even though at times with some uncertainty – for 83 objects. This work will be referred to later as "Schö".

Finally, in 1965, the list of the ACIC (1965) appeared, referred to in the sequel as ACIC. It was obtained by measuring forty original negatives obtained at the Observatoire du Pic-du-Midi, and also by means of the astronometric telescope of the U.S. Naval Observatory in Flagstaff.

2. The Individual Series of Comparisons of Observations

A. PRELIMINARY COMMENTS

What can we expect from a comparison of the seven lists of absolute heights on the Moon on the basis of the preceding evidence? Considerable differences – accidental as well as systematic – turned out, which many colleagues probably would not have thought possible. Some critical comments should be made here.

First, the use of the desk-type computing machines and especially of electronic computers, leads to a luxury in decimal points, which the inexperienced can interpret as an excess in accuracy of the results. For example, the four-decimal values of ξ and

η in Arthur's catalogue (ARTHUR, 1962) are quite sufficient, and the six-digit results of the AMS (1964) are unquestionably superfluous. Quite justifiably I, II and Schö have given the ξ's and η's to five digits and the more inaccurate ζ's only to four digits.

Above all, however, altitudes of AMS and ACIC cannot be accepted to the last meter as stated, in view of the fact that they are actually inaccurate by one to two kilometers. Of course, the computers can print innumerable decimals, but we should stop at 0.1 km. We find the same decimal-point luxury in the indicated mean errors (m.e.) of the individual coordinates. Mildly speaking, these figures are generally quite questionable. Anybody who has taken many measurements and not only calculated on a machine, knows that some observations in a series are in good agreement under entirely identical circumstances, whereas others do not agree, without the mean values actually being of different accuracy. We should point out here, as an example, the height indications for numbers 18 and 20 of the ACIC (1965). Only the average of all mean errors of a catalogue thus computed is suitable for an error discussion. It is not necessary to publish individual values.

To make the comparison it was necessary to establish a card index for the more than 1000 objects. The experiences made in doing this lead to a requirement for future publications: in I, II and Schö, the craters are listed by I.A.U. numbers (cf. BLAGG and MÜLLER, 1935) and their names are given, moreover ξ, η, ζ in addition to λ, β and the absolute height h. We should proceed similarly in the future. The ξ, η, ζ are the actual measuring results: the polar coordinates are derived from them. One group is used in a comparison with the Moon maps, especially at the telescope (for example, the *Orthographic Atlas* of the Lunar and Planetary Laboratory); the other group is necessary if we use the *Rectified Atlas* of LPL or the charts of ACIC. The names are desirable, because they familiarize us with the landscape better than coordinates. As long as the I.A.U. numbers (cf. ARTHUR, 1962) are not replaced by a more recent "official" catalogue, their lack in Ba and ACIC makes any use of these catalogues more difficult.

B. THE COMPARISON

The author – among other people – pointed out that, even with the Lick refractor, a determination of the absolute heights on the basis of libration effect is very difficult. For 6° libration, $\pm 0\overset{''}{.}05$ corresponds at the center of the disk to a height difference of only ± 1.0 km. However, this is the lower bound for reliable stellar parallaxes, on the basis of an average of 20 plates per star. The stars on the plates appear, however, much more sharply than many lunar objects. Therefore, the mean errors of 1 to 2 km must certainly be anticipated.

The first three columns of Table I reflect these indications; and additional ones will be explained later on. They refer to the five large catalogues. With regard to Sau and Schö, see below (pp. 287 and 288).

We now establish the distribution tables to calculate the systematic differences in height indications, the correlation coefficients, etc., for ten pairs of comparison possibilities, with intervals of 1.0 km (Col. 1 of Table II). Earlier investigations had

TABLE I

	Plate	No. of Objects	m.e. (km)	Reduction (in km)	Dispersion (in km)
I	5	150	± 1,17	0,0	± 1,78
II	12	140	± 1,78	+ 1,3	± 1,98
AMS	15	256	± 2,02	− 2,0	± 2,16
Ba	5	696	± 1,79	+ 0,4	± 1,88
ACIC	10	196	± 1,13	+ 0,1	± 1,20

TABLE II

X, Y	n	r m.e.	m.D. (km)	$X–Y$ (km)
II, I	60	+ 0,15 ± 0,13	± 2,27	− 1,36
AMS, I	58	+ 0,31 ± 0,12	± 2,09	+ 1,98
AMS, II	64	+ 0,05 ± 0,12	± 2,91	+ 3,98
Ba I	60	+ 0,00 ± 0,13	± 2,27	− 0,36
Ba II	73	+ 0,31 ± 0,11	± 2,13	+ 0,68
Ba AMS	106	+ 0,09 ± 0,10	± 2,75	− 2,00
ACIC, I	39	+ 0,76 ± 0,07	± 1,12	− 0,21
ACIC, II	29	+ 0,20 ± 0,11	± 1,79	+ 0,82
ACIC, AMS	31	+ 0,28 ± 0,16	± 1,83	− 1,58
ACIC, Ba	80	+ 0,48 ± 0,06	± 2,05	+ 0,25

shown that finer subdivisions would not have been justified because of a strong scatter. The second column shows the number n of the common object measured in two series, respectively; n being sometimes surprisingly small. Every author has made his own selection of objects necessary for intercomparison. Weimer suggested at the Bagnères conference (cf. KOPAL and FINLAY, 1961) a list of 140 craters as future third-order reference point, but this was followed only in II.

The strong scatter and differences in zero-point were the first surprise born out by our distribution tables. Divergences in the height data of the same crater of 4 and 5 km occurred very frequently, some reached 9 km (!!), without any certainty as to which is the "wrong" value. Table III gives as an example the distribution of the differences

TABLE III

Difference	Number
0,0 to 0,9 km	8
1,0 to 1,9 km	23
2,0 to 2,9 km	15
3,0 to 3,9 km	6
4,0 to 4,9 km	4
5,0 to 5,9 km	2
6,0 to 6,9 km	1
7,0 to 7,9 km	1
Total	60

in I and II, already after 1.0 km zero-point correction. From this follows a mean difference of ± 2.27 km. In an analogous manner, the values for the other nine comparisons were calculated in a similar fashion (penultimate column of Table II). From these values we can derive the true accuracies of the five series. Let us assume to clarify that we have any measurements A, B and C with the mean errors $x=3$, $y=4$ and $z=5$. The squares of the mean differences $(A-B)$, $(A-C)$ and $(B-C)$ are then

$$\overline{(A-B)^2} = 25, \qquad \overline{|A-B|} = \pm 5.0,$$
$$\overline{(A-C)^2} = 34, \qquad \overline{|A-C|} = \pm 5.8,$$
$$\overline{(B-C)^2} = 41; \qquad \overline{|B-C|} = \pm 6.4.$$

Conversely, we can obtain by such a "triangle comparison" in three series the true mean errors x, y, z from the mean differences.

Ten comparisons of five series are available in our case: i.e., the true mean errors must be derived by adjustment. Values in Column 4 of Table I resulted, which each have an uncertainty of ± 0.25 km. Obviously, I and ACIC are considerably more accurate than the other three series. On the average, the mean errors amount to ± 1.15 and ± 1.86 km, which is equivalent to a weight ratio of 2.6 to 1.

It is almost impossible to determine the origin of these weight differences; but it should be pointed out that glass copies were always measured in II, Ba and AMS. For the Lick refractor, ± 1.0 km in the center of the lunar disk corresponds to a $\pm 4.4\mu$, closer to the limb even less. It is sufficient to assume additional photographic errors in the copies of several microns to account for their greater uncertainty. For example, the heights based on Saunder's original plates are valuable, but not those obtained by Schöffl on the Lick copies. Unfortunately, Saunder's advice of 1907 (who already had had some bad experiences with copies) was forgotten. If, moreover, the old plates of I and Sau furnished such satisfactory results, the contributing factor may have been the fact that they (being relatively insensitive) required several seconds of exposure which helped to counteract the momentary blurring of the image. The quality of ACIC is in part due to the favourable climate on Pic-du-Midi and at Flagstaff; in part also to the large number of plates available for measurement.

Moreover, from the distribution tables for each of the five series we obtain the dispersion in height around their average. These averages can be found in the last column of Table I. They are, on the whole, only slightly higher than the mean errors in column 4. This could indicate that *all* measured heights are only the result of measuring inaccuracy, while the Moon actually is a sphere – a surmise which would be accurate if positive and negative values were distributed at random over the entire disk. However, all investigations made so far indicated that there are more or less pronounced clusters of signs, indicating the existence of actual areas of high and low elevation.

This is confirmed by the linear correlation coefficient r derived from the distribution tables and their mean errors (column 3 of Table II). They are all positive, i.e., the heights (depths) show up as such in all five series, only – except for one case

(ACIC-I) – the values of r are very small as a result of all kinds of measuring inaccuracies.

The last column of Table II shows systematic differences in the heights of the various series subject to our comparison. By smoothing, we obtain from this the amounts necessary to bring all series to the fundamental system I (Table I, column 5). In view of what is known of meridian circle measures, stellar parallaxes, proper motions, radial velocities, etc., such differences are not too surprising. The considerable amount for AMS goes mainly back to the fact that only the height of Mösting A was taken from I for level definition. The strong deviation of II has probably two causes: eight of the twelve plates were taken at a more or less large phase angle, so that there is the danger of shadow effects. On the other hand, the number of the connecting points from I is lowered on the phase plates.

Let us now, for completeness, turn to the two small series considered by Sau and Schö. Saunder (1907) measured hundreds of objects on four Paris plates which formed in 1912 the basis for his comprehensive catalogue (SAUNDER, 1911). Most of his ξ- and ζ-values are based on the measurements of three, or less than three, phase plates. Only for 38 craters close to the central meridian was it possible for him to derive absolute heights from all four plates. (An attempt made by me to derive usable heights from three plates failed.) Saunder augmented his plates in part by his own visual micrometer measurements, in part by those made by Barnard; but above all by the Breslau Catalogue made by FRANZ (1901). Accordingly, a greater zero-point difference with Sau-I was to be expected. The 38 points were, therefore, compared with the five large series (Table IV). Their columns show the reference catalogue X, the number of the common objects, the means of the heights according to Sau and X, and their difference. After showing the systematic differences between the individual

TABLE IV

Comparison	n	h_S	h_X	$S-X$	S	r	m.e.	m. diff.	m.e.
Sau–I	14	0.1	1.7	− 1.6	− 1.6	0.65 ±	0.15	± 1.11	± 1.17
Sau–II	11	− 0.8	− 1.1	0.3	− 1.6	0.25	0.28	1.58	1.78
Sau–AMS	14	− 0.6	2.5	− 3.1	− 1.1	0.18	0.26	1.76	2.02
Sau–Ba	27	− 0.7	0.2	− 0.9	− 1.3	0.44	0.16	1.98	1.79
Sau–ACIC	7	− 1.2	0.4	− 1.6	− 1.7	0.71	0.19	1.43	1.13

series and I (Table I, column 5) we obtain the difference between Sau and I from the five series which are in a surprisingly good agreement. Considering the weight of the individual series as calculated above, and the number of the intercompared points, the mean value becomes $S = -1.5 \pm 0.1$ km.

The linear correlation coefficients r in Table IV are higher than those in Table II, especially for the good series I and AC. This means that Saunder's heights are equally good. This is also borne out by the last two columns of Table IV, giving the mean differences. "Sau-I series" in comparison with the same values in Table I. Both columns are approximately equal, and their (weighted) averages are ±1.52 and

± 1.50 km, a clear indication of the heigh accuracy of the – unfortunately few – absolute heights of Saunder.

SCHÖFFL (1966) determined the ξ, η, ζ or λ, β coordinates (on the sphere) for 320 points close to the limb on the same twelve glass-copies of the Lick Observatory plates. Together with those of FRANZ (1913), they will be used to evaluate photographs of parts of the far side of the Moon. Also the absolute heights were calculated for 83 of these objects. Their mean errors are individually much higher than for objects located near to the center, as was to be expected. If we only take objects in which the computed mean error was less than ± 2.6 km, a similar comparison could be made with I. The following results were obtained: Schö-I $= 2.0 \pm 0.4$ km, and the mean difference, ± 2.55 km. If Schö must be counted among the less accurate series of the data, it is still of value to support our knowledge of lunar heights close to the limb, in areas which are sparsely covered by other series.

3. Plan for a 'Preliminary General Catalogue of Absolute Lunar Heights' and a 'Counter Map'

The preceding critical remarks lead to the following conclusions. If we want to know more about the geometric shape of the Moon, especially about the hypothetical tidal bulge, this cannot be achieved by using only *one* of the existing catalogues. The result would be just as uncertain as the various contour maps of FRANZ (1899), SCHRUTKA-HOPMANN (1955), BALDWIN (1963), or the AMS (1964). The same holds for the mathematical interpolation of heights in individual catalogues by three-dimensional trigonometric series (GOUDAS, 1963, 1965a, b).

Actually, the heliometer measurements of FRANZ (1889) should be repeated after more than 70 years in modern photographic form. Suggestions towards this end would exceed the scope of this address, and they were discussed by Arthur. The same net of approximately 200 points – for example, those of the ACIC – should be remeasured independently at *several* institutions with plates of *various* telescopes (Lick, Pic-du-Midi, Yerkes, etc.), until they result in a general catalogue. These 200 points – based on the same or different plates – should then serve as the reference points for the detailed selenodetic work.

The history of astronomy provides us with an exact analogy. The somewhat arbitrary selection of craters to be measured in every one of our seven series recalls somewhat the "picking up" of stars in the zones of Lalande, Bessel, Argelander, 150 years ago. The I.A.U. catalogue of BLAGG and MÜLLER (1935) corresponds to a degree to the "Bonner Durchmusterung" (1855). In stellar astronomy, some order was introduced by the zone catalogues of the "Astronomische Gesellschaft" (commenced in 1870). We had learned from its errors, and it was repeated in an exemplary manner from 1928 to 1932 under the direction of Küstner. The purpose was a systematically uniform determination of more than 100000 stellar positions. Towards this end, at first the so-called *Third Fundamental Catalogue of 1500 Stars* (FK3) was established from the material secured at – not one – but dozens of observatories by the "Astronomisches Rechen-Institut" in Berlin-Dahlem, where absolute meridian observations had been made since approximately 1850. Secondly, the positions of approximately 17000 reference stars for photographic plates were observed on seven different meridian circles from 1928 to 1932. The principle was that every star should be observed at least six times, and above all, by at least three different observatories. Moreover, 1200 stars distributed over the entire northern sky were measured in Babelsberg, Bergedorf and Bonn, in an endeavor to reduce the unavoidable sources of systematic errors. This distribution of work provided more accurate results than would have been obtained by observing the same star six times with the same instrument. Later on, the "Astronomisches Rechen-Institut" unified the partial catalogues from the seven observatories into one general catalogue. The 150000 photographed star positions were then added to this. The process was exactly the same in the second repetition of the "zone catalogues" which are now under way.

The suggestions for selenodesy, advanced above, are in harmony with these experiences. It seems most important: (1) to obtain as many plates as possible for judicious selection, (2) to obtain them by four or more different instruments, (3) to measure the same 200 craters at different institutes (not only one!) in accordance with common established guidelines.

However, since this project would require several years for its completion – if only because of the long periods of librations – it will be desirable to combine the material existing at present in the form of our seven series. This can be done most accurately by reducing the measured ξ, η, ζ – not the λ, β derived from them – to System I, or by determining the A, B, C, D by smoothing for the common craters in the equations $\xi_{II} - \xi_I = A\xi_I + B\eta_I + C\zeta_I + D$. The η and ζ coordinates must be treated in the same manner. Moreover, there are three adjustments to perform for the other five series. The ξ, η, ζ of all seven series on I can then be adjusted for all measured objects, and a general catalogue compiled for ξ, η, ζ. It should provide also the λ, β's and h, i.e., the absolute heights of the 200 points selected up to now. Unfortunately, the present author, who is 75 years old, has no collaborators in Vienna. He has an electronic computer at his disposal, but would like to finish some other studies first. The correct combination of all seven series must, therefore, be left to other investigators. However, the author (HOPMANN, 1967) was able to establish, along the lines indicated in the preceding sections, a simplified version of 'Preliminary Catalogue of Absolute Heights on the Moon' which certainly should be more accurate than the seven series considered individually. The already existing preliminary version of the card catalogue contains approximately 1051 objects. Out of these, 430 occur in one of the several "good" catalogues I, ACIC or Sau, or in at least two others. Only less satisfactory individual heights exist for 621 objects, mainly from Baldwin's long list (BALDWIN, 1963).

A preliminary enumeration of the "good" points resulted in Table V showing the extreme values +5.0 and −4.3 km. The positive values are strong by concentration in the region of Hipparchus etc.: the negative ones in various maria. On the average, however, deviation from the sphere seems to amount only to between 1.0 and 1.5 km.

It is further planned to plot these 1000 objects, with an indication of their accuracy, on a map on which the Moon radius should be equal to 50 cm, and to draw the corresponding contour lines. However, they should not be interpolated only between the individual points; but the morphology of the whole area must be taken into account with the aid of the photographic atlases of LPL, for example.

TABLE V

$h \geq$	3.0 km	9
Var. 2.9 to	2.0 km	33
	1.9 km–1.9 km	320
	2.0 km–2.9 km	31
$h \leq$	-3.0 km	7

The map on a scale corresponding to a lunar radius of 20 cm was drawn tentatively for the NW quadrant. This scale proved to be too small for the 250 points to be distinctly marked. However, one could already see that Mare Crisium is the deepest, with Mare Frigoris and Serenitatis already at the zero level: while Mare Vaporum belongs already to the continental shelf around Hipparchus (see below).

As an additional preliminary study, the map on a scale of $1 : 10^6$ of the Mare Cognitum (RLC 1) was investigated, which we obtained from ACIC in Manchester. Its area corresponds approximately to that of England (without Scotland and Ireland). My file contains 19 good and 4 less satisfactory absolute heights in this region, ranging from $+1.7$ to -3.0 km, with an average of -0.3 km. The scatter amounts to ± 1.1 km (i.e., is only slightly greater than the accuracy of measurements). By a comparison with the RLC map, which does not contain any absolute heights, we obtain the following groups (n=number, $\bar{h}$=average absolute height in km) (see Table VI).

TABLE VI

	n	$\bar{h}$
Mountainous Regions	5	+ 1.3
Flat Regions	5	+ 0.7
Mare Cognitum	11	− 0.6
Crater Floors	2	− 1.2

The effects of the local topography are, therefore, dominant. Nothing in the nature of a systematic drop is indicated, for example, between Guericke and the Riphaeus mountains, etc. However, the lower level of the maria is noticeable.

COMPARISON WITH ACIC AND AMS

R. W. Carder sent me a complete set of the maps published so far by the ACIC on the scale of $1{:}10^6$ and $1{:}5.10^5$. It is planned to compare the numerous relative heights given there with my 520 heights observed so far, etc. The maps are based on an extensive plate material, and supplemented by visual work of two observing teams in Flagstaff, who work with the aid of 20″ and 24″ refractors. The maps contain in part indications about absolute heights in the form of iso-hypses. I do not know how these latter were obtained. At the time of publication of the charts scrutinized below – 1961 and 1963 – only the scattered heights of I existed. Approximately at the same time, the U.S. Army Map Service published two countour maps, one with the western half of the Moon, one of the central parts to ±40° longitude and latitude. It seemed logical – again as a preliminary study – to compare the ACIC and AMS maps with my file of absolute heights.

The adjoining charts LAC 59 (Julius Caesar) and 78 (Theophilus) were selected for this purpose. They range in selenographic longitude from +10° to +30°, and in latitude from +16° to −16°. They contain, on the one hand, the highest points of the Moon at Cyrillus (according to the AMS maps), a part of the continental "bulge" around Hipparchus, etc., and, amounting to approximately 8% of the lunar disk together. Some figures bearing on the comparisons are listed in Table VII.

TABLE VII

	LAC	59	78	59 + 78
1	Number	40	37	
2	$\bar{h}\,(W)$	0.0	1.3	
3	$\bar{h}\,(L)$	3.7	6.3	
4	$\bar{h}\,(A)$	7.0	9.0	
5	W	± 1.41	± 1.48	± 1.45
6	L	1.31	2.04	1.68
7	A	0.95	1.07	1.01
8	$r\,(W\text{–}L)$	+ 0.53	+ 0.69	
9	$r\,(W\text{–}A)$	0.54	0.76	
10	$r\,(L\text{–}A)$	0.58	0.69	
11	$\overline{W\text{–}L}$	± 1.35	± 1.57	± 1.46
12	$\overline{W\text{–}A}$	1.21	1.00	1.11
13	$\overline{L\text{–}A}$	1.11	1.55	1.33

The first line indicates the number of points in the Vienna file. The known heights and individual observations are distributed approximately evenly for each sheet. Lines 2 to 4 give the mean values of these heights, in accordance with the Vienna file (W), the data on the LAC charts (L) in the second map of the AMS (A). It is immediately apparent that the terrain of chart No. 78 (Theophilus) is approximately 2 km higher than that of the chart No. 59 (Julius Caesar). Moreover, W, L, and A have different definitions of the zero points from which the altitudes are measured. Lines 5 to 7 give the corresponding standard deviations, which can serve as an indication of whether the three systems of lunar altitudes possess the same scale. The height variations on chart No. 59 exceed those of the mountainous region of No. 78, mostly at *L*; but the scale of *A* may be too small. On the first contour map of AMS, a point at Catharina is listed as being 9.3 km higher than the level of the zero-point Mösting A; on the second map, their difference was reduced to 5.2 km.

The correlation coefficients in lines 8 to 10 indicate only that the uncertainty of all these absolute heights are close to 1 km. In examining them we get the impression that the variation of the contour lines on the map is, on the whole, correct; but this need not be the case with all the individual numerical values as listed. Lines 11 to 13 in Table VII finally indicate the mean differences of the two chart systems. From the values last given we obtain by the "triangle comparison" (see p. 286) the external mean errors, i.e., for $W = \pm 0.90$, $L = \pm 1.16$, $A = \pm 0.66$. Accordingly, the system *A* seems to be the best. However, this might be due to its small scale height. If the values are adjusted to a standard deviation of $W = \pm 1.47$ km, the external mean errors for the different systems result as $W = \pm 0.90$, $L = \pm 0.96$ and $A = \pm 0.94$. In view of the limited size of sample material, all we can say at the present time amounts to this: except for the differences in zero-point and scale variations between *L* and *A*, the various height systems are approximately equivalent. So far, the absolute heights have an uncertainty of approximately 1.0 km.

The 40 + 37 altitudes were entered tentatively on square paper on the scale of the planned contour map, marking the certain and the individual values. The selection of the scale proved to be quite suitable. Subsequently, it was attempted to draw contour lines – or to mark especially high or low areas – using the corresponding pages of Kuiper's *Photographic Atlas* (Kuiper *et al.*, 1960). Unfortunately, absolute height measurements for Theophilus region do not exist. Moreover, a small high plateau 3 km in height shows at $\xi = +0.33$, $\eta = -0.24$; a second bulge seems to extend from $\xi = +0.32$, $\eta = -0.10$ to $\xi = +0.23$ to $\eta = +0.04$, 2 km in height. The Mare Tranquilitatis appears to lie approximately 1 km above the average Moon-level.

4. Problem of the Moon's Shape

The idea of a tidal bulge of the Moon goes back to Laplace (1800), who explained the observed synchronism between rotation and revolution of our satellite as a consequence of a difference of the three principal moments of inertia of a three-axial homogeneous ellipsoid elongated towards the Earth. In a study of now only historical interest, Hansen in Gotha assumed around 1860 that this tidal bulge was 60 km high. This seemed confirmed by an evaluation of two photographs made by Warren de la Rue, and in 1859 by Gusew in Pulkovo, leading seemingly to a tidal bulge of 83 km in height. Franz showed, however, in 1899 that this assertion was completely untenable. The observations made in the 19th and 20th century – in Schrutka-Rechtenstamm's investigation of 1955 – showed the ratio of the three moments of inertia to be 1:1.0002353:1.0006273; which, for a homogeneous Moon, would correspond to the differences of the three principal axes of inertia $a - c = 1.1$ km and $b - c = 0.7$ km. This would then be the "dynamical" shape of the Moon.

On the other hand, we can also derive for the "geometrical" shape as a first approximation the constants of a tri-axial ellipsoid with the aid of the equation

$$A\xi^2 + B\eta^2 + C\zeta^2 + D\xi\eta + E\xi\zeta + F\eta\zeta + G\xi + H\eta + I\zeta = 0$$

from the lists of absolute heights; where A, B, C indicate here the longitudes of its semi-axes: D, E, F their direction cosines: and G, H, I the coordinates of its center, which must not coincide with that of a sphere through the lunar limb. The origin of the rectangular coordinates ξ, η, ζ is the center of the sphere; the ζ-axis points towards the Earth; the η-axis towards the north, the ξ-axis towards the west.

In 1953, the author (HOPMANN, 1954) derived from the approximate values of the measurements made by FRANZ (1901) approximate absolute heights for 150 craters, resulting in an ellipsoid whose largest axis was 6.8 km longer than the mean axis, directed to a point in the SW quandrant. The values of G, H, and I were not determined here. On the basis of his re-reduction of Franz's measurements, SCHRUTKA-RECHTENSTAMM (1956) derived for the first time all 9 constants. Their mean errors remained unknown, except for $C-I=3.8\pm2.5$ km and $I=+5.1\pm2.0$ km, which, however, compensate each other in their action. The coefficients A to F were, therefore, calculated in a second adjustment without G, H, I, and they were then all found to lie within the limits of their mean errors. This indicated that the Moon was a sphere within an accuracy of ±0.5 km; any indication of a tidal wave could no longer be detected. However, the possibility of a small bulge of 2 to 3 kilometers height, which could be sharply bounded by the isohypses close to the center of the lunar disk, was not taken into account in these calculations. But even including it and taking into account Hayn's Moon profile (cf. SCHRUTKA-RECHTENSTAMM and HOPMANN, 1955), Schrutka-Rechtenstamm obtained $a=1738.7\pm0.4$ km, $b=1738.6\pm0.4$ km, $c=1737.6$ ±0.5 km, so that a sphere resulted within the mean errors. We can, of course, force through each of the five series such an ellipsoid – as it was done, for example, by RUNCORN (1966) with two different combinations of the ACIC values, or those of Potter (Leningrad), who gave a third solution from the same material. Later on, as soon as the contemplated 'Preliminary General Catalogue' is available, we can also make such an adjustment – then, however, with all 9 constants needed to describe the ellipsoid. This strongly levels off the many irregularities on the Moon's surface: but such an adjustment would be desirable, not only for a comparison with the dynamical ellipsoid, but also for studies about the internal structure of the Moon and its evolution.

The author (HOPMANN, 1954) indicated as early as 1953 on the basis of Hayn's lunar profiles the existence of bulges and depressions, which are clearly recognizable on Watts charts (cf. WATTS, 1963), especially on the figure on p. 949. Preliminary studies for the desired contour map also show such bulges; in addition to that mentioned above, close to the center of the disk, for Eudoxus, the Taurus mountains, between Mare Crisium and Serenitatis, etc. On the other hand, depressions are indicated in the regions of Mare Crisium, Imbrium, Humorum, and Nubium; whereas the Mare Tranquilitatis, Cognitum, and Vaporum appear to be elevated. If we enter the 150 points of A in a sketch of Lambert's equal-area projection, we see that this covers only less than 50% of the front side of the Moon. Approximately 60% are covered with ACIC. The central bulge amounts only to 2.5%. Accordingly, the results will remain unreliable in calculating the coefficients of an ellipsoid. If we solve the normal equations, the various unknowns can no longer be cleaıly separated. The situation can be improved – but still remains unsatisfactory – if Watts' Moon profiles are appropriately taken into account. A true answer with respect to the geometric shape of the Moon represented by a mathematical model will be obtained only when measurements on the back of the Moon are made with the same accuracy and uniformly added to the front. This will take a very long time.

However, one fact must be pointed out which transpires from the few photographs of the back side of the Moon available at the present: it shows almost no maria; therefore, to this extent the front and back side of the Moon are clearly unsymmetrical. Will perhaps the location of the center of a lunar tri-axial ellipsoid be found later on to be outside the ξ, η-plane? Is it not better, in view of such a possibility, to postpone calculation of ellipsoids (and harmonic expansions) at least for the time being? And should one not assume naturally that the Moon is a sphere with local deformations?

References

ACIC: 1965, 'Coordinates of Lunar Features', by D. L. Meyer and B. W. Ruffin, *Icarus* **4**, 513.
AMS: 1964, *Tech. Rep. No. 29, Part 2.* Corps of Engineers, U.S. Army, Washington, D.C.
ARTHUR, D. W. G.: 1962, *Comm. Lunar Planet. Labor., University of Arizona* **1**, No. 11.
BALDWIN, R. B.: 1963, *The Measure of the Moon.* Univ. of Chicago Press, Chicago.
BLAGG, M. A. and MÜLLER, K.: 1935, *Named Lunar Formations.* London.
FRANZ, J.: 1889, *Astr. Beob. Königsberg* **38**, Nr. 1.
FRANZ, J.: 1899, *Astr. Beob. Königsberg* **38**, Nr. 5.
FRANZ, J.: 1901, *Mitt. Sternw. Breslau* **1**, Nr. 1.
FRANZ, J.: 1913, *Nova Acta Leopoldina* (Halle **90**, Nr. 1).
GOUDAS, C. L.: 1963, *Icarus* **2**, 423.
GOUDAS, C. L.: 1965a, *Icarus* **4**, 188.
GOUDAS, C. L.: 1965b, *Icarus* **4**, 528.
HAYN, F.: 1904–1914, 'Selenographische Koordinaten', *Abd. K. Sächs. Gesell. d. Wiss.* (Leipzig) **27–32.**
HOPMANN, J.: 1954, *Mitt. Univ. Sternw. Wien* **6**, 13.
HOPMANN, J.: 1967, *Ann. Univ. Sternw. Wien* **27** (in press).
KOPAL, Z. and FINLAY, E. B. (eds.): 1961, *Problems of Lunar Topography* (Astr. Contr. Univ. Manchester, Ser. III, No. 90).
KUIPER, G. P. *et al.*: 1960, *Photographic Lunar Atlas.* Univ. of Chicago Press, Chicago.
RUNCORN, S. K.: 1966, Proceedings of the meetings of the NATO Advanced Study Institute, Newcastle (in press).
SAUNDER, S. A.: 1907, *Mem. Roy. Astr. Soc.* **57**, No. 1.
SAUNDER, S. A.: 1911, *Mem. Roy. Astr. Soc.* **60**, No. 1.
SCHRUTKA-RECHTENSTAMM, G. VON: 1956, *Mitt. Univ. Sternw. Wien* **7**, 151.
SCHRUTKA-RECHTENSTAMM, G. VON: 1958a, *Mitt. Univ. Sternw. Wien* **9**, 97.
SCHRUTKA-RECHTENSTAMM, G. VON: 1958b, *Mitt. Univ. Sternw. Wien* **9**, 251.
SCHRUTKA-RECHTENSTAMM, G. VON: 1966, *Ann. Univ. Sternw. Wien* **26**, Nr. 6.
SCHRUTKA-RECHTENSTAMM, G. VON and HOPMANN, J.: 1955, *Mitt. Univ. Sternw. Wien* **7**, 127.
SCHÖFFL, O.: 1966, *Ann. Univ. Sternw. Wien* **26** (in press).
WATTS, C. B.: 1963, *The Marginal Zone of the Moon* (Astr. Papers Amer. Ephem. Nautical Almanac, vol. 17).

OBSERVATIONAL UNCERTAINTIES IN LUNAR CONTROL SYSTEMS*

DONALD L. MEYER**

Abstract. Certain elements adversely affect a fundamental lunar control solution. Primary consideration is given herein to those distortions in the observations that have magnitudes greater than the desired increment of detection. An analysis is made of displacements in the recorded image due to atmospheric distortion, and methods of minimizing this effect through the use of sequential and long exposure photographs are discussed.

Introduction

The accuracy of three-dimensional coordinates derived for selected lunar features is limited by the observational data used. It is normal in a lunar control solution to assign a latitude and longitude within a lunar reference frame and an elevation above or below a mean spherical datum. The departures from a sphere imply the lunar shape.

Examination of the lunar limb, particularly as outlined in an annular eclipse, indicates general deviations of only 2 or 3 kilometers. It is reasonable to expect that undulations in the other regions do not exceed this by any appreciable magnitude. It then becomes apparent that observational uncertainties are extremely critical in the detection of small departures from the spherical form.

Angular Separation of Observations

Lunar control solutions are based on the concept that three-dimensional coordinates of a feature are implied as the Moon exhibits librations. The observation data consists of sets of measured plate coordinates for a group of features from photographs taken in different parts of the librational cycle. As such, they reflect a two-dimensional projection of the three-dimensional Moon as seen from slightly different viewpoints. Detection of the differential effects caused by departures from spherical form is dependent upon the angular separation. For example, Figure 1 shows the projected plate position of a lunar feature as seen on one photograph. This plate position can be the result of numerous positions (λ_1, λ_2, λ_3) and elevations above or below a mean sphere (h_1 and h_3). A series of photographs taken from this same view would

* The views expressed herein are those of the author and do not necessarily reflect the views of the U.S. Air Force or the Department of Defense.
** Aeronautical Chart and Information Center, St. Louis, Mo., U.S.A.

Kopal and Goudas (eds.), Measure of the Moon.

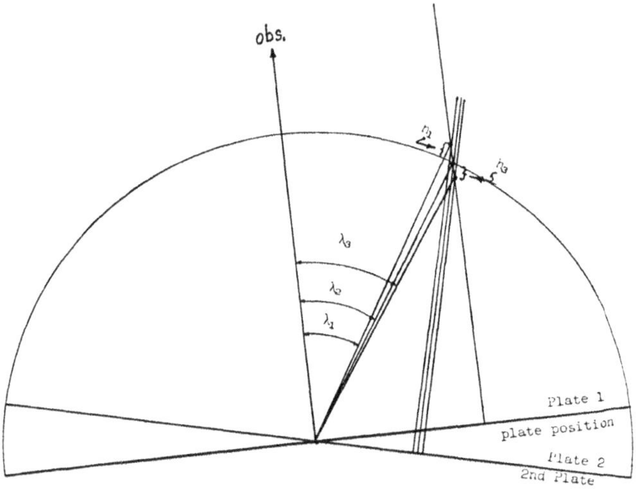

Fig. 1. Projected plate position of a lunar feature.

not resolve the actual location of the feature. Consequently, another photograph taken from a different viewpoint must be considered. There, the position of the feature on the second plate will be slightly different for each of the three possibilities shown. These differences in viewpoint are a function of the lunar librations.

Libration in latitude is fairly consistent with a maximum of about 6.7 degrees. Due primarily to evection, the maximum libration in longitude varies between 5 and 8 degrees. Thus, two observations taken from opposite quadrants, at instances of maximum librations, could result in an angular separation of 20 degrees. However, this is not a common occurrence, and it is more practical to expect librational differences of about 15 degrees.

Considering 15 degrees to be the expected value for the angular separations of observations, then some estimate can be made for the plate displacements caused by the undulations of the surface. The simplest case is in the center of the Moon with the feature being at the subterrestrial point for one observation. Figure 2 shows the displacement on the second plate caused by the feature being above the mean surface at P_1 rather than on the surface at P_2. The amount of this displacement is simply

$$d = h \sin \varphi$$

where d=the displacement in meters, h=height above the mean surface (meters), and φ=angular separations of the two observations.

This formula is not rigorously true due to a slight conical effect in the projection of the feature on Plate 2. However, this effect is negligible.

To convert the computed displacements into increments of the plate measurements, a photographic scale of 20 million is assumed. This scale is convenient for the refractors at the Pic-du-Midi and Lick Observatories, whose photographs have been

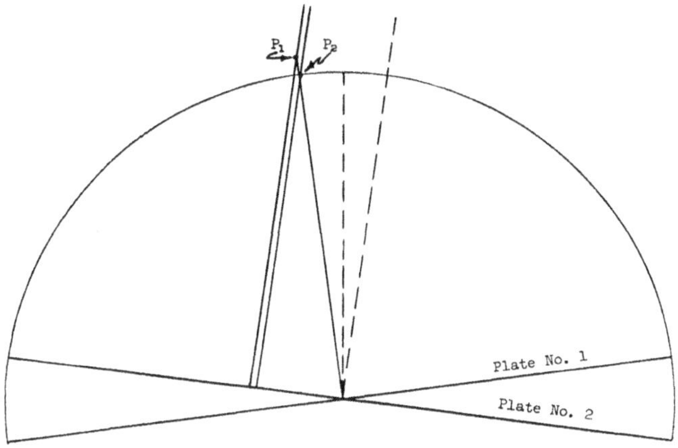

Fig. 2. Displacement due to undulations.

TABLE I

Plate Displacements in Microns
For Different Heights and Librations

Heighth Above Mean Sphere	Angular Difference in Librations					
	20°	15°	10°	6°	4°	2°
1 km	17	13	9	5	3	2
2 km	34	26	17	10	7	3
3 km	51	39	26	16	10	5

used in several lunar control studies. Table I gives the plate displacements in microns for various heights above the mean spherical surface and different angular separations between the librations.

The amount of the displacements decrease as the angular separation decreases. This continues until the angular difference is 0 degrees and the displacements disappear entirely. Also, displacements caused by a specific height increase away from center of the disc. The distance P_1 to P_2 no longer coincides with the height, and it continues to increase as the limb is approached. Figure 3 gives a simple version of the situation by placing the observer's positions and the features on the lunar equator.

Assuming the angular separation of 15 degrees ($\varphi_1 + \varphi_2 = 15°$), and a height above the mean sphere of 1 kilometer, then the plate displacements are 15 microns for a longitude of 20 degrees and 20 microns for a longitude of 40 degrees. This is for the assumed photographic scale.

The foregoing is not intended as a rigorous analysis of plate displacements caused by relief in the lunar surface, which is admittedly more complicated. In practice, some error must be expected in relating each photograph to the same origin and reference system. Also, the photographic scale changes with the lunar distance, and this causes

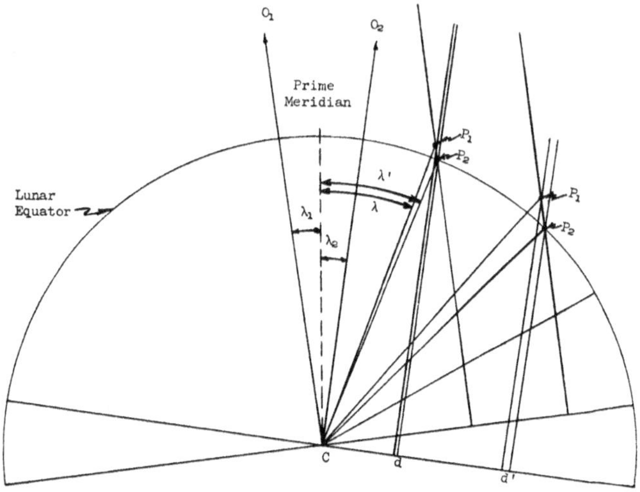

Fig. 3. Displacement increase.

slight variations in the computed displacements. The purpose is to determine a reasonable estimate of the displacements being sought in the observations.

Observational Uncertainties

Uncertainties in the observation can be placed in two categories. One is the actual measurement of the features. This includes the difference in interpretation of the feature on plates of different libration, phase and resolution, as well as the error in measurement. It could also include some physical parameters, such as the symmetry of the feature. The other category is the recording of the lunar image. This includes the atmospheric, optical, and photographic effects which disturb the true positioning of the feature's image.

Most uncertainties can be effectively reduced by established procedures and careful measurement. However, there is one source of positional error that is difficult to control or remove. This is the distortion of images at the telescope by atmospheric turbulence. Although image motion is readily observed in the focal plane, it is difficult to evaluate the amount of displacement. A lunar photograph records the moving imgage in some part of its random cycle, but the recorded positions may have geometric distortions of considerable magnitude.

The observational techniques employed by ACIC in the preliminary reduction of 196 points give some appreciation of the magnitude of "seeing" displacements. In order to reduce this effect, sequences of five photographs were measured by two operators. Exposure times were less than 1 second, and the sequences were generally taken in less than 2 minutes. The measurements on four plates are transformed into the center plate, and the final measured coordinates are a mean of these values.

This system includes 35 such photographs in seven sequences and required 28 plate to plate transformations to develop mean plate coordinates. The circular errors of the transformations varied from 28 to 39 microns with an average of 34 microns. However, various checks were made of the differences between the operators and their magnitude was about half the plate to plate errors.

The system also includes three original glass plates taken about 1 minute apart. These are long exposures taken with the 61-inch astronomic reflector of the U.S. Naval Observatory at Flagstaff, Ariz. The exposure times are in excess of 20 seconds, and the image excursions are expected to be averaged photographically. Plate to plate transformations were conducted in the same manner as with the short exposures.

The effects of seeing displacements can be seen in a comparison of plate to plate transformations between the two methods. One of the sequences of short exposures is a series of original glass plates taken at Pic-du-Midi. The average circular error of the plate to plate transformations is 28 microns. However, the same error for the long exposure plates in only 13 microns.

An extended test was made for seeing displacements using a series of 13 consecutive photographs taken at Pic-du-Midi Observatory. The exposure time is .2 seconds, and the lapsed time between exposures averages about 13 seconds. Table II lists the photographs and their Universal Time.

TABLE II

Times of Exposures Used In Seeing Test.
Pic-du-Midi Reference: 358.
Date: 25 August 1964

Photo Sequence No.		Universal Time
1		01 44 26
2		01 44 38
3		01 44 55
4		01 45 06
5		01 45 18
6		01 45 30
7	Center Plate	01 45 43
8		01 45 56
9		01 46 09
10		01 46 24
11		01 46 37
12		01 46 49
13		01 47 05

These are near full moon photographs with the evening terminator at slightly less than 70 degrees east longitude. They were selected so as to represent the seeing patterns over most of the lunar disc. Sixty-six well dispersed craters were selected for measurement. They are identified in Table III.

The purpose of this test was to demonstrate the distortions caused by seeing effects as opposed to measurement and interpretation errors. Using a series of photographs taken in a short period of time eliminates the effects of different phases and librations.

TABLE III

Seeing Test Points

No.	IAU No.	Name
1	1846 B	Galilei E
2	1835	Reiner C
3	1575	Bessarion C
4	1520	Hortensius A
5	1498	Gambart A
6	2933	Mösting A
7	1213	Bode G
8	1283 A	Eratosthenes B
9	1407	Pytheas A
10	1591 A	Diophantus C
11	1877	Naumann B
12	1614	Mairan E
13	1668	Harpalus E
14		Marpertuis L
15	1303	Leverrier
16	1298 D	Timocharis F
17	895	Aratus
18	632	Linné B
19	934	Cassini F
20	1124	Pico C
21	1356	Anaximenes
22	987	W. Bond B
23	677	Democritus A
24		N E of Hercules H
25	391	Cepheus A
26	645 A	Plana D
27	490 A	Posidonius F
28	281	Littrow D
29		E of Tralles B
30	275 A	Maraldi B
31	249	Maskelyne M
32	587	Tacquet
33		White Spot – Mare Vaporum
34	819	Silberschlag A
35	3667	Moltke
36	4287	Capella D
37		Taruntius B
38	4691	Langrenus D
39	4396	Biot
40	4359	Bohnenberger G
41	4157 A	Beaumont G
42	3727	Dollond E
43	3579 A	Albategnius C
44	3570	Argelander D
45	3791	Sacrobosco C
46		Rothmann K
47	4417 A	Reichenbach K
48	4442 A	Rheita P
49	4004	Nicolai A
50		Crater N W of Vlacg K
51	3901	Mutus B

No.	IAU No.	Name
52	3407	Cuvier C
53	3499	Aliacensis D
54	3066 C	Lippershey T
55	2856	Guericke C
56	2464	Euclides D
57	2447	Flamsteed D
58	2023	Rocca B
59	2417	Gassendi G
60	2419	Gassendi J
61		De Gasparis B
62	2325 A	Lacroix H
63	2598	Schiller A
64	2716 D	Longomontanus R
65		Ramsden A
66	2835	Lubiniezky F

Also, care was taken to select a series of photographs of fairly consistent quality. This was to prevent any significant differences in the ability to measure the various photographs.

Each of the 13 photographs was measured on a Mann Comparator in the orientation normally used by ACIC. This was with the x and y comparator readings increasing toward the lunar east and north. After all the points were measured, the plate holder was rotated 180 degrees, and the features measured again. This was to determine the uncertainties in the measurement and interpretation on each plate.

A series of least square transformations were performed on these measurements and the following equations were used:

$$X_1 = AX_2'' - BY_2'' + C_N,$$
$$Y_1 = AY_2'' + BX_2'' + C_N,$$

where

$$A = \cos\theta = \frac{\Sigma(X_2'X_1') + \Sigma(Y_2'Y_1')}{\{[\Sigma(X_2'X_1') + \Sigma(Y_2'Y_1')]^2 + [\Sigma(X_2'Y_1') - \Sigma(X_1'Y_2')]^2\}^{\frac{1}{2}}},$$

$$B = \sin\theta = \frac{\Sigma(X_2'Y_1') - \Sigma(X_1'Y_2')}{\{[\Sigma(X_2'X_1') + \Sigma(Y_2'Y_1')]^2 + [\Sigma(X_2'Y_1') - \Sigma(X_1'Y_2')]^2\}^{\frac{1}{2}}},$$

$$C_E = \frac{-A(\Sigma X_2'') + B(\Sigma Y_2'') + \Sigma X_1}{N},$$

$$C_N = \frac{-A(\Sigma Y_2'') - B(\Sigma X_2'') + \Sigma Y_1}{N}.$$

Scale $= 1 = [A^2 + B^2]^{\frac{1}{2}}$
X_1, Y_1 = Standard set readings
X_2'', Y_2'' = Set readings to be transformed
X_1', Y_1' = Standard set centroid readings
X_2', Y_2' = Centroid readings to be transformed
N = Number of craters measured.

The error of each transformation was determined from the residuals (dx and dy) and reported as a circular position error (C.P.E.). Where

$$\text{C.P.E.} = \sqrt{\frac{\Sigma dx^2 + \Sigma dy^2}{N - 1}}$$

The first transformations were concerned with each plate individually. Reverse plate readings were transformed into normal orientation and the circular errors noted. Then the normal plate measurements of all the plates were transformed into the measurements of the center plate (No. 7). A mean set of plate measures were derived for the craters by averaging all these values. Finally, each set of plate measurements was transformed into the mean set. Table IV is a listing of the circular errors of these transformations. It also gives a listing of the errors in transforming several plates into No. 9. Originally, this was to be the center plate in a larger sequence. However, the series was reduced in number due to a deterioration of the imagery after Plate 13. The errors of transforming to Plate 9 are included to show that the magnitude is not caused solely by No. 7.

TABLE IV

Standard Transformation Error (Circular) in Microns

				To Mean Coordinates	
Plate No.	0°–180° Orientations	To Center Plate 7	To Plate 9	0 Degree Orientation	180 Degree Orientation
1	16	31	30	21	23
2	16	38		23	23
3	15	33	33	20	21
4	16	31	29	18	21
5	17	30	30	18	21
6	15	31	31	17	20
7	17		34	24	24
8	17	34	33	23	23
9	15	34		23	24
10	15	30	31	20	18
11	16	31	31	20	21
12	16	35		23	21
13	15	33		19	22
Average	16	33	31	21	22

The significance of Table IV is that it is a comparison of the errors between plate orientations (0°–180°) as opposed to the errors between plates. It can be seen that the latter are about double the errors between orientations. The major factor introduced by going from plate to plate is not a difference in appearance or interpretation of the features but the random displacement caused by seeing effects. Since the random portion of the seeing and measuring error must be both accumulative and compensative, the seeing error must be of considerable magnitude to produce these differences.

Examples of the displacements, caused primarily by the seeing effect, are given in Figures 4 and 5. These are plots of the residuals in x and y for eight of the measured features. The residuals were developed by a transformation of each plate's measurements in the zero degree orientation into the mean measures derived for the whole series. In this manner, all plates were referred to a standard set of measures.

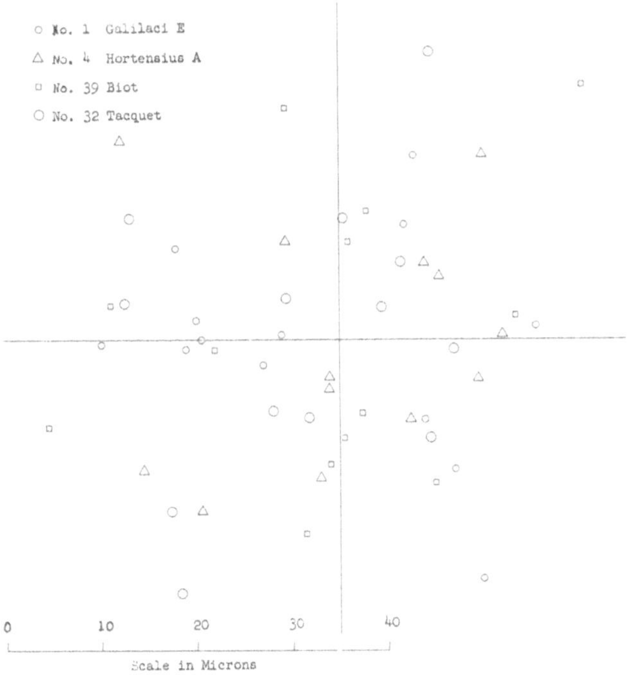

Fig. 4. Residuals in measurements after transformation to the mean value of the series.

The plots show the random behavior of a feature's position as recorded on different photographs. Differences greater than 40 microns can be observed between plate positions for every feature. In a few cases, the difference is greater than 60 microns. Also, the specific plates creating the maximum difference change from feature to feature and indicate the local nature of the seeing effect.

Next, a comparison was made of the ACIC method of using five sequential photographs assuming that the mean measures developed from the 13 sequential photographs represent the true positions of the features.

The center plate (No. 7), and the two plates before and after, were reduced by the ACIC method. A set of mean measures was derived for this short sequence of photographs. The circular error between these and the true set of measures was only ± 8 microns. The demonstration shows that significant distortions in a single photograph can be reduced to a usable magnitude with sequential photography.

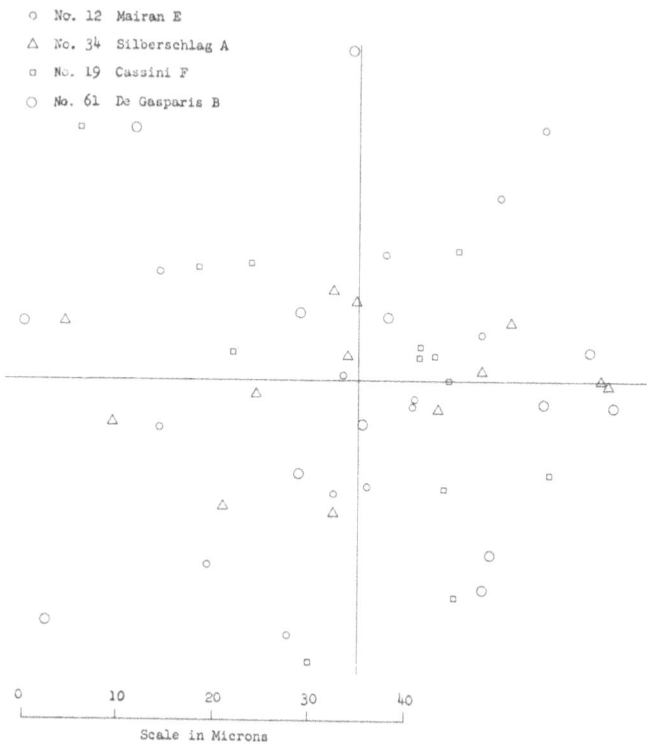

Fig. 5. Residuals in measurements after transformation to the mean value of the series.

Conclusion

A more precise lunar control solution from earth-based photography is possible, providing the observational uncertainties can be reduced to a usable level. The observation error can be reduced from ± 22 microns (which is the seeing and measuring error when only one photograph is used) to ± 8 microns when a sequence of five photographs are used. Additional photographs in the sequence yield a very small gain in accuracy. It is evident that the distortions in the lunar photograph are much larger than the effects of the small departures from a sphere. Plate distortions can exist as large as the total displacement being sought and larger than displacements caused by 1 and 2 kilometer heights. This places the major concern of a lunar control solution on the validity of the observations. Two methods are presently available to produce adequate observations. One is the use of a sequence of short exposures, and the other is an extended exposure time.

CONTROL INTEGRATION FOR LUNAR MAPPING

JAMES M. EIGEN AND JAMES D. HATHAWAY*

Introduction

The Department of Defense (DOD) Interim Control System 1966 has been established by the Selenodetic Branch of the Army Map Service (AMS), in cooperation with the National Aeronautics and Space Administration (NASA) and the Lunar and Planetary Branch of the Aeronautical Chart and Information Center (ACIC). Initial application of the new system will be in preparation of charts and maps for Project Apollo landing site selection. It is expected that the parameters descriptive of the lunar figure will be significantly improved from the harmonic analysis of the new control.

The results of three independent adjustments were combined to form the data for the DOD adjustment. Each of the three adjustments were reduced by the AMS processes to ensure equal treatment of all data. The three sources used to form the DOD system are the 256 craters of AMS-64 produced by AMS, the 196 crater reduction by ACIC, and the 484 crater reduction completed for NASA by AMS.

A least-squares solution was computed for 734 craters that formed the adjustment which contained 13966 observation equations and solved for 2886 unknowns. The control covers 74% of the visible surface of the lunar disk and is the largest adjustment ever completed by DOD.

Collection of Data

In preparing AMS-64, 15 photographs were selected from the Lick Observatory lunar photographs and copied on $\frac{1}{4}$ inch, microflat glass plates (cf. MARCHANT, HARDY, and BREECE, 1964). The copying was done with a Kel-Wat Photographic Dodger. The fluor-o-dodge system of contrast control was used in order that crater detail hidden by the high contrast range in the original negative could be made available for examination on the dodged copies.

The photographs are listed in Table I.

The 15 photographs described above were measured at Georgetown Observatory, with 24 to 204 crater images measured on each photograph. Measurements for each of 310 craters were made on one to eleven photographs, for the average crater on six

* U.S. Army Map Service, Washington, D.C., U.S.A.

Kopal and Goudas (eds.), Measure of the Moon.

TABLE I

Lick Designation	Universal Time	Date
M2	4^h 14^m 05^s	7 May 1938
M3	3 54 44	12 January 1938
M4A	8 03 25	22 October 1937
M5	13 00 40	24 October 1937
M7	12 47 19	20 August 1938
187	4 07 54	10 August 1940
190	3 39 08	7 April 1938
199	3 55 20	12 August 1940
202	4 19 06	8 December 1940
204	4 22 38	7 July 1938
206B	4 14 00	13 August 1940
209	12 48 30	22 August 1940
212	13 38 00	26 October 1945
214	11 19 34	6 September 1936
219	13 42 53	26 October 1937

photographs. Positioning of the plates in the measuring engine was facilitated through use of the image of a National Bureau of Standards reseau with 5 millimeter intervals, which had been overprinted on each of the plates during fluor-o-dodging. For the 0° orientation, horizontal grid lines on the plate were oriented parallel to the X-axis of the engine by reading coordinates of a series of points along one of the grid lines. The physical corrections to parallel were made by a rotation adjustment on the plate holder. After the features were measured in the 0° orientation, the plate was removed, rotated approximately 180° for the 180° orientation, replaced in the measuring engine and positioned in the same manner as described for the 0° orientation.

The coordinates of the geometric center of each feature were obtained by measuring and averaging the maximum and minimum ordinates and abscissas of the feature according to its orientation on the plate. The features chosen for measurements were identified by comparison of photographic plates with the I.A.U. lunar atlas.

The measurements were corrected so that the center of the coordinate system, to which the measurements were referred, coincided with the projection on the plate of the optical center of the Moon (MARCHANT, 1960). The measurements were rotated so that the axes, to which the measurements were referred, were parallel to the projection of the axes in the Moon's equatorial coordinate system. The components of the displacements due to refraction of the crater images on the photographic plates were corrected by refraction coefficients. The values of the semi-diameter of the Moon's image on each plate were obtained and measurements were converted from stereographically projected values to orthographically projected values.

In December of 1963, AMS received a contract from NASA to establish a control net about the Moon's equator. In order to produce a better control system the Corps of Engineers (OCE) authorized a similar polar strip to be established. After four iterations Group NASA contained 484 craters adjusted in a least-squares solution (cf. HARDY *et al.*, 1966).

In preparation for work in Group NASA eighteen plates were selected from the Lick Observatory lunar photographs and processed in the same manner as the AMS-64 plates. Fourteen of the original AMS-64 plates were used, the exception being plate M2. The four additional Lick plates are mentioned in Table II.

TABLE II

Lick Designation	Universal Time	Date
198	$4^h\ 46^m\ 55^s$	7 May 1938
M7A	12 57 18	11 September 1936
182	4 02 00	3 June 1938
192	4 22 37	20 July 1942

The photographs described above were measured at AMS with a total of 4774 crater images measured. An average of 265 craters were measured on each of the 18 photographic plates. For purposes of quality control, the designated craters for measurement on each photographic plate were randomly sorted into groups of 35 to 45 craters (HARDY and BOWLES, 1966). These groups were then of a size such that the entire measurement of an orientation, 0° or 180°, was accomplished in one measurement session. The measurement and reduction technique then proceeded in the same manner as previously described for the AMS-64 system.

The ACIC system used eight photographs on which each of the 196 craters comprising the system were measured. Two types of photography were used in the process (see Table III).

TABLE III

Identification of ACIC Photography

Reference Number		Date	Universal Time
Group I	89	21 December 1961	$22^h\ 14^m\ 29^s$
	195	9 February 1963	3 04 50
	118	21 April 1962	0 16 00
	358	25 August 1964	0 47 19
Group II	183	13 November 1962	22 39 43
	323	27 March 1964	22 12 35
	128	17 May 1962	21 52 35
	N-8	21 October 1964	7 37 10

Plate N-8 was taken at the U.S. Naval Observatory in Flagstaff, Ariz. It is a long duration exposure, 36 seconds, to minimize the effect of atmospheric distortion. The other seven photographs were taken at the Pic-du-Midi Observatory in France. These were taken using a modified K-22 aerial camera which photographed the Moon in sequences so that mean values were used for the observations (MEYER and RUFFIN, 1965).

The measurement procedure begins by placing the plate such that the selenographic coordinates of the observatory coincides with the center of rotation of the comparator stage. A continuous measuring operation is carried out on the 196 craters by two operators working independently. An arithmetic mean is then taken of the two operators' measures for each photographic plate in the sequences. The rotation and translation factors, computed by a linear transformation, transform all measurements to the reference system of the reference plate of the sequence. Several corrections are then applied, including corrections for refraction, and topocentric librations of the Moon (ALFONTE, 1966a).

Preliminary Work at AMS prior to DOD 66i

The ACIC meaned measures of 196 craters were reduced through AMS methods. Reference 89 was held fixed for orientation and scaling in the least-squares adjustment for simultaneous correction to the approximate plate constants and crater coordinates (ALFONTE, 1966b). The crater coordinates which resulted from the fundamental adjustment were then compared with other systems. Results of comparison among the three selenodetic control systems to be integrated showed that no one system disagreed from any other system by an excessive amount.

Weights

Under the direction of the Defense Intelligence Agency (DIA) a weighting procedure was developed to form the best solution for the combined data. The selection of the weighting criteria to be applied to the observation equation matrix was based upon the following:

(1) Librational Baseline (W_1). The strength of the analytical stereographic model is dependent upon perspective movement. The latter is derived from widely separated librational angles. This suggests assigning greatest weight to those sets of observations made under strongest librational conditions. This component of the observation weight was computed by forming the ratio of the average librational baseline for each crater to the maximum average librational baseline of any crater in the adjustment. The average librational baseline was formed by:

$$\text{average}\ \theta_{bk} = q^{-1} \sum_{i=1}^{q} \theta_{bki}$$

where $q=\frac{1}{2}n(n-1)$ and n=number of plates the crater was measured on. The librational baseline, the spherical distance between the librational centers of any two plates, was computed using

$$\cos\theta_b = \sin B_R \sin B_S + \cos B_R \cos B_S \cos(L_R - L_S)$$

where L and B represent librations of plates R and S.

(2) Observational uncertainty (W_2). A mathematical model is essentially a function of the measurements which describe the framework of the system. The uncertainty of that observation of greatest confidence provides a statistical measure for the determination of the strength of plate measurement. A 5:2 ratio was applied in favour of ACIC points as suggested by the five photos in the sequential process of ACIC and the two orientations in measuring at AMS. W_2 was computed as the ratio of the minimum observational uncertainty (σ_{ok}) of any crater in the adjustment to the σ_{ok} of the crater observed. The σ_{ok} were computed in previous fundamental adjustments of each data source independently and represent the plate agreements of the measures.

(3) Equal area distribution (W_3). An undistorted geometric model is best determined by assigning uniform density to the areas covered. Thus, a heavily populated region, e.g., a mountainous area, will not influence the determination of the best fitting surface by outweighing the contribution of the less heavily populated regions, e.g., maria. The visible surface of the Moon has been divided into areas, each equal to a 10° by 10° square at the equator. Each square was given equal weight and, therefore, the weight of each crater is inversely proportional to the density of identified features in its own square.

(4) Plate Coverage (W_4). In the event that certain systematic errors can be eliminated from sets of measurements, if those sets are measured in a homogeneous manner, the area covered by each plate was considered, using the maximum area covered by any plate as a statistical standard.

The resultant weights for the observation equations were then obtained as follows:

$$W_0 = \sqrt{W_1} \times W_2 \times \sqrt{W_3} \times \sqrt{W_4}$$

and then the normalized resultant weight obtained by $W_{n0} = W_0/\max W_0$. The effect of the weights will be analyzed and the results published at a later date. The orientation and scale of Reference 89, determined by a plate constant adjustment, using the coordinates of the 41 Schrutka-Rechtenstamm (S-R) craters contained in the ACIC system were held fixed in the fundamental adjustment. These 41 S-R craters used the *S-R* values as the initial approximations. All other craters used I.A.U. values as initial approximations.

The parameters of the datum point, Mösting A (I.A.U. #2933), as determined by KOZIEL (1964), were used to translate the internally adjusted network of control points to the real Moon.

$$\varphi = -\ 3°\,10'47''.0 \pm 4''.4$$

$$\lambda = -\ 5°\,09'50''.0 \pm 4''.5$$

$$H = 0''.40 \pm 0''.19 = (0.75 \text{ km} \pm 0.35 \text{ km})$$

$$R + H = 932''.98 \pm 0''.30 = (1738.75 \text{ km} \pm 0.55 \text{ km}).$$

The datum surface was considered to be a sphere of radius 1738 km$\pm$0.41 km (Newcomb, 1878).

The astronomic constants which are used to compute the librations (L, B), orientation (P), and scale (r) for each plate in the photogrammetric adjustment are:

$I = 1°32'01''.0 \pm 7''.1$	(Inclination of mean lunar equator to the ecliptic; KOZIEL, 1964).
$f = 0.633 \pm 0.011$	(Mechanical ellipticity of the Moon; KOZIEL, 1964).
$S = 932''.58 \pm 0''.23$	(Lunar semi-diameter at mean distance from the center of the Earth to the center of the Moon; Newcomb, 1878).
$\pi = 3422''.61$	(Equatorial-horizontal parallax of the Moon at the mean distance from the center of the Earth to the center of the Moon; Fischer, 1963).
$a_e = 6378.166$ km	(Equatorial radius of the Earth; Fischer, 1963).
$f_e = 1/298.3$	(Polar flattening of the Earth; Fischer, 1963).

Work prior to the FA

To prepare the data for the FA, plate constant adjustments were run to approximate corrections to the plate constants. The AMS reduction independently solved 171 least-squares solutions, which were iterated until the absolute values of the corrections were less than 10^{-7}. Each session had one adjustment that iterated for the plate constants. Table IV shows the number of sessions computed for each source.

TABLE IV

Computation	No. of Sessions
AMS–64	32
NASA	131
ACIC	8

The AMS-64 data started with 3492 observation equations that contained 987 unknowns. After analysis of this FA, 259 measurements and six craters were deleted because their residuals exceeded four times the rms uncertainty in the observation of unit weight.

A second iteration was then run and a second analysis was completed. Nineteen craters were deleted because their rms uncertainty in height exceeded ± 1945 meters. The third and final iteration contained 2842 observation equations with 909 unknowns. The results of this third iteration are described below. The rms uncertainties referred to the various coordinate axes are given in Table V.

TABLE V
RMS Uncertainty

in	$\pm$ meters
ξ	133
η	123
ζ	1378
W	896
N	628
H	858

With the completion of the third iteration, which contained 256 craters, AMS-64 was published. The knowledge gained from the three iterations helped to set the patterns for the Group NASA adjustments.

For Group NASA's first computation 9298 observation equations were formed which contained 2091 unknowns, 1569 for the craters, and 522 for the plates. A total of 276 observation equations were deleted after this computation.

A second computation using the remaining 9014 equations was performed and after analysis, 185 observation equations were deleted.

The third computation contained 8829 observation equations with 2010 unknowns, 1488 crater unknowns and 522 plate unknowns. A least-squares solution was performed and the standard FA analysis was applied. All the measurements were deleted for 12 craters, with a total of 805 observation equations deleted. Remaining were 8024 observation equations with 1974 unknowns, 1452 crater and 522 plate unknowns. Upon completion of this adjustment 484 craters remained to be used in DOD 66i. The rms uncertainties referred to the various coordinate axes are shown in Table VI.

TABLE VI
RMS Uncertainty

in	± meters
ξ	116
η	103
ζ	1121
W	636
N	600
H	718

The ACIC meaned measurements for 196 points were mathematically reduced using AMS methods. Corrections to the approximate plate constants and to the approximate crater coordinates were obtained by a least-squares adjustment with 3136 observation equations in 618 unknowns. One crater was deleted from this adjustment in two references because of apparent errors in the published data. No craters or equations were deleted from the solution on the basis of this adjustment. The rms uncertainties referred to the various coordinate axes are listed in Table VII.

TABLE VII
RMS Uncertainty

in	± meters
ξ	109
η	109
ζ	847
W	470
N	390
H	606

Fundamental Adjustment (FA) DOD 66i

The fundamental adjustment was accomplished in the following manner. A crater data tape was prepared to provide the necessary information. Data from ACIC consisted of 196 craters measured on eight plates. This source provided 3132 observation equations with 618 unknowns. Data from AMS came from two independent sources. The first source contained 256 craters. However, two craters were deleted because their heights exceeded 10 kms. It provided 2810 equations for 890 unknowns for 15 plates and 32 sessions. One session being the number of measurements one man could make in one day.

The second source from AMS were the Group NASA 484 craters which were measured on 18 plates and 131 sessions. This source consisted of 8024 observation equations in 1974 unknowns.

The total input for DOD 66i thus consisted of 734 craters and contained 13966 observation equations, for 2886 unknowns.

A least-squares solution was then performed and corrected crater coordinates and plate constants were determined. The rms uncertainties referred to various coordinate axes were computed using the following:

The weight matrix was formed and applied to the observation equations following Gauss' method. Let

C = diagonal weight matrix 13966×13966,
M = matrix of coefficients of observation equation 13966×2886,
X = matrix of unknown corrections to plate parameters and points 2886×1, and
L = differences between the errors in the adjusted measured coordinates and the errors in the approximate computed plate coordinates.

Then,

$CMX = CL$.

$\sigma\{\xi\}, \sigma\{\eta\}, \sigma\{\zeta\}$ = uncertainties in the coordinates referred to the selenocentric rectangular coordinate system.

$\sigma\{W\}, \sigma\{N\}, \sigma\{H\}$ = uncertainties in the crater positions referred to the westing (W), northing (N), and vertical axes (H) in the local horizon coordinates system at the crater.

The uncertainties found, are given in Table VIII, where $\sigma\{\text{hor}\} = [\sigma^2\{W\} + \sigma^2\{N\}]^{\frac{1}{2}}$. The rms uncertainty in the observation of unit weight is ± 16 microns at the average scale of the photographic plates.

TABLE VIII
RMS Uncertainties

in	$\pm$ meters
ξ	116
η	106
ζ	1218
H	751
hor	972

DOD 66i Compared With Other Systems

Upon completion of the FA, an error and coordinate analysis was conducted. This analysis was used to identify the anomalous craters that existed in the solution.

The error analysis consisted, in part, of the correlation between $\sigma_k\{H\}$, $\sigma_k\{\text{hor}\}$ and σ_{ok}, respectively, and θ_k, the spherical arc distance from the corresponding crater to the point of zero selenocentric latitude and longitude. $\sigma_k\{H\}$ being the uncertainty in the position of crater k referred to the vertical axis in the local horizon coordinate system at the crater. $\sigma\{\text{hor}\}$ is the uncertainty in the position of crater k referred to the westing (W) and northing (N) in the local horizon coordinate system at the crater. σ_{ok} is the uncertainty in the observation of unit weight of crater k. Studies were also made of the $\sigma\{H\}/\sigma_o$ and $\sigma\{\text{hor}\}/\sigma_o$ which is the $\sigma\{H\}$ and $\sigma\{\text{hor}\}$ with the observational uncertainty removed. A curve was fitted to the points, σ vs. θ with the equation $A\theta^2+B\theta+C=\sigma\{\ \}$ used in a least-squares solution to find the best curve for the given set of points. All craters whose sigmas were greater than one standard error were analyzed for possible deletions.

The coordinate analysis compared the height and horizontal references for the four data sources with the DOD 66i results. The results were compared with a system before and after an orientation and scaling was computed to adjust one system to the datum of the other. This orientation and scaling process rotates, translates and scales computationally the system to which the points are referred such that the two systems have the same coordinate system. The systems compared were the ones that comprised DOD 66i and in addition, the 104 S-R coordinates.

Results of comparisons among the four control systems are presented in Tables IX–XII. The rms distance is computed by:

$$\text{rms distance} = \sqrt{\sum_1^n \frac{(\Delta\xi^2 + \Delta\eta^2 + \Delta\zeta^2)}{n}}$$

where n = number of craters.

TABLE IX

RMS Differences in ± Meters Between DOD 66i and NASA for 484 Craters

Rms differences in	Before Adjustment	After Adjustment
ξ	173	246
η	377	206
ζ	1704	718
W	871	473
N	991	319
H	1155	541
horizontal	1319	570
distance	1754	786

TABLE X

RMS Differences in ± Meters Between DOD 66i and ACIC for 196 Craters

RMS Differences in	Before Adjustment	After Adjustment
ξ	421	171
η	674	236
ζ	717	439
W	433	200
N	646	295
H	735	386
horizontal	777	356
distance	1070	527

TABLE XI

RMS Differences in ± Meters Between DOD 66i and AMS-64 for 254 Craters

RMS Differences in	Before Adjustment	After Adjustment
ξ	613	574
η	556	192
ζ	2899	1603
W	1456	1112
N	1011	689
H	2346	1107
horizontal	1894	1308
distance	3015	1713

TABLE XII

RMS Differences in ± Meters Between DOD 66i and S–R for 104 Craters

RMS Differences in	Before Adjustment	After Adjustment
ξ	514	438
η	395	485
ζ	2218	1889
W	1266	1119
N	890	777
H	1716	1463
horizontal	1548	1362
distance	2311	1998

Charts were made to find craters whose height and horizontal difference, between two different systems, differed from the area in which they were included. These craters and the *H* contour chart of DOD 66i were then analyzed for possible deletions.

Future Work

With the fundamental net established, it is now possible to connect differential control to the base points. The 1049 supplementary ACIC points, the 1200 AMS intensification points, and the Franz measurements will serve as a good basis for this differential control.

The error analysis of the entire control system can be completed through the evaluation of the uncertainties in the system arising from sources other than the measurements (SCHWARZ, 1966). These sources include bias in the measurements due to phase effect and in the lunar constants upon which the mathematical models are based. The effect of each error source will be determined by strict differentiation with respect to the source.

The development of the topographic lunar figure can be progressive – three geometric models: sphere, ellipsoid of revolution, and triaxial ellipsoid – with the final determinations expressed in spherical harmonics (cf. CARLSON and MALLALIEU, 1966).

Reductions of data obtained from earth-based photography have probably reached the limits of practical usefulness. Tracking data obtained from the lunar orbiter will supplement the data presently used in selenodetic topographic models.

The dynamic and topographic figures will then be observationally related through simultaneous adjustment of the photographic and electronic data sources. The dynamic figure of the moon can be determined from reductions of the earth-based tracking data of the lunar orbiters (CARLSON and MALLALIEU, 1966).

References

ALFONTE, V.: 1966a, 'Control Integration for Lunar Mapping'. Preliminary Report. U.S. Army Map Service, Washington (NASA-T-37794(G)).

ALFONTE, V.: (1966b), 'Control Integration for Lunar Mapping'. Interim Report. U.S. Army Map Service, Washington (NASA-T-37794 (G)).

CARLSON, JR., A. E. and MALLALIEU, W. S. P.: 1966, 'Datum for Lunar Mapping'. Preliminary Report. U.S. Army Map Service, Washington (NASA-T-37794 (G)).

HARDY, Marian and BOWLES, Lawrence D.: 1966, 'Selenodetic Control for NASA Project Apollo'. Preliminary Report. U.S. Army Map Service, Washington (NASA-T-21657 (G)).

HARDY, Marian and HATHAWAY, James D.: 1966, 'Selenodetic Control for NASA Project Apollo'. Interim Report. U.S. Army Map Service, Washington (NASA-T-21657 (G)).

HARDY, Marian, HATHAWAY, James D., and BOWLES, Lawrence D.: 1966, 'Horizontal and Vertical Control for Lunar Mapping. Part Three: AMS Selenodetic Control for NASA Project Apollo – Group NASA)'. *AMS Technical Report No. 29*. Unpublished report. U.S. Army Map Service, Washington.

KOZIEL, Karol: 1964, 'Crater Mösting A as the First-Order Point of Triangulation on the Moon', *Life Science and Space Research* **2,** 141–144.

MEYER, Donald L. and RUFFIN, Byron W.: 1965, 'Coordinates of Lunar Features, Group I and II Solutions'. *Technical Paper No. 15*. Aeronautical Chart and Information Center, St. Louis.

Marchant, Marvin Q.: 1960, 'Horizontal and Vertical Control for Lunar Mapping. Part One: Methods'. *Army Map Service Technical Report No.* 29. U.S. Army Map Service, Washington.

Marchant, Marvin Q., Hardy, Marian, and Breece, Sam.: 1964, 'Horizontal and Vertical Control for Lunar Mapping. Part Two: AMS Selenodetic Control System 1964'. *Army Map Service Technical Report No. 29.* U.S. Army Map Service, Washington.

Schwarz, C.: 1966, 'Absolute Error Analysis for Selenodetic Control'. Preliminary Report. U.S. Army Map Service, Washington (NASA-T-37794 (G)).

THE MANCHESTER SELENODETIC CONTROL SYSTEM

G. A. MILLS AND M. E. DAVIDSON*

Abstract. Two methods in use at Manchester for determining absolute lunar coordinates, both based on the so-called "stereoscopic method" are discussed. The first method is dependent on determining the apparent coordinates of points on the lunar disc from photographs taken at different librations, which are eventually combined to form absolute coordinates. The second process makes no use of prior determinations of absolute heights. The organisation of the measuring and reduction processes is also discussed.

Introduction

The determination of the absolute coordinates of features on the lunar surface provides a control system for the preparation of accurate maps needed in the lunar space programmes, culminating in a manned landing. Also, a large number of absolute coordinates can be used to determine the overall shape of the Moon, and comparison with the Moon's moments of inertia would then yield information concerning the Moon's internal constitution. For this latter purpose it is unfortunate that only 59% of the Moon's surface is ever visible to an earthbound observer, and over substantially less than this is it ever possible to determine the absolute coordinates of lunar features.

Consequently, only when extensive data is available concerning the far side of the Moon, can any such deductions be made reliably. However, the determination of the absolute coordinates of features on the near side of the Moon is as easily done from the Earth as from anywhere else. Therefore, the problem has received some considerable attention in the past. However, examination of the methods and available materials of prior workers (most work being done early this century) revealed that more accurate data would certainly be forthcoming from a fresh attack on the problem.

At the commencement of the Manchester project only three extensive control systems were in existence. Intercomparison of these revealed that there were gross differences in the systems, both for individual points and for large areas of the lunar face. Clearly, a new approach to this problem was necessary. It was decided to measure about a thousand small features (splash craters, mountains, etc.) which show themselves on full-moon photography as local changes in albedo with diameters about five or six kilometres. These points were to be spread as uniformly as possible over the visible lunar surface and measured on as many different photographs as could be obtained. Their distribution is shown in Figure 1.

* Department of Astronomy, University of Manchester. Research supported by Grant EOAR 64–49 to the University of Manchester by the Terrestrial Sciences Laboratory, Air Force Cambridge Research Laboratories, throug the OAR European Office in Brussels.

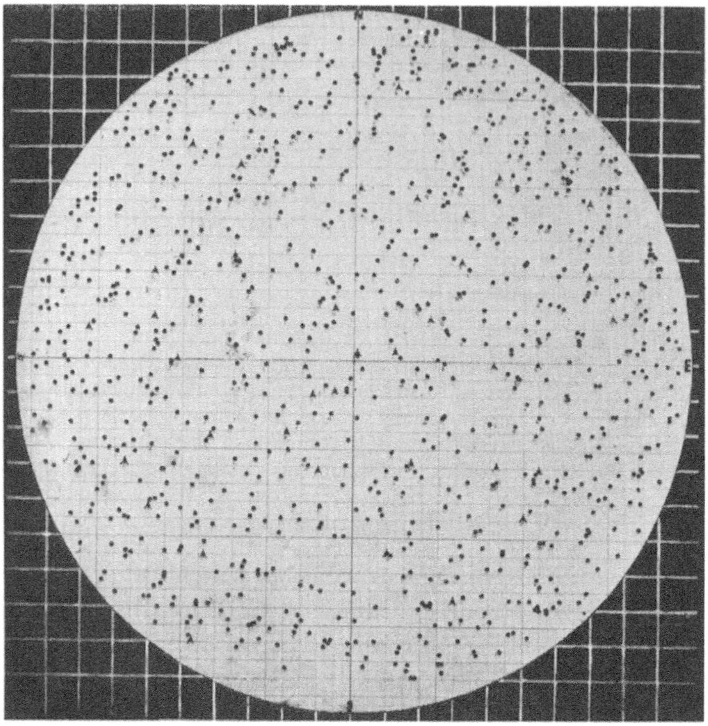

Fig. 1. Distribution of 960 points used.

It is hoped that in this way, any systematic errors in the final control system, arising from seeing displacements, libration errors, etc., will be eliminated. It was felt that the determination of the plate constants from measurements of the positions of features of "known" absolute coordinates might well introduce systematic error which could be present in the previous work.

It was decided to check on this by performing the reductions in two separate ways. The first is the usual method devised by SAUNDER (1905) and used by him and later workers. More recently, ECKHARDT and HUNT (1962) described a method which does not make use of any prior information, other than the plate constants of one plate and the coordinates of this plate in the lunar coordinate system.

The method, as originally put forward, allowed the remaining libration angles, the absolute coordinates of a number of lunar features and the remaining plate constants to be found. The input information would be the measurements of a large number of lunar features on a large number of plates. In experimenting with this method, some modifications have been made and these are discussed under Method 2.

Possible Sources of Systematic Error

It can be shown, as follows, that an ellipsoidal term in the Moon's figure is practically indistinguishable from a small error in the libration angles.

If the Moon is rotated through an angle about its y-axis the observed x-coordinate of the lunar feature is changed to x', where

$$x' = x\cos\theta + z\sin\theta .$$

If the Moon is spherical this can be written as

$$x' = x\cos\theta + \sqrt{1 - x^2 - y^2}\,\sin\theta . \tag{1}$$

On the other hand, if the Moon were ellipsoidal, this would be

$$x' = x\cos\theta' + c\sqrt{1 - \frac{x^2}{a^2} - \frac{y^2}{b^2}}\,\sin\theta' \tag{2}$$

For $a=b=1$, Equations (1) and (2) are equivalent if $c\sin\theta' = \sin\theta$, c being the elongation of the z-axis.

Thus an error of 0.°01 in the libration angle θ (taken as 10°) is equivalent to $c = 1.°001$ or an elongation of the z-axis of nearly 2 km. With a large number of different libration epochs, this source of error could be reduced; even so, it would be far better to determine the libration angles to 0.°002 or even better, beforehand.

The uncertainty in the libration angles after the work of ECKHARDT (1966a) now seems to be due to the uncertainties in f and I. If sufficient accuracy is not obtaining in these quantities, we can, through the use of a sufficient number of libration epochs, determine them from our own measurements, at the same time as we find the absolute coordinates. ECKHARDT (1966b) has, in fact, modified his reduction procedure to do this. However, in the past this important source of systematic error has been ignored, and it may well be that the discrepancies between the existing control systems could be due to this cause. Therefore, we perform the reductions for different values of f and I and examine their effect on the overall shape of the Moon.

A second source of error is that arising from the measurement of phase photography. The features measured are mainly small craters. At large phase angles they are partially filled with shadow and show contrast effects so that the apparent position of the crater is changed. Near the terminator, the effect is very obvious, however, the effect does extend further from the terminator than the observer can detect, and may still be serious. Although the estimation of the centre of a crater is largely personal, some estimate of the importance of this effect can be gained in the following way.

For simplicity, we regard our crater as being a paraboloid of revolution of depth $\frac{1}{3}$ of diameter. It is a simple matter to compute the apparent position of the shadow line to a terrestrial observer for various positions of the crater on the lunar disc. By taking the centre of gravity of the part of the crater not in shadow as the apparent position of the crater, we estimate the measuring errors due to the phase effect.

In Figures 2 and 3 we see the extent of these errors. Over a small area of the Moon, the Sun is entirely above the lip of our hypothetical crater, and over a much larger area, the effects of phase are negligible. At zero phase, of course, there is no phase effect. At 60° phase the effect extends over a large area of the visible surface. From this study we conclude that moon photography with phases greater than 20° would be of

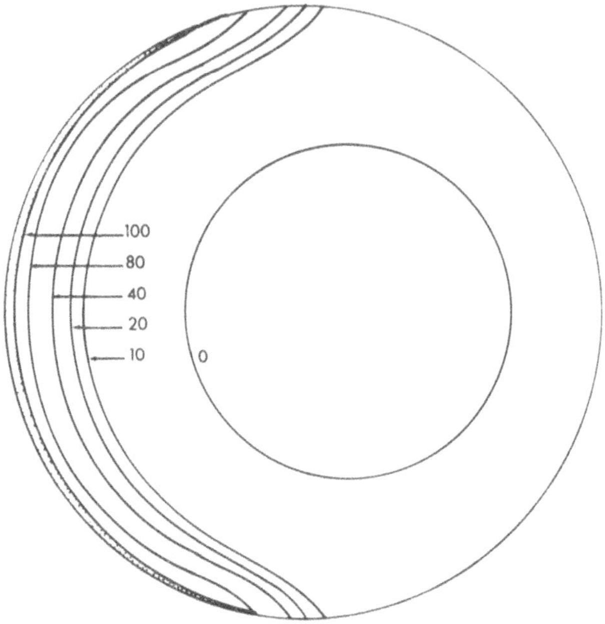

Fig. 2. The discrepancy between the real and apparent position of a hypothetical lunar crater at a phase of 10°.

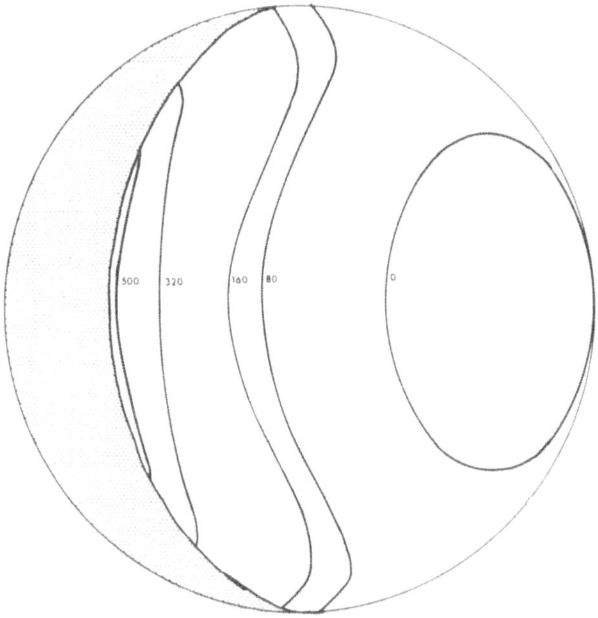

Fig. 3. The discrepancy between the real and apparent position of a hypothetical lunar crater at a phase of 50°.

dubious value except in the shadow-free area. It must be remembered, however, that in such a case the plate constants would be derived from measurements of some features outside this area.

The plate constants themselves, being derived from a small number of measurements, may give rise to systematic effects over the lunar disc. This is the most difficult effect to investigate and at the time of writing it has not been fully worked out. It will be published at a later date.

Plate Material

All photographs used in this project were obtained using the 24-inch refractor at the Pic-du-Midi Observatory. Only exposures made within a day of full moon have been considered as useful, because all 960 points can be seen on such photographs, but more important still, as we have already seen, the apparent position of a small crater changes with different phases of the Moon.

This full-moon-only restriction naturally limits the number of suitable librations, but as the Moon has been systematically photographed at the Pic-du-Midi Observatory since 1960, a sufficient variety of acceptable librations were found. The approximate Universal Times of each photograph are listed in Table I and the lunar librations corresponding to these time are plotted in Figure 4. Epochs 1 and 2 were recorded on Kodak P1200 plates, while the remaining epochs utilised Kodak S.O. 136 film. The exposure times range from 0.2 to 0.6 seconds.

TABLE I

Universal Time of Exposures of Photographs used

		Time			Epoch
		D.	H.	M.	
1960	Nov.	2	22	54	6
1962	Apr.	20	3	45	12
1962	May	17	22	46	4
1962	Oct.	13	2	56	5
1963	Jan.	10	0	30	9
1963	May	8	1	0	10
1963	Aug.	5	2	41	11
1963	Nov.	2	0	48	1
1964	Feb.	26	23	24	7
1964	Apr.	25	23	39	15
1964	Jun.	23	21	36	14
1964	Sep.	21	2	34	13
1964	Oct.	21	1	7	2
1965	Apr.	17	1	27	3
1965	Sep.	11	00	22	16
1965	Dec.	8	23	46	8
1965	Dec.	21	22	10	17
1966	Aug.	2	01	08	18

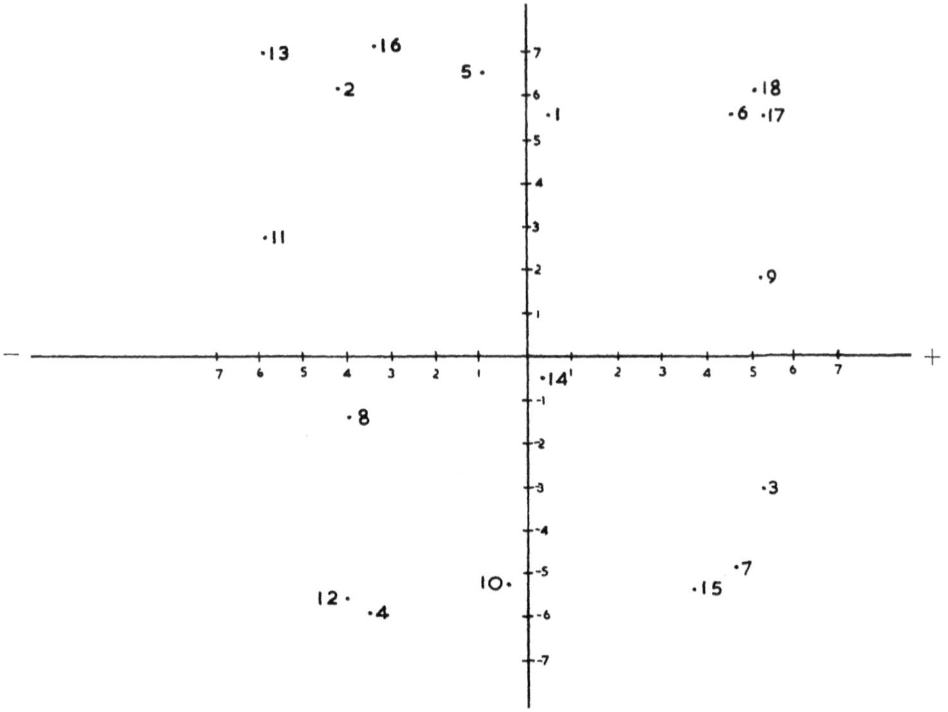

Fig. 4. Librations of epochs used.

It will be seen that some material has been included where the lunar libration in one coordinate is near zero, while that in the other coordinate is close to a maximum. We feel that such combinations help to minimize systematic errors, particularly in Method 2 discussed later. It is well known that the effects of "seeing" are extremely troublesome, particularly when one is using photographs of short exposure. The simplest way to overcome this problem is to use techniques of "sequential photography". Consequently, we have chosen six negatives from each epoch listed above, each one taken within a few seconds of the others, making a grand total of 108 negatives. We feel that such a number is sufficient to reduce seeing effects to the extent that they may be ignored.

Measuring Procedure

The measurement and reduction of all photography has, for technical reasons, been divided into three groups, each consisting of two plates from all eighteen epochs. At the time of writing, measurement of one group, that is 36 negatives, has been completed.

The Zeiss 3030 coordinate measuring instrument of the Department of Astronomy in the University of Manchester (Figure 5) has been used to measure all plates with the exception of the four plates from the first two epochs which were measured using a

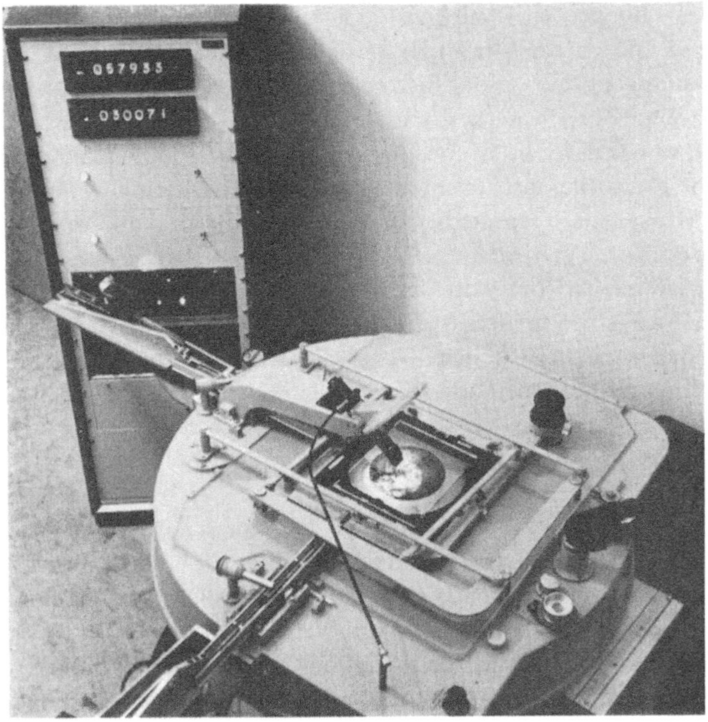

Fig. 5. Zeiss coordinate measuring instrument equipped with digital output.

Mann 422C comparator. The Zeiss instrument, which is housed in a temperature-controlled room, has recently been digitized and this has increased the speed of measurement by a factor of two.

It is virtually an impossible task to identify all our 960 points from memory, so the approximate positions of all points are predicted using a simple coordinate transformation. This is based on a plate constant determination from a preliminary measurement of a few control points. These predicted positions are sufficiently accurate to identify a point without ambiguity. Even so, each measurement is checked during the reduction processes to prevent any misidentification passing unnoticed. The original film negatives obtained at Pic-du-Midi are not entirely suitable for astrometry, so a certain amount of processing is necessary before measurement can be commenced.

Firstly, a positive copy of each negative is made, and this is used to produce a measurable negative by means of a Weston Fluorododge instrument. This effectively enhances the contrast of the programme points, enabling them to be measured more accurately. This treatment does not create any distortions in the image greater than four microns.

The main bulk of measurements have been made by Messrs. W. V. Garner and T. B. Ramspott, supplemented by observations from the authors and Mr. B. H. Bland. To eliminate effects of fatigue, no observer measures for more than one hour at a time. At each changeover, small check points, usually tiny bubbles in the glass plate, are measured and checked, in order to detect any possible movement of the plate during measurement. No serious discrepancies of this nature have occurred.

The first plate of each epoch is measured in both the forward and, using a rotating prism, reverse orientations by two measurers. The second plate of each epoch is placed in the measuring instrument at an orientation differing by 90° from that of the first plate. Measures are repeated again in the new forward and reverse directions, by two measurers.

All measurements from plates of one epoch are presented to the University of Manchester "Atlas" computer, where they are transformed to a common system and their internal consistency examined. The values are then stored on magnetic tape to await reduction by both methods described later.

Method 1

The absolute coordinates of a feature on the lunar surface may be conveniently determined from values of the apparent coordinates of the feature as measured on photographs taken at a sufficiently large number of different librations. The reduction of such apparent coordinates to absolute values is a relatively simple task, so the problem reduces itself to the determination, from a photograph, of the apparent position of a feature seen in projection against the surface of the mean lunar sphere.

If values for the scale and orientation for the photograph under examination were to hand, this calculation would be straightforward. However, the accurate determination of these two features is a difficult problem that has faced stellar astrometrists for more than half a century. It is invariably solved by determining the "plate constants" which are, in effect, parameters enabling one to transform measurements made on a photographic plate into celestial coordinates. When one is interested in the positions of stars these constants are determined by comparing measures of the positions of stars on a suitable negative with known values of their celestial coordinates. Thereafter the celestial coordinates of any other object on the plate may be determined by applying the plate constants to measurements of its position.

This method would be ideally suited to the determination of lunar coordinates, but it is by no means an easy task to photograph appropriate star fields simultaneously with the Moon. The most practical alternative is to use features on the lunar surface ("control points") whose absolute positions are known with sufficient accuracy.

When the project was initiated the most accurate list available was the re-reduction of Franz's data by SCHRUTKA-RECHTENSTAMM (1958). This list was later supplemented by a few values obtained at ACIC by MEYER and RUFFIN (1955) and is given in Table II. It was considered necessary that points included in the list should have well determined absolute coordinates and be adequately distributed over the lunar disc in order to provide reliable plate constants over all parts of the lunar images under investigation. The selection appears to have been most suitable and has been extensively used for the plate constant determination throughout this reduction system.

TABLE II
Data for Control Points used in Method 1

Control Point Reference	Long. λ	Lat. β	Height (km) Over 1738 km	Name of Feature
1	− 51.40	5.16	− 0.9	Reiner A
2	− 44.51	− 16.69	0.7	Gassendi G
4	− 37.20	37.81	− 2.1	Mairan E
5	− 36.08	7.16	0.3	Kepler A
6	− 31.08	0.19	0.2	Lansberg A
7	− 24.06	33.75	− 2.5	Carlini
8	− 11.52	− 11.51	0.7	Guericke C
10	− 10.00	− 52.33	2.4	Maginus H
11	− 8.06	− 1.77	1.7	Mösting C
12	− 6.86	− 5.56	1.5	Lalande C
13	− 3.16	− 4.97	3.3	W. Herschel C
14	− 3.06	8.75	1.6	Bode B
15	0.39	1.19	1.5	Bruce
16	− 5.16	− 3.18	1.4	Mösting A
17	6.82	1.63	2.5	Rhaeticus B
18	10.16	− 16.75	− 1.4	Abulfeda E
19	10.52	51.55	− 1.9	Egede A
20	11.80	27.73	− 1.1	Linné
21	13.22	6.95	1.4	Silberschlag A
22	19.19	16.63	− 0.5	Tacquet
23	21.00	24.74	− 0.7	Bessel A
24	23.62	− 42.43	− 0.6	Nicolai A
25	24.18	− 0.57	− 1.2	Moltke
26	26.16	− 17.01	2.9	Beaumont D
27	29.48	31.66	− 0.5	Posidonius A
29	33.77	− 26.10	− 0.5	Piccolimini L
30	34.87	− 17.82	2.1	Rosse
31	49.94	7.28	− 2.3	Taruntius A
32	46.00	14.23	− 0.3	Proclus F
33	50.26	27.26	− 0.1	Tralles B
34	1.07	29.20	0.0	
35	− 6.05	− 17.95	0.0	Alpetragius H
36	6.86	− 33.09	1.3	Aliacensis D
37	− 21.46	17.07	− 0.4	Draper C
38	− 25.34	− 12.37	1.3	Darney E
39	− 31.31	− 33.41	0.9	
40	41.05	− 24.46	− 0.5	Doppelmayer A
41	− 50.95	52.78	− 2.3	Harpalus E

Plate Constant Determination

The plate constant determination for each photograph under examination has been executed according to the following scheme: For each control point, the instantaneous direction cosines X, Y, Z, may be computed from its known selenocentric coordinates ξ, η, ζ by means of the equations:

$$
\begin{aligned}
X &= \xi \cos l' - \zeta \sin l', \\
Y &= -\xi \sin l' \sin b' + \eta \cos b' - \zeta \cos l' \sin b', \\
Z &= \xi \sin l' \cos b' + \eta \sin b' + \zeta \cos l' \cos b',
\end{aligned}
$$

where l' and b' are the topocentric librations at the time of the exposure, and may be conveniently computed by methods similar to that devised by ATKINSON (1951) or more rigorously, by the method suggested later.

We may now derive the projected coordinates X', Y', for each control point using

$$X' = X/(1 - Z \sin s'),$$
$$Y' = Y/(1 - Z \sin s'),$$

where s' is the Moon's topocentric semi-diameter.

These projected coordinates differ from the measured coordinates x, y of a control point because of differences of scale and orientation, translation of origin, and effects of atmosphere refraction. The relationship may be written

$$\left.\begin{aligned} X' &= a_1 + a_2x + a_3y + a_4x^2 + a_5xy + a_6y^2 + \cdots \\ Y' &= b_1 + b_2x + b_3y + b_4y^2 + b_5xy + b_6y^2 + \cdots \end{aligned}\right\} \qquad (3)$$

where $a_1\ a_2 \ldots b_1\ b_2$, etc are the plate constants.

It is not normally necessary to include terms involving x and y to any order above the first, providing the Moon is photographed at small zenith distances.

Thus if a set of, say, 30 control points is measured, Equations (3) may be solved by the method of least squares for the plate constants. For any other point on the plate whose measured coordinates are known, we may compute the projected coordinates using these plate constants. The instantaneous direction cosines can then be obtained by

$$\left.\begin{aligned} X &= X' - X'Z \sin s' \\ Y &= Y' - Y'Z \sin s' \end{aligned}\right\} \qquad (4)$$

Before these equations can be solved it is necessary to know the value of Z. This may be found by an iterative process or direct calculation from the geometrical configuration.

Lastly, the selenocentric coordinates can be found from

$$\left.\begin{aligned} \xi &= X \cos l' - Y \sin l' \sin b' + Z \sin l' \cos b', \\ \eta &= \qquad\qquad\quad Y \cos b' + \qquad Z \sin b'. \end{aligned}\right\} \qquad (5)$$

A good estimate of the astrometric quality of a plate and, hence, the accuracy of a position may be obtained by considering the probable error of unit weight, P.E. 1, as derived from the residuals of the least-squares solution for the plate constants. If the ith control point yields a residual v_i, then

$$(\text{P.E.}\,1)^2 = \sum_{i=1}^{n} v_i^2/(n-3),$$

where n is the number of control points used in the solution.

We have found that a good value for P.E. 1 is about 0.00025 lunar radii in each coordinate. Any plate with a value much greater than this should be considered suspect. It should be noted, however, that a high value can be caused by mis-identification of a control point. In this project, the computer programme examines each residual and in cases of high values for P.E. 1, repeats the least-squares solution without the control point yielding the highest residual. In this manner, the possibility of a fault remaining undetected is considerably reduced.

Absolute Coordinate Determination

If the procedure outlined above is applied to points appearing on photographs taken as vastly different librations, it is possible to combine the resulting apparent positions for each point to yield values for the absolute coordinate of that point.

The difference between the absolute coordinate ξ_A and the apparent coordinate ξ_i as derived on the ith plate may be expressed as

$$\xi_A - \xi_i = \frac{\sin l_i' \cos b_i' - \xi_i Z_i}{Z_i - \sin s_i'} h' = A_i h'$$

and

$$\eta_A - \eta_i = \frac{\sin b_i' - \eta_i Z_i}{Z_i - \sin s_i'} h' = B_i h', \tag{6}$$

where $h' = h/(1+h)$ and h refers to the height of the feature above a mean sphere, while the suffix i refers to the ith libration epoch. With a sufficiently large number of epochs, the above equations may be solved for ξ_A, η_A and h', the absolute coordinates of the feature.

It will be noticed that if this procedure is applied to the control points used in calculating the plate constants, then better estimates of their absolute positions will be obtained. These may be used for a recalculation of plate constants giving more accurate apparent positions for the 960 points.

Method 2

In lunar work it is usual to use three or four plate constants. These correct for orientation and translation of the plate in the measuring engine and the focal length of the telescope. In the present case atmospheric refraction is not necessarily the main source of distortion uncorrected by these constants. The film could possibly be stretched and also there has been found to be evidence of non-equivalence of the x- and y-scales in the measuring engine.

It is thought advisable to use a 6-constant transformation, as is the custom in stellar astrometry:

$$\xi = cx + by + a,$$
$$\eta = fx + ey + d.$$

As mentioned previously, we fix the libration angles before the reduction, otherwise we follow the method of Eckhardt and Hunt.

We measure the coordinates of the rth point (x_{rs}, y_{rs}) on the sth plate, assume an approximate focal length, f_s, and hence define a vector, $\mathbf{N}_{rs}$, in the local Cartesian system of the telescope;

$$\mathbf{N}_{rs} = (x_{rs}, y_{rs}, f_s)^T .$$

We operate on this with the plate constants matrix

$$\Delta_s = \begin{vmatrix} c & b & a \\ f & e & d \\ 0 & 0 & 1 \end{vmatrix} \quad \text{to obtain} \quad N_{rs} = \Delta_s \mathbf{N}_{rs} .$$

We define the unit vector $P_{rs} = N_{rs}/|N_{rs}|$, and rotate this vector through a series of angles (determined by the lunar libration angles, l_s, b_s) into the lunar coordinate system by operating with the matrix V_s to get Q_{rs}, so that

$$Q_{rs} = \begin{vmatrix} \cos l_s & -\sin b_s \sin l_s & \sin l_s \cos b_s \\ 0 & \cos b_s & \sin b_s \\ -\sin l_s & -\cos l_s \sin b_s & \cos l_s \cos b_s \end{vmatrix} P_{rs} = V_s P_{rs} .$$

In the lunar system the coordinates of the rth point are defined by the vector X_r and the coordinate of the sth observing site, by the vector

$$Y_s = (R_s \sin l_s \cos b_s, R_s \sin b_s, R_s \cos l_s \cos b_s)^T .$$

We define $W_{rs} = X_r - Y_s$, and $M_{rs} = W_{rs}/|W_{rs}|$.

If there were no observational errors $M_{rs} = Q_{rs}$, in fact $Q_{rs} = M_{rs} + \delta_{rs}$, where δ_{rs} is assumed small and normally distributed.

By an analogous treatment to that of Eckhardt and Hunt, we find the first and second derivatives of the log likelihood function, L. These are:

$$\frac{\partial L}{\partial X_r} = \sum_{s=1}^{m} \frac{Q_{rs} - M_{rs}}{|_{rs}| \sigma_{rs}^2 W},$$

$$\frac{\partial^2 L}{\partial X_r^2} = \sum_{s=1}^{m} \frac{Q_{rs} Q_{rs}^T - I}{|W_{rs}{}^2| \sigma_{rs}^2},$$

$$\frac{\partial L}{\partial \Delta_s^{ij}} = \sum_{r=1}^{n} \frac{\mathbf{N}_{rs}^j}{\sigma_{rs}^2 |N_{rs}|} \{(V_s^T M_{rs})^i - P_{rs}^i\},$$

$$\frac{\partial^2 L}{\partial \Delta_s^{ij} \partial X_r} = \frac{\mathbf{N}_{rs}^j}{|W_{rs}| \, |N_{rs}| \, \sigma_{rs}^2} \{(V_s^i) - P_{rs}^i Q_{rs}\}^T ,$$

where (V_s^i) represents the ith column of V_s, and

$$\frac{\partial^2 L}{\partial \Delta_s^{ij} \partial \Delta_s^{kl}} = \sum_{r=1}^{n} \frac{\mathbf{N}_{rs}^j \mathbf{N}_{rs}^l}{|N_{rs}^2| \sigma_{rs}^2} \{P_{rs}^i P_{rs}^k - \delta_i = k\} .$$

We then find the increments dX_r and $d\Delta_s$ to our initial approximate values by

$$\begin{vmatrix} dX_r \\ \\ d\Delta_s \end{vmatrix} = \begin{vmatrix} \dfrac{\partial^2 L}{\partial X_r^2} & \dfrac{\partial^2 L}{\partial X_r \partial \Delta_s} \\ \dfrac{\partial^2 L}{\partial \Delta_s \partial X_r} & \dfrac{\partial^2 L}{\partial \Delta_s^2} \end{vmatrix}^{-1} \begin{vmatrix} \dfrac{\partial L}{\partial X_r} \\ \dfrac{\partial L}{\partial \Delta_s} \end{vmatrix} \tag{7}$$

This method is a non-linear least squares solution for many variables. The use and validity of the method depends on the quantity of data available as the matrix of Equation (7) is very nearly singular. This can be seen when one realises that a focal length error is nearly equivalent to an increase in the mean lunar radius and also a plate translation error is nearly equivalent to a translation of the Moon along the z-axis.

This can be overcome by using a very large number of plates and points, or by fixing some of the variables by other means. In the latter case we are forced to use the data of previous workers, though not necessarily with high weights. Such a course of action, however, will be deferred until the time when the store space of the "Atlas" computer proves insufficient for the former task.

Preliminary Results

At the time of writing the results here presented are the only ones available: they are preliminary and subject to review. Calculations have been performed using the data taken from the ten libration epochs 3 through 12 (two plates each). The purpose of these calculations was to investigate the effects on the absolute coordinates through changing the parameters f and I. 94 lunar features were chosen from the 960 as being of consistently low measuring error and most evenly spread over the lunar surface.

The plate constants were calculated assuming that each point lay on the "mean sphere" of radius 1738 km. Then, assuming these plate constants, the absolute coordinates of the points were found. The residuals on each plate were then inspected for any systematic trends which could be attributed to an error in the libration angles, and this error was estimated in units of 10^{-3} degrees. Also the absolute coordinates were examined and the slope, B, of the graph of height versus the z-coordinate (Figure 6 is a typical example of such a graph) was found. This was repeated for various values of f and I. The root-mean-square estimated libration angle error was plotted against f for two values of I and is shown in Figure 7. The graph of the slope B versus f is shown in Figure 8.

The tentative conclusion to be drawn from these graphs is that I may well be higher than $1°32'.5$, and, in such a case the "bulge", B, may be very nearly zero – i.e., the Moon may be more nearly a sphere than hitherto supposed.

The intercept of the best line through the points of Figure 6, on the h-axis (i.e., at the limb) is -2 km. This is because the mean value of h for the points is 0 km – a consequence of our starting conditions and of the fact that the plate constants afford a partial compensation of any libration angle errors. We conclude that, in view of the present uncertainty in the values of f and I, any existing selenodetic control system is meaningless unless the values of f and I used in the reductions are quoted.

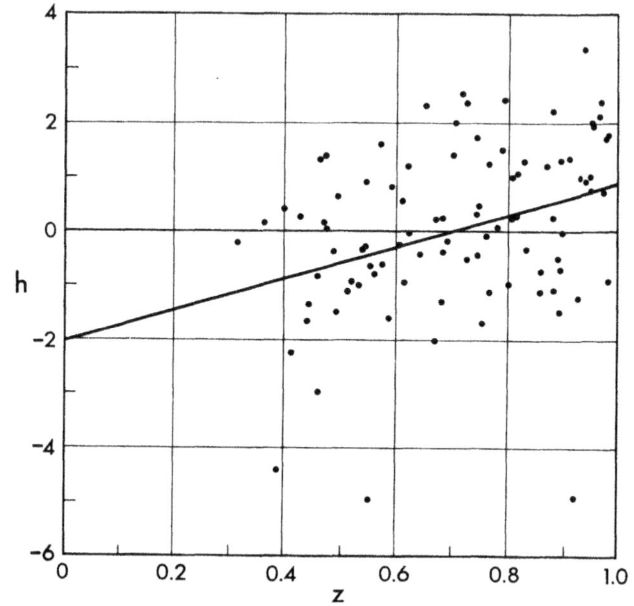

Fig. 6. A typical graph of the altitude h, in kms, versus the z-coordinate of the 94 points ($f = 0.64$, $I = 1° 32'$).

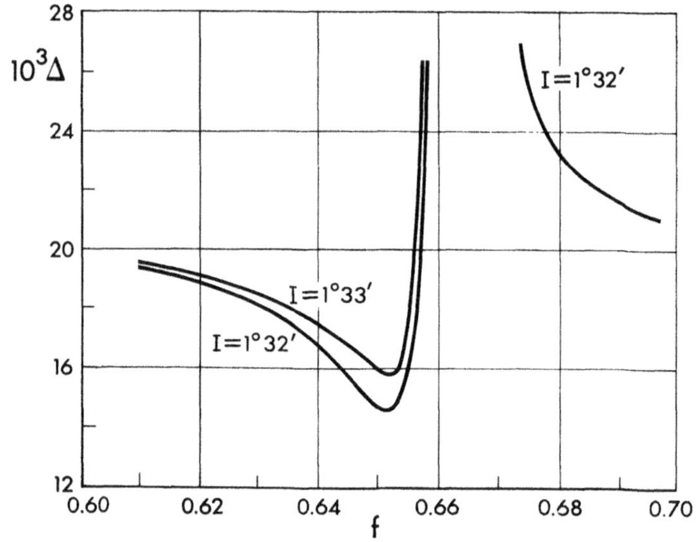

Fig. 7. Graph of the rms-error Δ of the estimated libration angle in longitude, in 0.001 of a degree, versus f (for $I = 1° 32'$ and 1° 3:25).

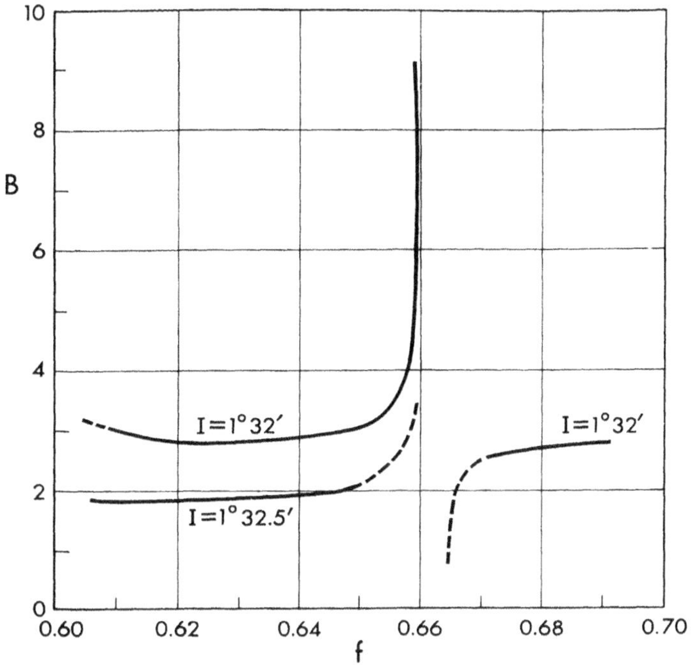

Fig. 8. Graph of the bulge B, in kms, versus f, for $I = 1°\ 32'$ and $1°\ 32'.5$.

References

Atkinson, R. d'E.: 1951, *Mon. Not. Roy. Astr. Soc.* **111,** 448.

Eckhardt, D. H.: 1966a, *Astron. J.* **70,** 466.

Eckhardt, D. H.: 1966b, in the present volume, pp. 40–51.

Eckhardt, D. H. and Hunt, M. S.: 1962, in *The Moon* (I.A.U. Sympos. No. 14) (ed. by Z. Kopal and Z. K. Mikhailov). Academic Press, New York and London, pp. 85ff.

Meyer, D. L. and Ruffin, B. W.: 1965, *A.C.I.C. Technical Paper No. 15.*

Saunder, S. A.: 1905, *Mon. Not. Roy. Astr. Soc.* **65,** 458.

Schrutka-Rechtenstamm, G. von: 1958, *Sitz. d. Österr. Akad. Wiss., Math.-Naturwiss. Klasse,* Abt. 11 **167,** 71.

THE LPL PROGRAM FOR SELENODETIC POSITIONS

Results and Comparison with Contemporary Selenodetic Work

D. W. G. ARTHUR*

Abstract. The precision of the LPL selenodetic measurements is discussed in the context of existing selenodetic triangulations. Attention is drawn to the possibility of appreciable errors in the fundamental and secondary positions, and to systematic errors of unknown origin in the photographic measures.

1. General

The basic data for a contemporary selenodetic triangulation are the position of the fundamental point Mösting A and the positions of twelve secondary points measured by Franz (1899) and Hayn (1904). A mean position for Mösting A has been determined by Schrutka-Rechtenstamm (1955) using modern rotation elements. The same author has applied these same elements (Schrutka-Rechtenstamm, 1956) to the secondary measures of Franz and Hayn to determine the selenodetic coordinates of the eight secondary points of Franz and the four secondary points of Hayn. These thirteen positions should therefore be in a uniform selenodetic system.

Nevertheless there is some evidence of systematic discrepancies. At present it is difficult to assess the four Ranger determinations of radii measured from the Moon's centroid to the surface. A simple interpretation is that mare surfaces around the center of face lie on a sphere of about 1735.4 km with the center at the centroid. I see no way in which this radius can be reconciled with the selenodetic radius of 1739.4 for the rim of Mösting A. Either the heliometer work leads to systematically high values of the radius of Mösting A, or the Ranger radii are systematically low.**

The evidence for errors in the secondary positions is internal. Since six of the eight Franz points lie on the continents, while three of the four Hayn points lie in the maria, we would expect the average height of the Franz points to be greater than the average height of the Hayn points, but in fact the selenodetic values indicate the reverse. On the average the eight Franz secondary values lie three kilometers below the mean sphere, while the Hayn positions lie about one kilometer below this sphere. The samples are small and obviously there are certain risks in the interpretation, yet I am certain that both sets are heavily affected by systematic errors. Thus the basis on which contemporary selenodetic triangulations must be erected is not above suspicion.

From the viewpoint of mere triangulation of the surface, the errors in the position

* University of Arizona Lunar and Planetary Laboratory, Tucson, Ariz., U.S.A.

** Cf. the subsequent paper by W. L. Sjogren (Eds).

of the fundamental point are not serious, since they are easily eliminated at some future date by a mere shift of the coordinate origin. However, I regard the errors in the secondary positions as so serious that normal positional work has been suspended at LPL since January 1966. The emphasis now is on measures on star trailed plates to determine a new set of secondary positions.

2. Systematic Errors in the Photographic Triangulations

Six three-dimensional triangulations of the lunar surface have been made from measures on lunar photographs. The first, of 55 points, was made by FRANZ (1899) using measures of five Lick photographs. The second, covering 38 points along the first lunar meridian, was made by SAUNDER (1905) using four Paris photographs.

More recently SCHRUTKA-RECHTENSTAMM (1956) has re-reduced the photographic measures by Franz at Breslau to obtain a net of almost 150 points. To avoid confusion we shall call this the Breslau triangulation. This triangulation is controlled by the Königsberg (Franz) secondary positions and one Hayn position. In later triangulations the tendency is to use the Breslau positions rather than the secondary positions. I have some reservations about the wisdom of this, even though it is known that the random errors of the Breslau positions are smaller than those of the secondary positions. We should not now be getting at the secondary data through a photographic triangulation of rather limited precision. The Army Map Service triangulation (BREECE, HARDY, and MARCHANT, 1964) contains 256 points and the triangulation of the Aeronautical Chart and Information Center (MEYER and RUFFIN, 1965) contains 197 points. The most extensive triangulation is by BALDWIN (1963) and contains 698 points.

The most conspicuous common feature of all these triangulations is the tendency to give high levels at the center of face trending decisively downwards towards the limb. Indeed for the Breslau, AMS, and ACIC triangulations it turns out that the height is statistically a linear function of the sine of the selenocentric arc from the center of face. If we write $R = \sin K$, where K is this arc, we get the regression

$$h = A - BR \qquad (1)$$

A least squares analysis was applied to the continental points of all three triangulations to obtain the results in Table I, for h in kilometers. The computations were repeated for the regression

$$h = A - BR + CR^2$$

with the rather peculiar result that although the dispersions about the regression line

TABLE I
The Regression $h = A - BR$

Triangulation	Points	σ	A	B
Breslau	46	± 1.25	$+ 3.82 \pm 0.48$	$+ 6.01 \pm 0.72$
AMS	130	± 1.78	$+ 5.00 \pm 0.55$	$+ 4.95 \pm 0.71$
ACIC	19	± 0.84	$+ 2.27 \pm 0.54$	$+ 2.70 \pm 0.79$

did not change by more than 0.01 km in σ for each triangulation, nevertheless the coefficient C was determined consistently as $+2.4$.

In the table σ represents the standard deviation of the height from the regression line, while the dispersions for A and B are also represented by standard errors.

The agreement in C troubles me, but the failure to diminish σ by the addition of CR^2 appears decisive. I must interpret the agreement as due to chance. Certainly if the term CR^2 is not real then *the three triangulations give no indication of an ellipsoidal elongation for an axis near the first lunar radius*, for this would correspond to a term $\beta(1-R^2)$ in the regression.

The three triangulations agree in a non-ellipsoidal elongation represented by (1), but with very different coefficients A and B. As indicated by the standard errors, there is some possibility that these represent experimental determinations of the same quantities. However, this figure is so strange, and the elongation (up to 6 kilometers) is so improbable and so at odds with the Ranger values, that I must conclude that A and B represent nothing more than some type of systematic error in the heights. The coefficients themselves are not always consistent with the idea that the regression (1) represents real altitudes. Since this is based on continental points only, and since the mean sphere is derived from the mean limb, which is largely continental, then the regression value of h should go to zero as R goes to unity. That is, for real heights we should have $A=B$. This holds only for the AMS triangulation, yet as indicated by Table I, this is the least precise of the three triangulations.

Even if (1) represents nothing more than systematic error, we can extract one piece of information from it, namely, the relative levels of continents and maria. The mare points were not treated by least squares but their deviations from (1) were noted for each of the three triangulations. These agree closely in an average negative deviation of 1.0 to 1.1 kilometers. Thus on the average the maria appear to be about one kilometer lower than the continents. This value is too small to be consistent with the results obtained from limb profiles measured at LPL. I make the central depths of Smythii, Marginis, Humboldtianum and Orientale to be 5, 2, 4 and 3 kilometers respectively, corresponding to average depths of half these amounts. Thus, the overall average depth is in excess of 1.5 km, which is appreciably larger than the selenodetic value. The discrepancy may be due to a sampling fluctuation arising from the location of the selenodetic points. GOUDAS (1966) asserts that the modern selenodetic results show that maria and continents are at much the same level; but it appears to me that the maria are on the average lower, although the difference is less than I expected.

If indeed A and B represent system, then the photographic method is in difficulties, for this non-ellipsoidal elongation has been present from the start. The small triangulation of Franz also gives a linear regression. The results obtained by Saunder are even more peculiar. This author was considerably puzzled by the characteristics of his prolate spheroid, whose longest axis proved to be shorter than the Moon's mean radius. Had he plotted his heights against R, Saunder would have obtained a curve which is strongly convex downwards. Thus, in his case ellipsoidal representation is quite inappropriate.

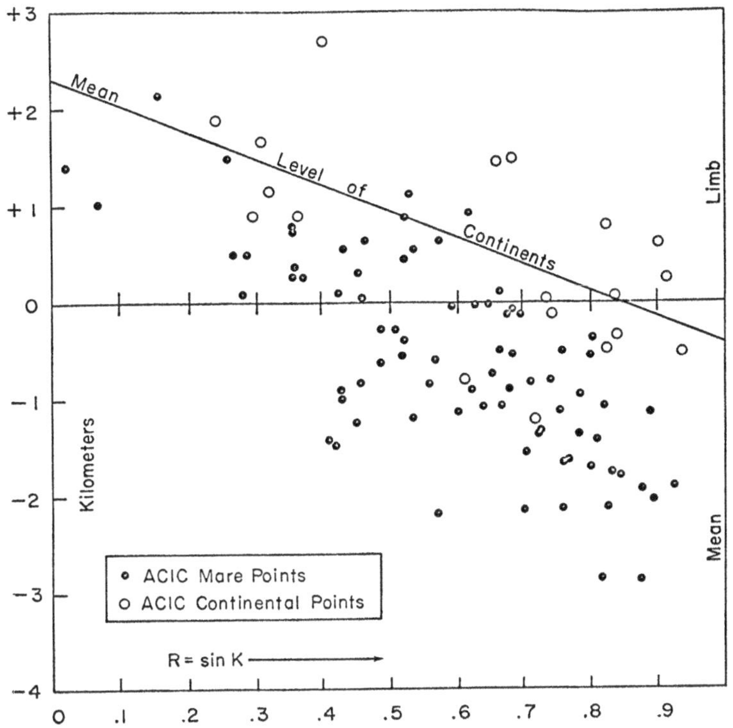

Fig. 1. Absolute altitudes for ACIC points.

The Baldwin triangulation results are not so accessible as the others and I have not completed the analysis. In this case, too, there appear to be systematic errors, and also a puzzling discrepancy between the height trends for the mare and continental points.

The source of the systematic height errors has not yet been identified. Indeed I cannot postulate any plausible system which leads to (1). If we simplify the problem and regard the photographs as orthographic projections (which they almost are) then the equations to determine the selenodetic coordinates (E, F, G) of a point, can be reduced to the form

$$x = aE + bF + cG,$$
$$y = eE + fF + gG,$$

where the coefficients $a, b, \ldots g$ represent the librations. The linearity of these assures us that the coordinate errors $(\delta E, \delta F, \delta G)$ are all linear functions of E, F and G, whether the errors arise from the measures (x, y) or from errors in the control values of (E, F, G). Since

$$\delta h = E\delta E + F\delta F + G\delta G,$$

this linear function leads to h as the sum of a first degree and a second degree function of E, F and G. This last is not directly comparable to (1) which represents the average characteristics of all vertical sections through the center of face. However, if we introduce a position angle P at the center of face by introducing the approximations

$$E = \sin K \cos P,$$
$$F = \sin K \sin P,$$
$$G = \cos K,$$

we obtain an expression for h with terms in $\sin P$, $\cos P$, $\sin^2 P$ and $\cos^2 P$. Replacing these functions of P by their average values we are left with an expression of the type

$$\delta h = A + B \cos K + C \cos^2 K. \tag{2}$$

The enquiry does not go deep enough to show which of the coefficients in this would always be negligible and which we might expect to be able to detect. Note that the last term simulates the effects of an ellipsoidal figure. It is clear that (2) bears no resemblance to (1), so the systematic errors we have in practice are not those we would expect. What is even more strange is that in the regression (1) there is virtually no trace of an ellipsoidal figure nor the systematic errors of a type we would expect.

The possibility that the photographic determinations have a persistent systematic error is of some importance, since I propose to use photographic measures to determine both fundamental and secondary positions. If indeed this error is present it is of the greatest importance to identify its source. The above considerations lead me to the conclusion that it does not come from errors in the secondary positions, but must in some way be generated in the photographic processes and measurements.

3. Scale Errors in the Plane of the Limb

One obvious source of systematic errors in the secondary and photographic points is incorrect scaling in the plane of the limb, that is, in the coordinates x and y. At LPL we take precautions to keep this source of error under close scrutiny and control, particularly since at least one of the above triangulations shows appreciable scale error.

For each measured plate we derive a scale from the Breslau positions and from measures on the limb itself. From the nature of the secondary measures one would not expect serious systematic scale errors in the plane of the limb itself, and this is in line with our experience. Thus far our "theoretical" radius, computed from the Breslau points, agrees to 0.00015 of the radius with the "observed" radius determined from several hundred measurements for the bright limb. Strangely enough Franz himself obtained discrepancies of ten times this amount in his reductions of the Lick plates. The discrepancy could be due to the defects of his non-rigorous reductions but I doubt this. This point appears to have been passed over by SCHRUTKA-RECHTENSTAMM (1958) in his re-reductions. From our measures to date I conclude that the Breslau triangulation has only very small systematic scale errors in the plane of the limb.

The same cannot be true of the AMS triangulation. The scale difference between triangulations can be tested as follows. For each common point the orthographic radius $R = \sqrt{E^2 + F^2}$ is computed and the scale difference between the two triangulations is then revealed by a regression of the type

$$\Delta R = mR,$$

from which the coefficient m is found as

$$m = \frac{\Sigma R \Delta R}{\Sigma R^2}.$$

For the difference AMS-Breslau I find

$$m = -0.0007$$

with a standard error of one tenth of this. Thus, since there is no evidence of an appreciable scale error in the Breslau triangulation it must be concluded that there is a serious scale error in the AMS results, amounting to one kilometer at the limb.

4. The LPL Selenodetic Measurements

The LPL methods of measurement and reduction have been discussed elsewhere (Arthur, 1965a, b), so here we give only those details which are necessary for the discussion of the results and their precision.

Each plate is measured in two orientations, differing approximately by 180°. The two rounds are united by a transformation of the type

$$\left.\begin{aligned} x_1 &= x_2 \cos\theta - y_2 \sin\theta + h \\ y_1 &= y_2 \cos\theta + x_2 \sin\theta + k \end{aligned}\right\} \tag{3}$$

The results published for each photograph give the mean values (x, y) and the semi-differences between the two rounds. The coefficients were formerly determined from the measures on the points themselves, but are now determined from two fiducial marks which can be bisected consistently to 2 microns. About 1000 points are measured on each photograph and the precision of the measuring process is estimated from a sample containing every tenth point.

Table II gives the standard errors of the mean x and y in microns and in units of the Moon's radius. These values do not include the effects of seeing or any other factor which distorts the photographic image.

TABLE II
Internal Precision of LPL Measures
on Yerkes Lunar Photographs

Plate	Standard Errors	
	x	y
1170	6.5 7×10^{-5}	6.1 6×10^{-5}
1269	5.0 5.5×10^{-5}	5.4 6×10^{-5}
482	7.8 9×10^{-5}	5.5 6×10^{-5}

To provide details of the external check on the precision a short listing of our reduction formulae is given. The Breslau points selected to control the reductions are converted to projected positions in the plane of the limb by

$$\begin{pmatrix} X \\ Y \\ Z \end{pmatrix} = \begin{pmatrix} a, b, c \\ e, f, g \\ i, j, k \end{pmatrix} \cdot \begin{pmatrix} E \\ F \\ G \end{pmatrix} \tag{4}$$

and

$$\left. \begin{aligned} X' &= X/(1 - Z \sin s'), \\ Y' &= Y/(1 - Z \sin s'). \end{aligned} \right\} \tag{5}$$

In these (E, F, G) are the rectangular selenodetic coordinates denoted by (ξ, η, ζ) by Schrutka-Rechtenstamm. The coefficients $a, b, \ldots k$ represent the topocentric librations and s' is the topocentric semi-diameter. These coordinates are referred to horizontal and vertical axes by

$$\left. \begin{aligned} u &= X' \cos\alpha - Y' \sin\alpha \\ v &= N' \cos\alpha + X' \sin\alpha \end{aligned} \right\} \tag{6}$$

where

$$\alpha = C' - Q' \tag{7}$$

and C' and Q' are the topocentric values of the position angle of the Moon's axis and the parallactic angle respectively. The horizontal and vertical plate coordinates affected by refraction are then

$$u' = u/\kappa', \quad v' = v/\kappa, \tag{8}$$

where $1/\kappa$ and $1/\kappa'$ are the vertical and horizontal refraction compressions respectively. These coordinates are connected to the measured coordinates (x, y) by the general transformation plane of rectangular coordinates, i.e.,

$$\left. \begin{aligned} u' &= px - qy + h \\ v' &= py + qx + k \end{aligned} \right\} \tag{9}$$

The coefficients p, q, h and k are found by the method of least squares treating u' and v' as independent observations of uniform weight. The residuals are computed as

$$\left. \begin{aligned} \delta u' &= u' - px + qy - h \\ \delta v' &= v' - py - qx - k \end{aligned} \right\} \tag{10}$$

The root-mean-square values of $\delta u'$ and $\delta v'$ are used to assess the external precision of the measure.

Table III gives the root-mean-square differences in u' and v' between the computed Breslau positions and the observed positions. Allowance is made for four degrees of freedom in the solution. It will be noted that the values are three to four times greater than the internal estimates, which perhaps emphasizes the perils of estimating dispersions in selenodetic work. The root-mean-square difference for E and F between the Breslau and ACIC positions is about 0.00015, and the deviations for these triangulations from the regression (1) suggests that the standard errors of the Breslau

TABLE III

External Precision of Selenodetic Measures on Yeskes Plates

Plate	Points	u'	v'
1170	85	$21 \times 10^{-5}\,r$	$25 \times 10^{-5}\,r$
1269	53	$22 \times 10^{-5}\,r$	$26 \times 10^{-5}\,r$
482	57	$18 \times 10^{-5}\,r$	$22 \times 10^{-5}\,r$
57	8	$24 \times 10^{-5}\,r$	$20 \times 10^{-5}\,r$

positions are about 1.5 those of the ACIC positions, which means that the standard error of the Breslau positions is roughly 0.00012 *r*. This implies a standard error of about 0.00018 *r* for the LPL measures on a single plate, that is, almost the entire amount in Table III. The disparity between the internal and external dispersions demonstrates very clearly that two measures per point on a normal lunar plate are quite sufficient. In our case the observer contributes a standard error of about 0.00007, while other sources contribute a further error with a standard value of 0.00016, that is, twice as much.

The progress in the position measurements is indicated by Table IV.

TABLE IV

Progress of Normal Position Measures on Yerkes Plates

Plate	Disk	Limb
Y. 1170	1099	439
Y. 1269	1467	403
Y. 482	995	447

Conclusions

The above is not much more than a superficial examination of the present status of selenodetic measurements, including those made at LPL. Even so it suggests that we should be rather more careful about the interpretation of these measures than we have been. There is a serious discrepancy between the Ranger data and the basic selenodetic position which will require the most thorough investigation. Even the internal evidence suggests that the secondary positions are not entirely satisfactory in some respects.

The LPL experience with normal lunar photographs suggests that these contain larger errors than those of the bisections themselves, that is, errors which are not diminished by any elaboration of the measures. I assess the LPL precision as not much superior to that achieved by Franz more than 50 years ago, on photographs with distinctly lower resolutions. The lower dispersions associated with the ACIC measures suggests that to make an appreciable advance we must use either serial photography or the long exposure technique developed by ACIC in connection with the astrographic reflector of the U.S. Naval Observatory, Flagstaff, Ariz.

The most disturbing feature of *all* the photographic triangulations is the evidence of systematic error from an unknown source. This demands the urgent attention of all who are interested in the precise triangulation of the lunar surface.

References

Arthur, D. W. G.: 1965a, 'The Reduction of Measures for Position on a Single Lunar Photograph', *Comm. LPL* **4**, No. 60.

Arthur, D. W. G.: 1965b, 'The Computation of Selenodetic Coordinates using the Librations', *Comm. LPL* **4**, No. 62.

Baldwin, R. B.: 1963, *The Measure of the Moon*. University of Chicago Press, Chicago.

Breece, S., Hardy, M., and Marchant, M. Q.: 1964, 'Horizontal and Vertical Control for Lunar Mapping'. *AMS Technical Report No. 29*, Part II.

Franz, J.: 1899, 'Die Figure des Mondes, IV: Heliometrische Bestimmung von 8 Fixpunkten auf dem Monde', *Astronomische Beobachtungen Königsberg* **38**.

Goudas, C. L.: 1966, *Icarus* **5**, 99–100.

Hayn, F.: 1904, 'Selenographische Koordinaten', *Leipzig, Sächsische Gesellschaft der Wissenschaften, I Abhandlung*.

Meyer, D. L. and Ruffin, B. W.: 1965, 'Coordinates of Lunar Features', *ACIC Technical Paper No. 15*.

Saunder, S. A.: 1905, 'First Attempt to Determine the Figure of the Moon', *Monthly Notices Royal Astronomical Society* **65**, 5.

Schrutka-Rechtenstamm, G.: 1956, 'Neureduktion der acht von J. Franz und der vier von F. Hayn gemessenen Mondkraterpositionen', *Sitzungsberichte der Österreichischen Akademie der Wissenschaften, Mathem.-naturw. Klasse*, Abteilung II **165**.

Schrutka-Rechtenstamm, G.: 1958, 'Neureduktion der 150 Mondpunkte der Breslauer Messungen von J. Franz', *Sitzungsberichte der Österreichischen Akademie der Wissenschaften, Mathem.-naturw. Klasse*, Abteilung II **167**.

ESTIMATE OF FOUR TOPOCENTRIC LUNAR RADII*

W. L. SJOGREN**

The Ranger lunar probes (6, 7, 8, and 9) which impacted the Moon at free flight velocities were tracked by three earth-based stations. These stations obtained accurate recordings of impact times and also range rate measurements for many hours prior to and within seconds of impact. Using this data, estimates of the dynamical center from four lunar surface points were determined. The range rate data determined the orbit relative to the tri-axial Moon and the recording of impact time triggered the proper radius vector (i.e., the selenocentric radius vector at the recorded time of impact, accounting for signal transit times).

The Doppler or range rate data for each mission consisted of approximately 2000 one-minute samples having a one-sigma accuracy of 1.4 millimeters/sec.† The data was fitted in a weighted least squares sense using a model with solar pressure, a tri-axial Moon and point masses for the Earth, Sun, Jupiter, Saturn, Mars, and Venus. Data near the Earth (prior to 170000 km from the Earth) were purposely ignored, so that the resulting orbit would be completely determined with respect to the Moon, and any erroneous ties with the Earth using data through the entire span would be removed. This was successfully shown in various computer computations in which the position of the Moon was perturbed by several kilometers and the resulting impact time was always the same. Also, perturbations in the Moon's potential, solar pressure, and data time tags were made, and always small resulting impact time changes occurred (i.e., .1 sec or less).

The discrepancy in matching the recorded impact time which was known to ± 5†† milliseconds was approximately 1.5 sec when ACIC charts were used for the radius of impact. Since the incoming normal velocity is approximately 2 km/sec, a decrease of approximately 3 km in the ACIC lunar chart elevation was necessary.

Comparisons of the Ranger tracking data results with the Army Maps and ACIC Charts is shown in Table I. Column 4 shows the actual Doppler data results which

* This paper presents the results of one phase of research carried out at the Jet Propulsion Laboratory, California Institute of Technology, under Contract No. NAS 7-100, sponsored by the National Aeronautics and Space Administration.

** Jet Propulsion Laboratory, California Institute of Technology, Pasadena, Calif., U.S.A.

† Data was weighted 13 mm/sec to cover various error sources, such as round off, refraction and integration error.

†† 400 μ sec for Ranger 8, and 7 μ for Ranger 9 relative to Doppler data time lag.

TABLE I

Lunar Radius Comparisons

Source / [a]Mission (1)	Lunar Elevation – km			Difference in Elevation		
	[b]U.S. Army Lunar Map (2)	[c]U.S. Air Force ACIC Lunar Chart (3)	[d]Ranger Flight Data (4)	Ranger-Army (km) (5)	Ranger-ACIC (km) (6)	ACIC-Army (km) (7)
Ranger VI	1743.2	1738.4	1735.3	− 7.9	− 3.1	− 4.8
Ranger VII	1737.7	1738.1	1735.5	− 2.2	− 2.6	+ 0.4
Ranger VIII	1741.6	1737.4	1735.2	− 6.4	− 2.2	− 4.2
Ranger IX	1738.1	1737.4	1735.7	− 2.4	− 1.7	− 0.7

[a]Impact latitudes and longitudes are RA VI 9.33°, 21.52°; RA VII − 10.63°, − 20.58°; Ra VIII 2.67°, 24.65°; RA IX − 12.83°, − 2.37° respectively.

[b]Army Map Service, Corps of Engineers, Washington, D.C., Topographic Lunar Map, 1 : 5000000. Compiled in 1963, printed June 1963, Edition 2–AMS.

[c]Aeronautical Chart and Information Center, U.S. Air Force, St. Louis, Mo., LAC60 printed September 1962, and LAC76, printed April 1964, both at 1 : 1000000.

[d]Technical Report No. 32-605: 'The Ranger VI Flight Path and Its Determination From Tracking Data', W. L. Sjogren, D. W. Curkendall, T. W. Hamilton, W. E. Kirhofer, A. S. Liu, D. W. Trask, R. A. Winneberger, W. R. Wollenhaupt. Jet Propulsion Laboratory, California Institute of Technology, Pasadena Calif., 15 December 1964.

Technical Report No. 32–694; 'The Ranger VII Flight Path and Its Determination From Tracking Data', W. R. Wollenhaupt, D. W. Trask, W. L. Sjogren, E. G. Piaggi, D. W. Curkendall, R. A. Winneberger, A. S. Liu, A. L. Berman. Jet Propulsion Laboratory, California Institute of Technology, Pasadena, Calif., 15 December 1964.

Technical Report No. 32–766: 'The Ranger VIII Flight Path and Its Determination From Tracking Data', W. R. Wollenhaupt, W. E. Kirhofer, E. G. Piaggi, F. Borncamp, A. S. Liu. Jet Propulsion Laboratory, California Institute of Technology, Pasadena, Calif. (to be published).

Technical Report No. 32–767: 'The Ranger IX Flight Path and Its Determination From Tracking Data', C. J. Vegos, W. E. Kirhofer, E. G. Piaggi, F. Borncamp, A. S. Liu. Jet Propulsion Laboratory, California Institute of Technology, Pasadena, Calif. (to be published).

have a one-sigma uncertainty of 300* meters. Column 6 shows a consistent decrease from ACIC charts of approximately 2.5 km, whereas Column 5 with Army Maps shows some inconsistency. In a private communication with Mr. Carlson of the Army Map Service he mentioned a possible error of 4 km in their evaluation of Rangers 6 and 8, for they are approximately in the same lunar area and this region is poorly defined by their control points. If this is true, then perhaps Army Maps would present the same picture as ACIC. Both ACIC and Army Maps derive their control from optical sources. Possibly this 2.5 km difference may be due to the fact that the optical center of the Moon along the Earth–Moon line is displaced from the dynamical center this much. Some independent work by Yaplee** using the lunar radar bounce data correlated with Army Maps should be directly comparable, but until relative elevations can be more firmly defined†, this check is weak. Future reduction of his data with ACIC charts or updated Army Maps could produce a consistent tie.

* Ranger 9 has a 700 meter one-sigma uncertainty since the last 11 minutes of tracking data before impact were not obtained.

** Yaplee, B. S., Knowles, S. H., Shapiro, A., Craig, K. J., and Brouwer, D.: 1963, 'The Mean Distance to the Moon as Determined by Radar', in *Proceedings IAU Symposium No. 21, Paris, France, May 1963.*

† Army Maps have been updated to a 1964 control since Yaplee's work.

ANALYSIS OF SATELLITE ORBIT PERTURBATIONS TO DETERMINE THE LUNAR GRAVITATIONAL FIELD*

W. M. KAULA**

Abstract. The selenodetic use of artificial lunar satellites entails a dynamical problem because the greater magnitudes and longer periods of disturbing accelerations lead to more significant non-linearities, and entails statistical problems because tracking from the Earth coupled with the variety of terms in the lunar gravitational field having similar effects leads to ambiguities and ill-conditioning.

A solution is proposed for the dynamical problem in which a von Zeipel transformation is applied to remove short period effects to obtain a non-linear intermediate orbit which can be integrated numerically with relatively long time steps.

The statistical problem is investigated by analyzing hypothetical tracking data incorporating the anticipated characteristics of the lunar gravitational field as well as instrumental error and ionospheric refraction. Regardless of tracking accuracy, it appears that a satellite of high inclination will be more valuable than a satellite of low inclination, and that the best solution is obtained by combining the two.

This paper is essentially a continuation of two previous papers (KAULA, 1963a, 1966a) on the use of artificial satellites to determine the gravitational field of the Moon, adapting techniques which have been applied to the geodetic use of earth satellites. The principal conclusions of these papers are: (1) If the shearing stresses implied in the Moon by the variations of its gravitational field are comparable to those implied by the Earth's variations, then the anticipated magnitude of normalized harmonic coefficients $\bar{C}_{lm}$, $\bar{S}_{lm}$ of the potential will be ± 2.0 to $3.5 \times 10^{-4}/l^2$. (2) The relatively slow rotation of the Moon and of the pericenter of a lunar satellite make it desirable to track a lunar satellite over several months to obtain the optimum results. (3) Carrying an orbit over such a long duration makes it preferable to integrate the orbit analytically. (4) The amplitudes of the perturbations by many of the harmonics in the lunar gravitational field may be some kilometers – large enough to be readily observable, but small enough to be treated as linear effects, similar to such perturbations of earth satellites.

Work since these papers has been principally on two aspects: (1) obtaining an orbital theory elaborate enough to be accurate and economical for arcs lasting a month or more, and (2) choosing the optimum tracking schedule and combination of satellite orbits. The remainder of the paper discusses each of these two problems, the dynamical and the statistical, in turn.

* Publication No. 502, Institute of Geophysics and Planetary Physics, University of California.
** Institute of Geophysics and Planetary Physics, University of California, Los Angeles, Calif., U.S.A.

Kopal and Goudas (eds.), Measure of the Moon.

The Dynamical Problem

Although the variations in the lunar gravitational field are expected to be somewhat larger in proportion to the central term than the Earth's, and the rotation of the Moon is much slower, it still should be feasible to calculate their perturbations as linear forced oscillations about a secularly precessing Keplerian ellipse. However, the much greater third body effect of the Earth, the much slower motion of pericenter, the probability of having to use satellites of appreciable semi-major axis and eccentricity, and the desirability of extending arcs over a much greater time-span indicate that the intermediate orbit to which these oscillations are added may have to be considerably more complex than for an earth satellite. Hence, to compromise between the flexibility of a numerical integration and the economy and insight of an analytical integration such as BROUWER's (1959), we propose to remove the short-period perturbations analytically and to integrate the resulting intermediary numerically – which may be possible to do in integration steps on the order of a day or so, since the highest frequency expected to have prominent effects is bimonthly.

Let us assume that the disturbances possible giving rise to significant non-linear long term effects comprise (1) the Earth, expressed as Legendre functions of degrees two and three; (2) the zonal harmonics of the moon J_2 and J_3; and (3) the equatorial ellipticity of the Moon J_{22}. Disturbing functions in terms of the osculating Keplerian elements of the satellite and thiıd body are given by KAULA (1961) for variations in the planetary gravitational field and by KAULA (1962) for the third body effects. However, two minor modifications are desirable in order to cope with the theoretical difficulties of obtaining a correct force function for the numerically integrated intermediary. Firstly, to perform the "book-keeping" associated with the mutual interaction of short period terms giving rise to long term effects, it is convenient to adopt a subscript convention such that all terms of like argument are combined with a positive coefficient on the mean anomaly term in the argument. Secondly, to account for the indirect effects of the perturbation of the Moon by Sun by using the lunar theory of BROWN (1905), either in tabular form or in an abbreviated analytical form such as that given by BATTIN (1964, p. 380), it is more convenient to express the position of the Earth in selenocentric inertial coordinates $\{r^*, \varphi^*, \lambda^*\}$ referred to the lunar equator and Cassini point at epoch.

The initial disturbing function R thus becomes:

$$R = \sum_{n=-3}^{3} \sum_{k=0}^{3} A_{onk} \begin{Bmatrix} \cos \\ \sin \end{Bmatrix}_{n-k \text{ odd}}^{n-k \text{ even}} \{n\omega + k(\Omega - \lambda^*)\}$$
$$+ \sum_{j=1}^{\infty} \sum_{n=-3}^{3} \sum_{k=-3}^{3} A_{jnk} \begin{Bmatrix} \cos \\ \sin \end{Bmatrix}_{n-k \text{ odd}}^{n-k \text{ even}} \{n\omega + jM + k(\Omega - \lambda^*)\}, \tag{1}$$

where, with subscript correspondences

$$\left.\begin{aligned} l &= [5 - (-1)^n]/2 \\ m &= |k| \\ p &= [l - (-1)^k n]/2 \\ q &= (-1)^k (j - n), \end{aligned}\right\} \tag{2}$$

the coefficients A_{jnk} are, from KAULA (1961, Eq. (27)) and KAULA (1962, Eqs. (2)–(3)),

$$A_{jnk} = \left[\frac{m^* a^l}{r^{*l+1}} P_{lm}(\sin\varphi^*)(2-\delta_{om})\frac{(l-m)!}{(l+m)!} H_{lpq}(e) + \left\{\delta_{2l}\delta_{2m}\frac{\mu}{a}\left(\frac{a_e}{a}\right)^2 J_{22} - \delta_{2l}\delta_{om}\frac{\mu}{a}\left(\frac{a_e}{a}\right)^2 J_2 - \delta_{3l}\delta_{om}\frac{\mu}{a}\left(\frac{a_e}{a}\right)^3 J_3\right\} G_{lpq}(e)\right] F_{lmp}(i) \tag{3}$$

In (1) and (3), $\{a, e, i, M, \omega, \Omega\}$ are the Keplerian elements of the satellite. In (3), m^* is the mass of the Earth times the gravitational constant, μ is the mass of the Moon times the gravitational constant; a_e is the equatorial radius of the Moon; the δ_{ij}'s are Kronecker deltas; $P_{lm}(\sin\varphi^*)$ is Legendre's Associated function; the inclination function $F_{lmp}(i)$ is (KAULA, 1961, Eq. (20)):

$$F_{lmp}(i) = \sum_t \frac{(2l-2t)!}{t!(l-t)!(l-m-2t)!\,2^{2l-2t}} \sin^{l-m-2t} i \times \sum_{s=0}^{m} \binom{m}{s} \cos^s i \sum_c \binom{l-m-2t+s}{c} \binom{m-s}{p-t-c} (-1)^{c-\kappa}, \tag{4}$$

where κ is the integer part of $(l-m)/2$, t is summed from 0 to the lesser of p or κ, and c is summed over all values making the binomial coefficients non-zero; and the eccentricity functions $H_{lpq}(e)$ and $G_{lpq}(e)$ are most succinctly defined as the Hansen functions:

$$\left.\begin{aligned} H_{lpq}(e) &= X_{(l-2p+q)}^{l,(l-2p)}(e) = X_j^{l,n}(e) \\ G_{lpq}(e) &= X_{(l-2p+q)}^{-l-1,(l-2p)}(e) = X_j^{-l-1,n}(e). \end{aligned}\right\} \tag{5}$$

The Hansen functions are valid for any $e<1$. The particular formulation we are using is that given by TISSERAND (1889, p. 256); an alternative formulation is given by PLUMMER (1960, p. 46).

To remove the short period terms on the second line of Equation (1) by the customary von Zeipel technique (BROUWER and CLEMENCE, 1961, ch. 17; KAULA, 1966b, Sec. 3.5), we have, using Delaunay variables,

$$\begin{aligned} L &= (\mu a)^{\frac{1}{2}} \\ G &= [\mu a(1-e^2)]^{\frac{1}{2}} \\ H &= [\mu a(1-e^2)]^{\frac{1}{2}} \cos i \\ h &= \Omega - \lambda^*, \end{aligned} \tag{6}$$

the force function:

$$F = \frac{\mu^2}{2L^2} + H\dot{\lambda}^* + R. \tag{7}$$

For the transformation

$$F(L, G, H, M, \omega, h) \rightarrow F'(L', G', H', -\omega, h) \tag{8}$$

we get

$$\left.\begin{aligned}
&\frac{\mu^2}{2L^2} + H\dot{\lambda}^* && = F_0' \\
&-\frac{\mu^2}{L^3}\frac{\partial S_1}{\partial M} + R + \dot{\lambda}^*\frac{\partial S_1}{\partial h} - \frac{\partial S_1}{\partial t} && = F_1' \\
&\frac{1}{2}\frac{3\mu^2}{L^4}\left(\frac{\partial S_1}{\partial M}\right)^2 + \frac{\partial R}{\partial L}\frac{\partial S_1}{\partial M} + \frac{\partial R}{\partial G}\frac{\partial S_1}{\partial \omega} && \\
&-\frac{\mu^2}{L^3}\cdot\frac{\partial S_2}{\partial M} + \dot{\lambda}^*\frac{\partial S_2}{\partial h} - \frac{\partial S_2}{\partial t} + \frac{\partial R}{\partial H}\frac{\partial S_1}{\partial h} && = F_2' + \frac{\partial F_1'}{\partial \omega}\frac{\partial S_1}{\partial G} + \frac{\partial F_1'}{\partial h}\frac{\partial S_1}{\partial H}
\end{aligned}\right\} \tag{9}$$

Substituting (1) in (9), and using the determining function S_1 to remove short period terms, we obtain for the force function F' of the intermediary to be integrated numerically,

$$F_1' = \sum_{n=-3}^{3}\sum_{k=0}^{3} A_{onk}\begin{Bmatrix}\cos\\ \sin\end{Bmatrix}\{n\omega + kh\}\begin{matrix}n-k\text{ even}\\ n-k\text{ odd}\end{matrix} \tag{10}$$

$$\begin{aligned}
S_1 &= \frac{L^3}{\mu^2}\int (R - F_1')\,\mathrm{d}M \\
&= \frac{L^3}{\mu^2}\sum_{j=1}^{\infty}\frac{1}{j}\sum_{n=-3}^{3}\sum_{k=-3}^{3} A_{jnk}\begin{Bmatrix}\sin\\ -\cos\end{Bmatrix}\{n\omega + jM + kh\}\begin{matrix}n-k\text{ even}\\ n-k\text{ odd}\end{matrix}
\end{aligned} \tag{11}$$

$$\begin{aligned}
F_2' &= \frac{1}{2\pi}\int_0^{2\pi}\left[\frac{3\mu^2}{2L^4}\left(\frac{\partial S_1}{\partial M}\right)^2 + \frac{\partial R}{\partial L}\frac{\partial S_1}{\partial M} + \frac{\partial R}{\partial G}\frac{\partial S_1}{\partial \omega} + \frac{\partial R}{\partial H}\frac{\partial S_1}{\partial h}\right]\mathrm{d}M \\
&= \frac{L^3}{2\mu^2}\sum_{j,n,n',k,k'}\left\{\frac{3}{2L}A_{jnk} + \frac{\partial A_{jnk}}{\partial L} + \frac{n'}{j}\cdot\frac{\partial A_{jnk}}{\partial G} + \frac{k'}{j}\frac{\partial A_{jnk}}{\partial H}\right\} \\
&\quad\cdot A_{jn'k'}\left\{(-1)^{n'-k'}\begin{matrix}\cos\\ \sin\end{matrix}\right\}\{(n-n')\omega + (k-k')h\}\begin{matrix}(n-n')-(k-k')\text{ even}\\ (n-n')-(k-k')\text{ odd}\end{matrix}
\end{aligned} \tag{12}$$

It would be more convenient to express F_2' in a form similar to F_1' in (10),

$$F_2' = \sum_{g=-6}^{6}\sum_{s=0}^{6} B_{gs}\begin{Bmatrix}\cos\\ \sin\end{Bmatrix}\{g\omega + sh\}\begin{matrix}g-s\text{ even}\\ g-s\text{ odd}\end{matrix}, \tag{13}$$

Such a form is particularly effective if $\sin\varphi^*$ in (3) is replaced by 0 and if r^* is replaced by a^* in these non-linear terms, since then the small coefficients B_{gs} will be for practicable purposes constant over the entire lifetime of some satellites, or at least constant over a time short – say, weeks – compared to a full cycle of pericenter ω for satellites of large eccentricity and semi-major axis. Working out the algebra of the subscript combinations yields:

$$\left.\begin{aligned}
&1)\text{ for } k' \leq k: && k' = k - s,\ n' = n - g \\
&2)\text{ for } k' > k: && k' = k + s,\ n' = n + g
\end{aligned}\right\} \tag{14}$$

So

$$B_{gs} = \frac{L^3}{2\mu^2} \sum_{j,n,k} \Bigg[\left\{ \frac{3}{2L} A_{jnk} + \frac{\partial A_{jnk}}{\partial L} \right\} \{ A_{j(n-g)(k-s)} + (1-\delta_{ok}) A_{j(n+g)(k+s)} \} + \frac{\partial A_{jnk}}{\partial G} \left\{ \frac{n-g}{j} A_{j(n-g)(k-s)} + \frac{n+g}{j} (1-\delta_{ok}) A_{j(n+g)(k+s)} \right\} + \frac{\partial A_{jnk}}{\partial H} \left\{ \frac{k-s}{j} A_{j(n-g)(k-s)} + \frac{k+s}{j} (1-\delta_{ok}) A_{j(n+g)(k+s)} \right\} \Bigg] \left\{ \begin{matrix} 1 \\ (-1)^{n-k+1} \end{matrix} \right\}_{g-s \text{ odd}}^{g-s \text{ even}} \tag{15}$$

Although the Delaunay elements are more convenient to derive the contribution to the long-term variations of the orbit from mutual interaction of short period terms, for purposes of integration of the intermediary the use of a disturbing function

$$R' = F_1' + F_2' \tag{16}$$

in the Lagrangian Planetary equations is equally valid. The form of the complete disturbing function R in Equation (1) is such that a representation in Keplerian elements is more convenient. Using (6) and the Kepler law $\mu = \mathbf{n}^2 a^3$,

$$\frac{\partial(a, e, i)}{\partial(L, G, H)} = \left\{ \begin{matrix} 2/\mathbf{n}a, & 0 & 0 \\ (1-e^2)/\mathbf{n}a^2 e, & -\sqrt{1-e^2}/\mathbf{n}a^2 e & 0 \\ 0 & \cot i/\mathbf{n}a^2\sqrt{1-e^2} & -1/\mathbf{n}a^2\sqrt{1-e^2}\sin i \end{matrix} \right\} \tag{17}$$

whence, also substituting from (6),

$$B_{gs} = \frac{1}{2\mathbf{n}} \sum_{j,n,k} \Bigg[\left\{ \frac{3}{2\mathbf{n}a^2} A_{jnk} + \frac{2}{\mathbf{n}a} \cdot \frac{\partial A_{jnk}}{\partial a} + \frac{\sqrt{1-e^2}}{\mathbf{n}a^2 e} \left(\sqrt{1-e^2} - \frac{n-g}{j} \right) \frac{\partial A_{jnk}}{\partial e} + \frac{1}{j\mathbf{n}a^2\sqrt{1-e^2}\sin i} ([n-g]\cos i - [k-s]) \frac{\partial A_{jnk}}{\partial i} \right\} A_{j(n-g)(k-s)} + \{1-\delta_{ok}\} \left\{ \frac{3}{2\mathbf{n}a^2} A_{jnk} + \frac{2}{\mathbf{n}a} \frac{\partial A_{jnk}}{\partial a} + \frac{\sqrt{1-e^2}}{\mathbf{n}a^2 e} \left(\sqrt{1-e^2} - \frac{n+g}{j} \right) \frac{\partial A_{jnk}}{\partial e} + \frac{1}{j\mathbf{n}a^2\sqrt{1-e^2}\sin i} ([n+g]\cos i - [k+s]) \frac{\partial A_{jnk}}{\partial i} \right\} A_{j(n+g)(k+s)} \Bigg] \cdot \left[\begin{matrix} 1 \\ (-1)^{n-k+1} \end{matrix} \right]_{g-s \text{ odd}}^{g-s \text{ even}} \tag{18}$$

In (15) and (18), the j summation is from 1 to ∞, while the other limits are:

$$\begin{gathered} \max(-3, -3-g) \le n \le \min(3, 3+g) \\ -3 \le k \le 3. \end{gathered} \tag{19}$$

The equations of motion applicable to the intermediary are (see, e.g., KAULA, 1966b, p. 29):

$$\left.\begin{aligned}
\frac{\mathrm{d}a}{\mathrm{d}t} &= \frac{2}{na}\cdot\frac{\partial R}{\partial M}\\
\frac{\mathrm{d}e}{\mathrm{d}t} &= \frac{1-e^2}{na^2e}\frac{\partial R}{\partial M} - \frac{(1-e^2)^{\frac{1}{2}}}{na^2e}\frac{\partial R}{\partial\omega}\\
\frac{\mathrm{d}i}{\mathrm{d}t} &= \frac{\cos i}{na^2(1-e^2)^{\frac{1}{2}}\sin i}\frac{\partial R}{\mathrm{d}\omega} - \frac{1}{na^2(1-e^2)^{\frac{1}{2}}\sin i}\frac{\partial R}{\partial\Omega}\\
\frac{\mathrm{d}M}{\mathrm{d}t} &= n - \frac{1-e^2}{na^2e}\frac{\partial R}{\partial e} - \frac{2}{na}\frac{\partial R}{\partial a}\\
\frac{\mathrm{d}\omega}{\mathrm{d}t} &= -\frac{\cos i}{na^2(1-e^2)^{\frac{1}{2}}\sin i}\cdot\frac{\partial R}{\partial i} + \frac{(1-e^2)^{\frac{1}{2}}}{na^2e}\frac{\partial R}{\partial e}\\
\frac{\mathrm{d}\Omega}{\mathrm{d}t} &= \frac{1}{na^2(1-e^2)^{\frac{1}{2}}\sin i}\frac{\partial R}{\partial i},
\end{aligned}\right\} \tag{20}$$

where, from (16), (10), and (13),

$$\left.\begin{aligned}
\frac{\partial R}{\partial\{a,e,i\}} &= \sum_{g,s}\left\{\frac{\partial A_{ogs}}{\partial\{a,e,i\}} + \frac{\partial B_{gs}}{\partial\{a,e,i\}}\right\}\left\{\begin{matrix}\cos\\ \sin\end{matrix}\right\}_{g-s\text{ odd}}^{g-s\text{ even}}\{g\omega + s(\Omega-\lambda^*)\}\\
\frac{\partial R}{\partial\{\omega,\Omega\}} &= \sum_{g,s}\{g,s\}\{A_{ogs}+B_{gs}\}\left\{\begin{matrix}-\sin\\ \cos\end{matrix}\right\}_{g-s\text{ odd}}^{g-s\text{ even}}\{g\omega + s(\Omega-\lambda^*)\}\\
\frac{\partial R}{\partial M} &= 0
\end{aligned}\right\} \tag{21}$$

While the evaluation of B_{gs} and its derivatives is laborious to program, as mentioned, it will not be time-consuming in the orbit calculation having to be done only once or at wide intervals.

Brown's theory gives the longitude, l and latitude, β of the Moon in a geocentric system referred to ecliptic and vernal equinox. To obtain the selenocentric position of the Earth referred to equator and Cassini point for use in calculating A_{jnk}, the signs are reversed and rotations from the vernal equinox to the Cassini point through the longitude of the node Ω_{M} plus about the Cassini point through the inclination of the ecliptic to the lunar equator, I, about $1°32'.1$:

$$\left\{\begin{matrix}\cos\varphi^*\cos\lambda^*\\ \cos\varphi^*\sin\lambda^*\\ \sin\varphi^*\end{matrix}\right\} = \mathbf{R}_1(-I)\mathbf{R}_3(\Omega_{\mathrm{M}})\left\{\begin{matrix}\cos\beta\cos l\\ -\cos\beta\sin l\\ -\sin\beta\end{matrix}\right. \tag{22}$$

where

$$\mathbf{R}_1(-I)\mathbf{R}_3(\Omega_{\mathrm{M}}) = \left\{\begin{matrix}\cos\Omega & \sin\Omega & 0\\ -\cos I\sin\Omega & \cos I\cos\Omega & -\sin I\\ -\sin I\sin\Omega & \sin I\cos\Omega & \cos I\end{matrix}\right\} \tag{23}$$

The range of latitude β is $\pm .09$ and of distance r^* is $\pm .05a^*$. Hence the neglect of these variations in calculating the non-linear coefficients B_{gs} should make a difference of about 1% in those coefficients. In the linear coefficients A_{onk} these variations are definitely significant and should, of course, be taken into account in the calculation of the A_{onk}.

The short periodic effects of the Earth involving the mean anomaly M of the satellite are quite effectively taken into account as linear perturbations with constant coefficients, in a manner similar to the tesseral harmonic perturbations.

The Statistical Problem

Sources of error in the use of artificial satellites of the Moon to determine the variations of its gravitational field include: instrumental error; refraction by the Earth's ionosphere; errors in the geodetic positions of tracking stations; omissions or errors in the lunar ephemeris; imperfections in the satellites attitude control system; and variations in the lunar gravitational field of a higher degree, of shorter wavelength, than those being determined in any particular analysis. Most of these effects are systematic, rather than random in character. Hence the ordinary sort of error analysis assuming randomness with respect to each other or errors at different times is likely to give extremely illusory results, particularly in view of the ill-conditioning of the situation. Hence it seems more prudent to construct a series of hypothetical observations incorporating the anticipated characteristics of the orbit and the tracking, and to analyze them by the same techniques to be applied to actual data. For economy, the model of the situation can, of course, be simplified in several ways: in particular, sources eventually removable by means other than tracking lunar satellites, such as station position and lunar ephemeris errors, can be left out. Also, although the geometry is different and the time scales are changed by factors of 5 and 27 respectively, it seems quite sure that the analyses for lunar zonal and tesseral harmonics will be best made separately, as they are for earth satellites, and that a distribution of observations adequate to determine the tesseral harmonics will be more than adequate to determine the zonal harmonics. Hence we shall assume the zonal harmonics known and make the error analysis for determination of tesseral harmonics only.

The technique of analysis applied was the geodetic satellite program SOA, adapted from the form applied to earth satellites (KAULA, 1963b). The principal parts of the program and their characteristics:

(1) SOA-P, the purpose of which is preliminary orbit and perturbation or partial derivative coefficient computation.

The general procedure applied in SOA-P:

(a) For each satellite in turn, determine by iterated least squares from a binary tape of observations orbital elements for a specified set of epochs;

(b) If so chosen, count the frequency distribution of various angles of the orbit;

(c) Calculate the amplitude factors and combinations of angles in the arguments of perturbations and partial derivatives under the assumption that they are representable as linear forced oscillations.

(2) SOA-F, the purpose of which is determination of spherical harmonic coefficients of the gravitational field from any number of arcs of any number of satellites by the technique of partitioned least squares.

The general procedure applied in SOA-F:

(a) Read from magnetic tape information generated by SOA-P common to all satellites (gravitational coefficients subscripts to be determined etc.); read as new input statistical data common to all satellites: initial values and a priori sigmas of parameters, etc.;

(b) For each satellite in turn (1) Read from magnetic tape information generated by SOA-P; read as new input information pertaining to final analysis: arbitrary polynomials to be used, if any; orbital arcs to be omitted, if any; satellite weighting factor; timing and observation variances; (2) For each orbital arc in turn, read observations from magnetic tape, form observation equations, increment normal equations; when arc end reached, subtract out from normals part corresponding to orbital constants of integration; (3) At end of satellite data, write normal equations with *unit* weighting factor on binary tape;

(c) After all satellites, read normal equations off tape and increment combination normals with weighting factors applied;

(d) Read in supplemental observation equations and their weights and variances, and increment normals;

(e) Solve normals for corrections to parameters.

Satellite orbit specifications to which these analyses were applied were:

(1) "R": the USSR satellite Luna 10, of semi-major axis a about 1.39 lunar radii, eccentricity e of 0.13, and inclination i of $72° = 1.26$.

(2) "L": the planned USA satellite Lunar Orbiter, $a = 1.544$, $e = 0.336$, $i = 15° = 0.262$.

(3) "G": a possible orbit of the planned USA satellite AIMP which is relatively good for selenodesy, $a = 2.10$, $e = 0.45$, $i = 160° \approx 2.80$.

(4) "F": a possible orbit of the planned USA satellite AIMP which is of average selenodetic value, $a = 4.42$, $e = 0.60$, $i = 2.90$.

(5) "P": a possible orbit of the planned USA satellite AIMP which is relatively poor for selenodesy, $a = 9.21$, $e = 0.75$, $i = 3.00$.

(6) "H": a hypothetical satellite of $a = 1.544$, $e = 0.336$, $i = 60° = 1.048$.

The variations which should be taken into account in calculating the dummy observation tape appear to be:

(a) Instrumental Error. We assumed:

1) G, F, P, and R:	$\sigma(r) = \pm 15$ m	$= \pm\ .9 \times 10^{-5}$	lunar radii
	$\sigma(\dot{r}) = \pm 1.8$ cm/sec	$= \pm 1.1 \times 10^{-5}$	lunar radii/1035 secs.
2) L, H:	$\sigma(r) = \pm 10$ m	$= \pm\ .6 \times 10^{-5}$	lunar radii
	$\sigma(\dot{r}) = \pm 0.2$ cm/sec	$= \pm\ .12 \times 10^{-5}$	lunar radii/1035 secs.

(b) Ionospheric refraction. We assumed for G, F, P, and R:

Error in Constant range effect at zenith $C_I = 50\ \text{m} = 0.29 \times 10^{-4}$
Error in Amplitude of diurnal variation $V_I = 34\ \text{m} = 0.20 \times 10^{-4}$
Random variation in time when sun at 90° hour angle $\sigma_R = \pm 125\ \text{m} = \pm 0.70 \times 10^{-4}$
Random variation in space when sun at 90° hour angle $\sigma_V = \pm\ 85\ \text{m} = \pm 0.50 \times 10^{-4}$

The systematic errors are 20% of the total magnitude of systematic variations. These values were used in the formulae:

$$\left.\begin{aligned} \sigma_R &= \left[1 + \frac{v_I}{C_I} \cos t_{\odot}\right] \sigma_R \\ \sigma_R &= \left[1 + \frac{v_I}{C_I} \cos t_{\odot}\right] \sigma_V \end{aligned}\right\} \tag{24}$$

in which $t_{\odot}$ is the hour angle of the sun. Then for the refraction at zenith at the instant of a particular observation

$$\left.\begin{aligned} \Delta R &= C_I + v_I \cos t_{\odot} + \text{Ran}(\sigma_r) \\ \Delta \dot{R} &= \{\text{Ran}(\sigma_v) - v_I \sin t_{\odot}\}\, \theta \end{aligned}\right\} \tag{25}$$

where θ is the rate of rotation of the Earth and Ran (x) is a randomly chosen variate from a Gaussian normal distribution of standard deviation x. Finally for zenith distance z,

$$\begin{aligned} \Delta r &= \Delta R \sec z \\ \Delta \dot{r} &= \Delta \dot{R} \sec z + \Delta R \frac{\mathrm{d}}{\mathrm{d}t} (\sec z) \end{aligned} \tag{26}$$

(c) Variations of the lunar gravitational field. Using the equal stress implication assumption, we made the normalized coefficients $\bar{C}_{lm}$, $\bar{S}_{lm}$ randomly chosen variates from Gaussian normal distributions with standard deviations $\pm 2.0 \times 10^{-4}/l^2$. Subscripts l, m were carried to 8, 5 for definitive runs. All satellites except "F" and "P" had at least nine harmonics causing perturbations of rms amplitude more than 1000 meters.

Hypotheses of tracking schedules tested for satellite "G" only were:

(1) The number of observations within a given duration. The duration should be at least the month, 27⅓ days, and preferably the common multiple of day and month, 82 days.

(2) Randomization versus regular intervals.

(3) Scale of randomization: i.e., the length of interval within which a prescribed number of observations should be made at randomly selected times.

(4) Spacing of observation sets: e.g., should tracking be intensive over one or two revolutions per day, or should it be spread out over the entire day?

Another hypothesis it was originally intended to test was time of day of tracking, to counter adverse ionospheric refraction. However, since the Moon moves only about 12°/day with respect to the Sun, this does not seem too practicable if we require a good distribution of tracking over the month. The basic premise of tracking over the month also permits always tracking with station which has the satellite closest to its zenith without sacrificing the advantages obtained by switching tracking between stations far apart on the Earth. Compared to a standard of uniformly spaced observations at a rate of 12 per day, nothing seems to help so much as simply more observations. By and large, schedules imposing some sort of regularity gave better results than those entailing some randomization. This situation is consistent with the principal source of error being systematic – such as the higher degree harmonics of the gravitational field – rather than essentially random, such as (presumably) instrumental error and ionospheric refraction variations.

Hence more extensive analyses extending over three months and combining different satellites were done using uniformly spaced observations. The results of these tests for some combinations are given in Table I for a rate of six observations/day. For these analyses the fictitious gravity field was extended up to $lm=85$.

The quantity $\Delta\sigma_l$ is defined as:

$$\Delta\sigma_l = [\sum_m (\hat{C}_{lm} - C_{lm})^2 + (\hat{S}_{lm} - S_{lm})^2]^{\frac{1}{2}} \tag{27}$$

The $\Delta\sigma_l$ is perhaps not too satisfactory a measure of the quality of a determination, since it is most influenced by the weakly determined coefficients; a more appropriate measure of the relative merit of two solutions might be simply the number of coefficients which are better or worse.

For some of the combinations, the calculations were repeated for a tracking rate of 24 times per day. The resulting improvements were moderate – less than 10%.

The following conclusions seem clear from Table I:

(1) The satellite of highest inclination and smallest semi-major axis, "R", the USSR satellite Luna 10, obtains the best solution. Higher inclination orbits are always much more sensitive to predominantly east-west variations in the field – (m/l near 1) and almost as sensitive to predominantly north-south variations (m/l near 0). The comparitive merit of satellite "R" given in Table I may be a little better than reality because it was assumed that the tracking was from stations more widely distributed than those actually being used, but this effect is probably slight over a duration as long as three months.

(2) The best solution is obtained by a combination of satellites of differing inclination.

In conjunction with the solutions in Table I, covariance matrices were calculated in the usual manner. Using nominal sigmas of about 100 meters in range and 10 cm/sec in range-rate, the resulting standard errors for the gravitational coefficients were on the order 1 or 2% of the actual errors. The correlation coefficients were low enough except between terms of the same symmetry and the same order, m, in which case the correlations persisted above 0.8 for solutions from any single satellite. The low inclination satellites, "L", "G", "F" and "P" were further inferior in being unable to distinguish strongly between symmetric (l-m even) and skew-symmetric (l-m odd) terms; even after three month's tracking, the correlations were high between

TABLE I

Results of Analyses of 81 Days of Various Combinations of Lunar Satellites Observed 6 Times per Day

Coefficient	"True" Value 10^{-6}	Satellites Used									
		G 10^{-6}	L 10^{-6}	R 10^{-6}	RL 10^{-6}	RG 10^{-6}	RF 10^{-6}	RLG 10^{-6}	LG 10^{-6}	LF 10^{-6}	LH 10^{-6}
C21	1.7	1.6	− 4.9	− 1.3	4.4	3.1	0.1	3.9	1.2	− 2.4	1.8
S21	− 37.0	− 34.1	− 54.5	− 31.7	− 35.0	− 40.4	− 34.6	− 37.3	− 47.8	− 49.1	− 32.0
C22	63.4	59.0	88.0	65.7	64.0	64.8	65.7	65.4	54.1	77.7	68.1
S22	93.8	71.0	104.0	92.4	90.0	96.9	92.3	95.4	101.0	101.0	108.0
C31	− 41.1	− 42.5	− 46.9	− 41.6	− 46.4	− 43.5	− 42.8	− 49.8	− 48.1	− 46.4	− 46.0
S31	6.6	0.8	11.3	3.9	8.7	2.4	3.3	5.7	8.8	11.3	8.4
C32	− 25.7	− 30.3	16.3	− 27.9	− 17.8	− 19.0	− 27.5	− 2.8	4.6	8.4	− 5.2
S32	32.6	22.5	52.7	36.6	35.0	37.8	36.4	33.2	31.2	49.1	44.8
C33	− 37.9	− 46.3	− 95.0	− 36.9	− 41.7	− 30.9	− 35.2	− 32.2	− 47.6	− 96.6	− 22.2
S33	43.6	90.5	109.0	46.2	38.2	39.6	42.1	37.1	111.0	68.9	20.1
C41	− 6.9	− 6.3	− 12.2	− 0.6	− 7.3	− 5.0	− 2.0	− 7.5	− 10.1	− 10.8	− 8.3
S41	1.4	9.4	− 14.5	− 6.1	− 2.9	2.7	− 3.1	− 1.3	− 9.2	− 11.6	− 1.0
C42	14.6	27.1	34.6	12.8	23.5	15.2	12.8	19.4	15.7	31.5	28.2
S42	− 1.6	− 30.1	− 3.1	3.4	18.3	− 9.7	3.9	4.8	5.9	2.2	11.2
C43	− 1.2	− 13.2	− 5.3	− 1.4	− 5.6	− 1.3	− 2.2	− 4.1	− 10.5	− 25.9	− 16.8
S43	− 5.8	22.3	− 0.6	− 5.7	− 4.0	− 7.5	− 6.3	− 8.1	− 20.0	− 3.2	− 11.1
C44	7.7	− 43.0	− 23.7	12.9	15.6	10.0	13.0	12.8	− 51.0	− 38.0	26.2
S44	− 11.4	60.2	− 153.0	− 16.7	− 20.0	− 16.8	− 16.7	− 19.3	− 42.9	− 145.0	− 35.6
$\Delta\sigma_2$		24.	26.	6.	5.	5.	4.	3.	16.	20.	16.
$\Delta\sigma_3$		47.	89.	6.	12.	13.	6.	26.	67.	75.	37.
$\Delta\sigma_4$		95.	148.	13.	26.	10.	11.	13.	70.	144.	39.

members of a set such as C_{22}, S_{32}, and C_{42}. The fact that "G", "F", and "P" are of retrograde inclination while "L" is of direct inclination is not of help to distinguish between terms of the same symmetry, since their perturbations have the same sign reversals. Satellite "L" may also suffer from its low pericenter causing poorer distribution of observations due to more frequent occulting.

Acknowledgements

The dynamical theory has benefited from discussions with Bruce Douglas of UCLA. The statistical computations were carried out at NASA – Goddard Space Flight Center with the assistance of W. D. Putney. A portion of this work was supported by NASA Grant NsG-216.

References

Battin, R. H.: 1964, *Astronautical Guidance*. McGraw-Hill Book Co., New York.

Brouwer, D.: 1959, 'Solution of the Problem of Artificial Satellite Theory without Drag', *Astron. J.* **64**, 378–397.

Brouwer, D. and Clemence, G. M.: 1961, *Methods of Celestial Mechanics*. Academic Press, New York.

Brown, E. W.: 1905, 'Theory of the Motion of the Moon; containing a New Calculation of the Expression for the Coordinates of the Moon in Terms of the Time, Part IV', *Mem. Roy. Astron. Soc.* **57**, 51–145.

Kaula, W. M.: 1961, 'Analysis of gravitational and geometric aspects of geodetic utilization of satellites', *R. A. S. Geophys. J.* **5**, 104–133.

Kaula, W. M.: 1962, 'A Development of the Lunar and Solar Disturbing Functions for a Close Satellite', *Astron. J.* **67**, 300–304.

Kaula, W. M.: 1963a, 'The Investigation of the Gravitational Fields of the Moon and Planets with Artificial Satellites', *Adv. in Space Sci. & Tech.* **5**, 210–226.

Kaula, W. M.: 1963b, 'Tesseral Harmonics of the Gravitational Field and Geodetic Datum Shifts derived from Camera Observations of Satellites', *J. Geophys. Res.* **68**, 473–484.

Kaula, W. M.: 1966a, 'Plans for analysis of lunar satellite orbits', in J. Kovalevsky (ed.), *Trajectories of Artificial Celestial Bodies as Determined from Observations*. Springer-Verlag, Berlin pp. 247–260.

Kaula, W. M.: 1966b, *Theory of Satellite Geodesy*. Blaisdell Publishing Co., Waltham.

Plummer, H. C.: 1960, *An Introductory Treatise on Dynamical Astronomy*. Dover Publications, Inc., New York.

Tisserand, F.: 1889, *Traité de Mécanique Céleste*, Tome I: *Perturbations des Planètes d'après la Méthode de la Variation des Constantes Arbitraires*. Gauthier-Villars, Paris.

LUNAR GRAVITY FROM ORBITER TRACKING DATA*

J. LORELL**

Abstract. A procedure for extracting gravity information from lunar orbiter tracking data is described with particular attention to the photographic orbiter to be launched shortly by the United States.

1. Introduction

In the near future, under the Lunar Orbiter Project, National Aeronautics and Space Administration (NASA) of the United States plans to place one or more satellites into orbit around the Moon. Although the primary mission of these satellites is to make pictures of the lunar surface, the tracking data will provide an unprecedented opportunity for study of the Moon's gravity field. A separate lunar orbiting project by NASA is the Anchored Imp (AIMP), a satellite in a loose orbit that will provide some information on the gravity field. Lunar Orbiter will be a more sophisticated vehicle than AIMP, however, and will orbit much more closely to the Moon. The intended pericenter and apolune distances are about 1784 km and 3588 km, respectively, which means a closest approach to the surface of some 46 km.

The use of Orbiter tracking data in describing the gravity field depends strongly on the type, precision, and coverage of the data. The data from Lunar Orbiter should be of a quality sufficient to determine 20 to 50 harmonic coefficients with 7 or 8 decimal digit accuracy.

To a certain extent, the results depend on the inclination of the orbit. Lunar Orbiter will have an inclination of some 15° to the lunar equator, which, though not optimum, can cover a large fraction of the gravity field. However, that part of the gravity field most sensitive to equatorial orbits will be mapped best.

Lunar Orbiter is expected to produce a tremendous quantity of tracking data. For selenodesy purposes, the radar range and range-rate will be the prime source of information. Data are sampled at 1-min. intervals, with continuous coverage during viewing periods. Over a 2-mo. span, some 100000 points of data could be accumulated; the selenodetic information will be extracted from these points. An efficient extraction process is, therefore, quite important.

* This paper presents the results of one phase of research carried out at the Jet Propulsion Laboratory, California Institute of Technology, under Contract No. NAS 7–100, sponsored by the National Aeronautics and Space Administration.

** Jet Propulsion Laboratory, California Institute of Technology, Pasadena, Calif., U.S.A.

In support of the Selenodesy Experiment, JPL is responsible for the precise reduction of the tracking data for selenodetic analysis. A description of the Lunar Orbiter Project and the Selenodesy Experiment was given by Michael and Tolson at the Geodesy Symposium in Athens (MICHAEL and TOLSON, 1965). This paper describes the computing methods for data reduction, illustrates the difficulties involved, and gives an estimate of what is hoped to be accomplished.

2. Information Content of Tracking Data

A. REPRESENTATION OF THE GRAVITY FIELD

To analyze the Moon's gravity field, a convenient mathematical description is necessary. This description is provided by the expansion of the gravity potential, Φ, in spherical harmonics:

$$\Phi(r, \phi, \lambda) = -\frac{\mu}{r}\left[1 + \sum_{n=1}^{\infty} \sum_{m=0}^{n} \left(\frac{R}{r}\right)^n P_n^m(\sin\phi)(C_{nm}\cos m\lambda + S_{nm}\sin m\lambda)\right]$$

in which $\mu = \mathrm{GM} =$ Moon's gravity constant, $M =$ mass of Moon, $r =$ radial distance, $R =$ mean lunar radius, $\phi =$ selenocentric latitude, $\lambda =$ selenocentric longitude, $P_m^n =$ associated Legendre polynomical, degree n, order m, and C_{nm}, $S_{nm} =$ harmonic coefficients.

If the Moon were a uniform sphere, and the coordinate system referred to its center, then all coefficients C_{nm} and S_{nm} would be zero. Thus the deviations from zero describe the deviations from a spherically symmetric gravity field about the origin.

In particular, the first order harmonics measure the position (x, y, z) of the center of mass. The second order harmonics measure the moments of inertial I_x, I_y, I_z. The third order harmonics measure the third order mass moments, fourth order harmonics measure the fourth order mass moments, etc., thus*:

$$\begin{array}{lll} RC_{10} = z & R^2MC_{20} = \frac{1}{2}(I_x + I_y) - I_z & \\ RC_{11} = x & R^2MC_{21} = I_{zx}, & R^2MS_{21} = I_{zy} \\ RS_{11} = y & R^2MC_{22} = \frac{1}{4}(I_y - I_x), & R^2MS_{22} = \frac{1}{2}(I_{xy}) \end{array}$$

The main problem in the Selenodesy Experiment is the accurate evaluation of as many harmonic coefficients C_{nm} and S_{nm} as appear important. Note that for a coordinate system fixed in the Moon with its origin at the center of mass, all harmonic coefficients are constant, and the first order coefficients are zero. Therefore, the basic coordinate system will be chosen as closely to this ideal as possible. However, since our best estimate of the center of mass of the Moon is that given by the lunar ephemeris, which may be in error by as much as 200 meters, it is possible that these ideal conditions do not hold to the desired accuracy. This possible error may be avoided by estimating values for the first order harmonics, and thereby correcting the ephemeris. Special care must be used, however, because such corrections can vary with time.

* The literature contains variations from this nomenclature, notably $J = -\frac{3}{2} C_{20}$ and $J_n = -C_{no}$.

B. THE TRACKING DATA

The Lunar Orbiter will be tracked by earth-based radar using NASA's Deep Space Network. It is hoped that range-rate data with a precision greater than 0.003 m/s, and range data with a precision greater than 5 m will be obtained. This information is extremely precise in contrast to telescopic data, especially for distance measurements, and, with it, the orbit should be determined with a precision greater than 0°.01 in the Euler angles, depending, of course, on the nature of the orbit and the tracking pattern.

A complete description of the data, how they are obtained, their format, and the essentials for processing them are available in a JPL Memorandum (LORELL, ANDERSON, and SJOGREN, 1965). The kind of accuracy that can be obtained is shown in Figure 1, taken from that Memorandum. This is a plot of S-band Doppler residuals from the tracking of Mariner IV. The noise is seen to be 0.00064 m/s.

The characteristics of a typical orbit for the mission are listed in Table I. Note that there are two orbits, the first of which will be held for a few days. The final orbit is scheduled to last for at least a year, 11 mo of which should yield quality tracking data for the Selenodesy Experiment.

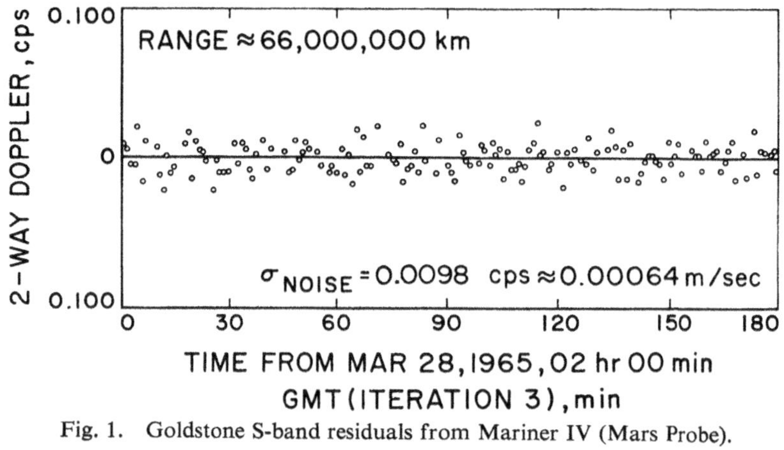

Fig. 1. Goldstone S-band residuals from Mariner IV (Mars Probe).

TABLE I

Lunar Orbiter Characteristics (Boeing Mission 110)

Orbit element	Initial orbit	Final orbit
a	2787.61 km	2686.04 km
e	0.286881	0.335791
i	13.4949°	13.4853°
Ω	46.6092°	45.9371°
ω	348.0953°	349.3650°
T	6/27/66	6/29/66
	4 hr, 4 min, 48.000 sec	1 hr, 33 min, 5.796 sec
Pericenter	250 km	46 km

The initial orbit, on the other hand, though available for only a short arc of tracking data, should yield very useful information. It will provide the very first estimates of the low degree harmonic coefficients. This information will be used to select the exact parameters that will ensure an adequate life of the final orbit.

The quantity of information on the gravity field that can be extracted from the tracking data depends on the kind of tracking plan used. Studies made in this regard show that a reasonable tracking plan will provide a more-than-adequate gravity analysis.

An arc of only two orbits, for example, is sufficient to yield appreciable information. Simultaneous two-station coverage yields improvement in the precision of the determination of the orbit by a factor of 40. Ranging data, however, improve the results by a factor of only two. Tables II and III give some of the details, showing the precision with which the state variables and some of the harmonic coefficients can be estimated.* The table was generated by a numerical simulation using as complete a model as necessary to represent pertinent effects. Thus the results should be quite realistic.

TABLE II

One-Sigma Statistics for Various Tracking Patterns

Case	A	B	E	G	EE
Coverage on two orbits	Two-way Doppler	Two-way plus ranging	Goldstone 5 hr; Woomera $2\frac{1}{2}$ hr	E with ranging	E with two orbits, day later
x (km)	2.37	1.37	0.069	0.036	0.011
$\dot{x}$ (m/s)	0.760	0.448	0.013	0.0076	0.003
$C_{20} \times 10^5$	2.47	2.18	1.25	0.702	0.01
$C_{30} \times 10^5$	8.96	7.93	4.20	2.25	0.23
$C_{21} \times 10^5$	5.71	5.24	2.16	1.31	0.11
$C_{22} \times 10^5$	0.251	0.155	0.180	0.089	0.036
$C_{32} \times 10^5$	2.33	2.11	1.03	0.543	0.064
$C_{33} \times 10^5$	0.030	0.015	0.023	0.011	0.009
$C_{42} \times 10^5$	0.286	0.210	0.167	0.079	0.010
$C_{44} \times 10^5$	0.0029	0.0024	0.0022	0.0017	0.0011
$S_{33} \times 10^5$	0.391	0.358	0.167	0.089	0.0155
$S_{42} \times 10^5$	0.813	0.725	0.323	0.177	0.027
$S_{44} \times 10^5$	0.0064	0.0059	0.0028	0.0020	0.0009

TABLE III

Statistics for the Kepler Elements (Angles referenced to the Earth Equator)

Case	a (m)	e ($\times 10^6$)	i, deg	Ω, deg	$(\omega + \nu)$, deg	ω, deg
A	34	6.2	0.30	0.42	0.36	0.36
B	29	5.8	0.18	0.25	0.21	0.21
E	14	2.5	0.0074	0.0074	0.0063	0.0063
G	9.5	1.8	0.0042	0.0043	0.0037	0.0038
EE	3.8	0.68	0.0010	0.0017	0.0015	0.0018

* Results obtained by William Sjogren of JPL.

Five cases are tabulated in Tables II and III: A) Doppler alone, B) Doppler plus ranging, E) Doppler with two stations, G) Doppler plus ranging with two stations, and EE) Doppler with two stations for two sets of data one day apart. Except for case EE, all the data are taken over two consecutive orbits having approximately a 7-hr span.

The data were used to fit both the Cartesian coordinates at injection, and eleven of the harmonic coefficients. Only the x- and $\dot{x}$-coordinates are tabulated, although all six coordinates were adjusted in the computation. The equivalent Kepler element values are shown in Table III.

C. DATA OVER LONG ARCS

One reason the Lunar Orbiter will yield so much gravity information is its proximity to the Moon for a long period of time. To take advantage of this fact, it is necessary to assimilate data that span long arcs (i.e., arcs of several hundred orbits). Figures 2 through 5 show the sensitivity of the orbit to the gravity field over such arcs. Changes in the Kepler elements are plotted against small changes in the first few zonal harmonics. Note that for node (Ω) and argument of pericenter (ω) the changes are appreciable. On the basis of a 0°.01 measurement capability for Ω and ω, it appears that the zonal coefficients should be measurable to one part in the eighth decimal.

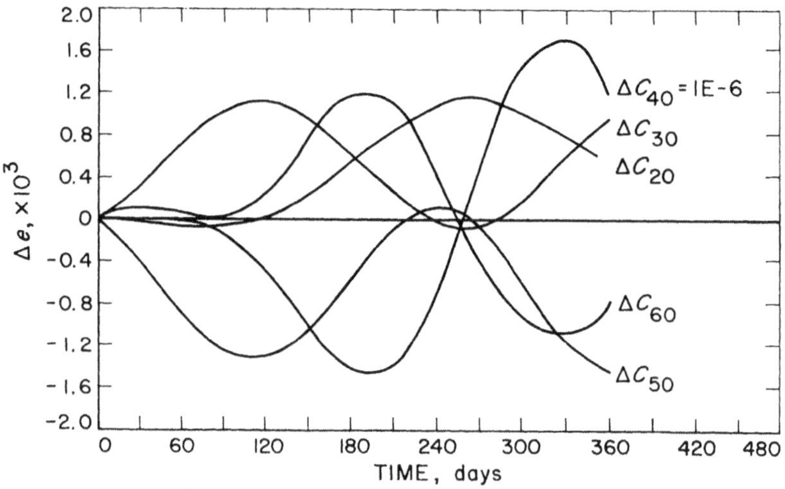

Fig. 2. Sensitivity of eccentricity to zonals.

Pursuing these computations, the sensitivity coefficients were combined with the simulated data to determine the precision of estimating the harmonics with various amounts of data. Tables IV and V give the results. These figures should be considered pessimistic because the data was compressed and represented by one Kepler element only (i.e., Ω). In the handling of the actual data, the data compression for long arcs will be more sophisticated, as described in Section 3 below.

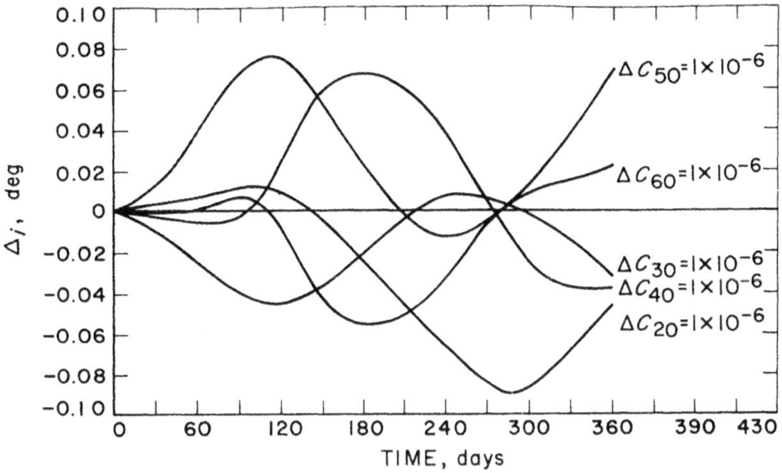

Fig. 3. Sensitivity of inclination to zonals.

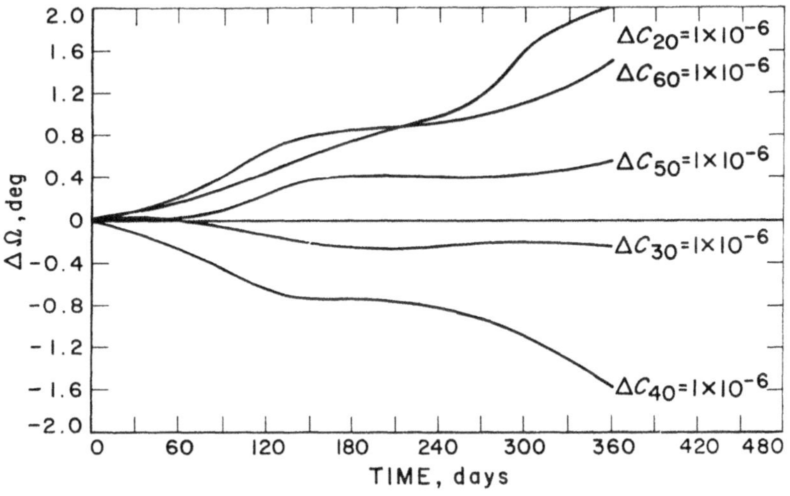

Fig. 4. Sensitivity of longitude of the node to zonals.

TABLE IV

Accuracy of Zonal Harmonics; Values of $10^6\,\sigma$, 22 deg orbit (Ω only data; $\sigma_\Omega = 0.01$ deg)

Harmonic	6 months	8 months	10 months	12 months
C_{20}	0.20	0.13	0.028	0.020
C_{30}	0.76	0.65	0.54	0.40
C_{40}	2.9	2.1	0.75	0.11
C_{50}	2.5	1.7	0.48	0.25
C_{60}	4.0	2.9	1.0	0.29

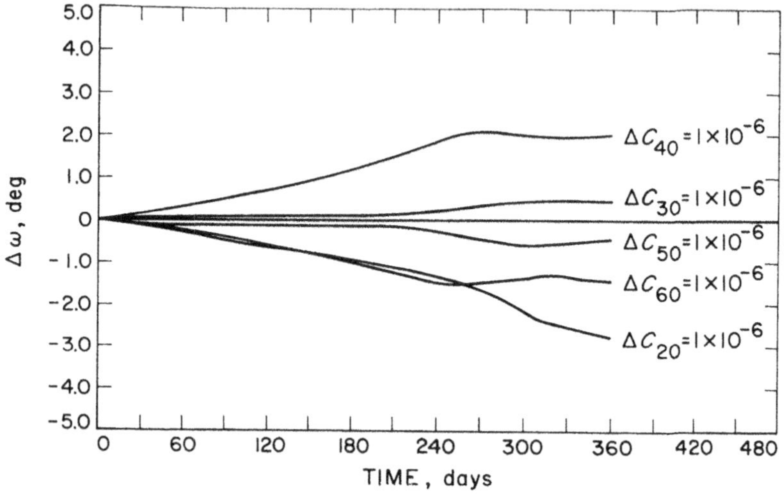

Fig. 5. Sensitivity argument of pericenter to zonals.

TABLE V

Accuracy of Tesseral Harmonics; Values of $10^6\,\sigma$, 15 deg orbit (Ω only data; $\sigma_\Omega = 0.01$ deg)

Harmonic	30 days	60 days
C_{21}	5.0	0.23
S_{21}	82	0.27
C_{22}	14	0.40
S_{22}	20	0.50
C_{31}	300	0.68
S_{31}	500	1.0
C_{32}	20	0.080
S_{32}	19	0.091
C_{33}	2.0	0.23
S_{33}	100	0.22

Table IV shows results using 12 mo. worth of data taken at 1-mo. intervals. Only the zonal harmonics were adjusted. In Table V, 60 days' worth of data at 2-day intervals was used, and all harmonics through the third degree were adjusted.

These computations are complete model simulations made without the help of a complete computer program, and so had to be limited in scope. Thus they should be considered only as indicative of what can be done, but, generally, the results must be on the conservative side. The amount of data in the real case will be much greater than that used here, and with the data points more closely spaced. Furthermore, all the Kepler elements will be used, and will be weighted statistically according to their covariance matrix.

3. The Computation Procedure

The tracking data for Lunar Orbiter are expected to cover at least 1 year's time in orbit, or about 2600 orbits (assuming approximately seven orbits a day). To tie this much data together is a major effort, even if the trajectory computation alone is considered. Excessive computer time prohibits direct computation.

Therefore, a two-stage data reduction procedure has been established. In the first stage, all the data over each short arc of a day or so is reduced and stored in the form of averaged Kepler elements and estimates of some of the harmonics. In the second stage, the totality of Kepler element sets from all the short arcs are used to fit a lifetime trajectory. This trajectory is computed without the short period effects, thus allowing a very rapid computation.

In each stage, the orbit-fitting procedure is computed by differential correction, using least squares on the residuals. The associated computer program is referred to as the SPS (Selenodesy Processing System).* Details of the information flow for the computation are shown in Figure 6.

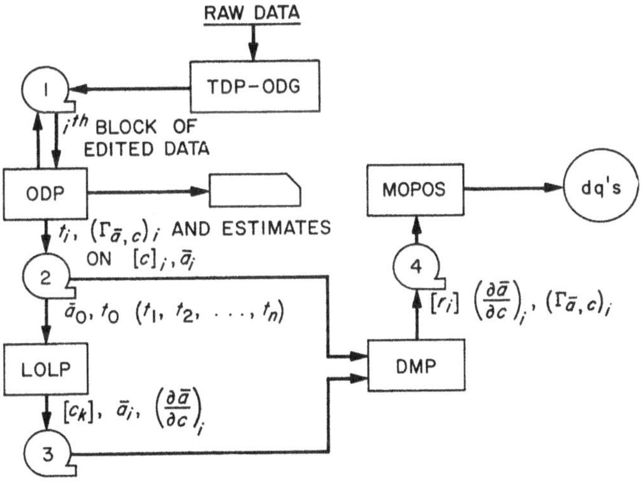

Fig. 6. Selenodesy processing system.

Raw data are first fed into the editing program's TDP (Tracking Data Processor) and ODG (Orbit Data Generator) where they are sorted, where bad points are discarded, and where the data are prepared for acceptance in the ODP (Orbit Determination Program). Here, by differential correction, a best orbit and, concurrently, best estimates are obtained on various constants, including gravity harmonics. A covariance matrix associated with these estimates is also produced.

* Dale Boggs of JPL has designed the computer programs for the SPS.

ANDERSON, NULL, and THRONTON (1964) provide a more detailed description of the orbit determination procedure and associated statistics used in the ODP. The philosophy of this procedure is to fit essentially all of the data by minimizing the sum of the squares of the residuals of the raw data. Six orbital elements plus a number of parameters (up to 11 dynamic constants in the present program) are adjusted simultaneously.

If A is the matrix of partial derivatives, W the weighting matrix on the data, and δz the column matrix of data residuals, the corrections to the parameters δx and the corresponding covariance matrix Γ_x are given by

$$\delta x = (J + \tilde{\mathbf{\Gamma}}_x^{-1})^{-1} (A^{\mathrm{T}} W \delta z + \tilde{\mathbf{\Gamma}}_x^{-1} \delta \tilde{x})$$

$$\Gamma_x = (J + \tilde{\mathbf{\Gamma}}_x^{-1})^{-1}, \quad J = A^{\mathrm{T}} W A$$

where superscript T denotes transpose, and the bold type denotes a priori values. W is limited to being a diagonal matrix, implying uncorrelated data. However, the corresponding weights for the Kepler elements used as data in the next stage of the computation are not so restricted.

Since the lifetime program operates on averaged Kepler elements, the ODP is designed to compute values of the averaged elements at epoch together with the appropriate covariance matrix. Data from several blocks of data, each block a few orbits long, are processed in this manner. The resulting information is accumulated and stored on tape, represented by "2" on Figure 6.

The next step is generation of a lifetime orbit using LOLP (Lunar Orbit Lifetime Program). Such an orbit is computed using the averaged elements from the first block of data.

The basis for this computation is the Kryloff-Bogoliubov averaging procedure described in BOGOLIUBOV and MITROPOLSKY (1961). It is applied here to the variational equations for the satellite orbit written in terms of the Kepler elements a, e, i, Ω, ω, and χ. In practice, because χ is not available as an observable quantity, and is not coupled with the equations for the other variables, it is not necessary to include the χ equation.

To perform the integration, the appropriate averaged gravity potential and its derivatives are necessary. For the satellite problem, some forms of this potential have been derived and published (LORELL and ANDERSON, 1964). A general form of the potential including short period terms has been published by Kaula (KAULA, 1963). However, for the present application, an independent development of the averaged potential seems to be the most useful. It can be written (omitting the Earth and Sun terms which are given in LORELL and ANDERSON, 1965) as follows:

$$\Phi = -\frac{\mu}{R}\left[\left(\frac{R}{r}\right) + \sum_{n=1}^{\infty} \sum_{m=0}^{n} (C_{nm} \bar{U}_{nm}^{C} + S_{nm} \bar{U}_{nm}^{S})\right]$$

where

$$\bar{U}_{nm} = \bar{U}_{nm}^{C} + j\bar{U}_{nm}^{S} = \left(\frac{1}{2\pi}\right) \int_{0}^{2\pi} \left(\frac{R}{r}\right)^{n+1} P_n^m(\sin\phi) \exp\left[jm(\lambda - \theta)\right] \mathrm{d}v$$

in which v is the mean anomaly, and θ is the angle of rotation of the Moon. This integral has been evaluated in a finite series that allows reasonable machine computation.

Averaged Kepler elements and partial derivatives are thus generated for the time points associated with each data block. The resulting information is fed into the DMP (Data Manipulation Program). Here the data is sorted, residuals are formed, and an input tape for MOPOS (Matrix Operator Program) is made. MOPOS performs the statistical operations required to produce corrections and covariance matrices associated with these corrections.

No provision is made for automatic iteration. In general, however, the corrections should be small enough to be in the linear range, so that iteration would be unnecessary.

4. Conclusions

Tracking data for the forthcoming Lunar Orbiter should yield detailed information on the Moon's gravity field, and include estimates of the harmonic coefficients up to the eighth decimal digit. In principle, there is no problem in effecting reduction of the data. There are, however, a myriad of practical problems that arise from (1) the high precision of the data, (2) the enormous amount of data, and (3) the long arc to be covered. The computer program described herein has been designed to handle these problems, and is expected to process the data quickly and efficiently.

References

Anderson, J. D., Null, G. W., and Thronton, C. T.: 1964, *The Evaluation of Certain Astronomical Constants from the Radio Tracking of Mariner II*, AIAA Series on Progress in Astronautics and Aeronautics.

Bogoliubov, N. N. and Mitropolsky, Y. A.: 1961, *Asymptotic Methods in the Theory of Non-Linear Oscillations*, Chapter IV. Hindustan Publishing Corporation, Delhi-6, India.

Kaula, W. M.: 1963, 'Satellite Orbit Analysis for Geodetic Purposes', in *Dynamics of Satellites* (ed. by Maurice Roy). Academic Press, New York.

Lorell, J. and Anderson, J. D.: 1964, 'Precession Rates for an Artificial Satellite', in *Proceedings of the XIII International Astronautical Congress – Varna 1962*. Springer-Verlag, Vienna and New York, pp. 451–461.

Lorell, J., Anderson, J. D., and Sjogren, W. L.: 1965, *Characteristics and Format of the Tracking Data to be Obtained by the NASA Deep Space Instrumentation Facility for Lunar Orbiter*. Technical Memorandum 33–230. Jet Propulsion Laboratory, Pasadena, Calif., June 15, 1965.

Michael, William H., Jr. and Tolson, Robert H.: 1965, 'The Lunar Orbiter Project Selenodesy Experiment'. Paper presented at the Second International Symposium on The Use of Artificial Satellites for Geodesy, Athens, Greece, April 27–May 1, 1965.

A DIFFERENTIAL ORBIT IMPROVEMENT PROGRAM FOR LUNAR ORBITER

GEORGE VEIS*

Abstract. The Smithsonian Astrophysical Observatory plans to expand the Differential Orbit Improvement program to determine lunar orbits as well. The orbits will be determined in a moon-fixed selenocentric reference system. Observations of directions, ranges, and ranger-rate (Doppler) from the Earth's surface as well as from the Moon's surface will be used. It is planned that observations of lunar satellites from a moon-fixed station will also be used. An ephemeris for the Moon will have to be given, but additional correction to those ephemerides will be determined the same way as correction to the observing station's coordinates are determined with earth satellites. The use of the phenomenological approach that has been so successful for earth satellites will have many advantages in the determination of selenocentric orbits, at least for the beginning of the lunar satellite era.

The Smithsonian Astrophysical Observatory has developed a computer program, based on a phenomenological approach, to determine orbits of artificial earth satellites from optical tracking data. The first version of this program – known as Differential Orbit Improvement program (DOI) – was prepared in the fall of 1958. Since that time the program has undergone many improvements, but the basic principle has not been changed. Detailed description of the program can be found in the literature (VEIS, 1960; VEIS and MOORE, 1960; GAPOSCHKIN, 1964).

The program assumes that the orbit of a satellite is a Keplerian one defined by the Keplerian elements which, however, are functions of time. These functions are specified with secular (polynomials in time), periodic, and hyperbolic terms. Any element b can be expressed in the form:

$$b = b_0 + \sum P_n (t - t_0)^n + \sum a_i \sin [b_i + c_i (t - t_0)] + \sum h_j e^{g_j [ln(k_j - t)]},$$

where t_0 is the epoch.

Accordingly, the orbit is defined with a set of parameters P_n, a_i, b_i, c_i, h_j, g_j, k_j for each element.

These parameters depend on the perturbing forces (zonal and tesseral harmonics of the Earth's gravity field, drag, etc.), and of course some of them are correlated. However, the program computes the values of the parameters that define each element by a least-squares fitting to the observations that ignores the physical meaning of each parameter.

An initial approximate value for each parameter is given, and the program computes a correction to each parameter by minimizing the sum of the squares of the

* Smithsonian Astrophysical Observatory, Cambridge, Mass., and National Technical University, Athens, Greece.

residuals to the observations. Analytical expressions give the derivatives of each observed quantity with respect to each parameter that describes the elements. The program can handle all forms of geometric observations, i.e., directions expressed in any system, and ranges. Doppler observations (range-rate) are included in the program. These derivatives constitute the elements of the matrix of the equations of conditions, which after normalization are solved to obtain the corrections to the approximate values. Since the observed quantities are not linear functions of the orbital elements, and the system is not linear, an iterative procedure is needed. The corrections are added to the approximate values and a new least-squares solution is obtained. The iteration is assumed to converge if the change of the standard deviation on two successive iterations differs by less than 1%.

The exact form and the number of terms for each element that one should use to describe the orbit depend on the satellite and the perturbations affecting its motion, and on the period the orbit is to cover. A rough orbit, however, can be determined using nine parameters, namely a constant eccentricity and inclination, a linear expression in time for the right ascension of the node and the argument of perigee, and a quadratic for the mean anomaly.

If the observations were perfectly distributed, the orbit to be determined would be a mean orbit. Since, however, the distribution of optical observations is far from perfect, modifications had to be made in the program to account for all short-period effects. The first step was to apply a correction for the short-period perturbations arising from the J_2 term. Furthermore, it was felt that the luni-solar perturbations should be included analytically in the program.

We can say then that the program in its present form can apply as corrections to the observations all perturbations which can be treated analytically and for the magnitudes of which we have a good knowledge. Actually, both zonal and tesseral harmonics can be treated analytically in the program.

This program is used not only very successfully for the determination of satellite orbits for prediction purposes and all routine tracking operations of the Observatory, but also as the core for orbit-analysis research, on the determination of the geopotential and the geocentric coordinates of the tracking stations, and on the study of the upper atmosphere.

The Smithsonian Astrophysical Observatory plans to expand the Differential Orbit Improvement program to determine lunar orbits. The orbits will be determined in a moon-fixed selenocentric reference system. Observations of directions, ranges, and range-rate (Doppler) from the Earth's surface as well as from the Moon's surface will be used. It is planned that observations of lunar satellites with respect to the Moon will also be used. An ephemeris for the Moon will have to be given, but additional corrections to those ephemerides will be determined in the same way that corrections to the observing stations are determined with the Earth satellites.

Since the perturbations arising from the Earth (and the Sun) will be very important and will distort considerably the orbit, an analytical treatment for those will be applied.

The reason that the DOI program was so successful in the determination of orbits of earth satellites since the early days of the space age was that there was no need either to make too many assumptions about the perturbations of the satellite or to have a complete and precise theory. Instead, a mean orbit was fitted to the observations phenomenologically and the analysis was made at a second stage. In the early days little was known about the gravity field of the Earth (e.g., the odd zonal harmonics) and about how far the upper atmosphere reached.

We expect that we will have to face a similar situation when the first satellites circle the Moon. We thus feel that such a phenomenological approach will have many advantages in the determination of selenocentric orbits, at least for the beginning of the lunar satellite area.

References

Gaposchkin, E. M.: 1964, *Differential Orbit Improvement (DOI-3)*. Smithsonian Astrophys. Obs. Spec. Rep. No. 161. 70 pp.

Veis, G.: 1960, 'Geodetic uses of artificial satellites', *Smithsonian Contributions to Astrophys.* **3**, 95–161.

Veis, G. and Moore, C. H.: 1960, 'Smithsonian Astrophysical Observatory Differential Orbit Improvement Program', in *Astronautical Information, Semiannual Proceedings*. JPL, Pasadena, Calif., pp. 161–184.

DISCUSSION

KOPAL: By this time, we have come to the end of the schedule for this morning and we can proceed to discuss the papers which were presented.

RÖSCH: First, I am sorry that I have to deny our American colleagues the right to claim that they have in Flagstaff the worst seeing in the world and assure them that with little difference, we could get at the Pic some pictures still worse than theirs.

Now, about the expression of sequential photography, I think it is a beautiful word which covers something which was not so well known until the word was discovered; because, after all, sequential photography is a notion which Lyot developed many years ago at Pic-du-Midi by taking a number of exposures of planets, then selecting the best ones among these negatives, and combining them in a composite unique image. We have done the same in studying the solar granulation by exposing thousands of pictures and selecting the best ones in order to produce movies showing the time-variations of position and size of granules. The short time-variations exhibited in these movies* for which the irregular atmospheric refraction is the sole cause look very much like what we have seen in this movie.

Now, some of you may already be aware that I am against this word "seeing" and the discussion here gives me a good example of the confusion which may arise from the use of this word because it covers two effects at the same time.

Suppose we consider the effects of so-called "seeing" on the image of a point source such as a star. There are two different effects, and I think the words blurring and wandering describe them with sufficient accuracy. I do not think that the word "seeing" can be used in the place of blurring, and thus we should separate blurring and wandering because they are indeed different. And now, I turn to the question as to how these effects can be treated over several plates. Let us suppose that the images are sharp, and that there is no blurring. Yet the points may suffer large positional displacement. This wandering over the disk of the Moon in completely uncorrelated, so one should be careful not to assume that by choosing a small number of reference points over the Moon and studying their wandering over a sequence of photographs that he can extrapolate and draw conclusions for all points.

Now, coming to the question of having long exposure photographs, it seems that if

* During this session Mr. R. W. Carder of ACIC presented a motion picture of the Moon taken through the telescope of the Flagstaff Observatory demonstrating dramatically the effects of "seeing"

we can guarantee that the guiding of the telescope is perfect, we shall have exactly the same results – with the difference that instead of having to measure many plates taken with short exposures and determining the average position, we obtain automatically the average by a long exposure. It should, however, be stressed that blurring can in this case be dangerously pronounced due to the finite dimensions of the lunar features which naturally introduce big uncertainties in their position.

MINNAERT: I intended to speak in the same direction as professor Rösch, but I wanted to add a few considerations too on the same point on the matter of long exposures. First, it is clear that the method has great disadvantages if you want to use plates for photometry. Also, it has disadvantages if you look at the resolving power, which you are lowering. Two close-by details may not be separated anymore, and this will be bad for position determinations. If I have a point around which the real distribution of light is symmetrical, then the averaging over a long time will preserve the symmetry, and the point will stay in its position. But, for points with assymetrical light distribution the effect of blurring will be a shift of the point of maximum light intensity and this, I would say, is a serious danger.

KOPAL: May I invite now comments by those of our colleagues who did not speak yet? Actually we should very much like to hear Dr. Weimer's comments on the subjects of this morning.

WEIMER: The first point that I wish to make is that there is a way of comparing plates taken sequentially which could be something like a blink measuring machine taking one plate as reference and measuring small differential shifts between this transparency and the others. And the second point is about the comment by professor Hopmann. I think that the list of reference craters should be reviewed again.

KOPAL: Would professor Hopmann like to comment?

HOPMANN: My dear colleagues, I am much impressed with the work done at Flagstaff, ACIC and elsewhere. We in Austria have now only one man interested in the Moon. Professor Schrutka is interested in the perturbation of some comets and so I have no assistant.

Perhaps I can satisfy the wish of Dr. Weimer to publish my already finished work on absolute heights. The other work on relative heights I will speak on tomorrow. Regarding the comments of Dr. Rösch on "seeing" I wish to say that I come from the School of Chrystal and I have learned to distinguish in our observing book not "seeing" but "Ruhe" and "Schärfe". How is it in English?

VOICE: Steadiness and sharpness.

HOPMANN: Yes, these are the right terms.

The last point I wish to make refers to the format of publication of the results. One should give the I.A.U. number of the crater, the names, xi and eta, zeta and lambda, beta and rho, and give finally the heights only to a tenth of a kilometer.

DOWNIE: I would like to go back again a little bit on this blur and wander. If we look at the basic phenomena causing these things I suppose we can resolve it into just a time-scale differential. Blur may be just like driving over a clobbestone road where you have high-speed turbulence whereas the wander is of a lower frequency.

RÖSCH: No, I am sorry, it is a different thing because blurring is due to the fact that the wave-front arising over your objective is not flat, and the pure wandering means that the wave-front is flat but it is moving. The blurring is due to the integration of wandering. It is a different phenomenon from the instantaneous blurring.

DOWNIE: It seems to me that one of the advantages or one of the possibilities of sequential photography is the statistical approach employed in taking a large number of points. You described the distance between two points as a stretched rubber pad, but then you say that we pick out the best pictures and limit ourselves to three or four or five, but have not we lost, in this way, the statistical approach of our technique? I mean, you can say that these five plates are the best, but somebody else can disagree. It may happen that the center of the image is correct on a slightly blurred plate, whereas on another plate the same image can appear very sharp, yet very far from its true position.

GOUDAS: First of all, the oscillatory character of the "seeing" suggests that you should expect the sharpest photographs when the formation is at its zero velocity positions, which are obtained when the formation is furthest from its mean position. Whereas, when the formation is at its near position, it moves faster and thus produces a blurred picture. It seems to me that the term "seeing" is representative of the phenomenon we discuss here, and I wish to ask professor Rösch if he has ever measured the blurring effect. We know exactly how to measure the "seeing" effects and we have obtained realistic estimates of their actual magnitude. How do you measure and what is the size of the blurring effects?

RÖSCH: Well, again I should say that there are two types of blurring; there is the blurring within a very short time and the blurring by integration of motion along a certain line. About the blurring in a very short interval of time I should say that there are some observers who have obtained exposures of one star in two milliseconds and obtained quite blurred images, changes of which we have measured.

GOUDAS: The question is, how much does it amount to it? We have for the wandering forty or even eighty microns in certain cases. What is the size of the blurring? I believe it is very small for sufficiently short exposures.

RÖSCH: It can be enormous.

KOPAL: I am sure that professor Rösch is a mine of information, but I suggest, Constantine, that you ask professor Rösch in private where we can gain all we want, and more.

GOUDAS: In regard to what Colonel Downie has said about the selection of suitable plates for measurements out of a sequence of many plates corresponding to the same libration, I wish to say that the method of "inspection" is unacceptable. It is not rigorous and I think that someone should develop a criterion of selection that will go deeper than the visual merit of the plate.

KOPAL: Would Mr. Carder like to comment on the preceding discussion?

CARDER: I hope that we at ACIC did not convey the impression that we really lay any claims on originating the concept of sequential photography. The point that we really want to make is that this is the first time that sequential photography has been made available. I think that ACIC was fortunate to have and make extensive use of this photography to improve the position measurements.

MEYER: I would like to answer two questions, one by Constantine about the statistical sampling. You must take the best images, because if you do not, you are going to introduce an enlargement of the measuring area and you have to have something that you can enter. Regarding the comments of Dr. Rösch on the use of long exposure, I wish to say that it looks as though a good set of sequential photography from Pic-du-Midi and selected long exposures may be compatible, but we have not rigorously checked this. It might be a good project for us when we get back.

KOPAL: Time is advancing and before too long we should adjourn for lunch. But before doing so, I should like, if I may, to take a minute or so to summarize some of the salient points which transpired from our discussion this morning.

I think there is no room for doubt in anybody's mind that the task of determining the shape of the Moon, when departures from a mean sphere amount to at best a few kilometers, is a task which taxes to capacity the best methods which photographic astrometry has developed so far. And, secondly, the task can be approached meaningfully only by a group effort. The brave work of early solitary investigators is barely more than of historical significance today; and it is obvious from what has been presented at this meeting this morning that, in order to obtain significant results, one must measure not only a few plates – like say, Franz or Schrutka or Baldwin have done – but a hundred or several hundreds, which necessitates a resort to new methods. We are fortunate enough at this stage – more so than our predecessors – that when these measurements are made, we can reduce them at high speed. The development of computers has relieved us completely of the drudgery of numerical computations which would have been utterly prohibitive to our predecessors. Moreover, the measurement and the reduction procedures are, I think, generally agreed upon and well in hand.

But the main area where agreement is as yet lacking is the technique of selection of the photographs. Therefore, we should take this task all the more seriously at the present time. It is a heavy responsibility. We know how to measure these points. Very interesting techniques to this end have been discussed here; and I am sure I am speaking for all of us when I express my admiration to Bob Carder and his colleagues for that movie which showed very vividly the effects of the – not "seeing" – but blurring and wandering. To see this movie would probably have horrified Franz and many of our predecessors. Now we at last face the problem realistically, and are in possession of adequate material; be it the sequential or long-exposure photography. We can reduce our observations with despatch, but we still do not use the same points; and, as a result, the systems published so far differ. It was very gratifying to learn this morning that before too long one Moon will emerge out of the the U.S. Dept. of Defense. I should like to congratulate our colleagues at ACIC and the AMS in advance to what will prove a satisfactory outcome.

However, their Moon will still differ from other people's Moons; and the parallel work that is going on (on a smaller scale) at Manchester or Vienna, still leaves much to be desired. Nevertheless, it seems to me that the task is of such a magnitude that forces should be pooled and an agreement made about a set of fundamental points. Or a decision reached which features should be used as fundamental points, which may then be measured and used by different groups in their own way. To try to arrive at an accurate global form of the Moon is still a task for the next decade. One step towards obtaining more meaningful results would be made if at least the different groups working in the field were agreed on the type of formations which should be measured and, if possible, the location and number of these formations. Smaller groups may not be able to measure 970 points but at least if they measure some of them the material will partly overlap.

The efforts of the individual investigators of the past ten years were neither of sufficient magnitude, nor did they overlap enough to make it possible to intercompare and reduce them to a common system – as astronomers would do with stellar catalogues. I suggest that, if time permits, any thought could be given to the type of formation which, in the experience of all selenographers here present, should be chosen for measurement as fundamental points, and which restrictions should be placed upon their size or topographic featuies. Mr. Davidson brought forth good reasons why it is inadvisable to undertake selenodetic measurements of lunar photographs at which the Moon is more than a few degrees from "full". It is desirable to arive soon at some kind of understanding – if not final conclusions – in this respect. Our lunar photographic material is now rapidly becoming extensive enough to enable us to select what we may consider as the optimal data; and not to have to take – like Franz or Saunder – any good plate which may come to our hands. We can choose now our libration angles at will, choose plates or sequence of films taken at the time of good seeing, and for such phase angles as we may want; the need is merely to spell out the optimum criteria to which all groups working in this field may wish to adhere for the sake of uniformity and ease of intercomparison. If the group of experts here present were to arrive at some specific recommendations of this type, these could prove of considerable assistance to all groups working in this field in the years to come.

. . . . . .

HUNT: I should also like to invite questions on the paper of Mr. Arthur, as well as on his paper presented yesterday, on photographic techniques for the determination of the Moon's rotation.

GOUDAS: The statement made by Mr. Arthur regarding the uncertainties in various control systems compare well with those of professor Hopmann, and I think we all are in substantial agreement on this subject. I should, however, like to stress that there are only two types of control systems: those which are based on sequential photography, and those which are not. These two kinds of control systems just are not comparable. For example, the data by Schrutka are based on a set of photographs where each libration is represented by only one photograph which was exposed under unknown "seeing" condition.

MILLS: I'd like to ask Mr. Arthur how many people measure each plate.

ARTHUR: I do not believe in mixing observers. An observer observes a plate and no one else goes on it. It is very dangerous to mix observers on plates, I think, unless if each measures the complete set.

MILLS: Yes. This is what we do, in fact, here. We have two people who measure the full set of points.

ARTHUR: I do not think there is any objection to a single person measuring a plate as long as he measures it forward and back, because he has a personal equation too.

MILLS: You measure it once in the 0° direction, rotate and measure it again?

ARTHUR: Yes. There is only one observer on a plate.

MILLS: Could I also ask how it comes that you have a plate in the machine for three months for the thousand points. This seems rather long compared with our material.

KOPAL: It has been our experience here that if you leave the plate on the measuring engine for much more than 24 hours, systematic shifts are likely to occur.

ARTHUR: Oh, I shall make that clear. We start with a pair of fiducial marks at the start of a session. We measure them at the end. In our very first session, these will not only shift, but they will shift secularly too; day by day, they will go further, but we get a pair of values for the secular points on the first day. We subsequently tie into those values throughout the measuring.

KOPAL: But this means that the error with which you can reset on the fiducial marks will go through!

GOUDAS: Regarding the relative elevation of the maria and continents I wish to say that it does not seem that all continents have larger absolute elevation from all maria on the lunar surface. For example, the absolute elevation of maria close to the optical center of the disc is larger than that of some continents which are close to the limb. The last slide shown by Mr. Arthur shows clearly the validity of this statement.

ARTHUR: Well, that is just a regression line.

GOUDAS: That is right.

ARTHUR: But is this real or is it generated in the measures?

GOUDAS: This is an essential question, but we have to depend upon what measures are available to date.

ARTHUR: This is the crunch. I say all this is fictitious and is generated in the measures.

KOPAL: How? Can you prove it?

ARTHUR: No, that is one of the baffling things about it.

KOPAL: Alright, fine; but why do you say so then, if you cannot prove it? Is it not a fact that what you say is a mere surmise on your part?

MEYER: Mr. Chairman, I wish to refer to the problem of measuring large craters where I think we have two cases. First, large craters sometimes lack symmetry and thus there is no way of getting a center by boxing or bisecting. Second, in Mr. Davidson's talk this morning about phase I noticed that he had some phase plates with big craters having massive walls. I think then in these cases we can expect strong phase shifts which can be noticed only when you compare plates of different phase.

HUNT: These last three papers on the lunar orbiter and selenodetic satellites are now open for discussion, and then perhaps Dr. Lorell will make a summary.

KOPAL: I should like to ask a question for Dr. Lorell or anybody who would care to answer. Is there any likelihood that the expansion of the potential with which you start is likely to converge?

LORELL: Well, I think you have to look at the use of that potential in the following way; it is a mathematical device to fit some data. You are only going to use a finite number of terms so you can consider these either as an asymptotic series or you take a finite number of terms and fit a certain number of data, just as if you had some data and fit to it a polynomial. What sort of polynomial would you fit?

KOPAL: In other words, you are treating this formula as a kind of Lagrangian interpolation where the addition of a new term may change all the proceeding ones.

LORELL: Yes.

KOPAL: But, if so, then these terms, or such empirical values as you may devise, will be wholly dependent on the extent of your series and weaken the physical meaning.

LORELL: Unfortunately, that is the case, and I think this has affected the results to some degree; and I think Dr. Goudas and Dr. Colombo will verify this. Also, I think that this has affected the results in the case of the Earth and exactly that sort of thing has been observed. Is that right?

COLOMBO: Yes.

GOUDAS: First of all, we cannot say whether it converges or not. What matters here is whether this expression is suitable. And the answer is yes, because if you have in the denominator the distance in growing powers which implies that the effect of the higher harmonics becomes smaller, we shall have a sufficiently accurate presentation of what is going on. The series may or may not converge, but we need not be concerned with it as far as the present practical problem goes. Regarding your second point about the change of previous harmonics upon addition of more harmonics, the answer is no, because the functions are orthogonal.

LORELL: Yes.

KOPAL: This is so only if you take the same number of terms.

LORELL: If you have an orthogonal expansion and you want to fit it to an infinite amount of data points, it is true that even when you truncate it, you will still not change the values of the previous terms. However, in reality you only have limited data points and hence you will not obtain the same result, i.e., fitting a few terms only will get different results than if you fit more.

KOPAL: I should like to add something that has been disturbing me for a long time, first with the Earth, and I think we will meet the same situation with the Moon. Why do we astronomers, always when it comes to tackling a potential, reach automatically for expansion in terms of spherical harmonics? I suppose we have been doing so automatically since the days of Laplace, because one can justify rigorously the procedure if the body in question is one in hydrostatic equilibrium. In the case of the Earth and the Moon, we deal with bodies that are not necessarily in such an equilibrium, and yet we still reach for the same mathematical machinery. The lesson that we got from using thus far such an expansion for the exterior potential of the Earth should be enough of a warning that we may be using the wrong mathematical tool.

In the future, when dealing with plastic bodies of the consistency of the terrestrial planets, we should perhaps no longer expand our potential in terms of spherical harmonics, but in terms of some other functions which also decrease with increasing inverse powers of the distance. Legendre polynomials of spherical harmonics are not unique in this respect.

GOUDAS: The Legendre polynomials as well as other orthogonal functions give the natural spectrum of the body. When you have a curve in two dimensions then you usually expand it in a Fourier series. This Fourier series expansion is unique; it is the natural spectrum of the curve, and it permits you to find all the existing frequencies in this curve. Now if you have a shape, a surface, in a three-dimensional space, you then find the natural frequency by expanding it into a series of spherical harmonics, thus obtaining the natural spectrum of this surface.

KOPAL: I think you can illustrate the case by taking the drastic example where the Moon is not a sphere or of a form close to it, but a cube. If you then try to explain the perturbations of a mass-point revolving around a cube by expanding its potential in spherical harmonics, this is mathematically possible on account of the absolute and uniform convergence of the potential. But I do not really know how many terms one should include before getting close to the true motion of the mass point.

MINNAERT: Mr. President, I would like to comment on another aspect of today's papers. I was very much impressed by the possibilities offered by the lunar orbiters and, as was left to be understood, the main aim is to determine the potential of the Moon. I wonder if it is possible to modify the method in such a way that the same orbiter will allow us to determine the shape of the Moon. Would it not be possible to add to the satellite an automatic range finder, signalling at each moment the distance between the point of the Moon vertically below it?

VOICE: This is being done.

MINNAERT: Oh, this will be done. And since the orbit is exactly known in space, the distance to the Moon would be known; also, we would have profiles of the Moon in all different angles.

LORELL: There is no instrumentation on the lunar orbiter satellite to measure the range to the surface of the Moon. However, there are cameras on board the lunar orbiter, and once the orbit has been established, it will, of course, be established inertially in an orientated coordinate system, then we can use the stereoscopic methods with a much wider base-to-height ratio, i.e., on larger librations to reduce the photography and find the positions of points on the surface of the Moon with respect to the spacecraft position in orbit.

KOPAL: What is the precision of this method in absolute heights?

LORELL: I do not have an answer for this at the moment.

COLOMBO: The physical libration is closely related to the orbital perturbation of a lunar orbiter, and we need a long range of observations in order to be able to detect the effect of the physical libration on the orbit. I do not think that this effect will be equal to zero.

LORELL: It is very small.

COLOMBO: It is very small, but if you observe its effect over a long time, you may detect it. The important question now is the length of time over which we must observe the orbiter in order to obtain a meaningful estimate of this effect.

LORELL: I'd like to emphasize that the accuracy with which this radar data can determine the orbit is such that even over a comparatively short arc, you can get good estimates of the long period effects. Of course, if you go to a longer arc – three months, six months, a year – then you get much better estimates of these effects which show up in the zonal harmonics. Now, the way that JPL has set up the computation is a two-stage procedure; the first stage to cover essentially the short arc calculation, and the second stage to cover the long arc and take care of the long period terms and so it does cover this area that you brought out.

COLOMBO: What about first-order harmonics?

LORELL: Well, the first-order harmonics represent, if you look at it, the displacement of the center of gravity from your origin of coordinates. Now we try to pick the origin of coordinates as at the center of gravity, but there is already some evidence that the center of gravity as given by the Ephemerides does not agree with the center of the figure of the Moon, and I am not sure that the center figure of the Moon is sufficiently well known in this regard.

COLOMBO: I suspect that the positions listed in the Ephemerides may not be the center of the focus of the osculating ellipses for the lunar orbiter.

LORELL: That is right, it might not be and therefore the origin of coordinates may differ from the actual center of the ellipse or the center of gravity by an appreciable amount. This will show up in the first order harmonics, but you have to appreciate that the discrepancy here is going to be time-dependent in a periodic fashion, and so if you are going to make an estimate of this discrepancy you can do so by using a short arc. I think we can make such estimates using arcs of one day.

KOZIEL: I think that all these new methods applied nowadays are very important; but the difficult term as far as I can see is the argument 2ω, in Hansen's notation, which in the forced libration term has a period of three years. And I am not sure whether meaningful results are possible if the observation time is shorter than three years. In addition, the free libration terms have period that is not determined yet. I think that we must try to get these terms from a long series of observations.

HUNT: How many years will the orbiter be observed?

LORELL: We have asked for one year.

KOZIEL: One year is one third of the period of the important term of the forced libration and there also is a big term with a period of one month too. In this way you have a possibility to obtain other terms with all good accuracy.

LORELL: The librations are going to be derived from the moments of inertia, therefore, it does not matter how long you observe. If there are free physical librations, their effect on the orbiter will be small. Once we have determined the moments of inertia, we can study the libration problem.

HOPMANN: Have I understood correctly, that you mean that the center of the visible Moon and the center of gravitation occupy different positions in the sky?

LORELL: There are three different points that we are talking about, not two. One is the center of figure, the second is the center of mass, and the third is a point defined by the Ephemeris. The position that the ephemerides tabulate and the center of figure differ by one kilometer.

HOPMANN: Half a second.

LORELL: But the center of mass – who knows where it is? We assume it is at the ephemeris point.

COLOMBO: The center of the Earth is to date known with an uncertainty of fifty or sixty meters, in spite of the many satellites we have around it.

KOPAL: One question for professor Koziel, please. Have you found in all your reductions, or any one of your predecessors, any component of free libration in latitude?

KOZIEL: You mean free librations in node and inclination?

KOPAL: Yes.

KOZIEL: They have been found to be small, but this does not mean that they do not exist.

KOPAL: But were terms representing such librations put in the equations of conditions and a simultaneous solution made for their coordinates?

KOZIEL: No, we have not made a simultaneous solution. The first estimate that I know of – namely that by Belkovich – is not made from a simultaneous solution.

LORELL: Can I ask professor Koziel one question? Supposing the orbiter comes out with a good estimate of the value of f. Will this make the study of the librations much easier, or how will it affect the work on the librations?

KOZIEL: If we obtain good estimates of the moments of inertia, or only the quantities β and f we shall be in a position to study the physical libration of the Moon with the exception of the free libration, the amplitude of which can only be established by precise observations.

KOPAL: If you feed extraneous f into these solutions eliminating it as an unknown, you will of course increase the weight of these solutions.

LORELL: Now, the clincher question: supposing we did get the moments of inertia very well defined by the orbiter data, and the value of f is appreciably different from the value 0.63 that you found: does this then mean that we are going to your considerations, Dr. Kopal, of a non-rigid Moon?

COLOMBO: The Moon definitely is non-rigid.

KOZIEL: No, excuse me, just a moment . . .

KOPAL: I shall attempt to answer this one, may I? Should it turn out that the value of f is very different from the one professor Koziel found, it would mean that the elastic property of the Moon differed very greatly from those of the Earth. You see, the f is in physical terms a ratio of the second harmonic rotational distortion over the second harmonic of rotational and tidal distortion. If these two distortions were hydrostatic, then at least for negligible mass ratio, you should get the value 1/4 while the actual value is closer to 2/3. The most direct interpretation of this fact, in my opinion, is this: the polar flattening of the Moon, small as it is (about a quarter of a kilometer in size) is essentially hydrostatic, but the tidal elongation is not; the reason probably being that the constant centrifugal force due to axial rotation is acting continuously, while the tidal attraction varies somewhat in magnitude as well as direction because of the eccentricity of the lunar relative orbit (which causes the angular velocity of revolution periodically to advance, or fall behind, the constant velocity of axial rotation). As a result, we may surmise that the tidal distortion of the Moon will not attain its present hydrostatic value, but will be smaller.

That this surmise may be correct is evidenced by a rather similar behaviour of our own Earth. The polar flattening of our globe is essentially hydrostatic (within better than 99% of its observed amount), presumably because the Earth rotates at an (essentially) constant rate over astronomically long periods of time; and, therefore, the centrifugal force never relents to allow any adjustment of the terrestrial ellipticity. On the other hand, the second-harmonic bodily tides – manifestly a dynamical phenomenon – are found to amount to only about one-quarter of their hydrostatic theoretical amplitude; and if the same defect were characteristic of the lunar globe as well, then you are led to anticipate almost exactly the same value of f as found by professor Koziel. I have but little room for doubt that this is indeed its correct explanation; and if, by any chance, the orbiter observations will lead to a drastically different picture, I shall be very much surprised.

KOZIEL: Dr. Eckert obtained recently a value of 0.638 for f – very close to my own value obtained in a completely different way. This makes me feel confident that the value I obtained is correct. I too do not expect any drastically different answer to be obtained by the lunar orbiter.

HUNT: The time has come when we should adjourn. Many thanks to all participants for their interesting contributions to this general discussion.

PART IV

TOPOGRAPHY AND MAPPING OF THE LUNAR SURFACE

PHOTOMETRIC METHODS FOR DETERMINATION OF LUNAR RELIEF

M. G. J. MINNAERT*

Morphological measurements of the relief of the lunar surface have been made by different methods, but mainly by *shadow* observations. This method is especially valuable if series of photographs are available, made at short intervals. In Manchester, Fielder, Rackham, Turner, Jones and others have developed this method and reached a high precision. We shall now consider in how far *photometric* measurements may be used for a similar purpose. This method has its own advantages, disadvantages and limitations, and it seems useful to know which are the possibilities.

The essential idea is this. Let the position of the Earth and of the Sun be given; consider a surface element of the Moon; its brightness will vary, according to its inclination. Thus, from a measurement of the brightness it must be possible to deduce information about the slope.

General Results of Lunar Photometry

Before we discuss the details of the method, let us first see what is known about lunar photometry. In selenocentric coordinates, let E represent the direction towards the Earth, S towards the Sun, P the normal on an element of the lunar surface (Figure 1). The brightness of a surface element P is given by the relation

$$B = B_0 f(i, \varepsilon, \alpha),$$

where i = angle of incidence, ε = angle of observation, and α = phase angle Earth–Moon–Sun.

The coefficient B_0 is the *albedo at full moon.*

The *photometric function* $f(i, \varepsilon, \alpha)$ describes the influence of the geometrical circumstances.

In general we first determine B_0 at full moon for all points in which we are interested. Then, at other phases, we divide B by B_0 and study the photometric function $B/B_0 = f(i, \varepsilon, \alpha)$. All observers agree that this function is almost the same for the whole lunar surface, as well for the mountain regions ("continents") as for the plains ("maria"). There are a few minor exceptions: e.g. the crater rays have a somewhat different function, apparently because the microstructure there is different.

* University Observatory, Utrecht, Holland.

Kopal and Goudas (eds.), Measure of the Moon.

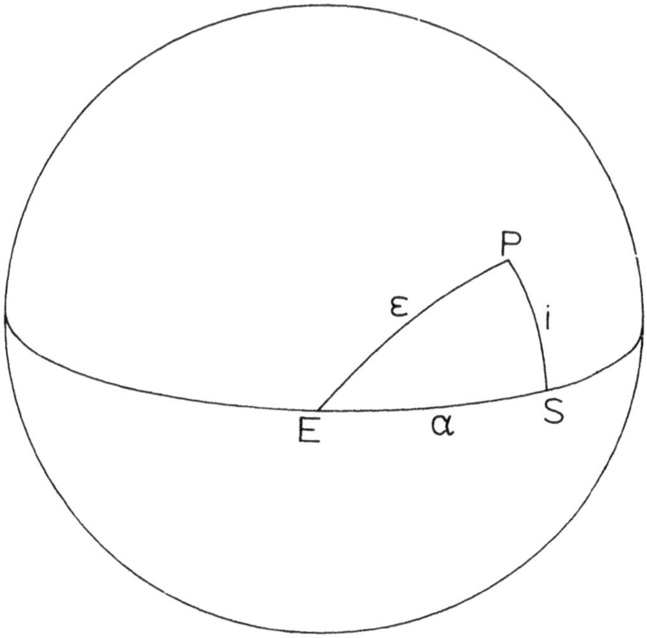

Fig. 1. Geometry of the lunar globe, illuminated by the Sun (direction *S*) and observed by the Earth (direction *E*).

(a) For our purpose the *synchronous photometry* is the most important. This concerns the brightness distribution on one single photograph, made at a definite phase $\alpha = \alpha^*$. Or on a series of photoelectric measurements, made practically at one and the same phase.

(b) *Successive photometry*, interconnecting definite lunar objects, observed at successive phases, is of course more ambitious and also more difficult, since variations of atmospheric transparency or instrumental sensitivity, effects of libration, etc., must be avoided.

Survey measurements have been made already earlier by several observers. The more modern results are obtained by photoelectric methods. There is an important series of Fedoretz, reduced and homogenized first by ORLOVA (1956) and MINNAERT (1961), later by BARABASHOV, EZERSKI, and PRISHLAK (1962), and finally again by BARABASHOV and EZERSKI (1965). There is another series of measurements by AKIMOV, not published in detail, but communicated by the author in the form of curves (1963) and formulae (1965). The extensive measurements of WILDEY and POHN (1965) at Mt. Wilson have not yet been combined into surface diagrams. The results of HERRIMAN *et al.* (1963), which came only recently to my knowledge, confirm very nicely the Russian results, except for phases near full moon, which are uncertain anyway.

One of the most important results is, that the full moon looks almost uniformly bright, a very interesting phenomenon, and only explained by the great porosity of the surface material.

For the other phases, Fessenkoff and I myself have found, that the isophotes on the lunar surface are approximately meridians; this has been confirmed even for phases near full moon by Wildey and Pohn. A considerable simplification thus becomes possible: in order to describe the brightness distribution over the lunar surface, it is sufficient to draw curves, giving B/B_0 along the equator for a number of values of α. For our purpose it seems appropriate to plot $\log B/B_0$ against the longitude λ; the arc ε then is equivalent to the selenographic longitude of the observed surface element and the arc i is $\lambda - \alpha$. If we are mainly interested in synchronous photometry, these logarithmic curves may be arbitrarily shifted in the vertical direction; for each lunar photograph we need only one of them.

Such sets of curves, derived from the series quoted above, are reproduced in Figures 2 and 3. The two series are very similar.

In order to represent these observations analytically, several formulas have been tried, among which the most generally known is that of Lommel–Seeliger:

$$\frac{B}{B_0} = \frac{\cos i}{\cos \varepsilon + \cos i}.$$

If different phases have to be interconnected, several authors simply multiply that

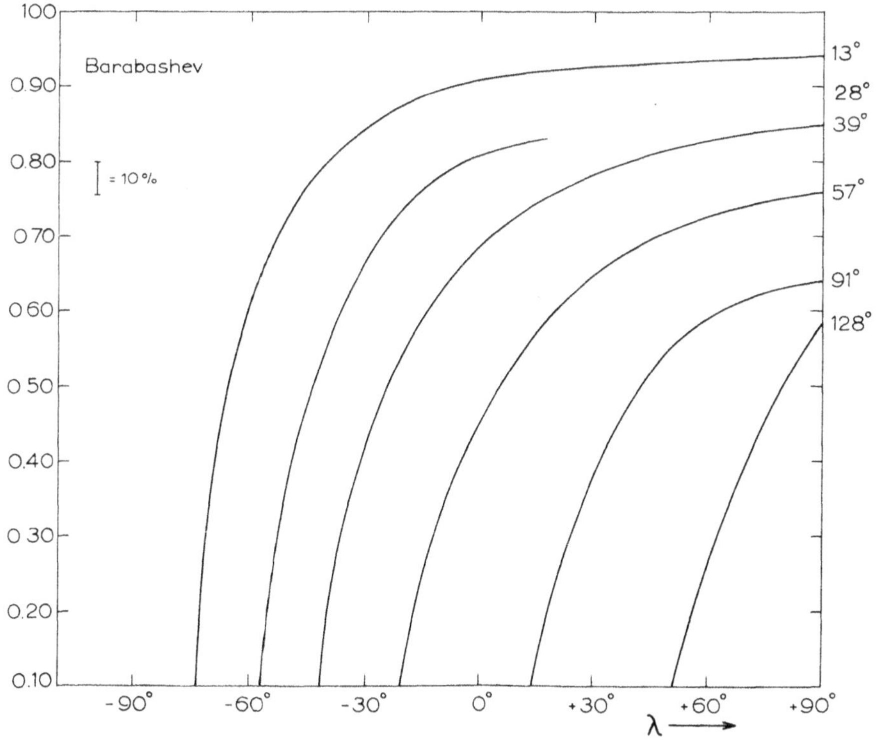

Fig. 2. Mean photometric function of the Moon according to Barabashov. The curves have been arbitrarily shifted in the vertical direction.

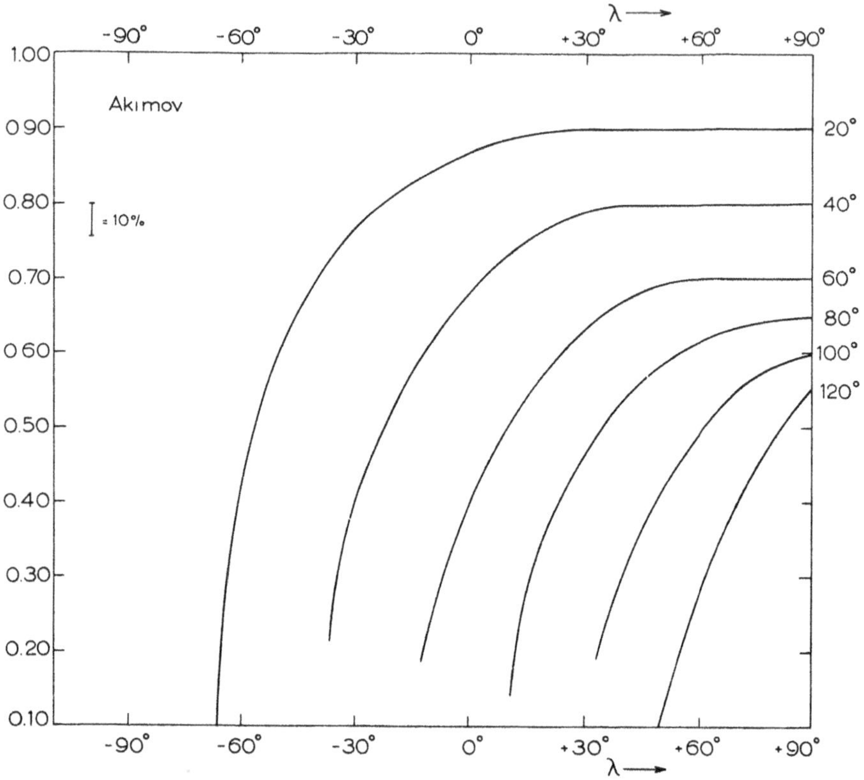

Fig. 3. Mean photometric function of the Moon according to AKIMOV (1963). The curves have been arbitrarily shifted in the vertical direction.

expression by a function of α:

$$\frac{B}{B_0} = \frac{\cos i}{\cos \varepsilon + \cos i} \cdot \phi(\alpha).$$

This is for example the type of formula proposed by HAPKE (1963). His function $\phi(\alpha)$ is rather complicated. These formulae are semi-theoretical.

Let us see in how far Lommel–Seeliger represents the observed photometric determinations. We consider only synchronous photometry and plot the calculated values of log B/B_0 as a function of λ, just as we have done first with the observed values (Figure 4). The differences between this set of curves and the observed ones are conspicuous; there seems no doubt that they are real, since they appear as well in the results of Fedorets as in those of Akimov and of Herriman; but a more detailed comparison shows that after all the discrepancies relate mainly to the region $90° > \lambda > 60°$. On the photographed lunar disc, this corresponds to the section $0.87 < r < 1$ of the equator: the bright limb region, opposed to the terminator. In that region Lommel–Seeliger gives a higher brightness than the observations quoted. As long as we agree to leave out that sickle, we may apply Lommel–Seeliger reasonably well.

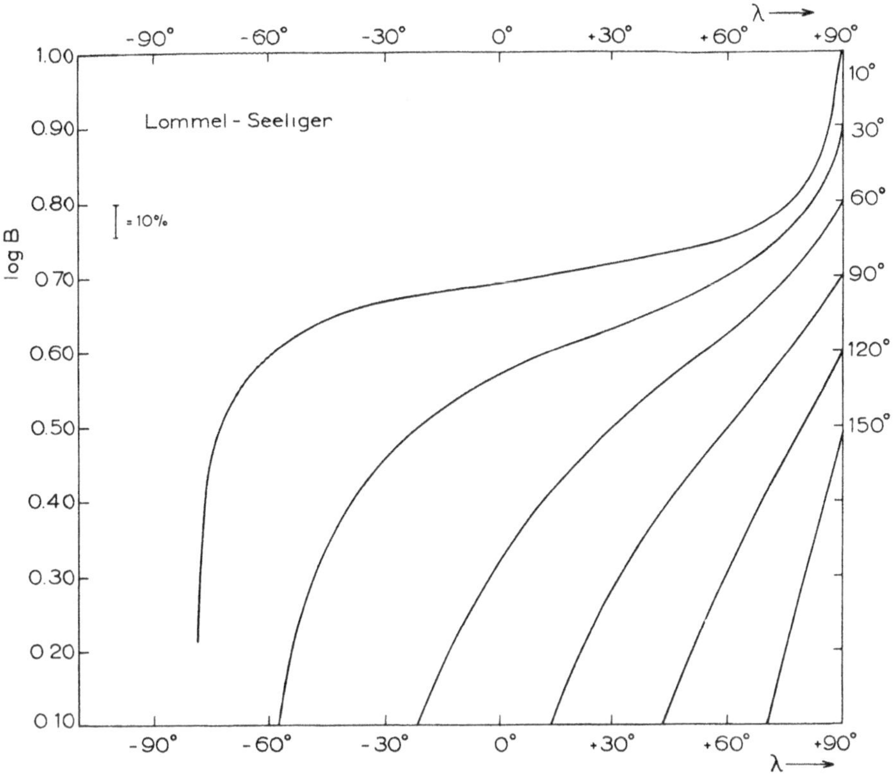

Fig. 4. Mean photometric function of the Moon according to Lommel–Seeliger's law. The curves have been arbitrarily shifted in the vertical direction.

Notice that near the terminator, where $\cos i$ can be neglected compared to $\cos \varepsilon$, the brightness becomes proportional to $\cos i$ (unless when the phase is near full moon).

Effects of Inclination

Let us now investigate the effect which we may expect if a surface element on the Moon is inclined over a certain angle with respect to the local horizon. If the normal is shifted along a meridian, this will have no effect; the surface element will behave as if its latitude were modified. But we have seen that the isophotes are meridians: so a change of latitude will not affect the brightness. Lommel–Seeliger's formula is in agreement with this.

Indeed (Figure 5), if β is the latitude, we have:

$$\cos \varepsilon = \cos \varepsilon_0 \cos \beta ,$$
$$\cos i = \cos i_0 \cos \beta ,$$

where ε_0, i_0 represent the equatorial components of the angles ε and i. By substituting in Lommel–Seeliger's expression, we see that the latitude cancels out.

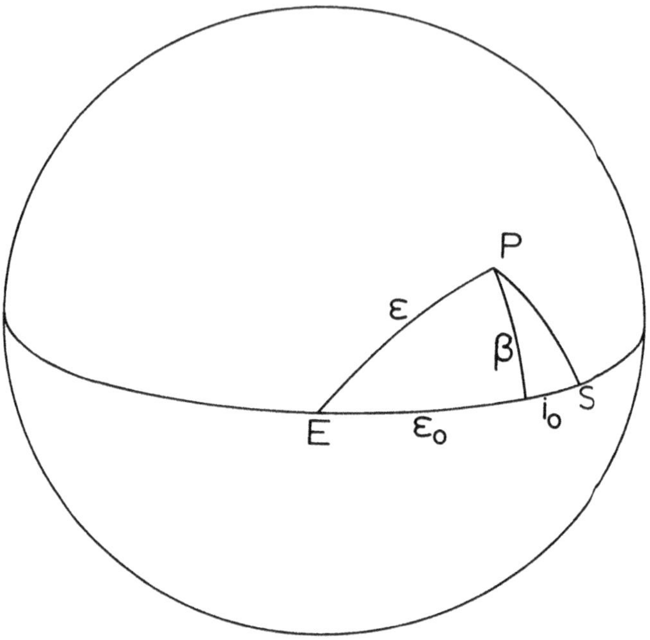

Fig. 5. Equatorial components of the angles ε and i.

Suppose now that the surface element has an inclination along the parallel. The brightness is modified, either increased or decreased, depending on whether the normal inclines towards the sun or away from it. An inclination is photometrically equivalent to a change in the longitude λ. Because we have plotted log B against λ, the slope of our curves gives directly the relation between the relative contrast $\Delta B/B$ and the inclination $\Delta\lambda$. This is precisely what we are interested in.

Take e.g. the curves of Barabashov (Figure 2), a phase angle of 57°, an area at longitude 30°. Let us assume that a contrast of $\Delta B/B = 0.10$ may be determined in a reliable way. That would correspond in the diagram to $\Delta \log B = 0.043$ and would require $\Delta\lambda = 19°$, which will seldom occur. The method will be applicable best in those parts where the slope of the curves exceeds 45°.

If we compute the contrast at one and the same point as a function of the inclination in longitude, we find that it increases almost linearly with the slope. An example is given in Figure 6.

We can also compute how the contrast, due to a local inclination, may be expected to vary in the course of the lunation. For illustration we consider a small area, inclined over 10° in the direction of the equator. In Figure 7 we have drawn as a function of phase the contrast, measured by log B'/B, of such an area with respect to the surroundings; the three curves correspond to areas located, respectively, at longitudes of 0°, 45° and 60°. The contrast is, of course, the strongest when the terminator approaches such an area. It is confirmed that the contrast changes sign when the phase

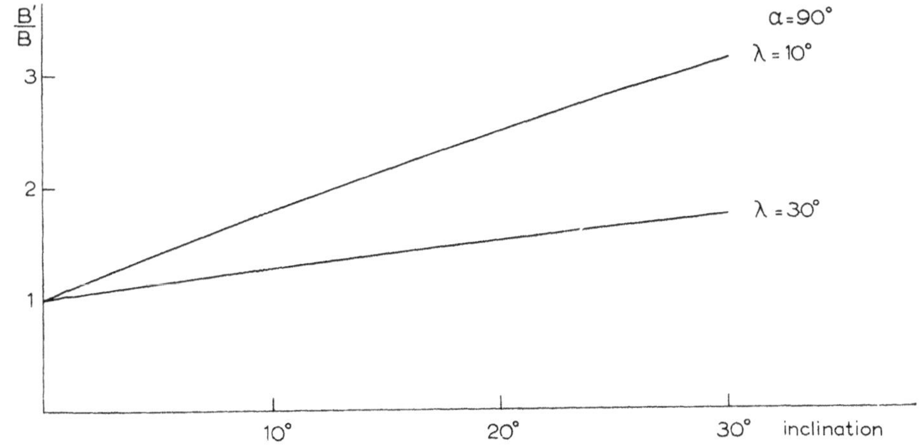

Fig. 6. Contrast corresponding to increasing inclination for the case $\alpha = 90°$, $\lambda =$ either $10°$ or $30°$.

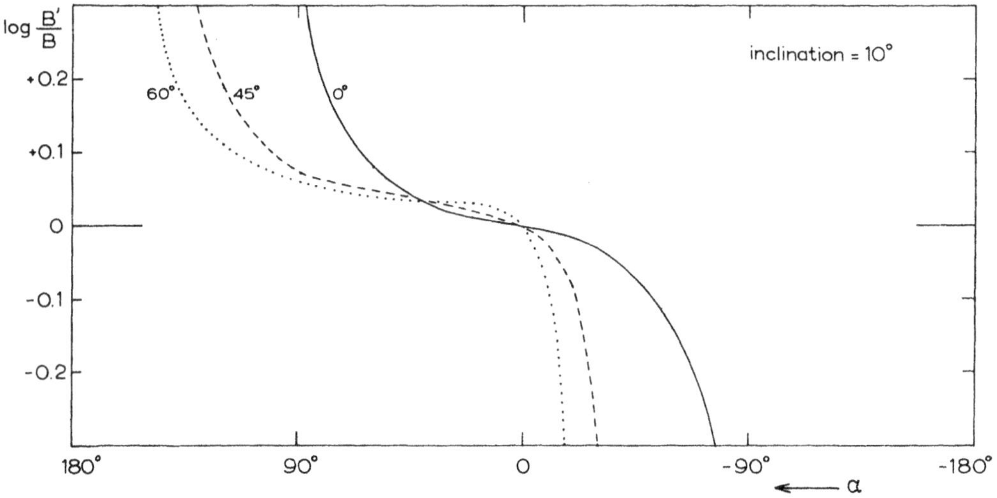

Fig. 7. Contrast as a function of phase between two neighbouring points, inclined over 10° with respect of each other (respectively for longitudes of 0°, 45°, 60°).

is reversed. All curves intersect at $\alpha = 0$. Such curves can help us to select the most favourable phase in order to study a given object.

Once we know the inclination at all points along a parallel circle, we may easily derive the whole profile, by a simple integration. This gives the heights along the parallel as a function of the *projected* distances; if we wish to know the heights as a function of the *true* distances, we must of course introduce in the integral a factor $1/\cos \lambda$.

A photometric study of the lunar surface detail gives a series of such profiles, all parallel to the equator and at successive values of latitude. There seems to be no possibility to interconnect them (Figure 8a). The only means is to assume that somewhere

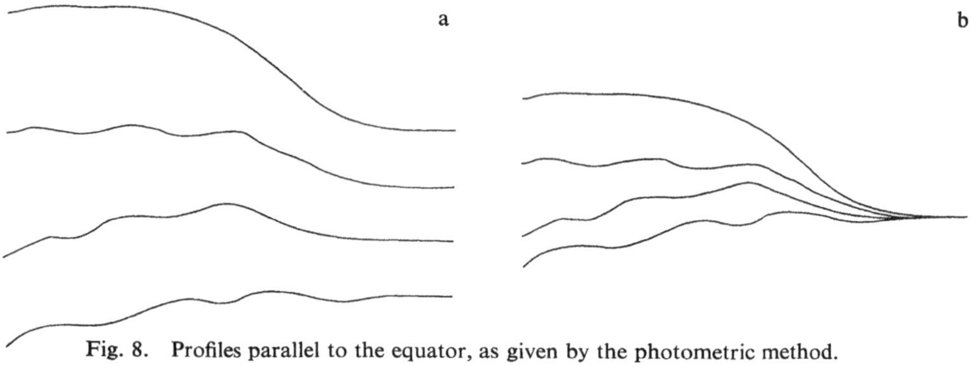

Fig. 8. Profiles parallel to the equator, as given by the photometric method.

the lunar surface is horizontal, and that all profiles can start from that common level (Figure 8b).

The circumstance that photometric measurements can only show the inclinations parallel to the equator, requires our full attention. This seems to be unavoidable, it is directly connected with the circumstance that the projection of the Sun is always on the equator. To show the effect of this preferential illumination, as an example, we compute the isophotes for a lunar crater which is supposed to have axial symmetry and of which the depth is described by $z = e^{-x^2 - y^2}$. The crater is assumed to be small compared to the radius of the Moon. The shape of the isophotes is simply determined by the locus of the points where $\partial z / \partial x$ has a certain value (Figure 9). At one side the

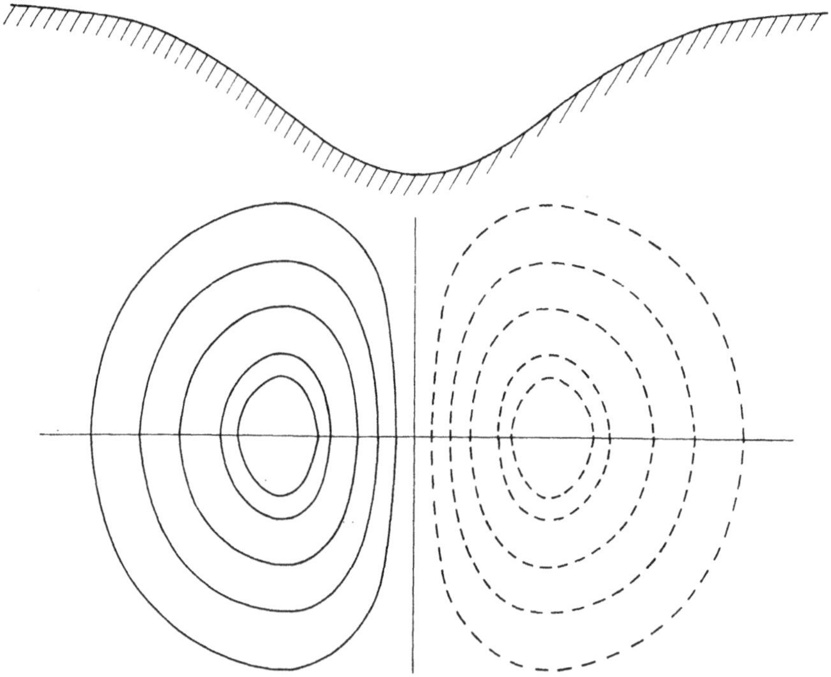

Fig. 9. Isophotes for an axially symmetrical crater.

slope is brighter than the surroundings, at the other side less bright. When the phase changes, the shape of the isophotes does not vary, but of course the contrast will be modified. The position of the crater will influence the shape of the isophotes only in so far, that nearer to the limb they are seen foreshortened by the projection. If the crater would be very big, then the inclination of its walls would combine with the spherical form of the Moon and the curves would be distorted.

This preferential illumination is very deceptive when we study lunar photographs. The low ridges which are seen in many plains of the Moon make the impression as if they were running mainly in the NS direction. Counts of these structures have recently been made by Dr. van Diggelen and corrected for the unilateral illumination. Only then are we able to look for a "tectonic grid"; it is really found to exist. Thus in studies about the lunar grid there is a double complication, which has to be avoided: the projection of the sphere on the plane of the photograph and the effect of unilateral illumination.

It would be most interesting if we could observe the same area, illuminated successively from two directions, in such a way that by the contrast structures appear, first in one direction, then in the other. This is only possible if we observe an area near the poles; for the preferential directions are meridians; and the only place where meridians intersect are the poles. We should therefore observe the poles by means of a satellite in a polar orbit, during the course of a lunation. Observations from such a satellite would, moreover, have the advantage to show the entire terminator region in its true size, not foreshortened; that means the region where our observations are the most precise, and which would also include strips on the back of the Moon. From the Earth we can only observe this terminator region conveniently near the central parts of the disc.

Other Effects influencing the Contrast

The strength of the photometric method is that it allows a determination of the surface relief already from one single photograph. On the other hand we must be aware that there may be disturbing effects, which can be eliminated completely only when we have several photographs, obtained at very different phases.

The most complete analysis of that kind was made by BARABASHOV and his collaborators in important publications (1962 and 1965). After having ascertained the mean photometric function for the Moon as a whole, they attribute all local deviations to the combination of two effects: (1) deviations due to the macrorelief, that means inclinations in longitude; (2) deviations due to a modified microrelief, e.g. to a locally greater or smaller porosity, characterized by a coefficient g after Hapke. They write

$$\Delta B = \frac{\partial B}{\partial \lambda} \Delta\lambda + \frac{\partial B}{\partial g} \Delta g ,$$

and from the theoretical photometric laws they derive $\partial B/\partial\lambda$ and $\partial B/\partial g$. Then, for different phases α, they take the observed deviations ΔB from the mean photometric function and, by least squares, they separate the local inclinations $\Delta\lambda$ and the variations in porosity Δg. By this procedure they find inclinations up to 11° and values of g, varying from 0.08 to 1.8. They are more interested in the variations of porosity than in the inclinations, because they find that the variations in porosity are correlated with thermal anomalies.

If we are primarily interested in the inclinations, there is a great advantage in studying only contrasts (brightness ratios) of neighbouring areas, and the way in which these ratios vary with the phase. For a photometric comparison of absolute brightness between records, taken at an interval of many days, is necessarily very difficult, considering the varying altitude of the Moon, the effects of atmospheric extinction, the changing sensitivity of the receptor, and the differences in libration. Moreover, we know that in broad lines a modified porosity affecting the whole field studied does not influence the brightness ratio between two areas of the field, since Hapke and many others assume $B/B_0 = A \cdot f(i, \varepsilon) \cdot \phi(\alpha)$. Consequently, if *a series of records at different phases* is available, *including full moon*, we can directly eliminate albedo differences, we are rid of a general, abnormal porosity, and our determinations of the relief could be vitiated only by local deviations of porosity. A check is obtained by testing whether the observed contrast behaves as a function of phase according to the curves of Figure 7; if not, local variations of porosity may be eliminated after the method of the Charkov astronomers.

If only *one record plus the full-moon photograph* are available, conclusions can be drawn almost in the same way, but the check disappears and local variations of porosity could not be found.

It has perhaps not been stressed sufficiently that, after all, there is no reason why in such considerations the full moon should play a preferential role! As soon as we have *two records at strongly different phases*, α_1 and α_2, we can always eliminate albedo differences by measuring the ratios B_1/B_2 for two neighbouring points, and by comparing the contrast to curves, analogous to those of Figures 2, 3, or 7.

Finally, if we have only *one photograph* at a certain phase, as is the case for the Ranger series, we shall provisionally assume that there are local differences neither in albedo nor in porosity. We may be confident that *general* deviations in albedo or porosity have no influence on the contrast. If the area studied is near the terminator, it is highly probable that the influence of inclination is the most important effect. Local differences in porosity are not very probable. As to local differences in albedo, from the morphology we may draw some indications. Moreover, the observed inclinations always appear in a preferential direction with respect to the Sun, while albedo differences are not subject to that preference.

A Comparison of the Shadow and Photometric Method

If we had an isolated mountain on an otherwise level area, one single exposure would be sufficient to determine the height by the shadow method. But as soon as we want to know the slope, or as soon as the relief is more complicated, a whole series of exposures at short intervals must be available. Even then, if the area is mountainous and irregular, the profile calculation is a complicated procedure.

On the contrary, for the photometric method one photograph gives already complete results, which become more reliable if one or two other records are available, taken at considerably different phases.

In both cases, the sensitivity is the greatest for areas near the terminator; one might say that the photometric method is practically only applicable up to, say, 40° longitude distance from the terminator.

Both methods yield profiles in planes parallel to the equator. In both cases we have no direct possibility of interconnecting them. In both cases for precise measurements some corrections have to be applied, among others for the finite angular diameter of the Sun, which produces a penumbra and diminishes slightly the contrasts in the marginal fringe near the terminator (see e.g. KOPAL, 1962). The instrumental blur and the scintillation will have a similar consequence. The elimination of such effects will be especially difficult in photographic work, as has been stressed by the work of the Manchester school; the procedure is simpler for photoelectric records.

There will also be effects of libration in longitude, by which one and the same region may be brought nearer to the terminator or further away from it. They should, of course, be eliminated carefully.

It would be most interesting to compare the merits of the shadow method and of the photometric method by applying both to the same objects. Obviously the selection of these objects should be fair.

Example of Application

Probably the first application of the photometric method was made by VAN DIGGELEN (1951). The object studied by him was very suitable: he investigated some low ridges in Mare Imbrium, photographed not far from the terminator. The inclinations are small, moreover one has the advantage that these ridges are surrounded by a flat area, which can be used as the fundamental level from which the successive profiles start. An example, studied by him, may be seen from Figures 10 and 11. As it will be noticed, the heights above the surrounding plain do not exceed 100 m and the slopes are at most 1%.

Later the method has been applied in various, not easily accessible publications, by DALE (1962), WILHELMS (1963), and MCCAULEY (1965). Quite recently the method has been adapted to the evaluation of the Ranger records and digitization was applied by WATSON (1966), RINDFLEISCH (1965), and NATHAN (1966). In this case, of course, the geometry is slightly modified. The direction of the straight line, connecting the camera with the lunar object, corresponds to a point C in selenographic coordinates (Figure 1), and now plays the role which the direction E (Earth–Moon) plays for ground-based photographs. The intensity equator is now defined by the plane CS through the centre of the Moon, and the isophotes are perpendicular to that circle. From geometrical considerations we find the direction of the parallels on the record,

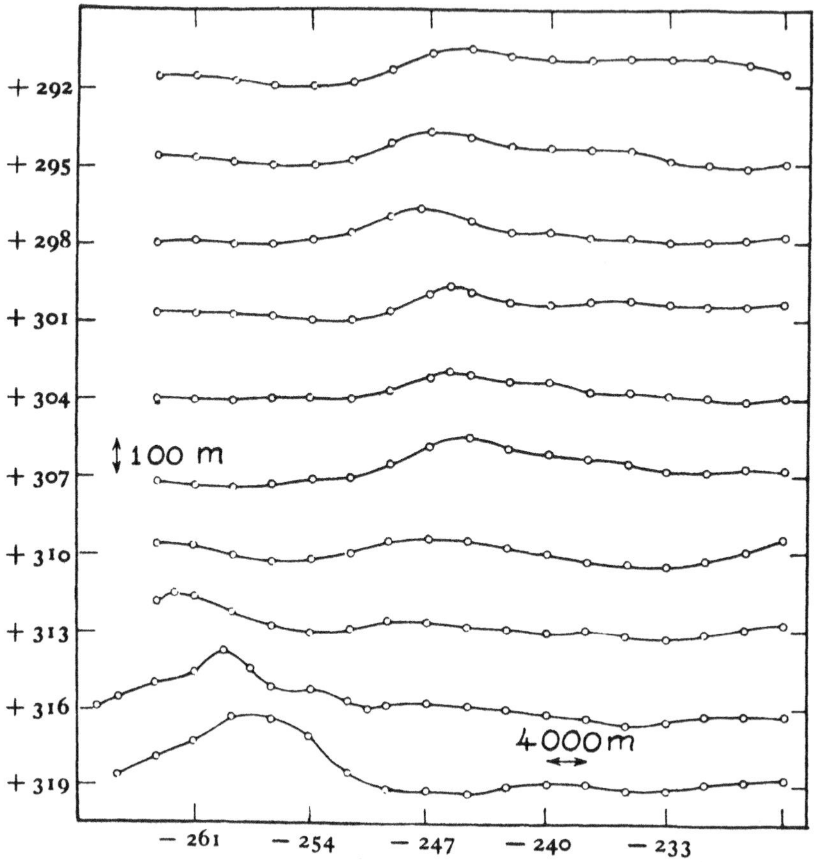

Fig. 10. Profiles parallel to the equator for a ridge in Mare Imbrium, after VAN DIGGELEN (1951). The heights are superelevated by a factor of 40.

along which the integration has to be performed. Some authors started from a mare which can be assumed to be level, others assumed that the average elevations are the same along each integration path.

Topography and Albedo Determination

The relation between topography and photometry is also important in cases where the unknown albedo has to be determined from photographs. Take for example the case of the Alphonsus peak, photographed surprisingly bright by Ranger VII. Kuiper attributes this to a deposit of salt, so as witnessed by himself on the Hawaii volcanoes. It is clearly of importance to determine the local albedo; but some assumptions about the topography will have to be made before an albedo determination is possible, since the inclination of that area may influence the brightness. It would be most interesting to have a record of this area at full moon on a sufficient scale.

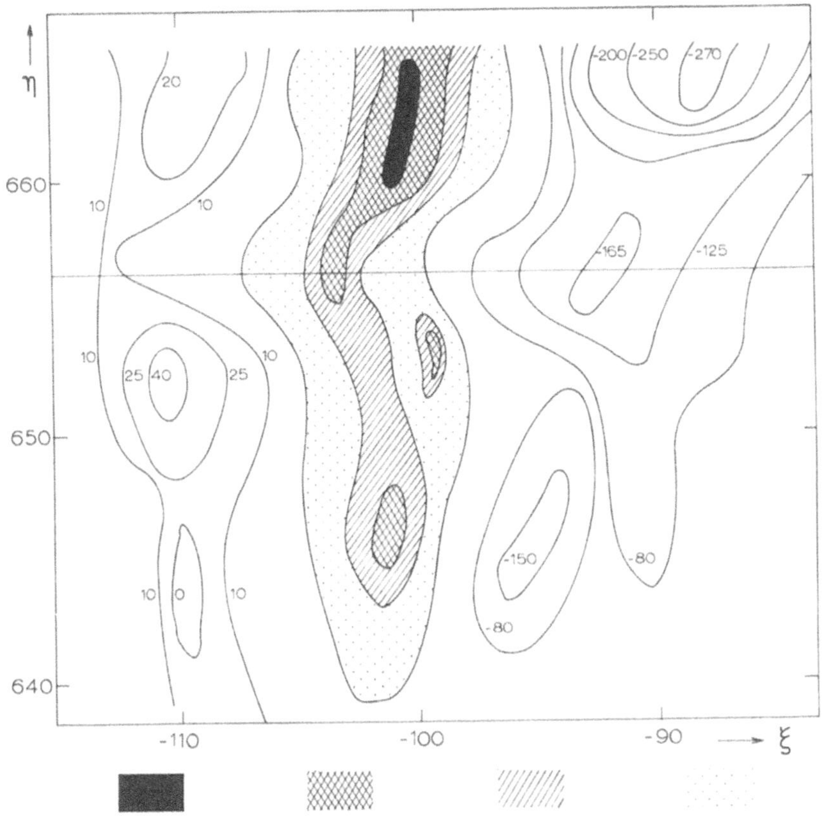

Fig. 11. Relief of a ridge in Mare Imbrium, after VAN DIGGELEN (1951).

References

AKIMOV, L. A.: 1963, *Tsirkular Observ. Kharkov* **26,** 43.

AKIMOV, L. A.: 1965, *Věstnik Khark. Univ.* **4**; *Ser. Astron.*, Vypusk **1,** 43–61.

BARABASHOV, N. P., EZERSKI, V. I., and PRISHLJAK, N. P.: 1962, *Tsirkular Observ. Kharkov* **25,** 15.

BARABASHOV N. P. and EZERSKI, V. I.: 1965, *Věstnik Khark. Univers.* **4,** 22.

DALE, E. D.: 1962, M. Sc. Thesis, Univ. of Manchester (unpublished).

FEDORETS, V. A.: 1952, *Trudy Observ. Kharkov* **2** (10).

HAPKE, B.: 1963, *Journ. Geophys. Res.* **68,** 4571.

HERRIMAN, A., WASHBURN, H., and WILLINGHAM, D.: 1963, *J.P.L. Techn. Rept. Nr. 32–384.*

KOPAL, Z.: 1962, in *Physics and Astronomy of the Moon.* Academic Press, New York and London, Chapter VII, pp. 262–264.

MCCAULEY, J. F.: 1964, *U.S. Geol. Survey*, open file report.

MINNAERT, M. G. J.: 1961, in Kuiper, *The Solar System*, III, p. 213.

NATHAN, R.: 1966, *J.P.L. Techn. Report. No. 32–877.*

ORLOVA, N. S.: 1956, *Russ. Astron. Journ.* **33,** 93.

RINDFLEISCH, T.: 1965, *J.P.L. Techn. Report. No. 32–786.*

VAN DIGGELEN, J.: 1951, *Bull. Astronom. Inst. Netherlands* **11,** 283.

VAN DIGGELEN, J.: 1966, *Bull. Astronom. Inst. Netherlands* **18,** 311.

WATSON, K.: 1966, *U.S. Geological Survey, N.A.S.A. Contract W-12*, 123.

WILDEY, R. L. and POHN, H. A.: 1964, *Astronom. Journ.* **69,** 619.

WILHELMS, D. E.: 1963, *Astrogeologic Studies, Ann. Progr. Report.*, Aug. 25, 1962.

SELENODESY FOR THE APOLLO PROJECT

JAMES H. SASSER*

Abstract. A review of the Apollo lunar landing mission is presented and areas indicated wherein lunar geodetic, cartographic and topographic information are currently deficient. Established programs to correct these deficiencies are described. Slope and protuberance frequency information at the scale of a few meters is presented for areas near the impact points of Rangers VII and VIII, and these data are compared with the predicted roughness of average lunar mare. Results indicate a rougher mare surface than previously anticipated.

Introduction

The Apollo lunar landing program presents unusual requirements for lunar geodetic, cartographic, and topographic information. In a few hours during a lunar landing mission the need for data on a planetary scale is supplanted by the need for regional cartographic data and topographic information at the scale of a few decimeters. The need for added information has given a strong impetus to lunar research programs in the United States in recent years, and selenodetic and selenographic programs have attained an importance heretofore unrecognized. The National Aeronautics and Space Administration (NASA) has established a cooperative program within NASA Centers, Universities and U.S. Government mapping agencies to provide the required information from observations made from Earth and a series of unmanned lunar probes. While describing the major elements and agencies participating in this program, a short description of the planned Apollo lunar landing mission proves useful to point out those phases of the mission wherein lunar geodetic, cartographic and topographic data are currently deficient and the steps being taken to remedy these deficiencies.

The Apollo Mission

The Apollo lunar landing mission will begin with the launch of the Saturn V launch vehicle from Cape Kennedy, on the eastern coast of Florida. The Saturn I-C, Saturn II, and Saturn IVB rockets will lift the spacecraft into an earth parking orbit. After a few orbits of the Earth, when it has been determined that all systems are functioning properly, the Saturn IVB stage will reignite and thrust the spacecraft into the trajectory that will provide arrival near the Moon some four days later. During this "trans-

* Space Science Division, National Aeronautics and Space Administration, Manned Spacecraft Center, Houston, Tex., U.S.A.

lunar" phase of the mission, the Command and Service Modules will separate from the Saturn IVB, perform a turn-around maneuver, and dock with the Lunar Module. This maneuver will free the Service Module propulsion system for use at lunar arrival. Navigational sightings will be made by the astronauts during the translunar phase of the mission prior to making any needed trajectory corrections. These sightings can consist of stellar observations, observations made of the limb of the Earth or Moon, and observations of recognizable earth or lunar landmarks whose positions are known in body-fixed coordinate systems. The geodetic positions of suitable landmarks almost anywhere on the Earth can be obtained to accuracies of a few hundred meters, but the high incidence of cloud cover makes the earth landmark sightings unreliable. Lunar landmarks, if properly selected, are always visible, but several problems arise in their use as the spacecraft approaches the Moon. The disadvantages consist of: (1) the large positional uncertainties in existing lunar control points – especially in the limb regions, (2) the large size of craters previously used as lunar control points, and (3) the uncertainty in the location of the Moon's center of mass with respect to the center of figure, which is used as the point of origin of existing lunar control systems. Some improvements have been made in (1) and (2) in recent work of the Army Map Service and the Aeronautical Chart and Information Center, but major improvements in the three areas can be obtained only from lunar photographic satellites, and lunar landing vehicles. Both types of spacecraft will be launched by the United States this year (1966). The Lunar Orbiter, a lunar orbiting photographic satellite, will provide medium resolution (8 meter) stereoscopic and high resolution (1-meter) monoscopic photographic coverage of large areas of the lunar surface. Five spacecraft are presently planned during 1966–1967. Program management for the Lunar Orbiter has been assigned by the NASA to the Langley Research Center. The Surveyor, a soft-landing, unmanned lunar probe, will provide measurements of lunar surface material properties and high-resolution television pictures of the lunar landscape visible from the landing spot. The Surveyor Program is managed for the NASA by the Jet Propulsion Laboratory of the California Institute of Technology.

When the Apollo spacecraft arrives at the proper position with respect to the Moon, the Service Module propulsion system will provide the velocity change necessary for the spacecraft to enter the selected lunar orbit. It is presently planned that this orbit will be nearly circular, at an altitude of about 150 kilometers above the surface of the Moon, inclined only a few degrees with respect to the lunar equator. During the lunar orbit phase of the Apollo mission navigational sightings can again be made: (1) to determine the parameters of the orbit achieved; (2) to determine the time-dependent position of the orbiting spacecraft with respect to the desired landing site, and (3) to determine the velocity change necessary for the Lunar Excursion Module to descend to the desired landing site. In addition to the required improvements in lunar geodetic control systems noted earlier, these computations require a better knowledge of the lunar gravitational field. This improvement will be obtained by analyzing radar tracking data from lunar orbiting satellites (MICHAEL and TOLSON, 1965). Lunar cartographic products, displaying navigational information, will be used by the astronauts during lunar orbit. Until more detailed photographs are obtained of the far side of the Moon, usable cartographic information will be available for only one-half of each orbit.

After several lunar orbits, two astronauts in the Lunar Module will separate from the Command Module, and descend to a desirable landing site on the lunar surface. Methods of determining the suitability of lunar landing sites will be discussed in the following section. After completing scientific activities and exploration in the vicinity of the landed spacecraft, the two astronauts will return to the Command Module in the ascent stage of the Lunar Module. The two spacecraft will rendezvous in lunar orbit. The astronauts will then enter the Command Module, and return to Earth with geological samples and photographic records of thier activities on the lunar surface.

Lunar Landing Site Analysis

The question of selecting suitable landing sites on the Moon may be divided, for purposes of discussion, in two parts; one dealing with the mechanical properties of the near-surface materials, and the other dealing with surface topography. Of the two, the former is potentially the most difficult, since direct measurements will be difficult to obtain to provide a basis for extrapolation. Based on his analysis of the Ranger IX photography, KUIPER (1966) has estimated a bearing strength of 1–2 kg/cm^2 for lunar materials in the floor of Alphonsus. On the basis of his analysis of Ranger VII photography, GOLD (1964) concluded that the Ranger VII impact area was covered by a loosely cemented dust layer several meters deep. Such analyses, made on the basis of photographic evidence alone, should be viewed with caution. KEENE (1965) has pointed out that image smear was a serious factor in Ranger VII photography, limiting the resolution and changing the appearance of the lunar scene.

The most promising approach to this problem seems to be to accumulate as much information as possible concerning lunar materials, including estimates of their genesis and morphology, and use this information to infer the surface mechanical properties. Studies of lunar material properties using radar, photometric, ultraviolet, infrared, and polarization measurements have recently intensified. SEBRING (1966) has reported on recent work and future plans for radar studies in direct support of the Apollo lunar landing program. Lunar geologic studies and studies of crater formative processes in terrestrial analogs of postulated lunar materials have also intensified in recent years, notably those performed by Gault of NASA Ames Research Center and Shoemaker, Moore and others in the U.S. Geological Survey. Assimilation of these data with that provided by lunar landing probes should allow rational estimates to be made of the degree of homogeneity and bearing properties of the lunar surface.

The second part of the question – that of describing lunar topography in meaningful terms – is of equal importance. If the lunar surface is homogeneously strong, the relative safety of potential landing areas can be evaluated only on the basis of topography. If the lunar materials are of heterogeneous strength, what might have appeared to be acceptable topography in the first case might well need to be analyzed in greater detail, and tolerable topographic relief reduced to allow for some surficial sinkage on landing. The conclusion is that an analysis of lunar surface topography will be the most valuable method of comparing the relative suitability of lunar landing areas. The methods to be used in analyzing lunar topography prior to a lunar landing mission are similar to those described below, pertaining to the analysis of Ranger data currently underway at the Manned Spacecraft Center.

RINDFLEISCH (1965) has described recent work at the Jet Propulsion Laboratory in the photometric analysis of data obtained from the Ranger series of spacecraft to obtain lunar topographic information. The method is a refinement of that first used by VAN DIGGELEN (1951). MCCAULEY (1964) used the technique to extract lunar topographic information from telescopic photography and prepare a terrain map of the lunar equatorial region. Additional work with this technique has recently been reported by WATSON (1966) and ROWAN and MCCAULEY (1966). NATHAN (1966) has developed techniques for mathematically filtering transmitted imagery to improve picture quality and provide a more faithful representation of the original scene. These methods will be used at the Manned Spacecraft Center to reduce the photographic data transmitted from the Lunar Orbiter spacecraft and to obtain digital topographic information.

In order to develop meaningful techniques of comparing the suitability of lunar areas for landings, digitized topographic information resulting from both filtered and unfiltered Ranger pictures was obtained on magnetic tape from the Jet Propulsion Laboratory. These data can be analyzed statistically in many ways, depending on the display desired. The most obvious method – and one of the more meaningful – is to subject the mathematically described surface to many landings of a hypothetical vehicle of the correct dimensions. This was performed for a hypothetical Lunar Excursion Module on data derived from the last few photographs taken before impact by the Ranger P1 and P3 cameras. The extent of the areas considered on the Ranger VIII data are indicated in Figure 1. The dimensions, resolution and other pertinent information concerning the input data are listed in Table I.

The *topographic* landing quality of each surface was evaluated in the following manner. (For these analyses, no vehicle-surface interactions were considered.) The

TABLE I

Nominal Dimensions and Resolution of last five Ranger VIII P1 and P3-Camera Exposures

	1001	1002	1003	1004	1005
Nominal East–West Dimension (meters)	67	149	232	314	396
Nominal North–South Dimension (meters)	47	106	164	222	280
Approximate Ground Resolution (meters)	.38	.82	1.3	1.7	2.2
Maximum Vertical Relief (meters)	2.7	10	10	18	15
	3001	**3002**	**3003**	**3004**	**3005**
Nominal East–West Dimension (meters)	141	386	631	875	1120
Nominal North–South Dimension (meters)	106	289	472	655	838
Approximate Ground Resolution (meters)	.74	2.0	3.3	4.6	5.9
Maximum Vertical Relief (meters)	5.7	14	30	35	48

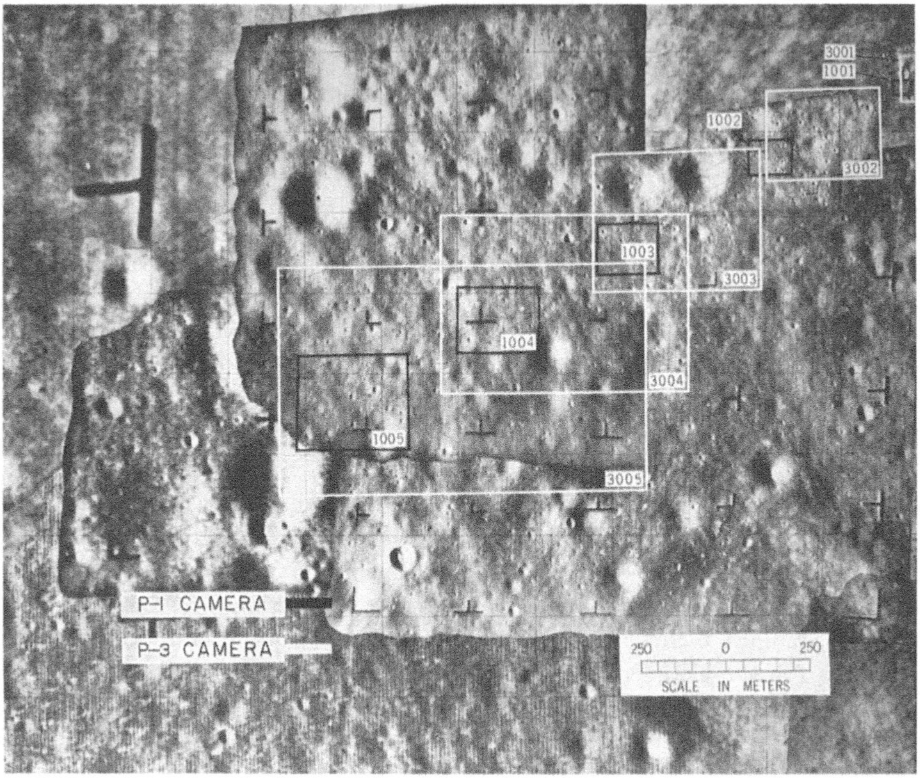

Fig. 1. Photomosaic showing Ranger VIII areas evaluated for slopes and protuberances at the scale of the Apollo lunar landing vehicle. Frame XOOY indicates the Y exposure before impact from the XP-camera. The photomosaic was produced by the U.S.A.F. Aeronatical Chart and Information Center.

first computer program was prepared to sample each elevation matrix with 10 000 random landings. Random number generators were used to locate the X–Y coordinates of the point to be considered within the digitized elevation data and the azimuthal orientation of the vehicle. When the four pads of the landing gear were within the boundaries of the elevation matrix, the landing was considered to be effective. The mean elevations of the two diagonals of the landing footprint were computed and the diagonal of highest mean elevation on the matrix assumed to contact the surface first. Because of the four-footed configuration of the vehicle, two possibilities remained for contacting the surface with a third foot. The lower elevation was chosen as the more stable position, and the maximum slope of the plane formed by the three feet in contact with the surface was computed. Additionally, the maximum protuberance above the plane – but within the vehicle bounds – was computed to determine the frequency of "bottoming". Cumulative frequency curves for the slope distributions of the effective drops are shown in Figures 2 and 3, and cumulative frequency curves for protuberances are shown in Figures 4 and 5. Such data will provide a rapid assessment of the

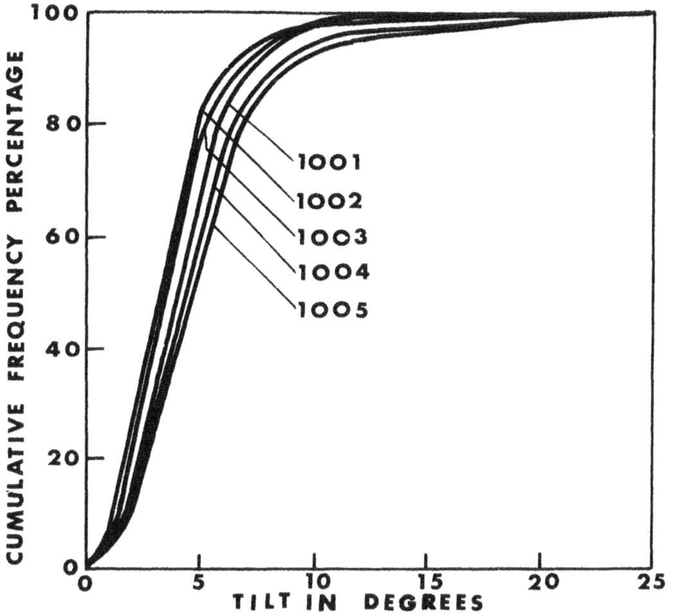

Fig. 2. Plot of cumulative slope frequency at the scale of the Apollo lunar landing vehicle. Percentage is based on 10000 random locations within each of the last five areas photographed by the Ranger VIII P1 camera before impact.

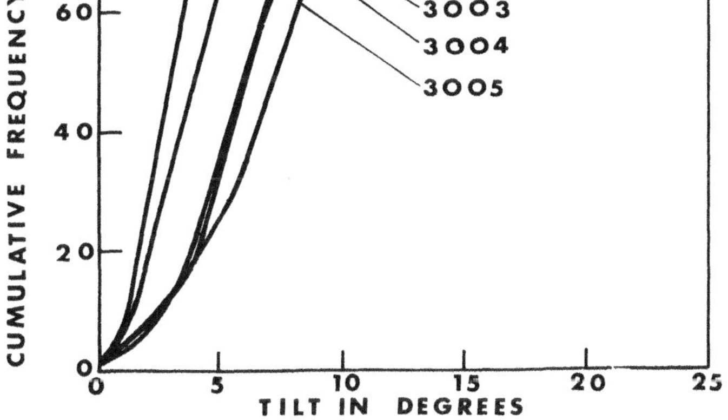

Fig. 3. Plot of cumulative slope frequency at the scale of the Apollo lunar landing vehicle. Percentage is based on 10000 random locations within each of the last five areas photographed by the Ranger VIII P3 camera before impact.

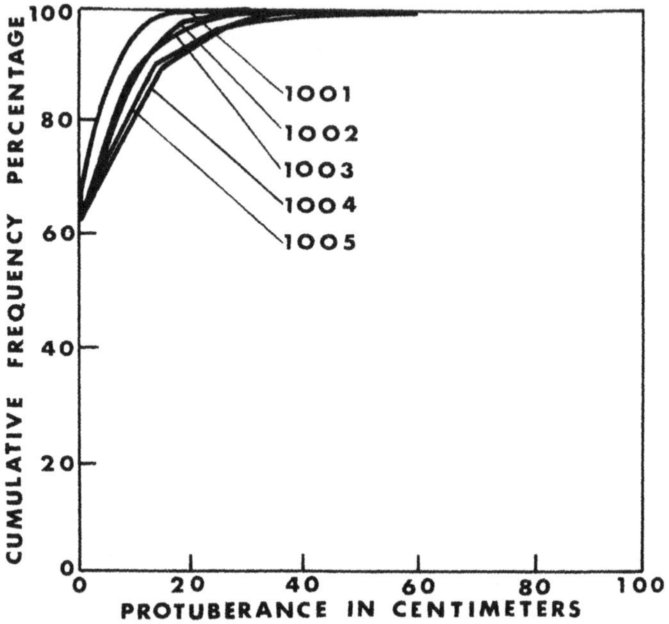

Fig. 4. Plot of cumulative protuberance frequency at the scale of the Apollo lunar landing vehicle. Percentage is based on 10000 random locations within each of the last five areas photographed by the Ranger VIII P1 camera before impact.

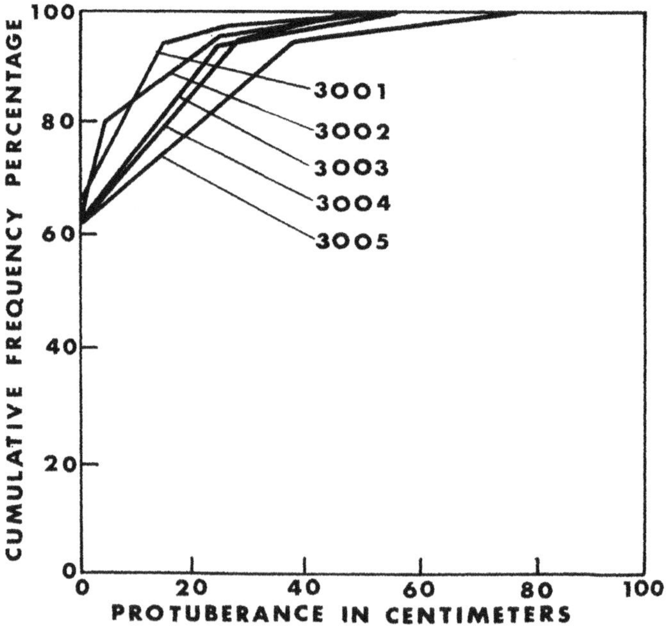

Fig. 5. Plot of cumulative protuberance frequency at the scale of the Apollo lunar landing vehicle. Percentage is based on 10000 random locations within each of the last five areas photographed by the Ranger VIII P3 camera before impact.

SAMPLE PORTION OF COMPUTER RUN

```
B B + O X X X X X X H H H X X
B B + O X X X O X H H X H X X
B B X O O O O X H H X X X X +
B B + O X I O X X X H H H H H
B B + X X O X X H X X H H + H
B B + X X X H H X X X H H H H
B B H O O O H X H H O H H H H
B B + H O X X X H H H X H H X
B B + I X H O H H H H X H X H
B B + + + + + + + + H H X X X
```

SYMBOL	ANGLE OF TILT IN DEGREES
B	BORDER
BLANK	0-2
—	2-4
I	4-6
O	6-8
X	8-10
H	10-12
+	OVER 12

Fig. 6. Example of computer print out for construction of isoslope chart shown in Figure 8.

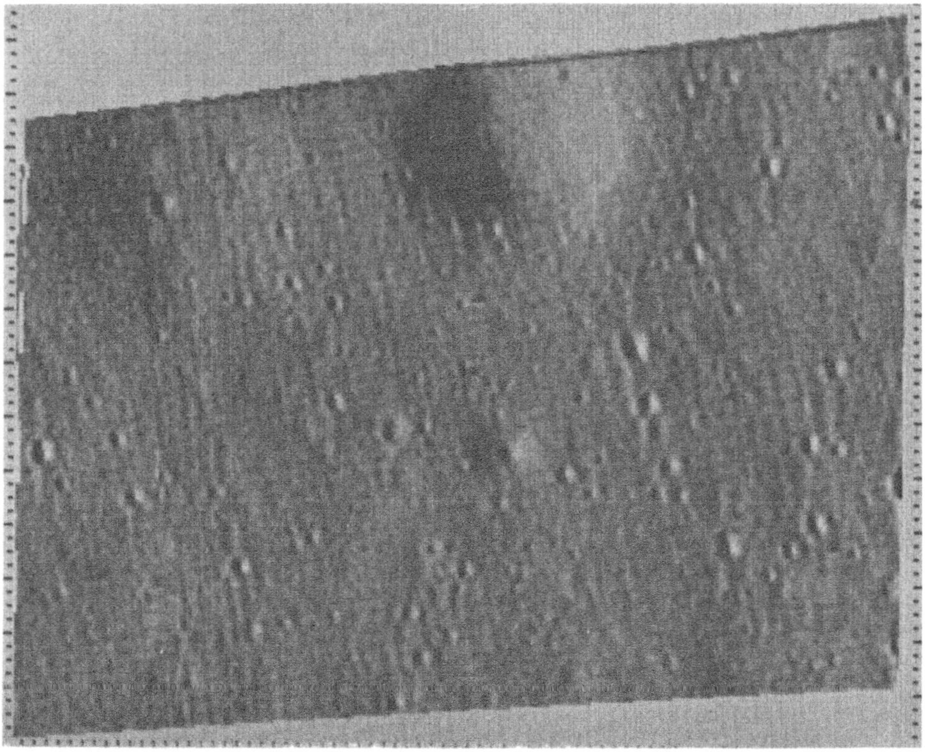

Fig. 7. Computer-generated photograph produced at the Jet Propulsion Laboratory. The sine wave filter enhances small detail not readily visible on the original Ranger VIII pictures. Exposure No. 3003.

suitability of areas for lunar landings, and allow quantitative comparison of the *topographic* suitability of potential lunar landing areas. However, this type of display tells nothing about the location of slopes and protuberances within the area analyzed.

This question was addressed by using a second computer program in the following manner. A regular grid of samples (120 columns × 55 rows) was analyzed for each elevation matrix. The only significance of this sampling interval was that this number of characters could be conveniently printed out by the computer on a single page. Random numbers again established the azimuthal orientation of the hypothetical vehicle, and the same procedure as before was used to compute slope and protuberance at each sample point. A character was then printed out to indicate the result at that point within the digitized topographic data. Figure 6 indicates the symbolization used for this process. It was then possible to contour the resulting tableaus and construct charts depicting isoslopes and the locations of protuberances within the terrain matrices. The similarity of these charts with associated photographs is quite good, as can

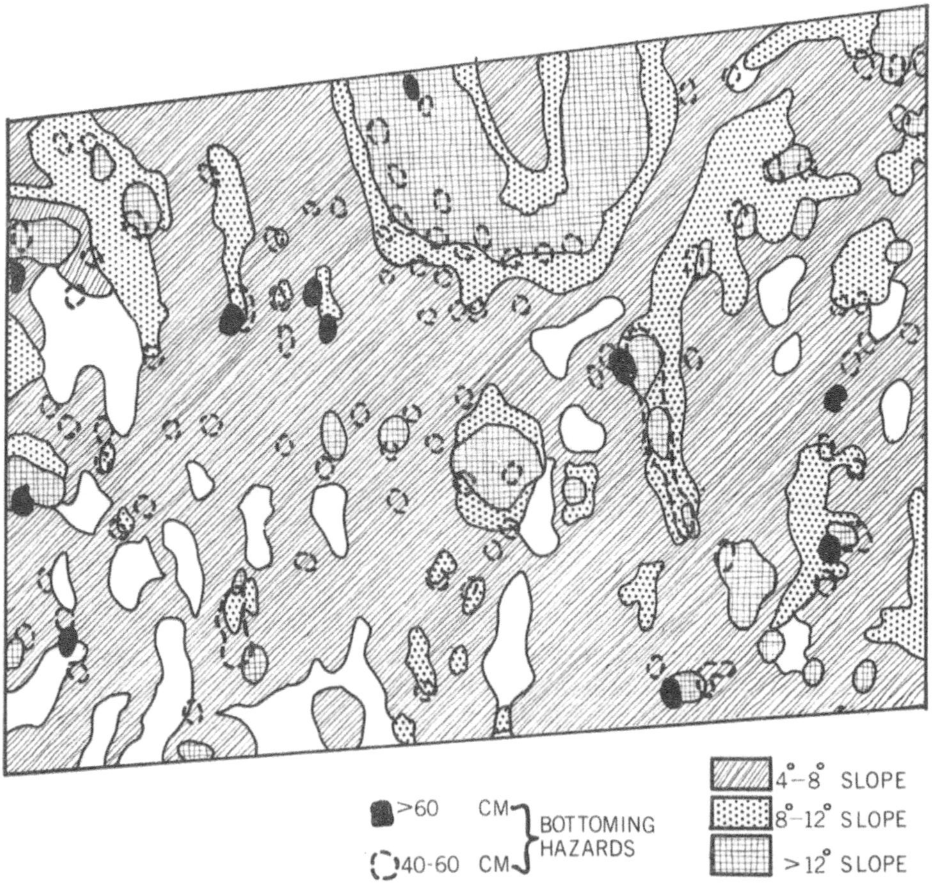

Fig. 8. Isoslope and protuberance chart generated from a slope and protuberance analysis, at the scale of the Apollo lunar landing vehicle, of the elevation matrix for Ranger VIII Exposure No. 3003.

be seen by comparing Figures 7 and 8. The charts allow immediate interpretation of the surface in terms of topographic suitability for lunar landings throughout the area processed.

The mean cumulative slope frequencies, at the scale of the Lunar Excursion Module, obtained from filtered and unfiltered Rangers VII and VIII data are shown in Figure 9. It is interesting to compare these results, pertaining to Mare Cognitum and

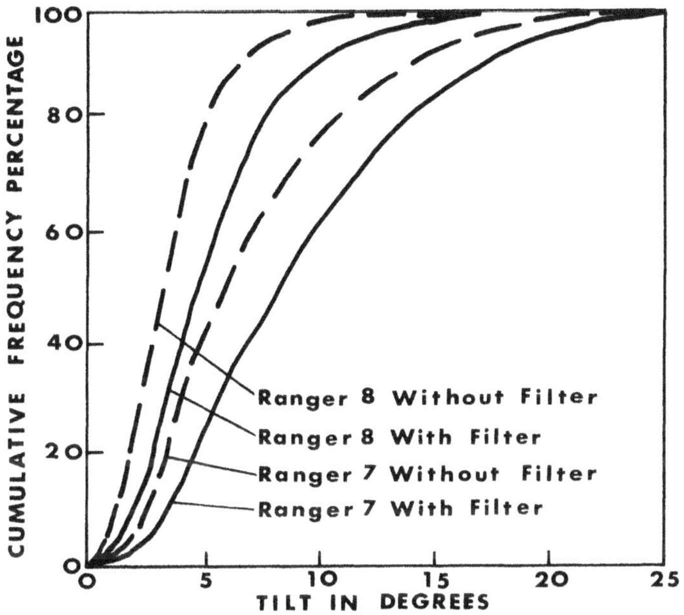

Fig. 9. Comparison of mean slope values at the scale of the Apollo lunar landing vehicle, of filtered and unfiltered data from Rangers VII and VIII.

Mare Tranquilitatis, respectively, with those predicted by McCauley (1964) and Rowan and McCauley (1966). The results of this comparison, shown in Figure 10, indicate a rougher mare surface than that predicted.

Conclusion

The improvements needed in lunar geodetic and regional cartographic information can be obtained from lunar photographic satellites and lunar landing probes. Obtaining detailed information on the surface mechanical properties and topographic relief in potential landing areas is a much more complex question involving many scientific disciplines. Programs now initiated to provide the information needed for a manned lunar landing mission will provide substantial improvements in scientific knowledge of many of the Moon's properties in this decade.

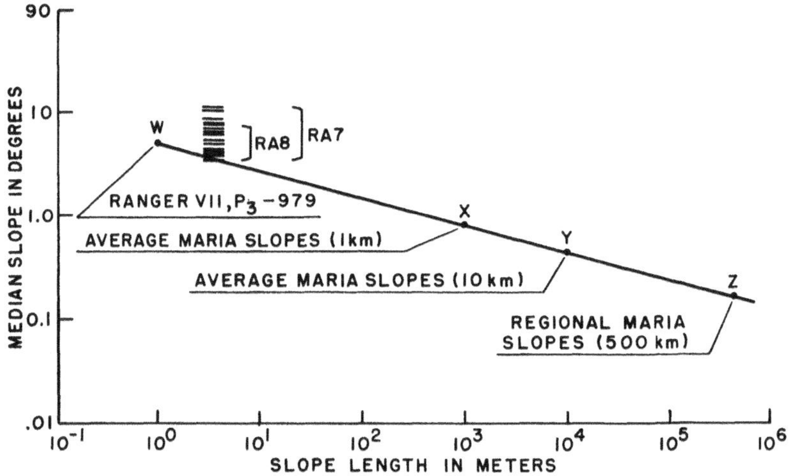

Fig. 10. Plot of median slope as a function of slope length for average lunar maria (ROWAN and McCAULEY, 1966). The dispersion of median slope values obtained from the analysis of the last five P1 and P3 camera photographs from Rangers VII and VIII is shown.

Acknowledgements

The author would like to express his appreciation to R. Nathan and T. Rindfleisch for providing the digitized Ranger topographic information used in these studies. In the areas of computer programing and statistical treatment, R. Rosencranz and J. Dragg are thanked for their extensive work in generating the descriptive lunar topographic information.

References

GOLD, T.: 1964, 'Ranger Moon Pictures: Implications', *Science* **145**, 1046–1048.

KEENE, G.: 1965, 'An Analysis of Photography from Ranger VII'. Paper presented before the Society of Motion Picture and Television Engineers, March 29, 1965.

KUIPER, G. *et al.*: 1966, 'Ranger VIII and IX Experimenters Analyses and Interpretations', *Jet Propulsion Lab. Tech. Rept. 32–800*, 153.

McCAULEY, J.: 1964, 'Terrain Analysis of the Lunar Equatorial Belt'. Preliminary report: U.S. Geological Survey, open-file report.

MICHAEL, W. and TOLSON, R.: 1965, 'The Lunar Orbiter Project Selenodesy Experiment'. Paper presented at the Second International Symposium on the Use of Artificial Satellites for Geodesy.

NATHAN, R.: 1966, 'Digital Video-Data Handling', *Jet Propulsion Lab. Tech. Rept. 32–877.*

RINDFLEISCH, T.: 1965, 'A Photometric Method for Deriving Lunar Topographic Information', *Jet Prop. Lab. Tech. Rept. 32–786.*

ROWAN, L. and McCAULEY, J.: 1966, in *Lunar Orbiter – Image Analysis Studies Report* (U.S. Geol. Survey Rept. 1966), pp. 89–127.

SEBRING, P. *et al.*: 1966, *Radar Studies of the Moon.* Mass. Inst. of Tech. Lincoln Lab. Rept., February 15, 1966.

VAN DIGGELEN, J.: 1951, 'A Photometric Investigation of the Slopes and Heights of the Ranges and Hills in the Maria of the Moon', *Astron. Inst. Netherlands Bull.* **11**, 283–289.

WATSON, K.: 1966, in *Lunar Orbiter – Image Analysis Studies Rept.* (U.S. Geological Survey Report, 1966), pp. 1–19.

TERMINATOR PHOTOGRAPHY IN OBLIQUE ILLUMINATION FOR LUNAR TOPOGRAPHIC WORK*

ZDENĚK KOPAL**

Abstract. Advantages to lunar topography of ground-based photographs taken under oblique illumination – especially in the penumbral zone of the lunar surface illuminated by the only partly risen Sun – are discussed. It is pointed out that such photographs require, in general, exposure times from 10 to 100 times longer than would be conventionally regarded as "normal"; and advantages of such "over-exposed" ciné photography in minimising by rapid projection the film-grain noise are stressed.

Illustrative examples of such photography underline the advantages of this method. In particular, a comparison of two photographs of the crater Ptolemy – one taken with the 43-inch reflector of the Observatoire du Pic-du-Midi, the other with the FB camera of Ranger 9 from an altitude of 1406 kms above the lunar surface – discloses that while the linear resolution of the Ranger camera at this altitude exceeds that of the 43-inch telescope more than ten times, the vertical resolution of the ground-based but oblique-illumination photograph exceeds that secured from the Ranger to an almost equal extent.

It is pointed out that photographs taken from terrestrial observatories of the lunar surface illuminated by the partly risen Sun can furnish vertical relief data that are generally unobtainable from spacecraft unless the latter, while overflying the terminator, can secure sufficiently short exposures at the light level prevailing in the penumbral zone.

The need for resolving the maximum amount of horizontal or vertical detail on the lunar surface, in support of the current mapping programmes to precede manned landings on the Moon in the next few years, should scarcely require any explanation or emphasis at this time. Most information available so far on the topography of the near face of the Moon has been obtained with the aid of ground-based telescopes from the distance of the Earth. The recent contributions by spacecraft of the Ranger or Orbiter type have increased the linear resolution of details on the Moon (albeit over only very small fraction of its surface so far) by a factor of 100–1000 in excess of the best resolution attainable from the distance of the Earth (of the order of 10^6 in the immediate proximity of the TV-cameras of a soft-lander); and, in the future, their supremacy will remain unchallenged as far as the *horizontal* resolution is concerned. However, it is also possible to show that, in so far as the *vertical* resolution (or determination of surface roughness) is concerned, ground-based photography from the Earth can, under certain conditions, exceed that from a spacecraft operating in the proximity of the Moon by a considerable margin. The aim of the present contribution will be to

* Work sponsored by the Aeronautical Chart and Information Center under Contract AF 61 (052)–829 with the Department of Astronomy, University of Manchester, through the European Office of Aerospace Research, U.S. Air Force.
** Department of Astronomy, University of Manchester.

establish the circumstances under which this happens to be true, and to document our case with appropriate examples.

As is well known, the most powerful method for the determination of relative altitudes (or level differences) on the Moon is based on a triangulation of shadows cast by the respective eminence on the surrounding landscape in the rays of the illuminating sunlight. The efficiency of this method is, in turn, at its best at the time of the sunrise or sunset, when the respective terminator passes through the centre of the apparent lunar disc (i.e., the locus of minimum foreshortening). Along the limb of the Moon, the limiting precision with which one can determine the altitudes of lunar mountains is given by the resolving power of the telescope employed; which for a 24-inch aperture in yellow light is (by Rayleigh's criterion) close to 500 metres on the lunar surface at the mean distance of the Moon; and about 300 metres for a 43-inch telescope (the largest aperture used for systematic lunar photography so far).

However, near the centre of the apparent lunar disc, the relative altitudes can be greatly magnified by the shadows cast in the oblique rays of the Sun. Simple geometry reveals that, at the time when the lower limb of the Sun just clears the lunar horizon, the ratio of the length of the shadow to the height of the object casting it is equal to $\cot 0\overset{\circ}{.}5 = 115$ (and exceeds 115 in the penumbral zone illuminated by partly risen Sun); though when the Sun's centre has risen 2° above the lunar horizon, this magnification ratio has diminished to 27. If, therefore, photographs can identify on the Moon at sunrise (or sunset) shadows (say) one kilometre in length, these should indicate the presence of vertical obstacles less than 10 metres in height; and even smaller obstacles (or uneveness of ground) could be detected in the penumbral zone (which extends for less than $0\overset{\circ}{.}5$ in longitude along the lunar "equator of illumination" and, as such, is not more than approximately 14 kms wide).

Have any such objects (or ground irregularities) been actually recorded on the Moon from the distance of the Earth? Unfortunately, most photographs of the lunar surface obtained so far are ill-suited for such a search; for their exposures – approximately correct for the centre of the entire field of the negative – were usually much too short to bring out any detail inside their relatively narrow twilight strip. It may perhaps come as a surprise to many readers to hear that most photographs of the Moon he may have seen show a "false terminator", marking the region where the Sun has already risen several degrees above the lunar horizon (for example, cf. Figure 1 in Kopal and Rackham, 1963); for exposures long enough to reveal details in the twilight zone would have left the rest of the field largely burned out.

Examples of the increased definition of the real terminator with increasing exposure times were published by the writer before (cf. Kopal, 1965, Figures 2–6 and 2–7); and reproduced below are two pairs of typical new photographs to illustrate further this point. Figures 1a and 1b show sunset over the western plains of Mare Imbrium north of the Carpathian Mountains, with craters Lambert and Pytheas near the centre of the field. Both photographs were taken with the 43-inch (*f*/15) reflector at the Observatoire du Pic-du-Midi on 1964 August 3 at 03.34 U.T. – both within the same minute – but one with the exposure of one-thirtieth of a second (Figure 1a); while the other was

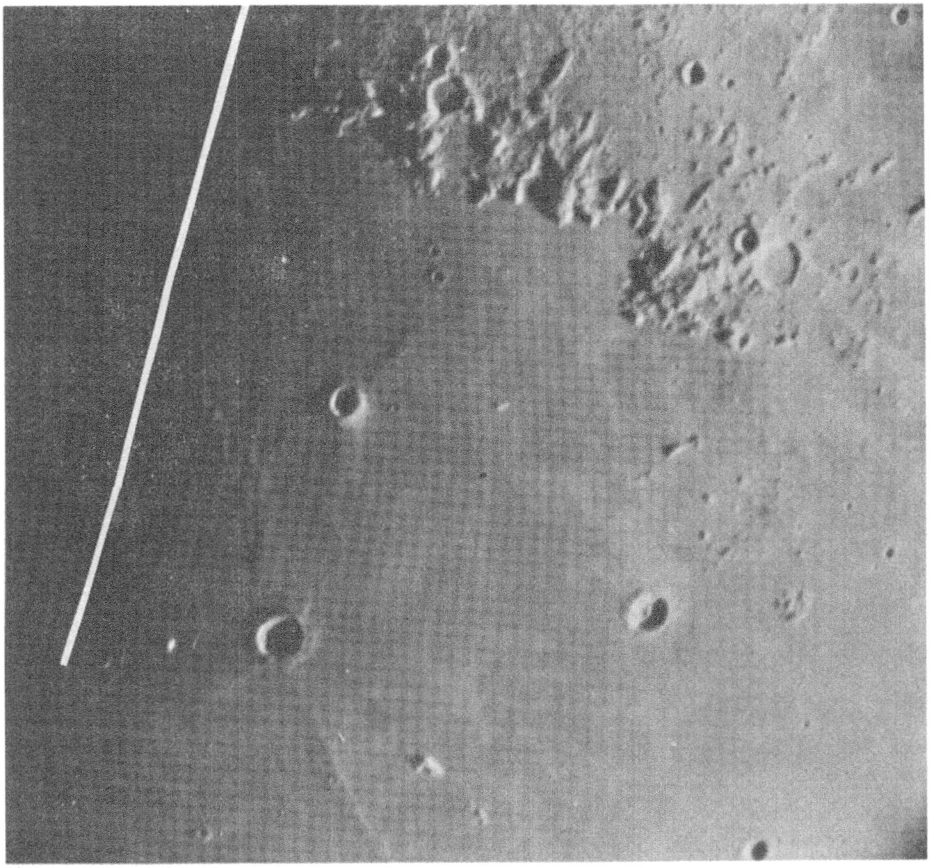

Fig. 1a. Sunset over the Western plain of Mare Imbrium, north of the Carpathian mountains, with craters Lambert (below) and Pytheas (above) near the terminator, as photographed on 1964 Aug. 3, 3.34 U.T. with the 43-inch reflector at the Observatoire du Pic-du-Midi. The exposure time at $f/15$ was $^{1}/_{30}$ sec; the position of the actual sunrise (where the upper rim of the Sun's disc just becomes visible above the lunar horizon) is marked on the left with a white line.

exposed one-quarter of a second (Figure 1b). The former exposure was appropriate for the field as a whole, but left the sunrise zone completely underexposed. An eight-fold increase in exposure for Figure 1b brought out details in the terminator zone, at the expense of over-exposing all regions where the Sun stood more than 2–3 degrees above the lunar horizon.

However – and this is essential – the terminator region on Figure 1b constitutes a strip where shadows cast in the rays of the rising Sun are of greatest help in measuring relative altitudes or any kind of ground irregularities. The actual line of the sunset terminator (marked by a white line on Figure 1a) moves westward at a rate of $14\frac{1}{2}$ km per hour. In view of the fact that the "penumbral zone" of the Moon (illuminated by a partly-risen Sun) is approximately 14 km wide, it follows that any region of the lunar surface can be studied under these conditions for only about one hour each

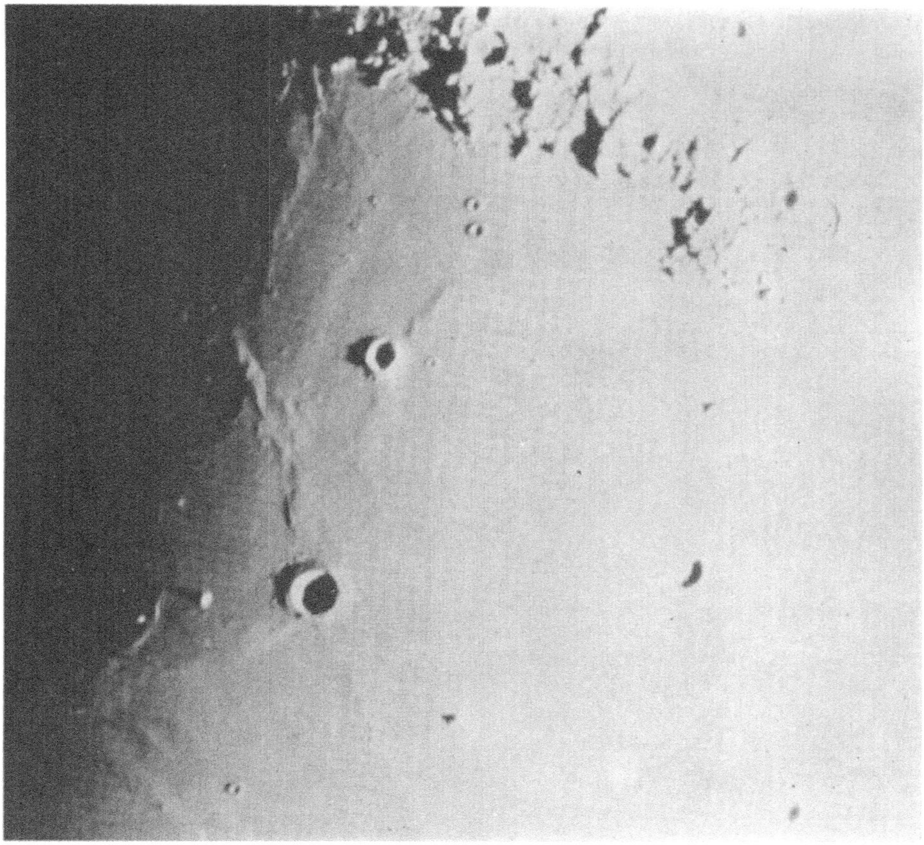

Fig. 1b. A photograph of the same region of the lunar surface as shown on Figure 1a, taken at the same time (within the same minute) with the same instrument, but with an exposure time of 0.25 sec. With this prolonged exposure, the punumbral zone along the terminator emerges from the darkness and reveals details invisible on Figure 1a.

month. But during that time the shadows are veritably racing over the lunar surface – moving through a distance of 220 metres (equal to a theoretical angular resolution of the 43-inch telescope, in yellow light, at the mean distance of our satellite) in less than one minute of time!

Cine photographs of sunrise or sunset on the Moon – suitably exposed to reveal details in the penumbral zone of the respective terminator – should look quite spectacular when projected on the screen at an accelerated rate (and such inspection would, incidentally, be the quickest way to distinguish genuine lunar surface details from the emulsion noise). The principal reason why no such movies are available so far goes back to the difficulty in guiding a large telescope, with the requisite precision, for sufficiently long time. Ordinary guiding errors tend to bring successive frames of a ciné reel (taken with a time lapse of a minute or so) out of positional adjustment; so that the projected image on the screen would exhibit a high-frequency "random walk" caused by mechanical imperfections of guiding mechanism together with the effects of atmospheric refraction anomalies. However, it should be possible to obtain a smooth film by subsequent re-positioning of each individual frame and re-copying on a new roll; but this, too, remains yet a task for the future.

The second pair of illustrations accompanying this text is of an entirely different kind. Figure 2a shows an enlargement of the Ptolemy-Alphonsus region of the Moon,

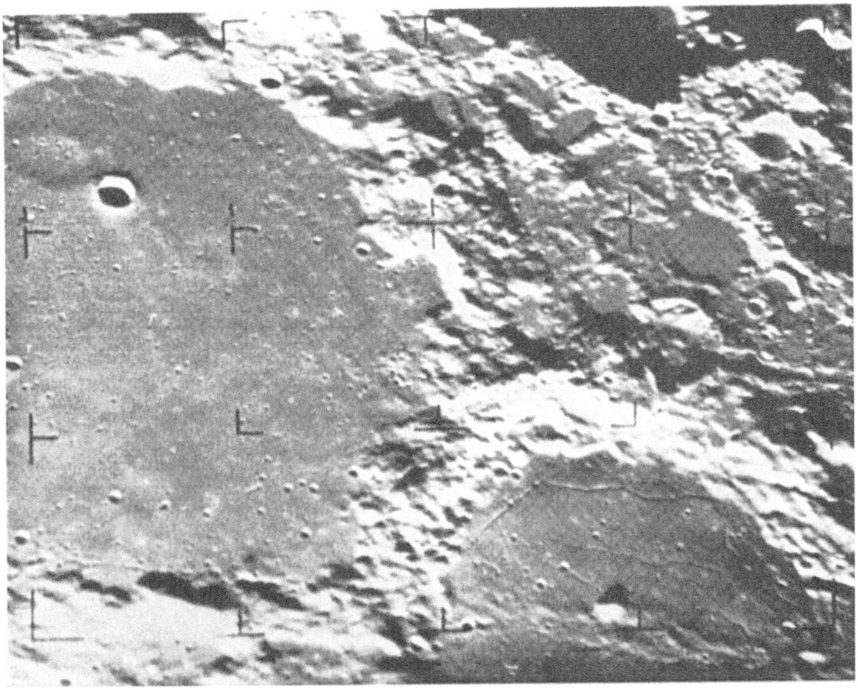

Fig. 2a. A view of the Ptolemy-Alphonsus region of the lunar surface, secured by Ranger 9 FB Camera (Frame No. 193) on 1965 March 24, 13h 57m 44.7 sec U.T., from an altitude of 1405.6 km above the lunar surface (635.3 sec before impact). The altitude of the Sun above the lunar horizon was 10.°4; and ground resolution close to 300 meters.

taken by the Ranger 9 FB camera (Frame No. 193) on 1965 March 24th at 13 h 57 m 44.7 sec U.T. from an altitude of 1406 kms above the lunar surface (from closer proximity, this region drifted out of the field of view). Its ground resolution of close to 300 metres on the floor of Ptolemy exceeds by about a factor 2 that of Lyot's best photographs taken from Pic-du-Midi in 1945, which were later (after Lyot's death) published by Dollfus (1962; cf. Figures 4 and 14). It shows incomparably more detail than a photograph reproduced on the accompanying Figure 2b, taken by Titulaer (cf. Rackham, 1967) with the 43-inch (*f*/30) telescope at Pic-du-Midi on 1966 August 9, at 1 h 35 m U.T.; but there is a good reason to account for it. While the photograph reproduced on Figure 2a was exposed a small fraction of a second (to arrest the motion of the spacecraft) outside the atmosphere, Titulaer's photograph on Figure 2b was exposed for 6 seconds to bring out the details along the terminators. Intervals of con-

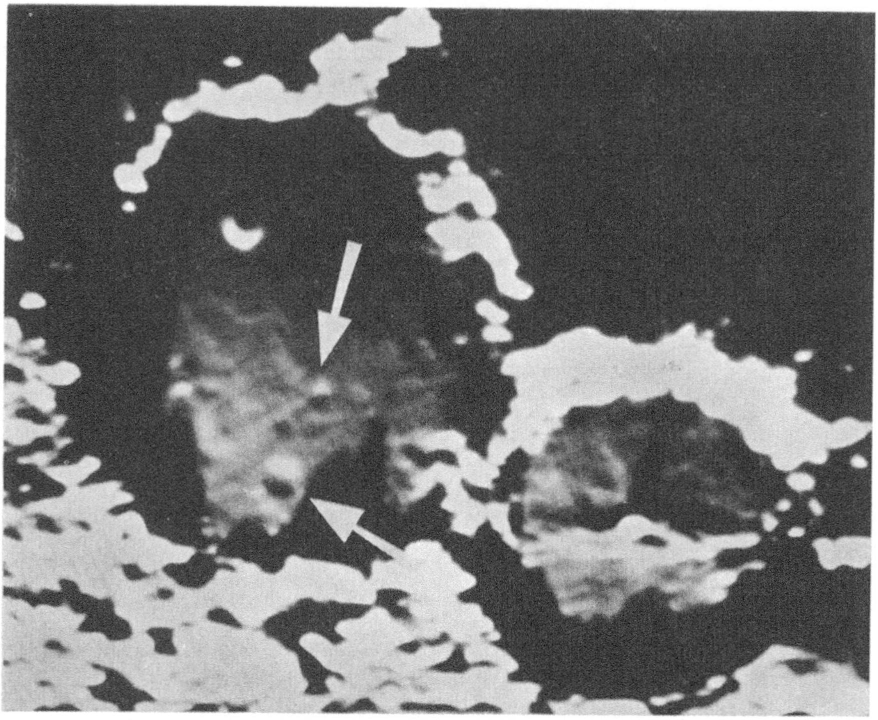

Fig. 2b. A photograph of the same region of the Moon, taken by Titulaer on 1966 August 9, at 1 h. 35 U.T. with the 43-inch telescope of the Observatoire du Pic-du-Midi. The altitude of the Sun above the lunar horizon ranged between 0–2.°7; and an exposure of 6 seconds was necessary (at $f/30$) to bring out details of the penumbral zone. A comparison of this photograph with the one reproduced in Figure 2a reveals the existence, on the floor or Ptolemy or Alphonsus, of many shallow features invisible on the Ranger photograph (in spite of the favourable vantage point in space from which it was taken) because of the higher altitude of the Sun above the lunar horizon. The most prominent circular depressions, 9 and 15 kms in diameter, unseen by the Ranger, are marked with white arrows.

tinuous steady seeing which last for 6 seconds do not come by often; and Titulaer's photograph paid its price in resolution to the vagaries of fluctuating quality of the image.

However, its relatively long exposure enabled it to record sufficient detail in a zone where the altitude of the Sun above the lunar horizon ranged between 0°–2°.7; while at the time when the Ranger photograph was taken, it ranged between 9°.0–11°.5 in the same zone. In spite of a manifestly lower horizontal resolution (by a factor of more than ten) of the photograph reproduced on Figure 2b, the low altitude of the Sun more than made up for it in vertical resolution; for a diminution of the obliquity of solar illumination from 10° to 1° revealed on the floor of Ptolemy (or the neighbouring Alphonsus) the presence of shallow craters and other features (the most conspicuous of which are marked on Figure 2b by white arrows), the existence of which would be quite unsuspected from the Ranger evidence alone. A triangulation of their shadows measurable on Figure 2b renders the depth of these formations to lie between 100–

200 metres; and their dimensions (9 and 13 kms across) are such that the slopes of their inner walls are close to only about 1°.

This example – of many others that could be invoked – demonstrates that earth-bound lunar photography at low Sun can furnish vertical relief data on the Moon that are unobtainable, not only from ground-based photographs taken with more conventional exposures, but also from spacecraft operating in much closer proximity to the lunar surface – unless the latter are on missions that permit them to overfly the lunar terminator regions and are, at the same time, able to photograph these regions with exposure times sufficiently short to arrest the motion of the spacecraft at the prevailing level of illumination. This was not true for all the Rangers (whose TV-cameras required light flux obtaining when the Sun was not much less than 10° above the lunar horizon); nor is it true for the Orbiters whose predominantly transverse motion requires even higher light intensity of the image.

At the terrestrial observatories, an extension of the photographic coverage of the Moon to the terminator illuminated by the partly-risen Sun requires, in general, exposures 10 to 100 times longer than have so far been considered "normal"; and this, in turn, rules out the use of slow optical systems with large focal ratios. If, however, we are to retain sufficient angular resolution, the focal length of the system employed for this task must remain long enough to give us the desired scale. These two requirements evidently restrict us to the use of large telescopes only – such as the 43-inch reflector at the Observatoire du Pic-du-Midi. A systematic programme of lunar terminator photography will, we hope, be commenced with this instrument in the near future, to supplement space-born photography carried out from closer proximity to the Moon. Only when this has been done we can regard the photographic contributions to the study of lunar topography to have exploited the methods and facilities now at our disposal.

References

Dollfus, A.: 1962, in *The Moon. I.A.U. Sympos. No. 14* (ed. by Z. Kopal and Z. K. Mikhailov). Academic Press, London and New York, pp. 325–342.
Kopal, Z.: 1965, *Space Science Reviews* **4**, 737–855 (cf. in particular pp. 773–774).
Kopal, Z. and Rackham, T. W.: 1963, *Icarus* **2**, 329–333.
Rackham, T. W.: 1967, *Icarus* **6** (in press).

MEASUREMENTS AND REDUCTIONS OF RELATIVE LUNAR ALTITUDES

T. W. RACKHAM*

Abstract. This paper describes the results of experiments conducted to evaluate the residuals and root mean square residuals of control point measurements made by five members of the Manchester Lunar Group. Each member was asked to measure identical control points on a given lunar negative on the Mann Comparator. The resultant rms residuals are compared and their effects are noted on the results of the relative height reductions. It is shown that the reduced heights are not sensibly affected by personal errors.

Introduction

Since early 1959, techniques for the measurement of lunar heights have been developed and pioneered in the Department of Astronomy at Manchester, and this work continues under the leadership of professor Kopal. In the same period great advances were made in the field of astronautics, and man's first venture into space in 1961 made it certain that a manned lunar landing would be attempted within the decade.

So the task of charting the Moon, which had formerly been of academic significance to a few selenologers, rapidly grew into a work of formidable proportions that had to be completed with extreme urgency. It is therefore not surprising nor coincidental that the Mancunian effort has been tightly meshed with the work of the Aeronautical Chart and Information Center of the U.S.A.F. in St. Louis, Missouri, where lunar charts have been made since 1960, and during the last two years several thousand measurements of lunar heights have been sent from Manchester for inclusion in the LAC 1:1 000 000 series.

Operations in Manchester have been tailored to suit the requirements of ACIC, which is another way of saying that schedules have to be prepared and deadlines met, which further means that calender dates take on sinister significance.

The measurement of relative lunar heights by the shadow method calls for a precise knowledge of the position of any chosen shadow-casting point on the lunar surface and, in addition, the length of the shadow must also be defined accurately. These quantities are easier to discuss than to obtain for, notwithstanding the perfection of the related geometry and the purity of the mathematical expressions by which the reductions are computed, the marriage between the theoretical and the practical is not a simple alliance. It is, however, one that must be attempted if any sort of chart is to be produced.

* Department of Astronomy, University of Manchester.

Four shadow-measuring techniques have been developed and used in the Manchester Department of Astronomy and these are briefly described below. Three of the methods may be bracketed together because they are parallel in the sense that each produce a shadow length that can be used for a height reduction as long as the position of the shadow-casting point can be read from an existing gridded chart. The fact that this position has to be obtained in this way imposes certain limitations arising from inaccuracies in determining the exact position on the chart and from reading coordinates that may also be subject to positional errors. In addition the work of reading off the positions of several hundred points can be tedious and time-consuming and is a task that does not occur in the fourth method, described later.

The first three methods may be described as follows:

(1) *Microdensitometric.* An enlarged microdensitometric tracing of a shadow is made on graph paper and is subsequently measured. If careful attention is given to the sensitometry of the film the evaluation of densities corresponding to half-intensities values can result in the determination of shadow lengths to a high degree of precision.

(2) *Isodensitometric.* An isodensitracer provides a two-dimensional appraisal of a selected area of lunar negative but instead of single scans of shadows this instrument produces a line-by-line "raster" of the image. Coding of the raster lines into patterns comprising full, dotted and blanks gives additional information, that can be identified with incremental density differences on the negative. Two-dimensional density curves of lunar shadows are thus drawn by the instrument and, if a known contour of density corresponds to the half-intensity isophote in the image brightness, the shadow lengths can be quickly measured. Since it is usual to enlarge the scale of the isodensitracing by 50 or 100 times the work of measuring the shadows can be accomplished with a graduated rule.

(3) *Measurement of Stable-based Prints.* Although not possessing the high accuracy of the two previous methods the measurement of stable-based enlargements of lunar images is a simple technique that can be used with some success to provide rapid reductions of lunar heights.

The fourth method differs in several important respects from the three methods outlined above, for it must be conceded that these, in principle, are so very similar as to constitute variations on but a single theme. Let us therefore call the fourth method the method of control points.

(4) *Method of Control Points.* Much of the time of the Manchester Conference on Selenodesy has been devoted to discussion of the properties of control points on the lunar surface, and in general terms we can assert, if we use a few dozen precisely determined control points scattered fairly evenly over the visible hemisphere, that we can fit any global photographic negative of the Moon to these and in so doing supply an accurately known frame of reference that will facilitate the computation of the lunar coordinates of any other point on the visible lunar hemisphere.

The Mann Comparator (on loan to the Department of Astronomy from the U.S.A.F.) has been successfully used for this and associated work for more than two years. Without going too deeply into the technicalities of the task of measuring the lunar images on this machine we can appreciate that we require a means of transforming the x, y machine coordinates so that the points they describe fit exactly into a three-dimensional hemispherical coordinate system. Just how accurately this can be done depends partly on the choice of the lunar control points and how well their respective positions are known. Much depends also on the number of points used, for, the larger the number, the more anchorage points there will be to tether a given lunar negative.

Before we even get started we find ourselves on one horn of a dilemma, for if we are to have the best possible "fit" we require to make large numbers of measurements of large numbers of control points, but, if we are to work on a realistic production basis to satisfy fairly tight schedules, we find that we cannot afford the additional man-hours involved. The gnomon of time may therefore be recognised as the other horn of our dilemma. Obviously some form of optimisation must be aimed at if these difficulties are to be minimised: to do this we must try to ascertain how thinly our control point net can be spread and attempt to determine the least number of measurements that can be made on each point without sacrificing accuracy. To this end the Manchester Lunar Group has conducted some experiments which will be described within these pages.

Where teamwork is involved we naturally have to contend with variations in the methods of working of each member of the group. In the identification of control points we immediately encounter problems because no two operators place their cross-wires of the measuring engine on exactly the same point. Each member has his own "personal equation" which is not easy to assess since it includes optical aberrations of the eye as well as his cerebral reactions to what he sees. Because of personal errors involved in measuring the basic net of control points on a given negative as a prelude to the work of relative height measurement and reduction, and in view of possible misreadings of the measuring dials of the Mann Comparator, it has been our practice to have a minimum of two members of the team to provide the control point data. If this part of the work is not completed correctly the rest of the reductions will be adversely affected.

So as to provide reasonable coverage of the lunar disk we have used up to 41 control points of which 30 are common to the ACIC system and whose positions are known with a proportional accuracy of 10^{-4} in latitude and longitude. This list of control points is included in the contribution of Drs. G. A. Mills and M. E. Davidson in this volume, and the reader should turn to these pages.

Having taken into account the prevailing librations of the Moon, at the instant of a given exposure, one can then proceed to convert the control point coordinates λ, β and h into ξ, η, and ζ.

This yields sufficient information to calculate the instantaneous direction cosines, X, Y, Z, which may readily be converted to plane coordinates by using the formulae:

$$X' = X + XZ \sin s',$$

$$Y' = Y + YZ \sin s'.$$

The expression sin s' embodies the horizontal parallax of the Moon; the semi-diameter and the topographical position of the telescope.

The task of the computer programme that Dr. G. A. Mills has devised is to compare the X', Y' coordinates with those obtained by direct measurement on the Mann Comparator. The computer is also instructed to determine which of the machine measures corespond to longitude and which to latitude, and to make corrections that result in increasing x, y machine coordinates conforming to increasing longitudes and latitudes respectively. Using expressions including x, y, and the plates constants, the computer solves for the plane coordinates X' $Y'$and forms residuals for each control point. After this, the rms residual for X' and Y' is computed from all the control point values and individual residuals are compared to these. The computer programme rejects any individual point that produces a residual $2.718(x)$ rms residual and repeats the process until all values come within the accepted limits.

In the first of the experiments conducted by members of the Lunar Group in Manchester, five men were asked to measure 27 control points on a gibbous phased negative exposed at the Pic-du-Midi Observatory on 20th July, 1964 at 20h. 56m. U.T. (for its reproduction, see Figure 1). Each control point was measured 15 times per person, making a total of 405 pairs of measurements per person which can be reckoned to be more than one day's work. The purpose of this task was as follows:

(1) To amass sufficient data as to allow meaningful comparisons to be made between the five sets of residuals of all individual control points.

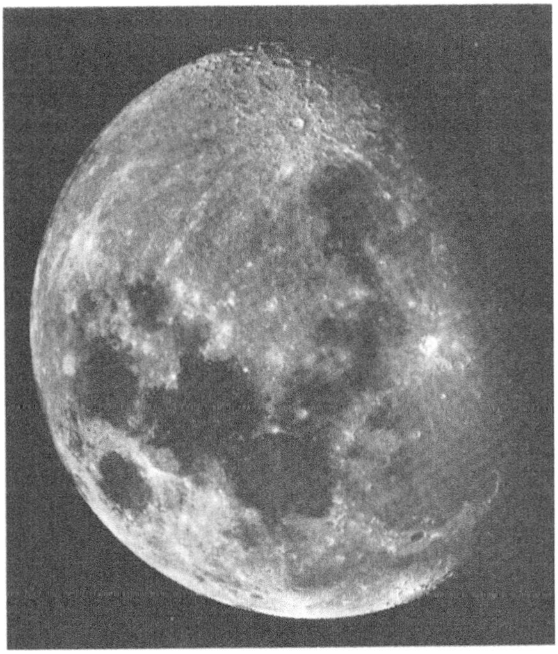

Fig. 1. The Moon, 20th July, 1964 at 20h 56 min. Observatoire du Pic-du-Midi photograph, taken with the 24-inch refractor.

(2) To compare each operator's rms residuals in order to see what variations there were from one operator to another.

(3) To determine the relative height differences caused by 1 and 2 using constant relative height input data.

Because of the large amount of numerical data involved, the author has abandoned tabulated material in favour of graphical representations which show essentially how the mean machine coordinates, x, y, for each control point vary from one operator to another. In Figures 2 and 3, each operator is identified by the letters A, B, C, D and E. Vertical lines indicate the total spread in microns for each point and short horizontal lines crossing them show where the mean position for each operator's measurement is located. Figure 2 gives this information for x and Figure 3 displays similar information for y. In both figures the abcissae represent the control point reference numbers.

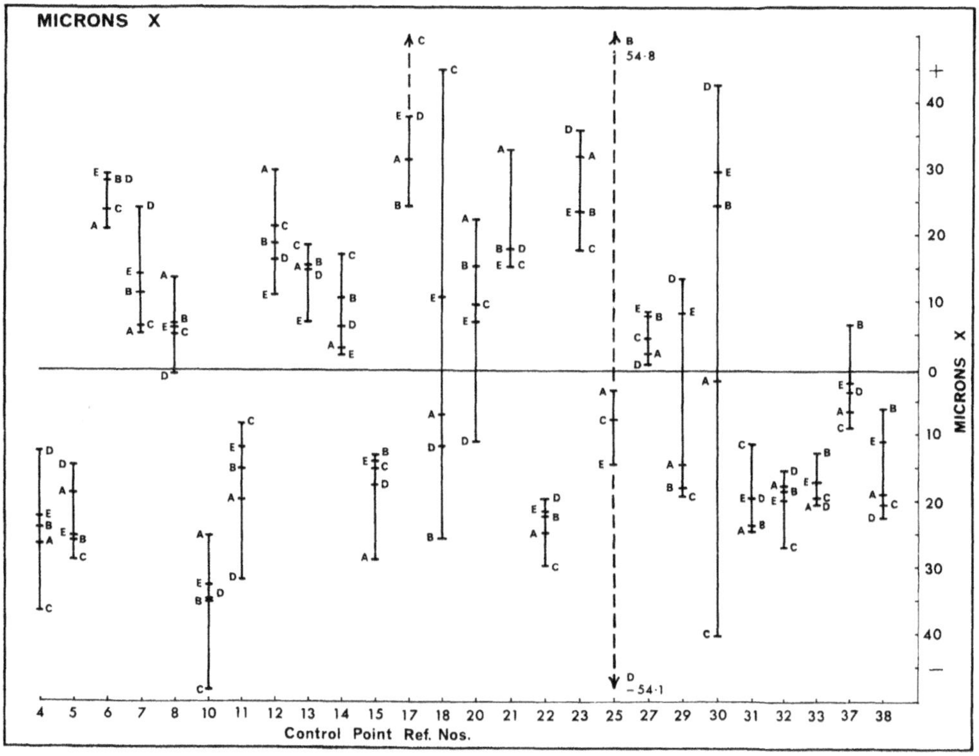

Fig. 2.

These figures show that in the main the agreement between all five operators is reasonably good. On the other hand, there are one or two points that show considerable variation, and it is fairly certain that identification errors, or possibily, errors arising from misreading the instrument scales may have played some part in these few isolated instances.

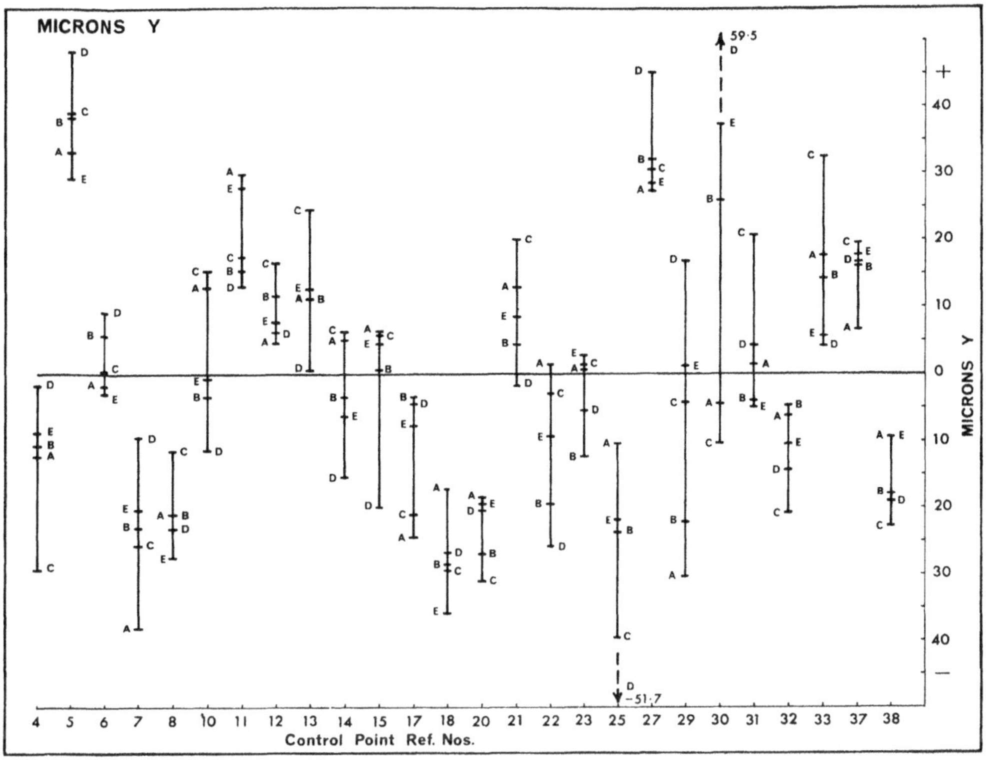

Fig. 3.

Using all the information contained in Figures 2 and 3 we can obtain the rms residuals for all 27 control points that were measured by the five operators from the Pic-du-Midi negative of 20 July, 1964, 20h. 56m. U.T. (Figure 1). This information is given in Table I. Also given are the equivalent values in terms of lunar radii, kilometres and seconds of arc.

TABLE I

RMS Residuals for 27 control points measured on negative secured at Pic-du-Midi on 20th July, 1964, 20h. 56m. U.T.

	lunar radii		kms		microns		sec/arc	
A	0.00028	0.00023	0.5	0.4	22.1	17.7	0.25	0.20
B	0.00028	0.00023	0.5	0.4	21.9	18.4	0.25	0.21
C	0.00037	0.00028	0.6	0.5	29.0	21.8	0.33	0.25
D	0.00032	0.00031	0.6	0.5	25.1	24.1	0.28	0.27
E	0.00024	0.00023	0.4	0.4	18.6	17.9	0.21	0.20
	0.00030	0.00026	0.5	0.4	23.3	20.0	0.26	0.23
A	0.00002	0.00003	0.0	0.0	1.2	2.3	0.01	0.03
B	0.00002	0.00003	0.0	0.0	2.4	1.6	0.01	0.02
C	− 0.00007	− 0.00002	− 0.1	− 0.1	− 5.7	− 1.8	− 0.07	− 0.02
D	− 0.00002	− 0.00005	− 0.1	− 0.1	− 1.8	− 4.1	− 0.02	− 0.04
E	0.00006	0.00003	0.1	0.0	4.7	2.1	0.05	0.03

It is relatively easy for us to inspect and compare the five sets of rms residuals, and a glance at Table I reveals that operators A and B made measurements that resulted in very close agreement. By comparison, C's residuals are on the high side and D's are also larger than the average values. Operators E's rms residuals are fractionally better than those of A and B. Even allowing for the larger values of C and D the rms residuals of about 0.5 kilometre, or between 0.20 and 0.25 seconds of arc implies dispersions of measurements not much in excess of the diameter of an Airy disk (20 μ) in the principal focal plane of the 60-cm refractor at the Pic-du-Midi Observatory.

No attempt will be made to explain the differences in the rms residuals of operators A, B, C, D and E but, at the same time, we may pursue the problem to see what effects these differences produce on the relative heights since this is really the "raison d'être" for the measurement of the control points in the first instance. What then happens to the end-product, the relative lunar heights?

First of all, we have to define the shadow length which means, in the case of the control-point method, that we have to measure the machine coordinates of the shadow-casting point and those of the shadow tip. For the purposes of this paper only a few values have been given from the large number that was available – the author has no intention of frightening the reader away by asking him to peruse vast amounts of computer printout.

Let us turn to the few results shown in Table II, which gives samples of the height printout data for points 395–405 inclusive. To produce this table identical shadow-point and tip coordinates were used, and the variations between the heights and errors (the latter printed below each height value) are entirely due to differences arising from personal errors in each set of individual control point measurements. Columns A, B, C, D and E contain the computed heights and errors in metres. The LAC area under scrutiny was in this case LAC 24, so points 395–405 refer to heights in the Sinus Iridum region of the Moon.

As might be anticipated, the results follow the same general trends as the control point data. For example the heights and errors arising from the use of relative height input data with control point data A and B are remarkably similar. The differences amount to only a negligible 1 metre for points 400 and 401. All the rest have exactly the same values, and the errors differ by no more than 4 metres in one point (403) – the rest are closer than this. The results from C and D control-point data reveal heights that are within a metre or two of the mean values. Seldom is there a difference of as much as five metres between any of the five columns. The larger residuals of data C and D are reflected in the larger errors. For point 398, for example, the error for column C is twice that of column E. Only in points 400 and 401 do we see a reversal of this situation, but here the errors are too small to be meaningful.

Table II represents a small section of a large amount of output data that shows the same trend. It is comforting to note, despite the variations arising from the "personal equations" of different observers, that the computed relative height values are sensibly constant and only the errors displayed below them show variations that reflect the quality of the rms residuals.

TABLE II

Height in Metres

Point No.	A	B	C	D	E
395	2689 30	2689 30	2693 43	2692 36	2694 25
396	2956 37	2956 37	2960 49	2958 42	2961 32
397	2344 41	2344 42	2347 54	2346 47	2348 37
398	717 77	717 80	718 127	718 104	718 62
399	769 33	769 34	770 58	770 46	771 25
400	1775 2	1776 2	1779 1	1778 2	1781 3
401	1580 1	1581 1	1583 3	1583 0	1586 2
402	513 32	513 33	514 58	514 47	515 24
403	632 83	632 87	633 129	633 109	634 71
404	2035 12	2035 12	2038 17	2037 13	2040 10
405	559 33	559 35	560 56	560 46	561 27

There are other ways in which we can obtain information. In the above we have changed the control-point data so that each operator's measurements have been used consecutively with identical values of shadow input data. What then happens if we retain the best set of control point data (that of operator *E*) and vary the number of measurements of each shadow casting and shadow-tip point? So as to evaluate these quantities the computer was given relative height data in batches of 3, 6, 9 and 12 individual point coordinates for each shadow-casting point and tip and the results for heights 395–405 inclusive are given in Table III.

From this table there is a trend that suggests that the computed height values fall as the number of measurements of coordinates are increased, and this difference is emphasised between the values taken from the 3 and 6 measurement columns. There is a greater stability about the values that occur in the 9 and 12 measurements columns which might be interpreted as meaning that nine or twelve measurements per point

TABLE III

Point No.	No. of Shadow input Measurements				
	3 Height	6 Height	9 Height	12 Height	Mean Height
395	2694 25	2690 28	2680 28	2678 24	2685
396	2961 32	2907 34	2901 30	2899 25	2917
397	2348 37	2324 25	2297 25	2282 22	2313
398	718 62	661 129	567 109	635 79	663
399	771 25	757 40	743 39	739 32	752
400	1781 3	1778 2	1771 2	1767 3	1774
401	1586 2	1578 3	1568 3	1569 3	1560
402	515 24	493 44	482 35	484 28	493
403	634 71	590 66	564 63	569 49	589
404	2040 10	2040 3	2052 3	2045 2	2044
405	561 27	546 26	554 23	548 20	552

are sufficient for this type of reduction. Three measurements per point are undoubtedly insufficient to yield reliable information about the true positions of the shadow-casting and shadow tip points. The percentage errors are, of course, larger for the shorter shadows since the errors of measuring remain approximately the same regardless of the shadow length.

Conclusion

So many variables are involved in the work of shadow measurement that it is difficult to make hard and fast rules about the conduct of the work. The control-point method is rapid and is the only one that can be used effectively to reduce the large amount of information that is needed to fill a lunar chart as large as the ACIC LAC areas in a comparatively short time. If our computations presented within these pages are typical, we can state the following:

(1) Personal errors in the manipulating of the Mann Comparator and the reading of the dials do not adversely affect the height reductions, although the magnitude of the residuals is reflected in the errors given with the heights.

(2) Twelve or fifteen measurements of each pair of control point coordinates make a reliable fabric upon which to base the relative height reductions. The work of measuring the control points should be split between two or more operators.

(3) Not less than six measurements – nine if sufficient time is available – should be combined to obtain reliable coordinate positions for shadow-casting and shadow-tip points.

Acknowledgements

The author would like to thank Dr. G. A. Mills, who devised the Atlas Computer programme from which these results were obtained, and also Mr. B. J. Temple, who collated much of the material during the author's absence at the Pic-du-Midi Observatory in April 1966.*

* Work reported in this paper has been sponsored by the Aeronautical Chart and Information Center under Contract AF 61(052–829 with the Department of Astronomy, University of Manchester, through the European Office of Aerospace Research, U.S. Air Force, in Brussels.

HIGH-RESOLUTION LUNAR PHOTOGRAPHY

P. V. SUDBURY*

Abstract. Recent improvements in the resolution of earth-based lunar photography are discussed. Procedures for the assessment of photography and its contribution to the lunar mapping programme are outlined.

Introduction

The continuing programme of lunar full disk photography carried out with the 24-inch Pic-du-Midi refractor by Manchester University's Department of Astronomy has been described previously. Recently the work has been supplemented by routine high resolution photography of small fields with the new 1-metre telescope at the same site. The nature of the photographic subject, the characteristics of the telescope, the proven excellence of the site, and the requirements of the programme placed conflicting restrictions on the design of the camera. Results indicate that a successful compromise has been achieved.

The Photographic Subject

The aim of high resolution photography is to detect changes in brightness over elements of the lunar surface whose size corresponds to the theoretical resolution of the telescope.

The most common representation of the photometric properties of the lunar surface is a combination of the Lommel–Seeliger law, a phase function and the local full moon albedo. The photometric brightness of an element of lunar surface is then given by

$$B = A\phi(\alpha)\frac{\cos i}{\cos r + \cos i},$$

where i denotes the angle of incidence; r, that of reflection, $\phi(\alpha)$ is function of phase angle α, and A, the local full moon albedo.

It is evident that, for a single photograph, the phase angle α is everywhere constant, and need not be considered further.

The full moon albedo A has not yet been determined photoelectrically for elements of the lunar surface smaller than 5″ in diameter. On this scale the range of albedo

* Department of Astronomy, University of Manchester.

seems to be 0.06–0.18, giving a maximum subject brightness contrast of 3:1. At much higher resolution Rangers VII, VIII and IX gave no indication that any increase in albedo contrast can be expected when smaller elements of the surface are considered. Full moon photography, aiming to detect local albedo differences of 1%, must record a contrast of 1.2:1.

The dependence of the brightness on the angles of incidence and reflection allows the detection of changes in slope between adjacent elements in the lunar surface. The simultaneous variation of i and r with slope makes the analytical expression of the brightness differences rather complicated. Numerically we can obtain figures for the

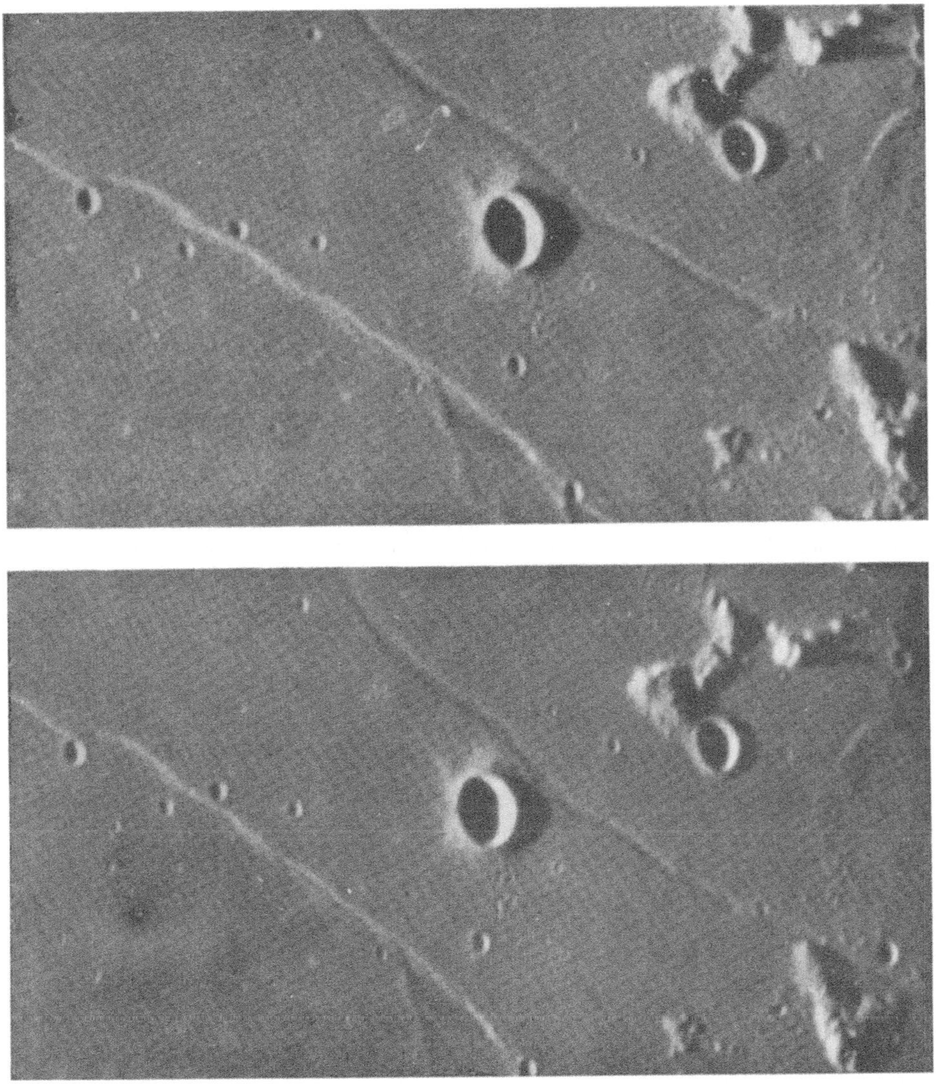

a

b

Fig. 1. Cauchy. (*a*) 1-metre telescope $f/15$ $\frac{1}{8}$-second exposure; (*b*) 1-metre telescope $f/35$ $\frac{1}{2}$-second exposure. The resolution advantage at $f/35$ is shown by improved visibility of the smaller craters.

favourable case (see, in this volume, Minnaert's paper, Figure 6) of $r=10°$, $i=80°$ and a change in slope of 5° degrees. This gives a brightness contrast between adjacent elements of about 1.6:1, and once again the aim must be to detect small contrasts to the resolving limit of the telescope.

It is only in the terminator zone, where shadows exist, that normal photographic contrasts are encountered. Unfortunately, much of the contrast in the subject is softened by the atmosphere and the optical system, and while the contrast of the coarser details that survives until the image plane, consideration of smaller elements at the edge of shadow detail shows that the image contrast drops from about 50:1 for elements of about 10000 times the area of the Airy disk to zero as the limits of seeing or resolution are reached with elements a few Airy disks in size. To detect shadow detail to that limit low contrasts must again be recorded satisfactorily by the emulsion.

The Light Path

Although Rayleigh scattering of moonlight produces a background which is small compared with lunar intensities, small angle scattering by water droplets and ice in clouds (which are sometimes present on nights of excellent seeing) will often reduce contrast considerably, and requires a compensating increase in image scale (see Figure 2c).

Dispersion by the atmosphere is also of considerable importance. Table I gives the dispersion per 100 Å at different wavelengths for the range of zenith distances encountered in lunar photography from the latitude of Pic-du-Midi. The figures may be

TABLE I

Dispersion in Seconds of Arc per 100 Å as a Function of Zenith Distance and Wavelength

z \ λ	3000 Å	4000 Å	5000 Å	6000 Å	7000 Å
20°	0″.088	0″.037	0″.019	0″.011	0″.007
30°	0″.138	0″.059	0″.030	0″.017	0″.011
40°	0″.202	0″.083	0″.044	0″.025	0″.016
50°	0″.284	0″.114	0″.060	0″.035	0″.022
60°	0″.412	0″.173	0″.088	0″.051	0″.032
70°	0″.650	0″.272	0″.140	0″.081	0″.051

compared with the 0.15 second of arc theoretical resolution of the 1-metre telescope. The permissible bandwidths are given in Table II. The figures are for sea-level and the situation for Pic-du-Midi (altitude 2862 m, mean barometric pressure 540 mm Hg) is rather more favourable.

The phenomena known collectively as "seeing" may conveniently be subdivided into two components. The first is associated with small scale spatial variation of refractive index which break up the incoming wavefront and give rise to blurring of the

TABLE II

Bandwidths Giving Less than 0″.1 Dispersion

z \ λ	3000 Å	4000 Å	5000 Å	6000 Å	7000 Å
$z < 40°$	50 Å	170 Å	330 Å	600 Å	900 Å
$z < 70°$	15 Å	40 Å	70 Å	120 Å	200 Å

images. The second is associated with time variations of refractive index and causes displacements of the image.

A night of good seeing is one in which the blurring is small and the time variations are slow. The problem of selecting the best moments of seeing is made easier if exposure times are short ($\sim\frac{1}{25}$sec) but this is not normally possible in lunar photography if faint detail is to be recorded.

The Optical System

In its present dome the 43-inch reflector can only be used as an $f/15$ Cassegrain. Some form of secondary magnification is therefore required, and this is now accomplished with an $x2$ Barlow lens, achromatised for d and a (5876 Å and 4358 Å) and designed to give a 5-inch diameter field when used with the $f/30$ Cassegrain.

Other elements in the light path are colour filters and a mylar diagonal respectively 20 and 10 cm in front of the focal plane. These components are believed to be within $\lambda/4$ of constant optical thickness over the required areas of their surfaces.

In considering the smallest crater detectable with the system, intensity variations within the first diffraction minimum of the telescope must be considered. Table III

TABLE III

Intensity Variations
within the First Diffraction Minimum

x	0	0.24	0.48	0.72	0.96	1.22
I	1	0.92	0.71	0.43	0.16	0.00

gives these intensities in terms of distance from the geometrical image centre. Under low angles of illumination a lunar crater has an intensity distribution which approximates to a bright circular annulus. In the image plane of a perfect optical system such an annulus will be broadened by diffraction but will have a totally dark centre if it has a diameter twice the radius of the first diffraction minimum. A slightly smaller annulus, one and a half Airy disks in diameter would be imaged with a central intensity 60% of the annular intensity. The resulting brightness contrast of 1.6:1 would make photography of the ring form very difficult. On the Pic-du-Midi 1-metre this corresponds to a crater diameter of 0″.23 and probably represents the ultimate limit that could be achieved. The smallest craters resolved to date are about 0''.35 in diameter, which corresponds to a spread function with half intensity width of about 0''.20.

The Camera

Requirements for a high resolution lunar camera for the *f*/15 Cassegrain focus can be listed as follows:

(1) Secondary magnification optics.

(2) Filters to limit the passbands in accordance with Table III.

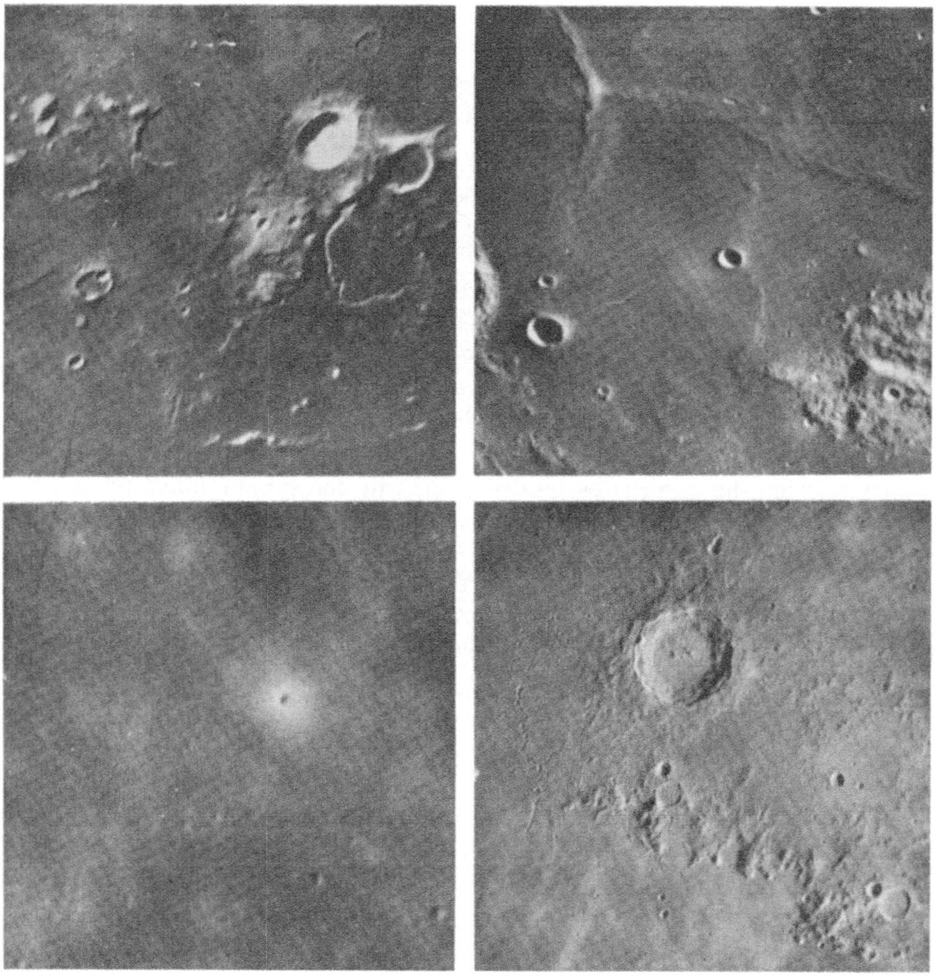

a b c d

Fig. 2. (*a*) Aristarchus, November 16th 1964, 1-metre telescope *f*/15; (*b*) N. E. Mare Serenitatis, November 24th 1964, 1-metre telescope *f*/35; (*c*) Linné (dia. 2 kms), January 10th 1965, 1-metre telescope *f*/32; (*d*) Copernicus, August 20th, 1965, 1-metre telescope *f*/32.
(*a*), (*b*) and (*c*) on Ilford Pan F, (*d*) on Kodak Panatomic-X. (*c*) illustrates the detection of albedo variations and fine detail through cloud with large image scale.

(3) Continuous visual monitoring to determine the best moments to initiate the exposure. Monitoring throughout the exposure will allow the observer to assess the probable quality of each frame as it is taken.

(4) Reflex focusing to allow changes in focus to be followed during long periods of photography.

(5) Focusing and squaring on to 40 microns. Film vibration during exposure less than 3 microns.

(6) Rapid shutter response.

The secondary magnification is supplied by a 3-inch diameter 24-inch focal length Barlow lens. The lens was designed to give a 5-inch diameter field at $x2$ with the $f/30$ Cassegrain. Aberrations are negligible over a $2\frac{1}{2}$-inch diameter field when used at $x2$ with the $f/15$ Cassegrain.

The filter normally employed is a Chance OY2. This gives a 900 Å passband (5500 Å – 6400 Å) when used with normal panchromatic emulsions, and limits chromatic dispersion to about 0''.1 of arc at a zenith distance of 40 degrees. For extreme zenith distances a Chance OR2 used with an aerial emulsion of extended red sensitivity gives a better passband (6400 Å–7100 Å). Photography is also carried out in blue light near full moon.

A mylar diagonal of 20% reflectance gives a reflex focus above the camera. This system allows accurate framing of the region to be photographed, precise focusing in light of the same colour as the photography and visual monitoring of the images through an eyepiece. The last aspect is of great importance and means that exposures are only initiated when a moment of good seeing appears imminent or has already begun, and the quality of the frame can be judged at the end of the exposure by the appearance of the image and its motion on the cross wire grid during the exposure. The importance of this degree of observer control over the conditions of photography cannot be emphasised too strongly and has resulted in a greatly increased output of excellent frames. The situation can be illustrated by the typical observation, at the camera in good seeing conditions, that only once in five minutes will seeing and telescope motion achieve simultaneously the required 0''.1 of arc stability for a $\frac{1}{2}$ second exposure time. The chance of recording that moment by random photography is less than 1 in 500. Experiments with monitored photography give a corresponding figure of about 1 in 5.

Photography is normally taken on 70 mm perforated film, in 50 exposure cassettes. The cassettes are run through a Linhof magazine with a $2\frac{1}{4} \times 2\frac{3}{4}$-inch format area. The camera also has provision for 4×5-inch plates.

Emulsions

The choice of emulsion and the image scale must be considered together. During experimental work four image scales were used: 13''.2, 7''.6, 5''.7 and 3''.8/mm, corresponding to focal lengths of 15, 30, 45 and 60 metres. The considerations given earlier show that detail at the resolving limit of 0''.15 will have contrast less than 1.6:1, and

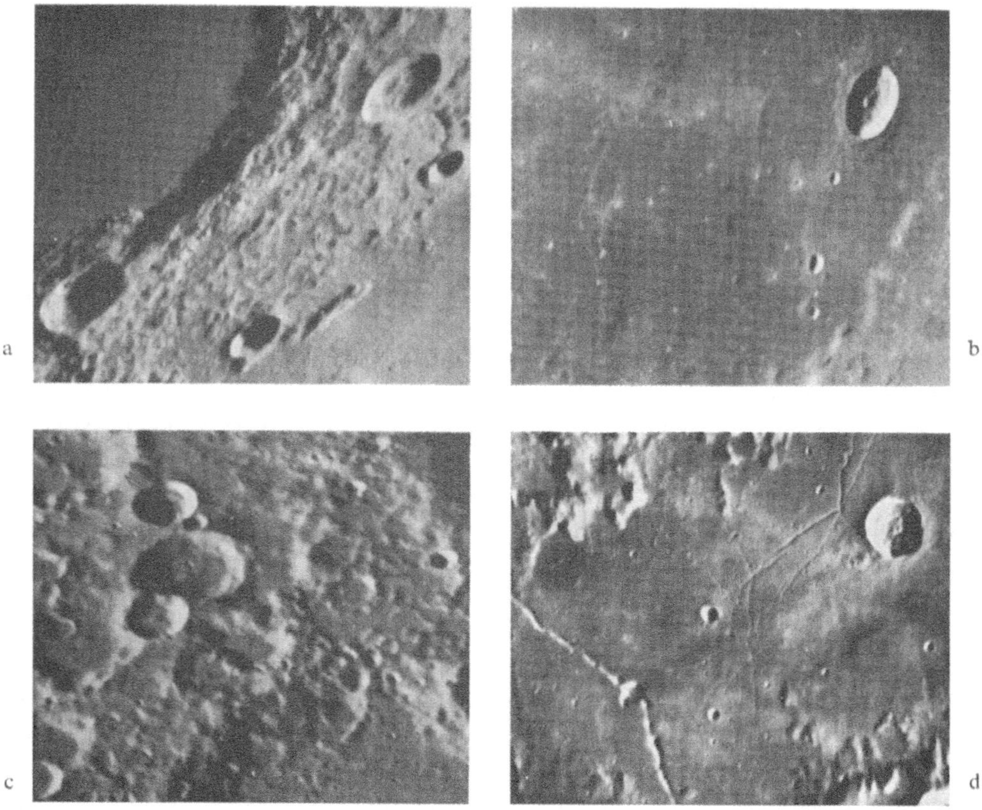

Fig. 3. Some Results from the Routine Programme. (*a*) September 20th 1966, 04h 12m; (*b*) January 4–5 1966, 00h 36m; (*c*) March 30–31st 1966, 22h 31m; (*d*) July 9th 1966, 03h 02m.

the emulsions must be at least matched for resolution at that contrast, and preferably over matched by a factor 2. The required resolutions at test object contrast 1.6 were, therefore, 88, 44, 33 and 22 lines/mm.

Table IV gives the resolutions and film speeds of some Kodak emulsions. The combination of image scales and emulsions that have achieved resolutions close to 0.″2 of arc is listed in Table V. In each case the emulsion resolution to test object contrast 1000:1 exceeded the telescopic resolution by a factor 6 or more.

In spite of the apparent desirability of reducing exposure times to lessen the effects of image movements during photography, it seems certain that the best lunar photography has been taken with exposures in excess of one second.

Two other considerations influenced the final choice of image scale. The first was the desire to use 35 mm, 70 mm or 120 roll film and take advantage of the choice of emulsions and precision-camera components available to the commercial market. The second was the need to cover fields at least 5′ of arc square if reasonable coverage was to be obtained. The difficulty of obtaining adequate control for the measurement of small high-resolution fields was evident early in the work, and has impaired the

TABLE IV

Resolution of Some Kodak Emulsions

Emulsion	Resolution TOC 1000:1	Resolution TOC 1.6:1	Speed ASA	Focal Length Required (metres)
649–0	2000			
SO 243	465	205	1.6	15
SO 132 (4404)	475	200	1.6	15
SO 226	336	112	6	15
V–0	225			
IV–0	136–225			
SO 136	186	65	20	30
Pan X (4404)	170	65	20	30
Plus X (8402)	135	52	64	30
Plus X (4401)	105	40	80	45
III–0	96–135			
II– 0	69–95			
Double–X	85	38	125	45
Super XX	75	30	100	45
Tri–X	71	22	200	60
103–0	55–68			

TABLE V

Examples of the Matching of Image Scale and Emulsions in High-Resolution Lunar Photography.

Telescope	Image Scale	Emulsion Resolution T.O.C. 1000:1	Emulsion
120-inch Lick	2″ /mm	80/mm = 0″.025	II–G
61-inch Flagstaff	13″.2/mm	2000/mm = 0″.006	649–F
1-metre Pic-du-Midi	6″.2/mm	186/mm = 0″.033	SO136

usefulness of much high-resolution material and prevented its use in the determination of relative altitudes.

The final choice of image scale was 6″.2/mm (32 metres effective focal length) to be used with Kodak Panatomic-X or SO-136 emulsion. The field is 7′.2 × 5′.5 corresponding to the $2\frac{1}{4} \times 2\frac{3}{4}$ format of the Linhof magazine. A complete terminator can thus be covered in five overlapping fields.

Results

Each night worked is graded from one to five, according to the quality of the best negative from the night. The grades may be interpreted approximately in terms of the resolution of craters and an angular resolution

Grade	Smallest Crater	Resolution	Remarks
5	0.9 km	0″.4	Excellent
4	0.9–1.2 km	0″.5	Very good
3	1.2– 2 km	0″.8	Fair
2	2 – 4 km	1–2″	Poor
1	4 km	2″	Bad

The results fall into three sections: (a) Tests carried out with a 35 mm camera November 1964–January 1965; (b) Trials with the 70 mm camera July–August 1965; (c) Subsequent routine work with the 70 mm camera:

Nights Worked	Quality				
	1	2	3	4	5
(a) 14	1	2	2	6	3
(b) 15	2	2	5	5	1
(c) 38	4	4	10	7	10

On this basis about 12 excellent nights a year are to be expected. Whole disk coverage to resolution better than 0″.4 may be completed in about two years.*

* Investigation supported by the Terrestrial Sciences Laboratory, Cambridge Air Force Research Laboratories, under Grant EO AR 64–49 to the Department of Astronomy, University of Manchester, through the European Office of Aerospace Research, in Brussels.

ISODENSITOMETRIC AIDS TO LUNAR CHARTING

M. T. JONES*

Abstract. Two possible applications of the recently developed Technical Operations Isodensitracer to lunar charting are discussed:

(i) The determination of the selenocentric coordinates of photographed lunar features requiring densitometric analysis, with particular reference to relative altitude determinations through the medium of the shadow-projection method;

(ii) The charting, by photometric methods, of lunar regions registered on photometrically controlled photographic images obtained from spacecraft based camera systems, with particular reference to the Orbiter series of space stations.

Introduction

The introduction of spacecraft-based camera systems to lunar surface studies has added much to our knowledge of the surface's small scale structural relief and should, in time, enable the accurate techniques of stereoscopy and photometry to be applied to a large-scale charting of the whole lunar surface. However, as the immediate aim of present space programmes is to select suitable targets for a manned lunar landing, the use of such spacecraft observing platforms for surface mapping is somewhat restricted to areas within close range of the lunar equatorial belt and to regions where large scale surface irregularities are absent. Consequently, a full topographical analysis of the small-scale surface relief will, for some time at least, continue to require the reduction of photographic data collected by terrestrially based telescope systems. Stereoscopic and photometric methods have a limited application to such data and the most suitable reduction process is given by determining relative altitudes from the length of shadows cast by illuminated surface features. This method was originally applied by Schröter as far back as 1791, but its experimental techniques are being continually brought up to date by the Astronomy Department at Manchester University, under the direction of professor Kopal. The most recent development in this field has been in the application of isodensitometry for determining the selenocentric coordinates of shadow casting features.

The conventional shadow projection method for lunar charting is somewhat restricted in that it can only provide relative height values between pairs of shadow points and leaves much of the available photographic data unanalysed. It is, thus, unlikely to provide an efficient method of analysis when spacecraft-based camera

* Department of Astronomy, University of Manchester.

systems are developed to obtain photometrically controlled photographic coverage of the lunar surface. With the probability of such coverage being obtained in the near future from the NASA Orbiter programme, it is essential that attention should also be paid to the application of the photometric method to a large-scale charting of selected areas of the lunar surface. It is proposed that the Technical Operations Isodensitracer (I.D.T.) may be adapted for such an analysis and that it will fulfil most of the requirements of an accurate photometric analysis system.

The Shadow Projection Method for Use on Photographic Images obtained through Terrestrially Based Telescope Systems

GEOMETRICAL ASPECTS

The basis equation for the shadow method relates the height, h (in terms of the mean lunar radius k) of a shadow-casting eminence above its corresponding shadow tip, to the angular selenocentric shadow length, ω, cast under a solar altitude, v, as follows:

$$\frac{h}{k} = \frac{\cos(v - \omega)}{\cos v} - 1 .$$

A full analysis of this equation in terms of suitable measureable parameters may be found by reference to KOPAL (1962). Once the parameters of observation, relating the relative configuration of the observer-sun-moon system at the time of photographic exposure have been interpolated from the appropriate *Almanac*, the sole requirements for a solution for h are determinations of:

(a) the selenocentric latitude and longitude, β_p, λ_p, of the shadow-casting eminence, and

(b) the shadow length, L, projected in the focal plane of the observing telescope.

The reliability of determined heights must obviously depend on how accurately these two parameters can be measured, and the fractional error introduced by inaccuracies in these measurements can be shown to be:

(a) From inaccuracies $\delta\lambda_p$ and $\delta\beta_p$ in λ_p and β_p:

$$\frac{\delta h}{h} \simeq C \tan(\lambda_\odot - \lambda_p)\,\delta\lambda_p - C \tan\beta_p \cdot \delta\beta_p ;$$

C being a corrective term to allow for lunar surface curvature and having the value

$$C = \left(1 - \frac{r \tan\mu \cos^2 v}{2k \sin v \sin\theta}\right)^{-1},$$

where $\lambda_\odot$ = selenocentric solar longitude, f = telescope focal length, $\mu = L/f$ if f is large, r = distance from observer to shadow-casting point, and θ = angular separation of the Sun and the observer at the shadow-casting point.

(b) From inaccuracies δL in L:

$$\frac{\delta h}{h} \simeq \frac{r \cos v}{h \sin\theta} (\tan^2 v - 2h/k)^{\frac{1}{2}} \frac{\delta L}{f} .$$

Errors in determined heights are particularly sensitive to inaccuracies in L where short shadows are being measured, while errors in λ_p tend to dominate as the shadow lengthens.

DENSITOMETRIC ANALYSIS OF SHADOW BOUNDARIES

In view of the existence of penumbral zones at the ends of photographed lunar shadows, caused by (a) the finite angular diameter of the Sun as viewed at the lunar surface and (b) resolution factors controlled by the apparatus function, direct measurement on visual viewing machines is of limited accuracy in a location of the required shadow points. Accurate shadow-point identification requires a full photometric analysis to be made of individual density scans taken across the shadows being measured. The most suitable output for these is given by instruments capable of automatically recording density scans on calibrated sheets from which the shadow length may be measured, and a full account of the necessary techniques and measurement criteria may be found by reference to JONES (1966).

For lunar charting, three distinct methods of scanning present themselves:

(1) *Single microdensitometer tracing.* This is obviously limited in its application and can only provide scant information on the lunar terrain.

(2) *Shadow progression tracing.* This approach utilises the natural progression of shadows over lunar terrain, as the Sun rises or sets, as a means of analysing that terrain. In its original form (MCMATH *et al.*, 1937) this method involved a systematic repetition of single tracings over a sequence of photographs. This was somewhat limited by difficulties in obtaining a precise repetition of the scanned shadow and was liable to systematic inaccuracies if any undetectable progression of the shadow-casting point occurred as the solar altitude changed. A more systematic approach is offered by analysing shadow progression with the Raster technique of scanning.

(3) *Raster tracing.* This requires a series of closely spaced parallel tracings from a single negative, and makes the fullest use of available data, providing information in a North–South direction as well as East–West. Its greater negative coverage leads to a more accurate identification of the relative positions of analysed points, and consequently a more accurate determination of their selenocentric coordinates.

SELENOCENTRIC COORDINATE DETERMINATION

Due to librational effects and the influence of any deviations of the lunar surface from a mean sphere, an accurate determination of the selenocentric coordinates of any lunar point ideally requires that a complete grid system be built onto the photographic image under analysis from available lunar control points. However, this necessitates the use of high-precision measuring machines which rarely accommodate a microdensitometric analysis system with which to perform an accurate identification of shadow points. Furthermore, as it is usually impractical to extend microdensitometric scanning to cover a region containing sufficient control points, selenocentric coordinates must be determined by interpolation from existing lunar charts.

Until recent years, the most serious limiting factor of the shadow method had been the unsuitable presentation of lunar coordinate systems. For a long time, the accepted reference was the map prepared by Blagg and Müller and published by the I.A.U. in 1935. This map was portrayed by a simple cartographic sketching technique but this, by its very nature, produced only a very subjective skeleton outline of lunar surface features (see Figure 1). A much more accurate cartographic presentation is now given

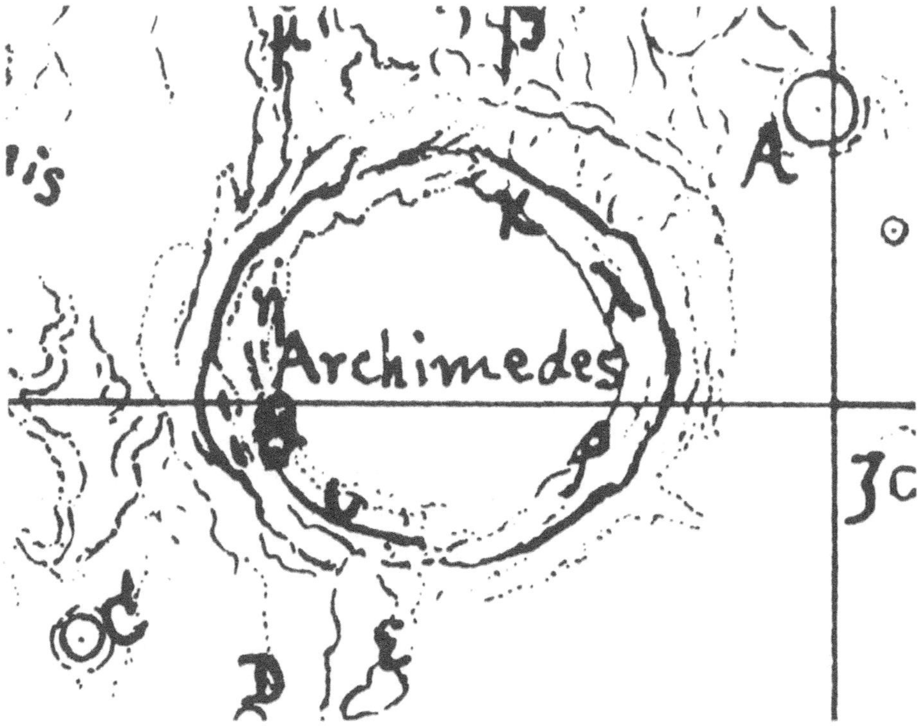

Fig. 1. Archimedes as portrayed in the Blagg–Müller Atlas. South is to the top, and the grid lines ξ, η are drawn at 0.1 lunar radii intervals.

by the Lunar Charts prepared by the United States Air Force on a scale 1:1 000 000 (Figure 2). Unfortunately, conventional cartographic presentations introduce identification uncertainties in the transference of coordinate systems to lunar negatives blurred by resolution factors. Indeed, the only satisfactory medium for the presentation of such coordinate systems for this use is photography itself, where the surface representation is accompanied by the tone variations, in terms of resolution and surface slope, that occur on the negatives used for height determinations. At present the best use of the photographic medium is made in the *Orthographic Atlas of the Moon* compiled by Kuiper, Arthur and Whitaker and published as a supplement to the *U.S.A.F. Lunar Atlas* (Figure 3). It incorporates the positional data of Franz and Saunder, supplemented by Arthur's figures, and presents an orthographic coordinate

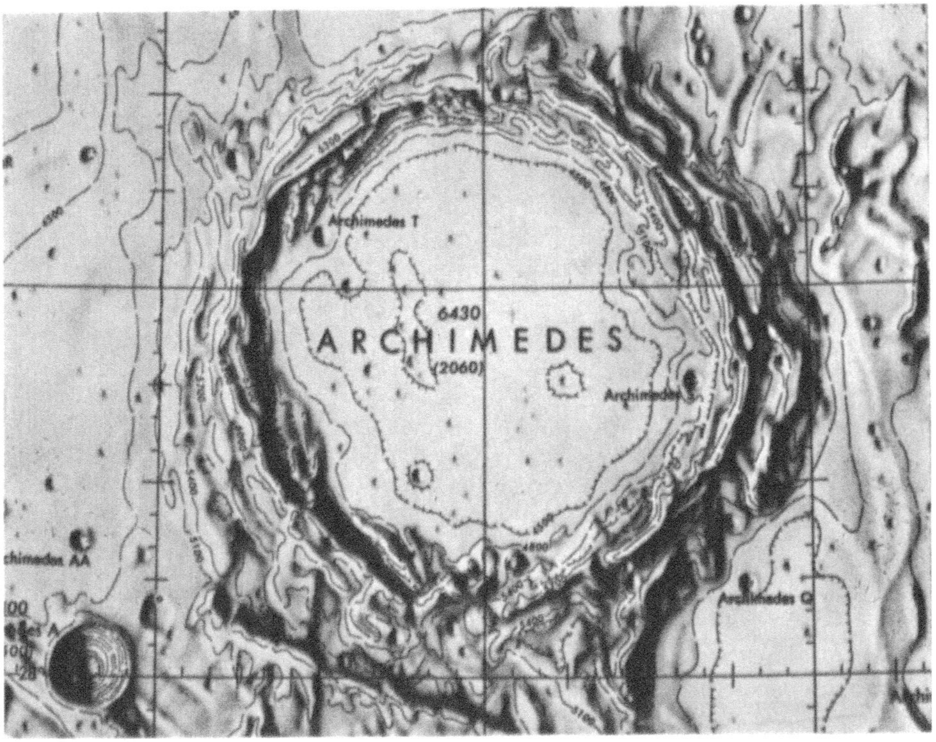

Fig. 2. Archimedes as portrayed on the U.S.A.F. Lunar Charts. North is to the top, and the grid lines λ, β are drawn at 2° intervals.

system portrayed on an orthographic projection. The telescope image is, except for small libration effects, virtually that of the portrayed orthographic projection, and so the system greatly facilitates identification. The coordinates are non-dimensional, being effectively direction cosines, and apart from scale factors and small corrections depending upon the projection, they can be easily transferred to any photographic images under analysis.

A point of selenocentric longitude λ and latitude β at a distance R from the centre of the lunar sphere has orthographic map coordinates of

$$X = R\xi = R \cos\beta \sin\lambda$$
$$Y = R\eta = R \sin\beta$$

but, for simplicity, Kuiper's *Atlas* treats all points as lying on a mean sphere of radius 1738 kms$=k$, so that its orthographic map coordinates ξ_k, η_k are given in units of this mean figure. It is readily observed that any deviations $\pm\ \delta R$ from this mean sphere will then result in inaccuracies in ξ, η of

$$\delta\xi = \pm \frac{\delta R}{k} \xi_k$$

$$\delta\eta = \pm \frac{\delta R}{k} \eta_k$$

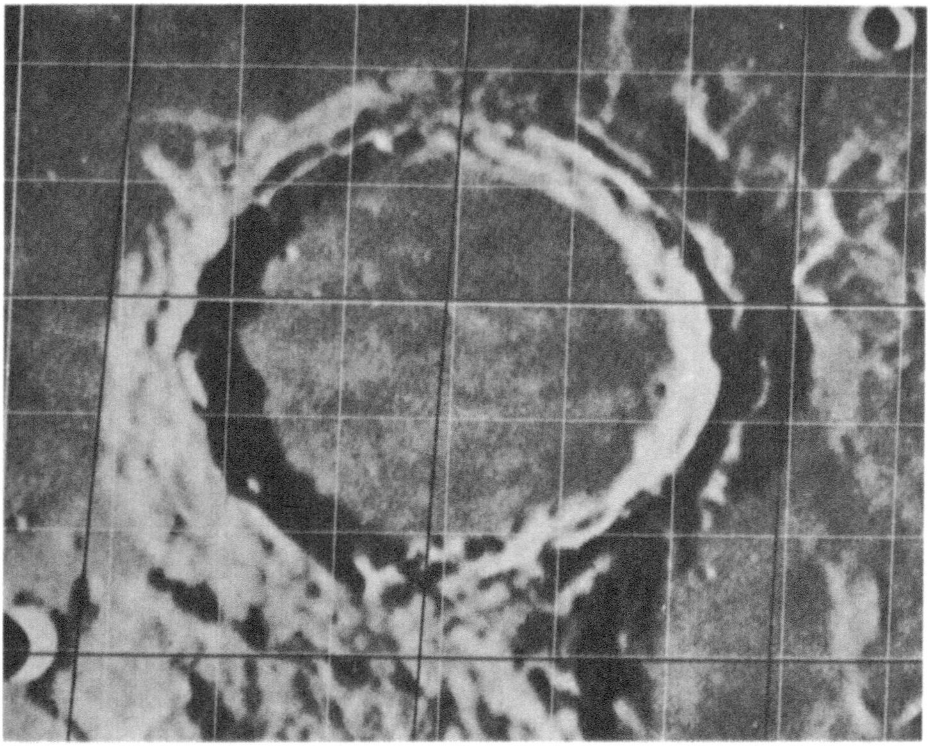

Fig. 3. Archimedes as presented in Kuiper's *Atlas*. North is to the top, and the grid lines λ, β (black) are drawn at 2° intervals, while the ξ, η (white) lines are drawn at 0.01 lunar radii intervals.

Furthermore, when the orthographic coordinates of a shadow casting eminence are read off a grid transferred from Kuiper's *Atlas*, they will be read off as a projection of the casting eminence against the local lunar surface and will not necessarily represent the true selenocentric position of the eminence. If the projected peak coordinates are determined as λ_B, β_B, the appropriate corrected coordinates λ_p, β_p of the peak may be derived from:

$$\sin\beta_p = \sin\beta_B \cos a + \cos\beta_B \sin a \sin b$$
$$\sin(\lambda_B - \lambda_p) = \sin b \sin a \sec\beta_p$$

where $\sin b = \sin(\lambda_B - \lambda_E) \cos\beta_E \sec(v + \theta)$, and $a = \pi/2 - v - \theta - \sin^{-1}(R/(R+h) \cdot \sin(v + \theta + \pi/2))$, λ_E, β_E being the selenocentric coordinates of the terrestrial observer.

The accuracy of selenocentric coordinate determination by reference to the *Atlas* is obviously limited to that of the *Atlas* itself, which claims an accuracy of at least 0.001 in either orthographic coordinate in the central area of the lunar disk. It has been found by computational checking that a factor 3 improvement on the 0.001 figure may occur in certain regions (as in the Archimedes region analysed), but these figures cannot hope to be representative of regions well away from the disk centre. For regions approaching the limb, the above corrections may be applied to minimise errors although it must be appreciated that a large uncertainty may be present in the final derived height values.

GRID TRANSFER TECHNIQUES

To reduce errors in the determination of the selenocentric positions of shadow edges to a minimum, great care must be exercised in utilizing the material of Kuiper's *Atlas*, and this necessitates an accurate transference of its coordinate system to the microdensitometric data. As microdensitometer tracings reduce the two-dimensional negative into one-dimensional scans they are, in such a form, unsuitable for accurate coordinate determinations and the two-dimensional scene must be recreated, together with the photometrically determined shadow-casting and shadow-tip points. It is for this reconstruction that isodensitometric techniques become invaluable. For each scan, trace density levels at suitable intervals are projected to the scan abcissa and these intersecting positions are transfered in single line to graphical paper. Successive scans are linearly represented at separations chosen to incorporate the magnification factor used in their production and points of equal density are connected so as to provide an isophote diagram representative of the original negative area analysed. Shadow-casting and shadow-tip points are also transfered to this diagram. Unfortunately, with single scan microdensitometers an isophote construction is long and tedious, usually providing only a skeleton reproduction of the original scene (Figure 5). A more suitable output for such a construction is presented by the newly developed I.D.T., which

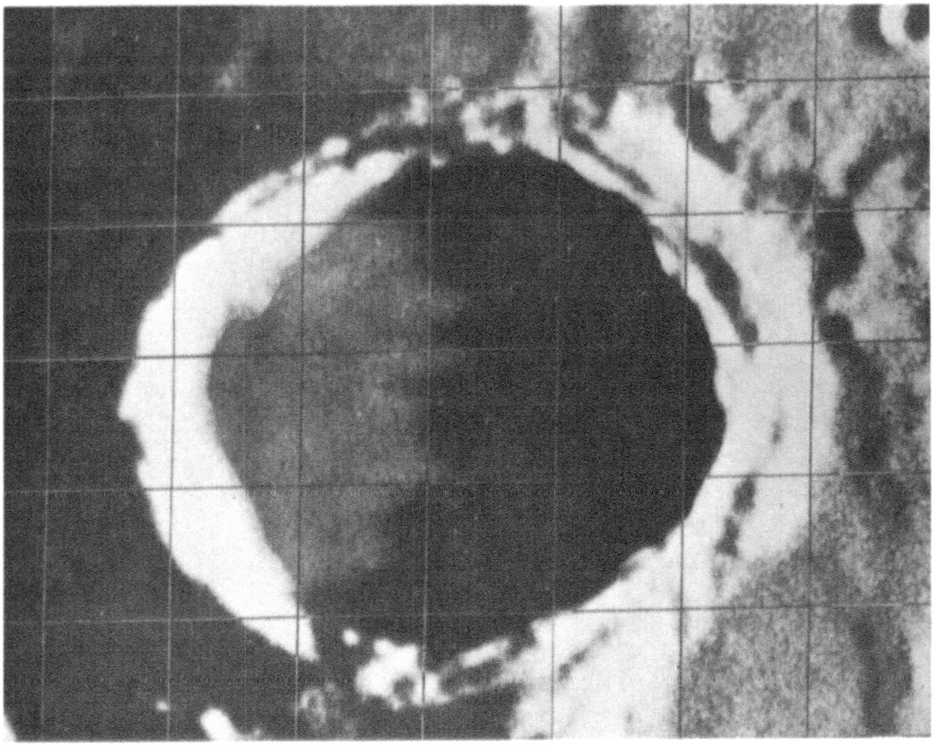

Fig. 4. A typical intermediate grid transfer from Kuiper's *Atlas* to a photograph of the crater Archimedes. North is to the top, and the grid lines ξ, η are drawn at 0.01 lunar radii intervals.

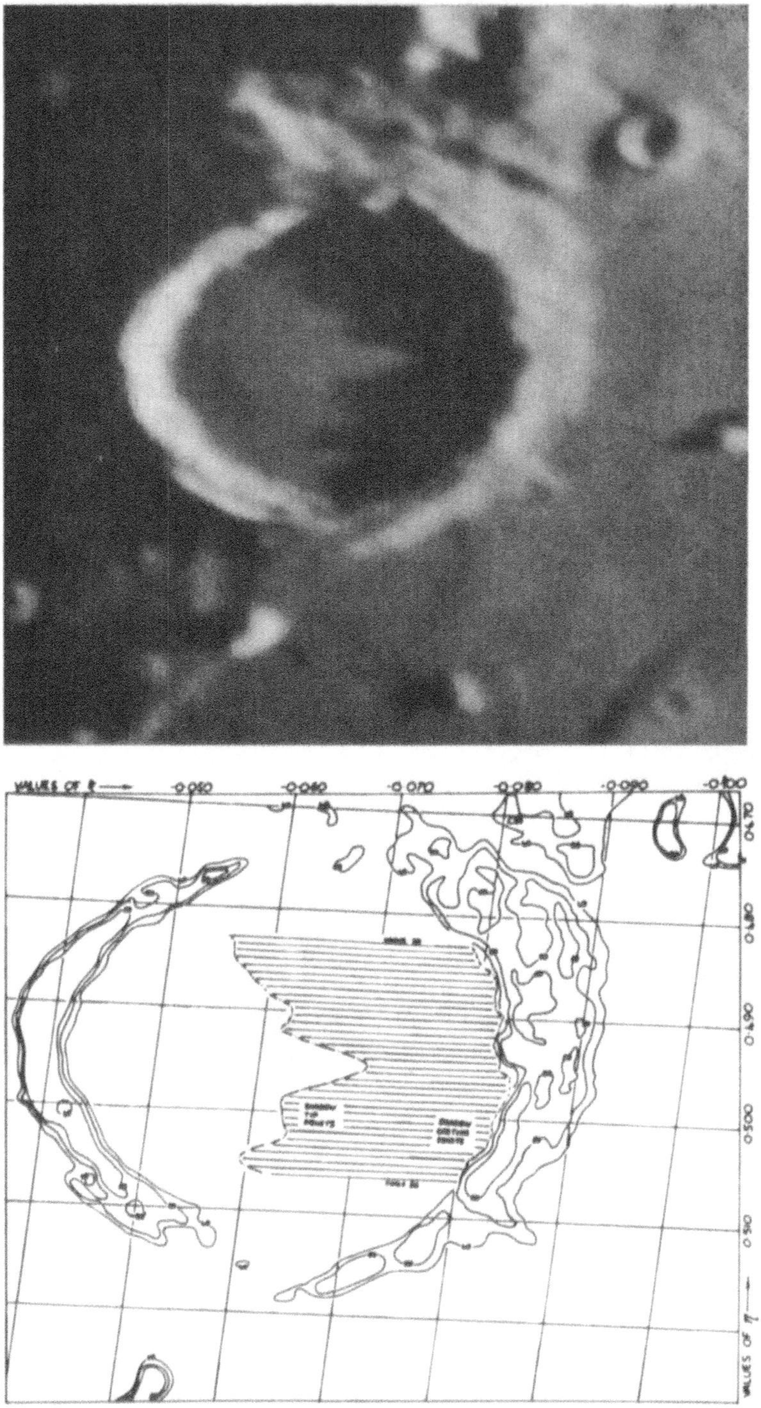

Fig. 5. A typical grided isophote diagram constructed from a manual correlation of successive microdensitometer scans. Above is shown the photographic image of Archimedes from which the analysis was made.

automatically scans and measures density, plotting the values as a quantitative two-dimensional map of the scanned area, thus eliminating the tedious manual correlation of data from successive traces. With its dropped line technique of output it is also possible to locate shadow boundaries directly on the isophote diagram (Figure 6).

Although fully covering the visible lunar disc, Kuiper's *Atlas* only portrays each lunar region under a single angle of illumination. This does much to complicate grid transfer when areas are analysed from photographs taken under widely differing phase conditions, and may lead to interpretation inaccuracies when the isophote diagram is grided by direct reference to the *Atlas*. Instead, an intermediate grid transfer to an enlarged positive print of the negative used in the analysis of a given region becomes necessary. This follows from the fact that it is easier to allow for the effects of phase changes, on the apparent position of grid lines, by use of a photographic image than by a lined isophote diagram. For grid transferring it should be noted that, on the standard orthographic map, grid lines appear as two mutually orthogonal sets of parallel straight lines, thus defining a uniform grid of squares. Librated photographs distort this grid into two families of ellipses. However, for small lunar surface regions (the permissible area being dependent on the region's position on the disk and the libration) it is possible to assume straight grid lines, although the two sets are not necessarily mutually orthogonal.

The intermediate grid transfer is accomplished by transfering, to the positive print, the intersection points of the atlas grid lines with various peaks and craterlets near the analysed region and, over a small area, a straight line construction through these points recreates the grid system (see Figure 4). Once a grided print is available under the same lighting conditions as the isophote diagram, a grid transfer to the diagram can commence. Diagrams built on the normal accepted magnification of 50 restrict the area onto which grid lines may be transfered, and consequently the number of suitable transfer reference points is greatly reduced. This normally requires that the isophote grid has to be constructed from a least squares computation on the available points. Once this has been completed, the orthographic coordinates of shadow points on the isophote diagram can then be determined by linear interpolation and finally converted to selenocentric values.

The values obtained for the orthographic coordinates ξ_p, η_p of the shadow-casting point may be checked by comparing the coordinates ξ_T, η_T of the shadow tip, as measured off the grided isophote diagram, with those computed from the measured shadow length and the determined values of ξ_p and η_p. This computational check was used by the author on all shadow-tip coordinates obtained from an extensive experimental analysis of the walls of Archimedes and Plato, and the agreement between each pair of derived values was found to be satisfactory. Differences in excess of 0.0005 lunar radii were rare and, in general, a figure of 0.0003 lunar radii or less was more representative for those areas analysed. These figures illustrate that, providing it is used carefully, the developed isodensitometric aid is a most suitable one for minimising errors in the determination of the selenocentric coordinates of lunar surface points analysed by shadow-projection techniques.

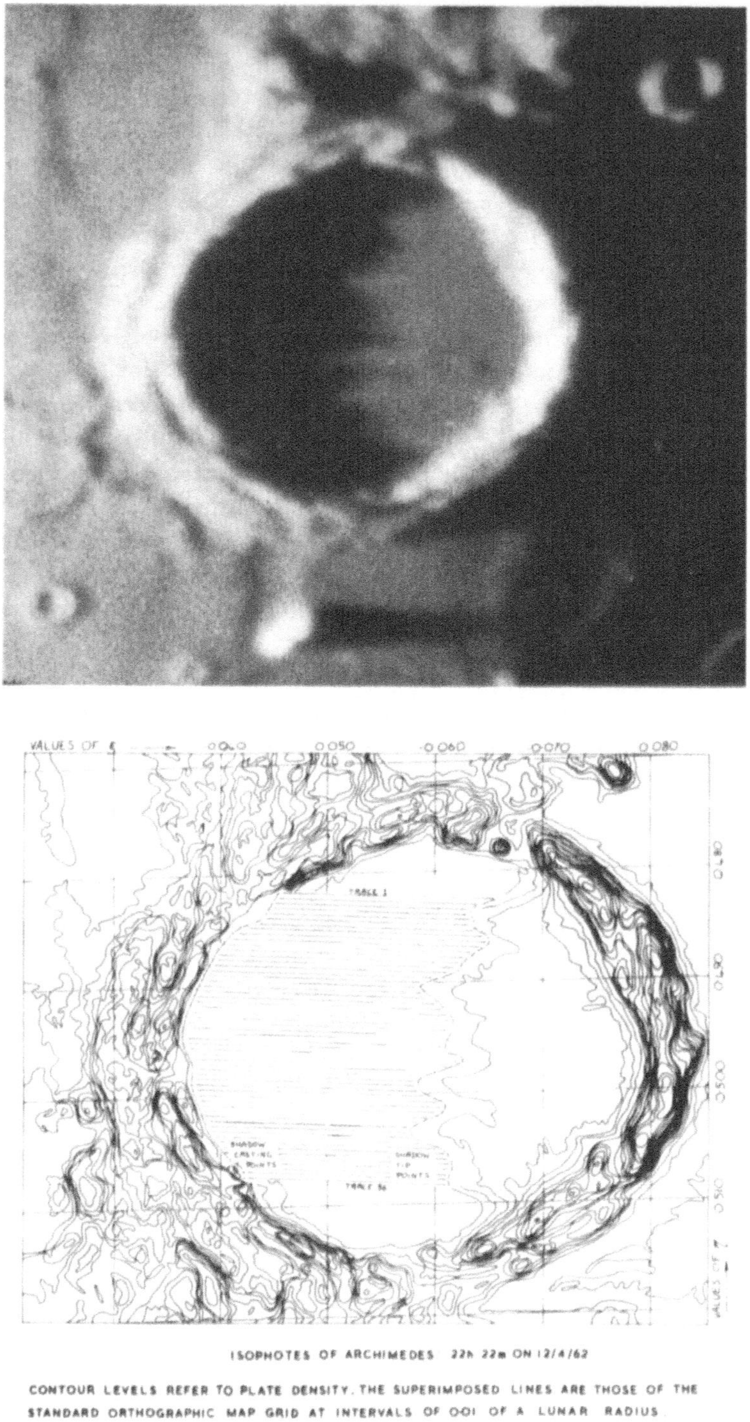

Fig. 6. A typical grided isophote diagram constructed by an I.D.T. automatic correlation of successive density scans. Above is shown the photographic image from which the analysis was made.

The Photometric Method for Use on Photometrically Controlled Photographic Images obtained through Spacecraft-Based Camera Systems

GENERAL PRINCIPLES

The basic principle of the photometric method is that the light intensity scattered in a given direction from any discrete element of the lunar surface may be considered as being characteristic of the surface slope of that element. However, before any practical application of this principle can be made, some consideration must be paid to the form of the lunar surface scattering law.

On experimental evidence, VAN DIGGELEN (1959) suggested that a single photometric function ϕ of the angles of incidence (i), observation (e) and phase (g) could be found to represent the effect of illumination variations on the light scattered from any lunar surface element, i.e. the intensity (light flux/unit projected area/unit solid angle) scattered to an observer by a surface element at P could be given by

$$I_p = A_p \phi(i, e, g) \qquad (1)$$

where A_p is the normal albedo (light flux scattered/unit projected area/unit solid angle for normal illumination and normal observation) of the element at P. Thus, for a given photographic image, assuming for the present that the imaging system is perfect, i.e. the transparency transmittance is a direct function of the intensity distribution of the input scene, the recorded intensity I'_p at any corresponding image point P' can be represented by

$$I'_p = A'_p f(\lambda_p, \beta_p, \alpha_p) \qquad (2)$$

where λ_p, β_p = the selenocentric coordinates of the object point P, α_p = relative surface slope, at P, to the lunar sphere, and A'_p = the instantaneous albedo at P.

As Equation (1) was suggested by experimental evidence from earth-based observations, there is a possibility that it may not be sufficiently exact for photometric mapping. The instantaneous albedo A'_p is introduced in Equation (2) such that it is equal to the normal albedo A_p if Equation (1) is exact but, if this is not strictly true, it contains a correction factor depending on i, e, and g. However, for the method to be presented below, an actual knowledge of the phase dependence of A'_p is not essential.

To overcome present uncertainties in the exact functional nature of the lunar surface scattering law, and to determine instantaneous albedo values, the Orbiter space stations are designed to give the surface dual camera coverage with fine-resolution shots nested within those of a coarser resolution. It is intended to provide suitable stereoscopic pair coverage on the medium-resolution shots so that a correlation of slope and photometric values on such shots may yield information on the illumination constants, for use in a photometric analysis of the fine-resolution shots. For each pair of simultaneously taken single shots from each camera there will be an accompanying test card in order that the photometric properties of the whole system may be evaluated. This then enables the density D'_p of each photographically registered surface element to be converted to a value I_p' representative of the intensity I_p received from that element by the spacecraft station.

In order to obtain a suitable photometric expression for slope evaluations, a series of values of x_p, y_p and D_p' with corresponding slope values α_p, determined stereoscopically, are required off the medium resolution calibration imagery. The values x_p, y_p are taken to represent the positions of each analysed image point P' relative to a Cartesian coordinate system projected within the boundaries of the photographic image. With this series of values it is then possible to construct a photometric law in the form

$$I_p' = f(x_p, y_p, \alpha_p). \tag{3}$$

It is initially assumed that the photographed surface area has a constant uniform albedo A_m and that a best fit relationship in the form of Equation (3) can be found averaged over the whole analysed region. If variations in instantaneous albedo A_p' do occur they can be deduced in units of the mean albedo from the relationship

$$A_p'/A_m = I_p'/I(x_p, y_p, \alpha_p) \tag{4}$$

where I_p' = actual recorded intensity at the image point P', and $I(x_p, y_p, \alpha_p)$ = recorded intensity expected at P' from the best fit relationship (3) for I_p'.

A reiteration between (3) and (4) should then be performed to obtain an expression for (3) that is convergent in the form

$$I_p'' = f(x_p, y_p, \alpha_p) \tag{5}$$

where $I_p'' = I_p'/A_p'$, and until convergent values are reached for each A_p'.

The plate constants can then be derived in the form

$$\alpha_p = f(x_p, y_p, I_p'') \tag{6}$$

so as to be independent of albedo variations, by a best fitting of α_p to the values of x_p, y_p, I_p''. With the plate illumination constants and the instantaneous albedo known, any photograph nested within the photometrically analysed shot of the stereopair, and taken at the same instant, can be charted by simply feeding the x, y, D values of each point into a computer for the appropriate numerical reductions. This assumes that the numerical reductions accomodate any geometrical and sensitivity variations between the two cameras and that the albedo figures derived on the medium resolution scale can be applied on the fine-resolution scale, i.e. in the case of Orbiter photographs, that microalbedo variations on a metre scale are small.

THE ISODENSITRACER DENSITOMETRIC ANALYSIS SYSTEM

As a collosal amount of data is likely to be collected by each Orbiter mission it is imperative that an efficient system of analysis is used to determine x, y, D values. It is obvious that an individual manual reading of the density of each element is of little use, and that an automatic scanning of the photographic image is essential.

The densitometric analysis system of the I.D.T. (see Figure 7) is essentially that of the Joyce Loebl Microdensitometer (M.D.M.). In it, an optical arrangement enables

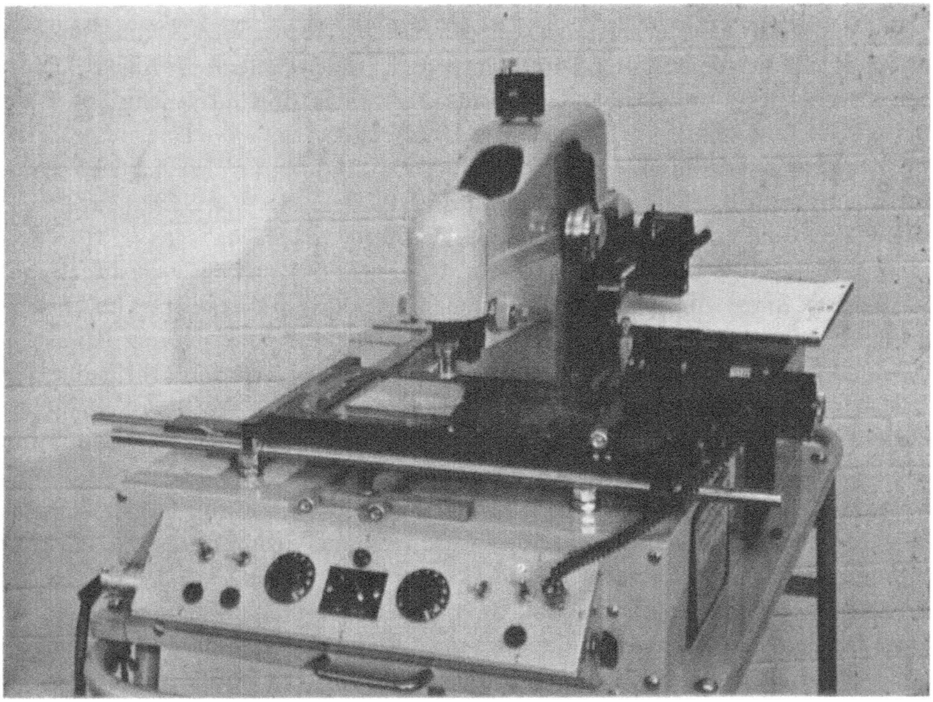

Fig. 7. The densitometric analysis table of the Tech/Ops. isodensitracer.

a collimated beam of light to be split into two beams. One is passed through the negative under analysis on the specimen table, refocussed onto a scanning slit and directed onto a photomultiplier cathode. The other beam passes through a calibrated continuous density wedge and is then directed to the same photomultiplier cathode. A rotating chopper between the two beams enables the photomultiplier to view each in turn, and should they not be equal in intensity, a servomotor adjusts the position of the wedge until balance is obtained. Attached to the wedge is a pen capable of recording the wedge's position and hence the density of that part of the negative viewed by the scanning slit. A mechanical lever system (with a range of magnification factors) connects the specimen table to a recording chart table so that, when the chart table is driven electrically, density scans are obtained across a given ribbon of the negative. The advantage of this two-beam balancing system is that the instrument's output is independent of its own parameters and so the system may be run for long periods of time without being affected by drift problems, i.e. a complete reproducability of record is possible. Unlike the M.D.M., which only scans individual ribbons and requires a manual resetting for successive scans, the I.D.T. is capable of automatically scanning successive parallel ribbons, printing the results on a dropped line output in a three-symbol code. The I.D.T. in effect samples the density variation in a scan and changes the output symbol as the density changes by a discrete preset interval and it is possible to identify the individual density level without ambiguity.

The I.D.T. permits continual densitometric analysis to be achieved with a minimum amount of supervision but, in its standard form, it requires a tedious visual evaluation to be made of the printed data. This form of output is obviously inefficient if large numbers of I.D.T. charts are to be analysed, and a great saving in time and effort can be achieved by digitising the I.D.T. so that the x, y, D output figures may be recorded directly on punched tape during the densitometric analysis. The output can then be fed directly into a computer system for the necessary reductions.

The design of the I.D.T. is such that the x, y, D values can be automatically read in a digitised form with only a few instrumental modifications. During the analysis the negative is arranged to move under a fixed reading head in both x- and y-directions. Assuming scans are taken along lines of constant y, then motion in the x-direction, over the negative under analysis, follows identically that of the recording table by means of the precision ratio arm. Due to the high precision of this ratio arm in producing scans on a magnified scale it is proposed that the x-coordinate pick off be actuated by the recording table with a suitable moire fringe reading head attachment. The automatic stepping of the I.D.T. to each parallel scan is accomplished by a stepping motor that is actuated by pulses obtained from the I.D.T. program unit (Figure 8). Thus, the y-coordinate can be obtained by a simple counting system that would enable these pulses to be interpreted by the digitiser unit and transferred to punched tape. Once the density levels to be recorded by the I.D.T. are set within the

Fig. 8. The Tech/Ops. isodensitracer with its programming unit.

instrument's programme unit then, as the density of the analysed specimen region reaches one of these levels, the status of the electrical signals applied to the pen write out mechanism changes. For digitisation, these signals may be read directly from the I.D.T. programmer circuitry and, again by a suitable counting technique, transformed by the digitiser unit to density levels on punched tape. Thus, with this proposed digitisation, the x, D-values would be punched out at each preselected density level and the y-value punched out at the commencement of each scan.

THE ANALYSIS OF I.D.T. OUTPUT

a. *Calibration of Plate Constants and Instantaneous Albedo Evaluations*

For a given fine-resolution shot (FR(A)), photometric calibration requires the corresponding medium resolution shot (MR(A)) of the frame pair, A, together with a further shot (MR(B)) for stereoscopic evaluation of the lunar terrain covered. After the stereoscopic evaluation of the lunar terrain on shots MR(A) and MR(B) has been completed to provide a set of x_p, y_p, α_p values, the MR(A) shot is then subjected to an I.D.T. analysis for a determination of the I_p' values so that the form of Equation (6) may be derived. The I.D.T. analysis should be made to accommodate the FR(A) coverage in the field of shot MR(A) and also to cover an intensity range of at least that given on the FR(A) shot. Great care must be taken in correlating stereoscopically determined slope values, α_p, with their corresponding photometrically determined intensity values I_p' at x_p, y_p, to allow for instrumental averaging effects and for the fact that the intensity is only received in the sun-observer-analysed point plane.

To determine the most efficient numerical analysis system for calibration some idea is required of the probable amount of available data.

Single-shot fine-resolution coverage at periselenium passage is expected to be 16.6×4.15 kms of lunar surface for the Orbiter shots. If this is analysed with a slit width of 50×50 m relative to the surface (a figure chosen to give an approximately 50 micron square view of the MR image within the spacecraft) and the scale of density variations is of the order of 50 m relative to the surface, then approximately 30000 sets of values of x_p, y_p, D_p' will be available for analysis. In decimal digit output this corresponds to approximately 700 yards of punched tape output and it would be an inefficient process to try to read this into a large store off-line computer for reduction. As this large amount of data is far in excess of that required for plate constant reductions, it is preferable to systematically sample the MR(A) image density instead, and present a much smaller tape to the off-line large store computer. However, a complete analysis of the fine-resolution coverage of FR(A) shown on MR(A) is still required in order that instantaneous albedo figures over that region can be determined and one is still confronted with the problem of handling large tapes. To overcome this difficulty, it is essential that a small store on-line computer is used in conjunction with the digitised I.D.T. Each set of x_p, y_p, D_p' values could then be fed into it direct and, with an edited phased input of x_p, y_p, α_p values, suitable programming and $I_p' \vee D_p'$ calibration, could enable instantaneous albedo figures to be punched out for each scan.

TABLE I

Flow Diagram of a Proposed Orbiter Data Analysis System – Phase A
The Determination of the Plate Constants and Instantaneous Albedo figures

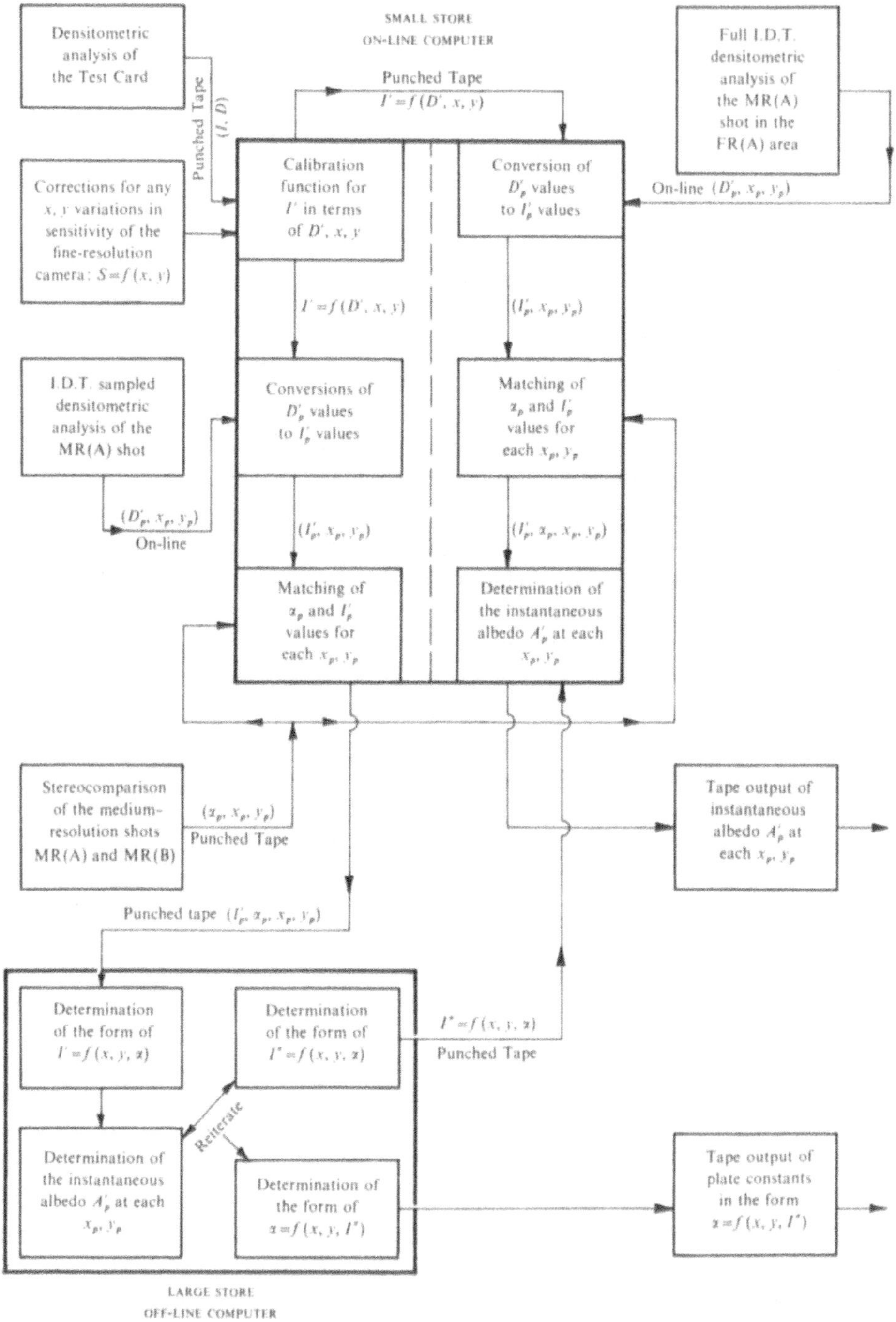

TABLE II

Flow Diagram of a Proposed Orbiter Data Analysis System – Phase B. The Determination of Slopes on the Orbiter Fine-Resolution Shots

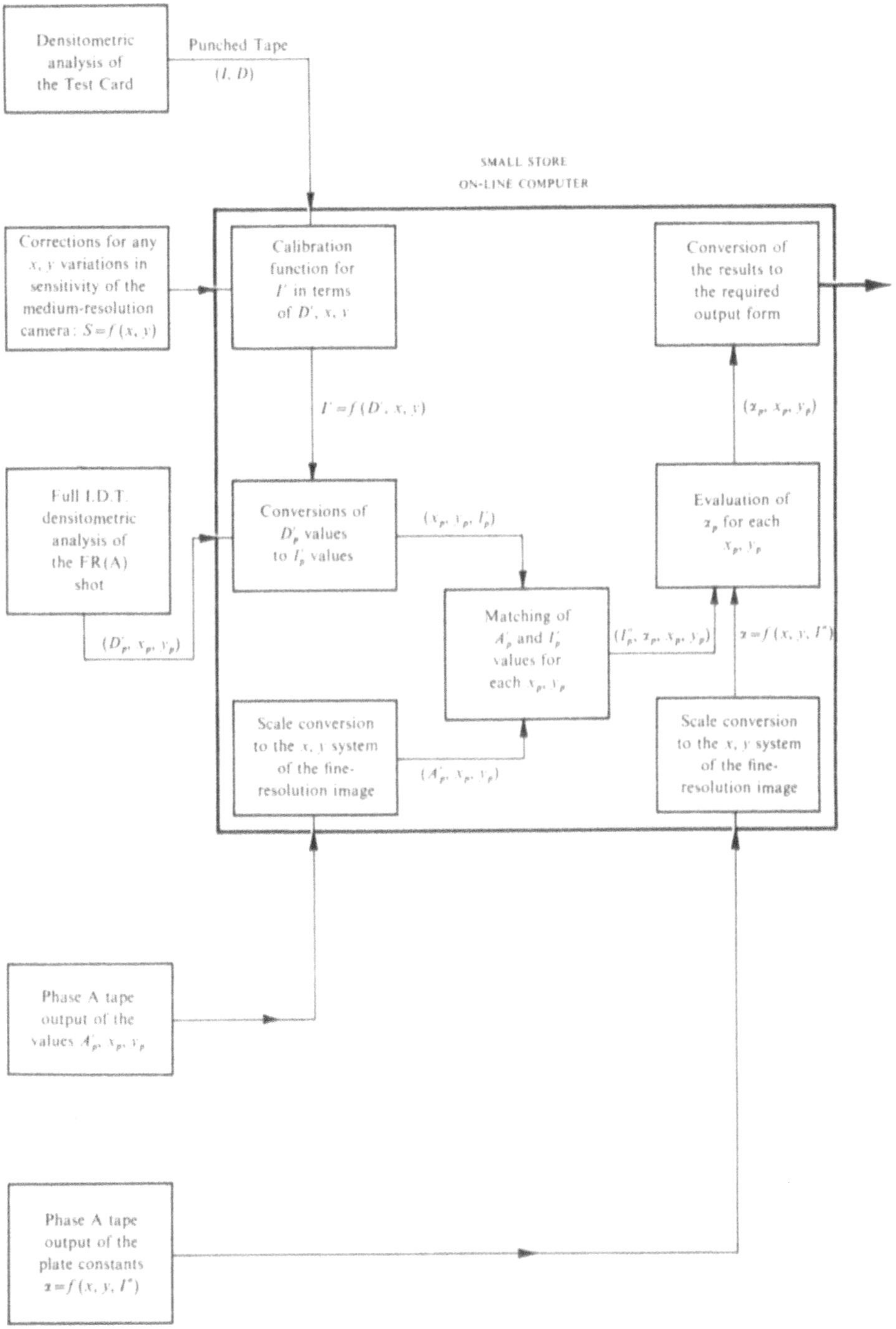

The x, y, D store could be cleared for the next scan and the complete tape of albedo figures could be returned in stages to the on-line computer when required for the analysis of the fine-resolution shots.

b. *Mapping of the fine-resolution shots*

As demonstrated in the previous section, the amount of data available is, in the fine resolution coverage registered on the corresponding medium resolution shot, excessive for a full analysis by an off-line large store computer and such an approach would be grossly inefficient for use on the fine-resolution data itself. As each set of x_p, y_p, D'_p values needs to go through the same standard reduction procedure for a determination of slope α_p, the use of a small store on-line computer again becomes essential for an efficient analysis of data.

Values of x_p, y_p, D's are read directly into the on-line computer and stored for each I.D.T. scan. The instantaneous albedo figures should be fed into the computer in stages to deal with given sets of scans as they are analysed. Then, with the necessary internal programming for geometrical and photometric scale conversions between the fine- and medium-resolution cameras, use of the determined relationship (6) will enable slope values to be determined for each analysed point in the scan. These can be punched out and the x, y, D store cleared for the following densitometric scan.

It must be stressed that slopes determined by the above method are expressed as component values in the plane containing each analysed point, the Sun and the spacecraft observing station. From photometric methods it is not possible to obtain information in other planes and this should be borne in mind for any map construction from the determined slope values.*

References

JONES, M.: 1966, *Astronomical Contributions from University of Manchester*, Series III, **140.**

KOPAL, Z.: 1962, in *Physics and Astronomy of the Moon*. Academic Press, New York and London, chapter VII.

MCMATH, R. R., PETRIE, R. M., and SAWYER, H. E.: 1937, *Pubs. Obs. Univ. Mich.* **6,** 67–76.

VAN DIGGELEN, J.: 1959, *Recherches Astronomiques de l'Observatoire d'Utrecht* **14,** 2.

* Investigation supported by the Terrestrial Sciences Laboratory, Cambridge Air Force Research Laboratories, under Grant EO AR 64–49 to the Department of Astronomy, University of Manchester, through te European Office of Aerospace Research, in Brussels.

DETERMINATION OF THE MOON'S SHAPE BY THE PHOTOMETRY OF ITS TERMINATOR

ODILE CALAME*

The method of determination of the Moon's shape which I wish to discuss goes back, in its rudimentary form, to RITTER (1927), and is based on the fact that the geometrical shape of the terminator depends on that of the lunar globe. If the latter were a sphere, the beam of sunlight intercepted by our satellite, as seen from the Earth, would constitute half of an ellipse, to which we refer as the theoretical topocentric terminator. A comparison of the form of the terminator as actually observed with the theoretical one (depending solely on the phase of the Moon) should, in principle, enable us to determine the deviations of the actual lunar surface from the mean sphere along the arc of the terminator.

1. Relation between the Altitude and the Inclination of a Surface Element

In actual fact, the elevation h of any element of the lunar surface above or below the mean sphere is not the only parameter influencing the shape of the terminator; for it is also necessary to consider the angle of incidence of solar radiation on the Moon; and in order to do so we must take into account the local angle i of inclination of the respective surface element with a plane tangent to the mean sphere at the respective point. If, however, the actual form of the lunar surface deviates but little from a sphere, it should be legitimate to retain only the first powers of h and i in the equations of our problem, and to ignore all higher powers and cross-products.

2. Equations of the Problem

Let us adopt a rectangular system of selenocentric coordinates ξ, η, ζ, defined so that the ζ-axis is oriented towards the observer; and the ξ-axis to the Sun. If, moreover, φ denotes the topocentric phase of the Moon, and $\Delta\xi = \xi - \xi_c$, where ξ and ξ_c are the measured and calculated coordinates of a point P or P' observed or calculated in the terminator, the altitude h of this point above the mean lunar sphere should be given by the equation

$$h = \frac{\Delta\xi\sqrt{1-\eta^2}}{\cos(\alpha+\varphi-i)} + (1-\eta^2)\left\{\frac{\cos\varphi}{\cos(\alpha+\varphi-i)} - 1\right\} \tag{1}$$

* Observatoire du Pic-du-Midi, France.

Kopal and Goudas (eds.), Measure of the Moon.

if the mean radius of the Moon is taken as our unit of length, and the angle α follows from the auxiliary equation

$$\sin\alpha = \frac{\sqrt{1-\eta^2}}{\sqrt{\eta^2+D^2}}, \qquad \alpha < 2'', \tag{2}$$

where $D = MS$ (see Figures 1 and 2).

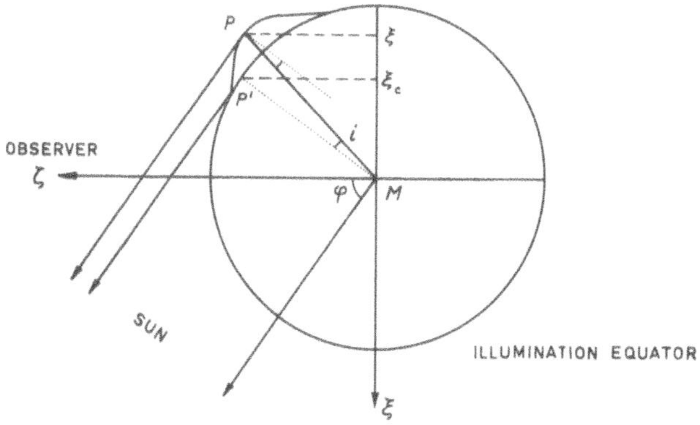

Fig. 1.

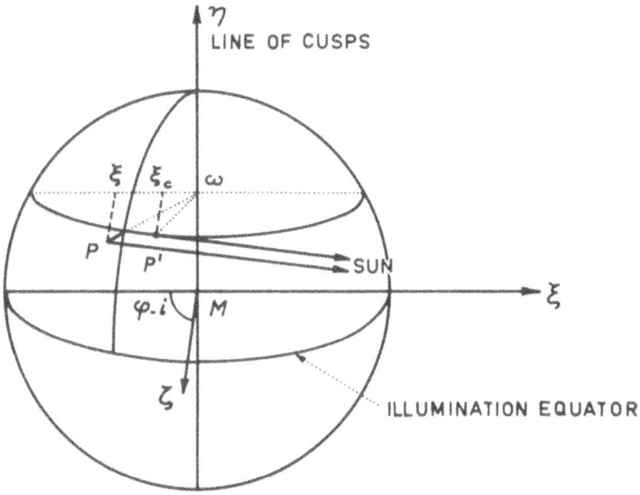

Fig. 2.

The need to introduce the angle i in Equation (1) reveals that absolute values of h can be obtained with its aid only if i can be determined from an independent relation; and this can be done as follows. By definition, the angle i represents one between the selenocetric radius-vector to P and the normal to the actual surface, along each parallel of illumination and arc of the terminator. In consequence, the parameters

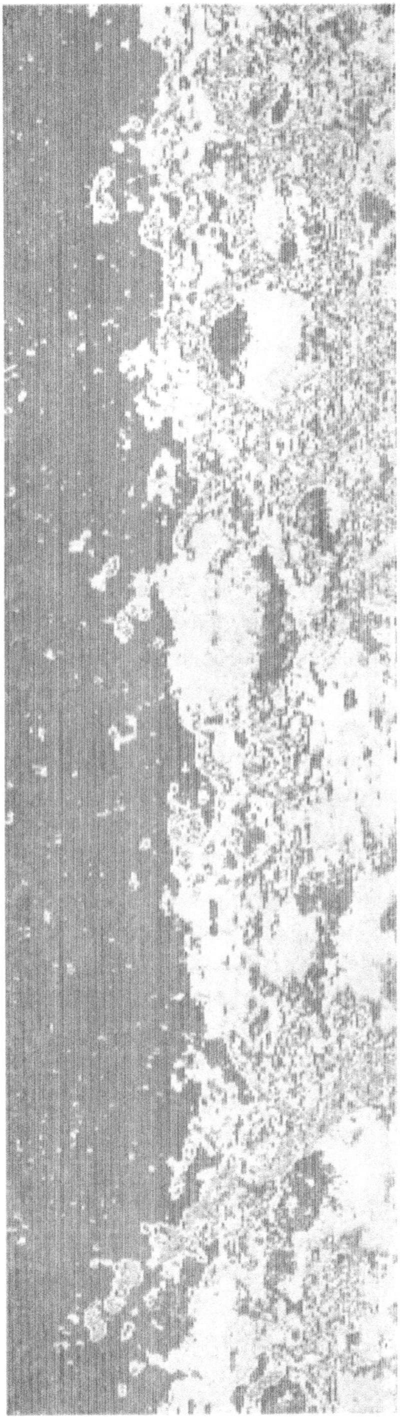

Fig. 3.

h and i are further related by the differential equation

$$\tan i = \frac{\mathrm{d}h}{\mathrm{d}(\varphi - i)} \frac{1}{1 + h - \eta^2}; \tag{3}$$

and the problem is to solve the Equations (1)–(3) simultaneously by numerical integration and successive approximations. A large number of observations can enable us to eliminate certain errors by averaging (for example, for two complementary phases).

3. Method of Measurements

The instrument used for obtaining the lunar photographs necessary for this purpose was the refractor of 38 cm free aperture and approximately 6 metres focal length at the Observatoire du Pic-du-Midi. One problem is to photograph the instantaneous position of the true terminator; for by definition (and the finite sensitivity of photographic emulsions) the points of zero illumination (defining the true terminator) are invisible. In order to circumvent this difficulty, photographic plates used for this purpose were uniformly pre-exposed; and it was found that, on pre-exposed plates, the form of the terminator was no longer dependent on the exposure time, by virtue of the compensation for the sensitivity threshold.

The utilization of such plates proceeded in the following two steps:

(1) to refer the position of the terminator with respect to neighbouring craters by means of photometry; we have utilized to this end the Joyce-Loebl isodensitracer which traces automatically the isophotes of the regions under investigation (cf. Figure 3).

(2) to determine subsequently the positions of these neighbouring craters with respect to 196 reference points of ACIC by means of a Zeiss comparator – i.e., by a micrometric method.

One of the difficulties of such work arises from the need to employ two different methods of reference, which imposes a limit to the attainable precision. In fact, another method is being experimented with: namely, to use a camera of the Markowitz type, which should permit to take photographs of the Moon superposed on a star field. Stellar images can lend themselves for the purpose of reference more easily than the images of the craters, the apparent positions of which can be influenced by shadow effects making micrometric settings difficult. These experiments should permit us to compare the efficiency of these methods, of application also to other astrometric problems.

In conclusion, the method of terminator tracing should be primarily applicable to hypsometric studies of flatlands (maria), and may lead to their absolute height determination. The process offers a greater precision in regions of increasing distance from the centre – a fact which should make it possible to compare direct height determinations in the marginal zone with those resulting from the observed terminator contours.

Reference

Ritter, H.: 1934, *Astr. Nachr.* **252**, 157.

ISODENSITOMETRIC MEASUREMENTS OF LUNAR SLOPES FROM THE RANGER PHOTOGRAPHS

A. RIFAAT*

The aim of the present communication will be to obtain by photometric methods the elevation profiles for seven fields of the Moon photographed in March 1965 by Ranger 9. We will assume, first, that the sunlight scattered by the lunar surface obeys a photometric law of the form $I \equiv I_0 f(\alpha, \varepsilon, i)$, where I_0 denotes the albedo of the respective element of the lunar surface; α, the phase angle; i and ε are the angles of incidence and reflection. In order to minimise the number of unknowns in our problem, it will hereafter be assumed that the albedo of all surface elements under consideration remains constant. The angle of incidence i can also be considered constant for each frame of the Ranger photographs, in the last stage of its flight since its variation over the field of view of the respective camera is very small. Within the framework of these approximations, the distribution of brightness in the field of view should depend only on the angle ε of reflection which, in turn, depends on the slope of the surface. Therefore, if the angle of incidence i can be deduced from the observations by photometric measures, and if the respective positions of the Sun, the Moon and the spacecraft are known at the time of observation, we should be in a position to determine the direction of a normal to the respective element of the lunar surface at each point of the field, and thus its inclination to the horizontal. A numerical integration of these derivatives of the lunar surface curves, if sufficiently densely spaced, can then enable us to determine the actual warping of the local surface.

The mathematical formula relating the shape of the surface with its photometric properties observable at a distance were deduced already by RITTER (1934); and the reader of the present volume is referred to the preceding contribution by Mlle CALAME (1967) for their formulation appropriate for our problem. In our interpretation of the measurements we shall, moreover, follow MINNAERT (1961) in his conclusion that lunar isophotae are essentially identical with the 'meridians of illumination' and do not depend on the latitude. This, in turn, renders hypsometric curves along each parallel of illumination independent of each other in zero point of the altitude – i.e., while an integration of surface derivatives along any parallel of illumination can, in principle, be used to determine the respective warping of the surface, no such continuous curves can be obtained along the meridians of illumination.

In order to adjust the zero points of adjacent hypsometric curves along the

* Dept. of Astronomy, University of Manchester. On leave from Helwan Observatory, Cairo, Egypt.

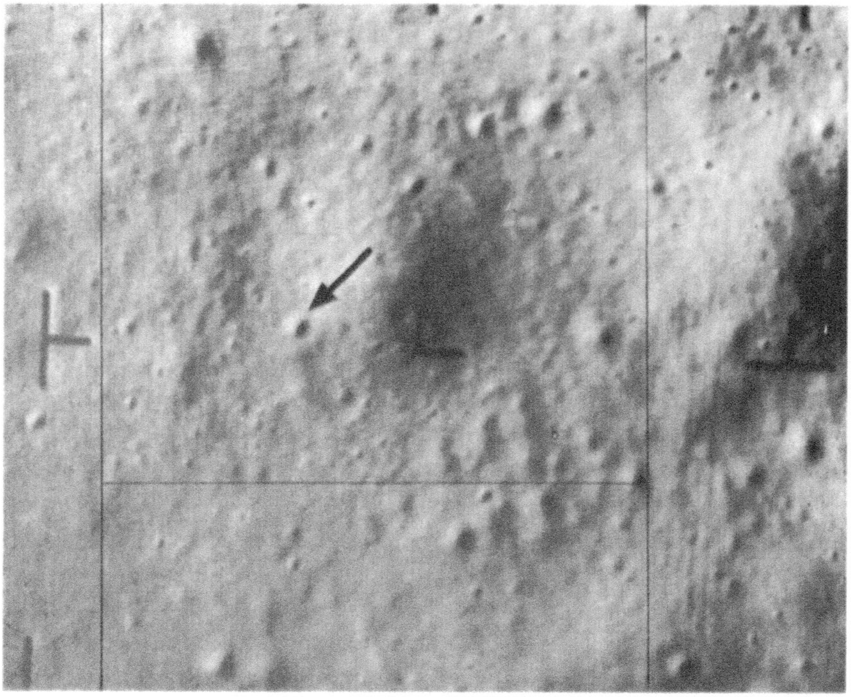

Fig. 1a. A sector of frame number 440, camera FA of Ranger 9.
Altitude 7.1956 km, 2.972 seconds to impact.

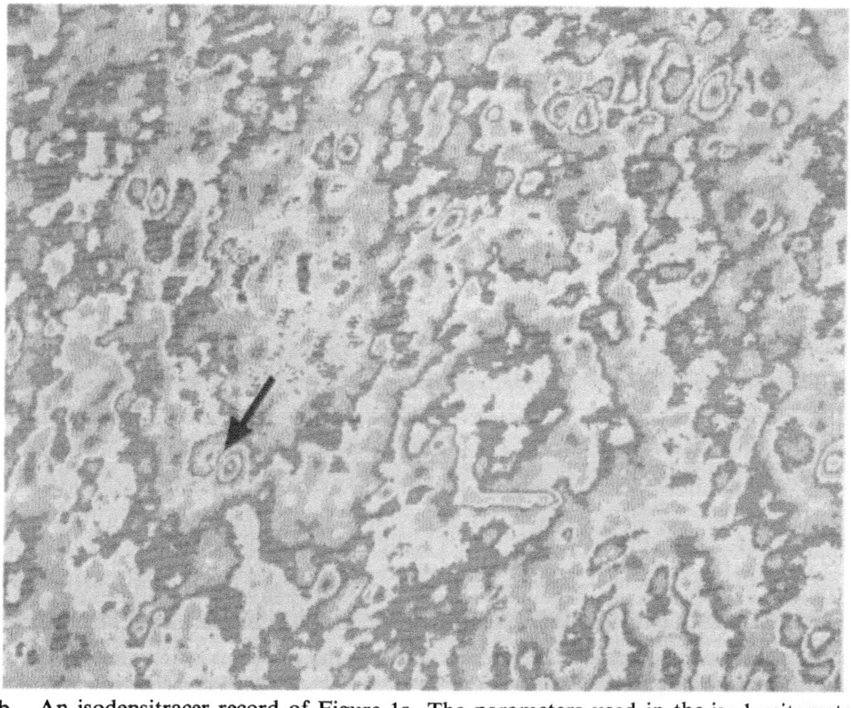

Fig. 1b. An isodensitracer record of Figure 1a. The parameters used in the isodensitometer are: spot height = 12.5μ, spot width = 12.5μ, objective $\times$ 20, wedge number D176, ΔD increment = 0.02, differential control = 5, and pen damping = 10. The resolution is 0.5 mm.

Fig. 2a. Frame number 439, camera FB of Ranger 9. Altitude 13.3851 km, 5.532 seconds to impact.

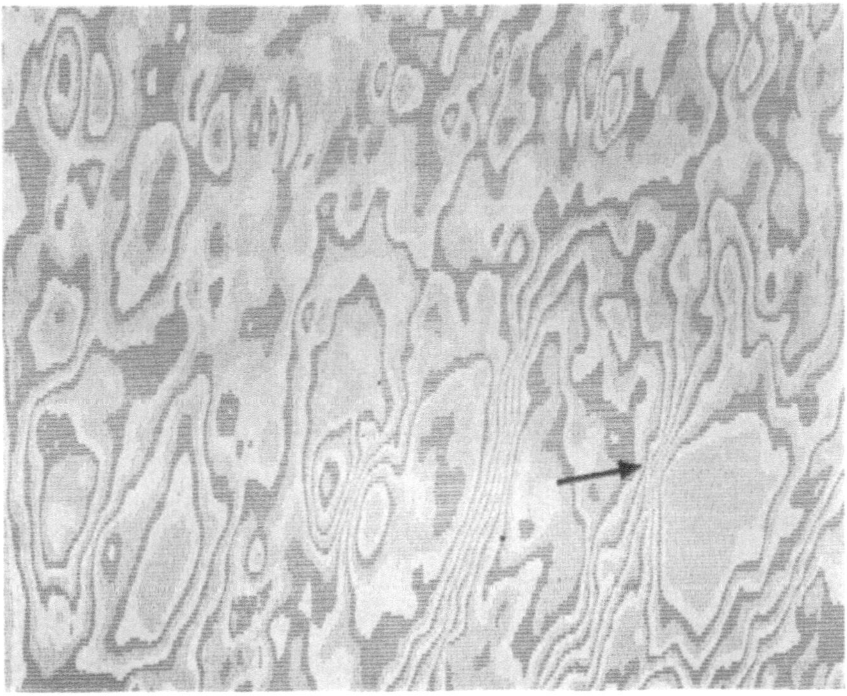

Fig. 2b. An isodensitracer record of Figure 2a. The parameters used in the isodensitometer are: spot height = 150μ, spot width = 25μ, objective $\times$ 10, wedge number $D176$, ΔD increment = 0.04, differential control = 5, and pen damping = 10. The resolution is 1 mm.

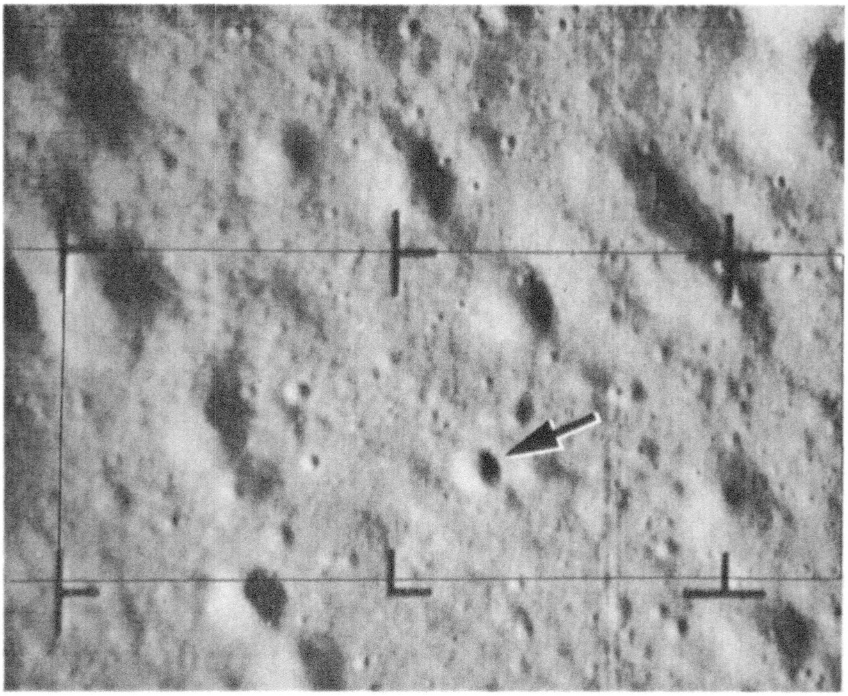

Fig. 3a. Frame number 438, camera FA of Ranger 9. Altitude 19.5729 km, 8.092 seconds to impact.

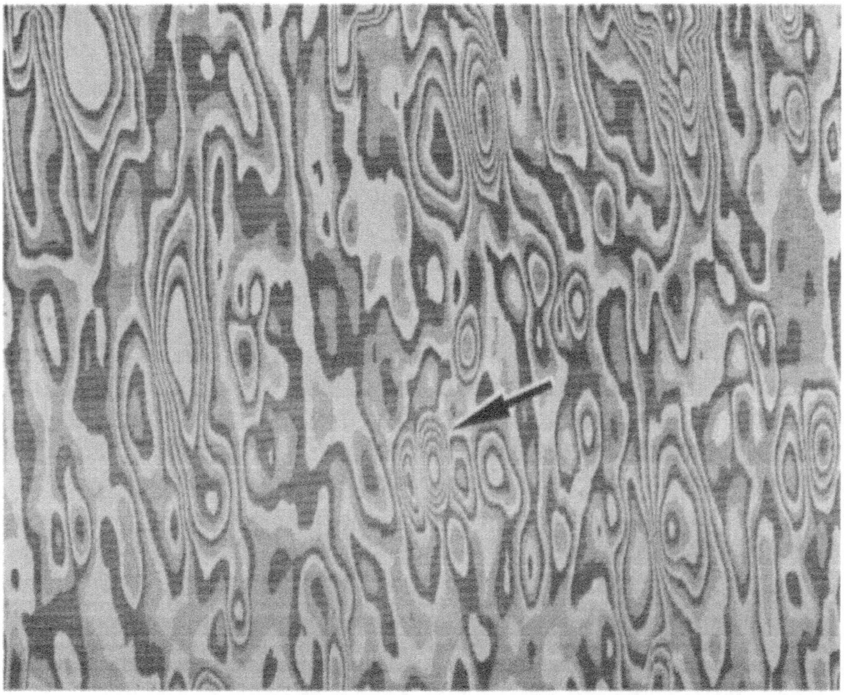

Fig. 3b. An isodensitracer record of Figure 3a. The parameters used in the isodensitometer are: spot height = 50μ, spot width = 25μ, objective × 10, wedge number $D176$, ΔD increment = 0.04, differential control = 5, and pen damping = 10.

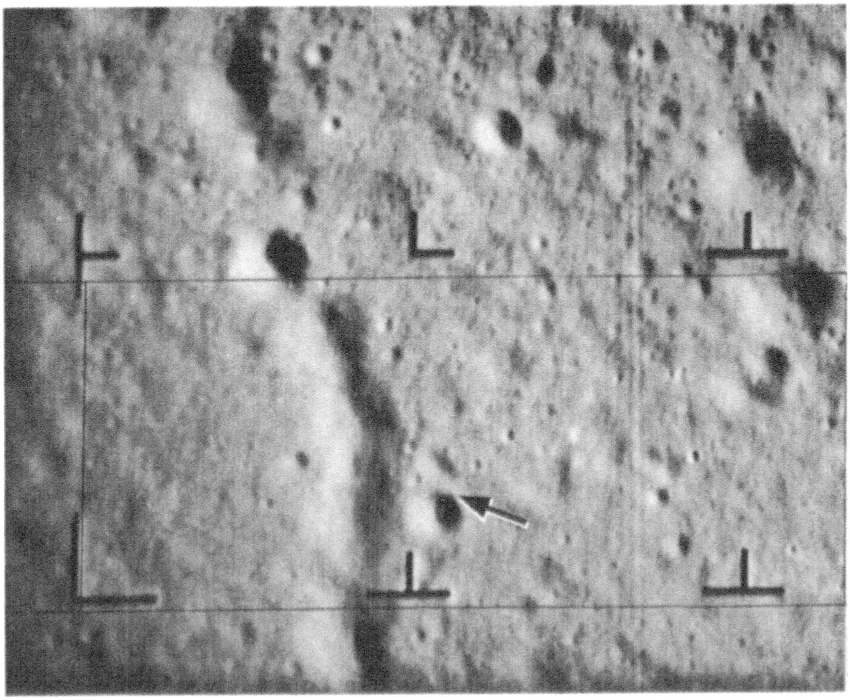

Fig. 4a. Another area of frame number 438.

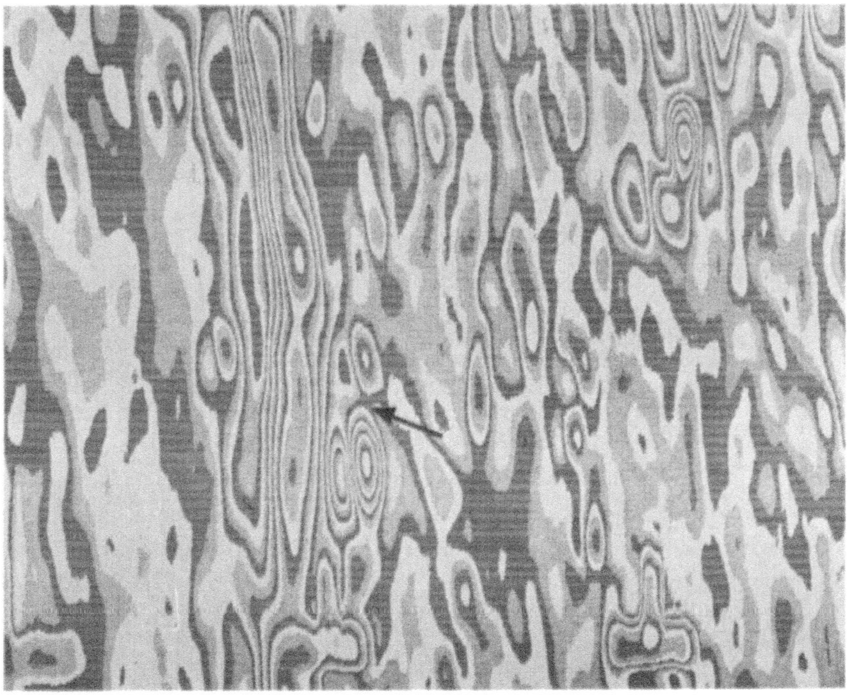

Fig. 4b. The same parameters used as in Figure 3b.

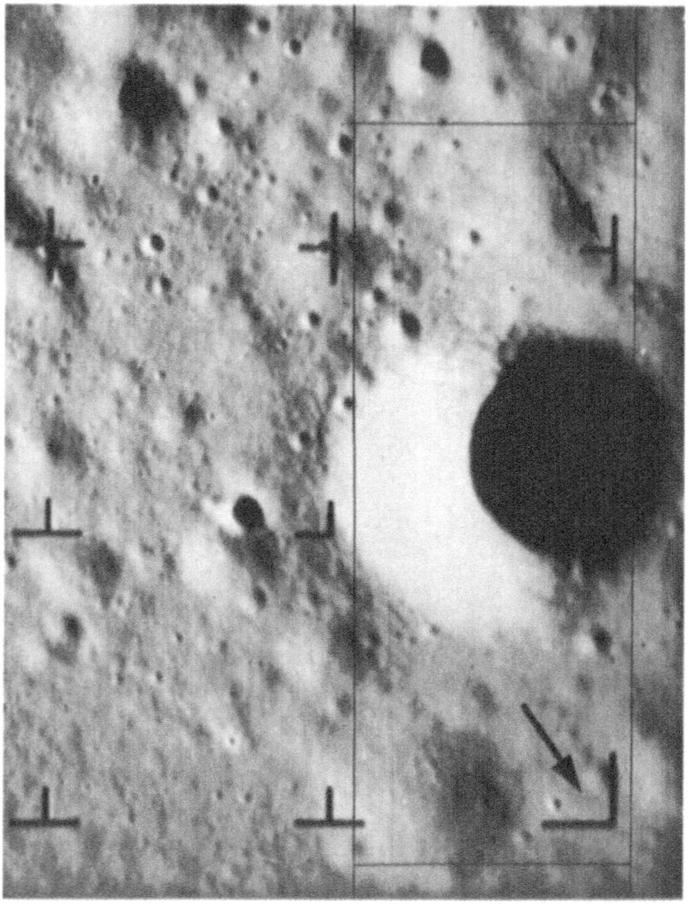

Fig. 5a. Alphonsus G. B. in frame number 438, camera FA of Ranger 9.

parallels of illumination, several methods of approach suggest themselves. If, for instance, the surface region under investigation appears to be reasonably flat and devoid of shadows, it is reasonable to follow RINDFLEISCH (1965) in his assumption that the mean altitude remains the same along each strip; and to adjust their zero points accordingly. Or, if any feature within the field of view exhibits a distinct shadow the length of which can be measured, its geometry (for a given altitude of the Sun above the lunar horizon) should permit us to determine by standard methods the difference in height between the top (or rim) of the shadow-casting obstacle and the tip (or the outline) of the shadow – and thus the mean inclination of the respective hypsometric curve can be calibrated. In addition, if the shadow outline intersects several adjacent hypsometric curves, their respective zero points can be adjusted accordingly.

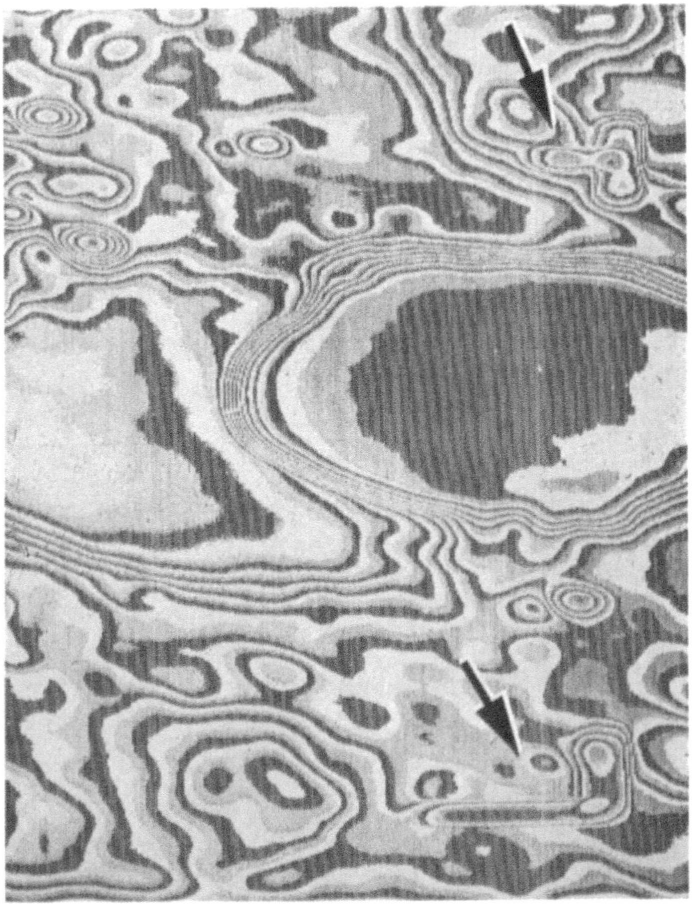

Fig. 5b. The same parameters are used as in Figure 3b.
Altitude 19.5729 km, 8.092 seconds to impact.

Application of the Method

Particular areas recorded by Ranger 9 have been selected for photometric analysis performed with the aid of the Joyce Loebl-Tech. Ops. iso-densitracer of the Department of Astronomy, Manchester University. Figures 1b–5b illustrate typical isodensitometric transcript of direct photographs reproduced on Figures 1a–5a.* The optimum parameters used in obtaining such records accompany each illustration. The spot height, which has been adjusted to reduce film graininess without integrating over too large a surface element, equals the slit height times the magnification of the objective. The discrete density increment ΔD is equal to the grey wedge length per

* The actual field photometrically analysed on each b-figure is marked on the corresponding a-figure by a rectangle (the proportions on each pair of plates are not necessarily the same). North is on top of each figure.

commutator division (in cm) times the density (per cm) of the individual grey wedge. The arrows indicate the same areas in each corresponding pair of plates.

The reductions of the photometric measure for hypsometric profiles of the selected regions are under way; and their results will be ıeported in a subsequent communication.

References

Calame, O.: 1967, in the present volume, pp. 451–454.

Minnaert, M. J. G.: 1961, in *The Planets and Satellites* (ed. by G. P. Kuiper and B. M. Middlehurst). University of Chicago Press, Chicago, Chapter VI, pp. 213–248.

Rindfleisch, T.: 1965, Jet Prop. Labor., Calif. Inst. of Technol., *Technical Report No. 32-786.* Also *Photogrammetric Engineering* **32** (1966), 262–276.

Ritter, H.: 1934, *Astron. Nachr.* **252,** 157.

DISCUSSION

Kopal: I should like to express our sincere thanks to professor Minnaert for so lucid a survey of such fascinating field as represented by the photometric methods in the service of lunar topography; and to make two brief remarks. First, the shadow method does not permit us to measure altitude differences between any two points of the lunar surface, but only between those which are situated on a "parallel of illumination" which passes through the shadow-casting obstacle; and only within such fanwise spread of it as is afforded by the libration in latitude. The second point which I should like to make concerns the need of separating local albedo and inclination in lunar terrain studies. I agree with professor Minnaert that this is impossible as long as observations have been carried out in only one colour. However, colorimetric studies of the Moon indicate that a change in local albedo is usually (if not always) accompanied by a change in colour index which, small as it is, can nowadays be measured well within the limits of observational precision. Therefore, the best way of ensuring that what we measure as an albedo change should be interpreted as a change in slope, and not in reflectivity, would be to verify that change in albedo is not accompanied by a change in colour index; for only when this is the case can we surmise to our satisfaction that it is the inclination, and not reflectivity, which has undergone a local change.

Minnaert: I am quite in agreement on this last point; only in most cases we may not have a possibility of making colorimetric measurements on the photographs. I should like to ask what is simpler: to measure small difference in colour, or to take a photograph of a full moon?

Kopal: The colorimetric measures can be made simultaneously in three of six colours, at the same time, virtually speaking, (i.e., at the same hour or within the same hour) as when the plate is taken. And that will have the advantage that you will never have to combine data which may have been secured under different seeing conditions, with different extinctions as would be obtained, say, between a plate at a particular time and an adjacent full moon or some other full moon. I was very happy to see professor Minnaert underline the need of a concerted attack on the shape of the lunar surface by more than one method. Shadow methods, the photometric method and photometry in several colours should be combined; and, if possible, pursued at the same time (or within the same evening) because the Sun fortunately rises very slowly above the lunar horizon. It is on the basis of such records that the shape of the lunar surface can finally emerge from an analysis. We are anxious in Manchester to make such a consolidated attack, and shall report what has been done in the course of this afternoon.

MINNAERT: I am very gratified that we fully agree that such a concerted attack will be most profitable and useful, but as to the colorimetric measurements, it is quite true that, for example, from a Ranger you could arrange to have simultaneous photographs in two colours. However, I think we should remember that the differences in colour with which we are concerned are very small. You probably remember that up to five years ago in the Soviet Union there were two groups of workers, one saying that there were no measurable differences in colour, whereas the other one was finding some small differences, which would imply that observations from a Ranger would not be easy.

KOPAL: I should only like to mention that, in this matter, we have progressed beyond the work of the Russians. The differences, small as they are (the differences in colour index amount to, at most, a few hundredths of a magnitude), this can now be measured with a satisfactory significance by simultaneous scanning techniques of and formation of the ratios of intensities measured through two or more different spectral channels simultaneously. These techniques have been developed, and those of you who are going to visit the Pic over the weekend will see this photometer – though not in action, because we shall be there during daytime. It was designed and built by Mr. Roberts, one of our graduate students, for precisely this type of work. We can now trace by means of ratio-meter the ratio of output in three or six different spectral channels and establish small colour differences very significantly. Several of our colleagues in the United States have done likewise.

The only advantage we can claim in this colorimetric work is the adoption of the scanning technique. We can do so simultaneously and do not have to compare colours measured at different times. I think this is very essential.

KARANDIKAR: What are the spectral windows selected for such purposes?

ROBERTS: One is about 40 Å at 7889, the other 40 Å at 6500, and the last one is 40 Å at 5500.

KOPAL: The photometer we have in Egypt, which operates at the Kottamia telescope which professor Samaha described, has three more bands.

ROBERTS: Yes, further into the blue.

GOUDAS: I should like to ask professor Minnaert if he would like to comment on how these techniques could be used for high-resolution photographs. High resolution photographs taken from a lunar orbiter, taken from a distance, say 50 to 1000 km over the surface, will, hopefully, soon be obtained. Is there anything that you could tell us on such applications?

MINNAERT: This case is more or less the same as the reduction of the Ranger photographs with the difference that the Orbiter could perhaps be made to photograph areas close to the terminator. The Ranger photographs were at a certain distance from the terminator where we have greater contrasts due to the inclination. The differences in albedo will be more negligible compared to the differences in inclination and we will have the best results. Most of what I have said refers to possibilities of measurements from a rocket or an Orbiter near the Moon. It is highly desirable to have the same areas photographed also at phases and at full moon. That will be difficult from such an Orbiter, but we would already be extremely happy to have photographs or photoelectric images of the terminator region as observed by an Orbiter, but the whole method could be applied.

Mr. President, I was very impressed to hear about the Apollo plans. Now, as far as I understand, this will be a photographic record which afterwards can be transmitted. It will then be necessary for density calibrations to make photometry afterwards because now the density curve of the photographic plate will be unknown. Will you rely on the fact that you can reproduce the curve afterwards in the laboratory?

SASSER: The long focal range camera will be calibrated and also on the film there will be density wedges that will indicate any errors that may have occurred in the transmission and reproduction of the film, so we do feel that we have a long focal length camera, photometrically calibrated.

RIFAAT: How were the slopes measured?

SASSER: By the photometric method, completely by a computer, for each Ranger photograph; and the method has been described in a communication by JPL.

ARTHUR: Just one thing – I would warn anyone against using estimates of correctness from the rectified atlas. You get a rather complicated printing process here which is used to enhance contrast and I am afraid it has played the devil with tones. I just would not trust it.

HOPMANN: Thank you, Mr. Arthur, I am glad to hear it.

MINNAERT: These measurements along the fault near Cauchy were made on the assumption that the albedo was everywhere the same. Something in favour of that would be that in a similar case Dollfus has found that the polarization does not change along such a slope, but this may support the assumption of constant albedo.

RACKHAM: Professor Hopmann has talked about Luna 9 landing. Could he amplify a bit more on the luminosity he mentioned of the lunar surface?

HOPMANN: The luminosity – I do not remember. This will perhaps be better known after they produce a complete atlas.

WEIMER: I do not think I need to emphasize the fine work on lunar mapping done at ACIC, and I hope that the observers will have still more success.

HOPMANN: I am most delighted that at Flagstaff you work so intensively. This is absolutely necessary. I also wish that all workers on the problems of the Moon should see the Moon itself by a telescopic magnification of about 400 or 500. You have a better impression of the whole thing than with the best photographs.

KOPAL: I should like to make a few remarks concerning professor Hopmann's earlier communication about the use of visual observation for shadow measurements instead of making measurements on photographs. With all due respect to the achievements of all our predecessors, I think the visual methods are now out of date. Visually the shadow method is at its best when the shadows are longest – i.e., when the ratio of a shadow to the height is at its maximum. But when this happens, the shadows are not sharp: they are bordered by penumbra due to the fact that the illuminating source (i.e., the Sun) is half a degree in diameter. The difficulty in determining where the shadow begins and ends has been the principal source of error in visual work in the 19th century and up to the present time. The width of the shadow band when the Sun is rising above the lunar surface may be as large as 30 km; and when you are dealing with visual micrometry, you must make up your mind where this shadow begins and ends. It is, however, next to impossible to make up one's mind where to set systematically on the given part of the shadow curve. I think the only way to do so is the microdensitometric one, setting at either in the inflection point of the density curve or some other well defined point on it. The difficulty is lessened when the Sun gets higher and the shadows are shorter but precisely at this time the technique is losing its advantage and its errors grow. This cause, which has not been realised till in the past seven or eight years, was the principal source of the errors of the 19th-century observers in trying to determine heights on the Moon.

The second point I should like to make regarding the use of photographs is that most photographs are worthless for shadow measurements unless the exposure time is known at least to a minute. Again, if you wish to get the best results, you choose the photograph with a terminator when the shadows are low; but the shadow is racing on the lunar surface at this time fairly rapidly; and unless you know at least within a minute when the photograph was taken, you commit errors in the computed height which exceed the errors of the measurement itself. This is the one fact which disqualifies for practical use a large part of the earlier photography of the Moon, good as it was, because the observers (such as those of Mount Wilson who produced those beautiful plates 30 or 40 years ago) did not take the trouble to record their times of exposure. They recorded the evening but did not trouble to record the hour and minute because at that time they did not see the point. For such plates – and many of these are reproduced in the literature – so long as the exposure time is not known, and I should say known within a minute, errors of height measurements from the shadow may become unacceptably large.

ARTHUR: Against the photographic method I would put, though, the existence of systematic errors. Now I really cannot speak with authority about modern plates; but some years ago I measured some Paris negatives, and it turned out that the shadow boundaries on those photographs had been moved through 0.12 mm. Now this was proved by the fact that they gave systematic errors that got larger and larger as you approach the terminator. When you subtract 0.12 mm then craters of the same size give consistently the same depths, and this has imperiled some of the older thick-emulsion photographs.

KOPAL: But these negatives are now about 60 or 70 years old and almost any amount of shift may have occurred. This again is a good warning against too much emphasis on the remeasurements of old plates. With modern emulsions – and I think Mr. Carder can bear me out on this – such shifts cause negligible errors in comparison to errors arising from other sources that cannot be avoided. Modern emulsion bases are so good that no shifts larger than a couple of microns or so possibly occur in the course of subsequent processing.

ARTHUR: I have always been astonished that Saunder and Franz could get such good positions as they did with these kind of shifts of the emulsions available at the time.

KOPAL: The positions, especially of heights, of the early observers have been given to more significant figures than they really deserve; and this makes them to look nice on paper, but if we re-measure them at the present time, large discrepancies often are discovered. I should like to warn – if any warning is needed – that greatest caution should be exercised in the use of the old material. Schmidt, the greatest producer of lunar heights by visual methods in the 19th century, when reducing his measures he confused often the points measured and took the means of observations of two different heights, quoting such a mean to one meter. Some of these mixups have later been disentangled by Miss Blagg and other investigators in this field. I wish to stress that all such measures as you may find in the books of the 19th century and even older vintage, are essentially only of historical interest.

LORELL: Did I hear you mention pre-exposure of plates?

KOPAL: Dr. Rösch did so. And I added that in our particular case the lunar landscape not reached by direct sunlight is illuminated by the Earth.

SASSER: What is the estimated horizontal dimension of these features one or two meters in height?

KOPAL: Close to a kilometer, between half a kilometer and a kilometer. The prints which are being circulated are not the best, inasmuch as they were not taken at the time of the best seeing. I think probably Pat Sudbury could tell you that the seeing was not much better than average.

KOZIEL: Any more questions?

HOPMANN: Dr. Richter at Altenburg has studied a lot the method of equidensitometry and built an isophotometer generating the isophotes of the different nebulae and other objects. Is it also possible to use this for your purposes?

KOPAL: It can also be done by means of other apparatus, which we have at our disposal here. Already it is being done in France by Mlle Calame working with professor Rösch. Mlle Calame is rather shy of speaking English, but I am sure that if you ask her, she will be very glad to show you some of the results of precisely the kind of work which you mention.

HOPMANN: Yes, I can speak with Mlle Calame at the Pic tomorrow or sometime, because we have made some first experiments in this direction at Vienna.

GOUDAS: Could you possibly comment on the combination of the long exposure with sequential photography, or is there any necessity to do so?

KOPAL: We merely take sequential photographs at a slow rate to follow the gradual development of shadows. There is no reason why a cine-camera could not be made to work at the speed of, say, 4 or 5 frames per minute. I think if one had such a cine run and if one could only run it as such, the result would be spectacular.

VOICE: This, however, would require very precise guidance.

KOPAL: Oh, but I think that over short intervals of time we can guide better than to 0.″1. The guiding of the 43-inch on the Pic is very good. However, the guiding errors, such as may be, would simulate the wandering of the image: the image will be jumping by microscopic amounts back and forth, but still the racing of the shadow across the surface of the Moon should be clearly discernible; because when the tip of the shadow of a mountain or obstacle just happens to touch tangentially the lunar surface, the speed at which the shadow moves across the lunar surface is about 30 kms per hour.

ARTHUR: I was wondering how this compared with pre-exposure, because we tried pre-exposure very briefly; and it is our experience that it does bring out detail around the terminator. There is also a drop in contrast unless you control this pre-exposure very carefully and, of course, the drop in contrast means a drop in resolution.

KLAGES: When you pre-expose like this, much of the image would build up on the silver grains and start making the figure look larger. How much of a shift does this give you in the position you are trying to determine?

KOPAL: We do not know as yet. This work is only beginning. We did not spare very much time from our programs for this experimentation, and we do not have enough data at the moment to answer your questions. We should very much like to know it, and in due course we shall.

ARTHUR: There is one point that applies to the question of long shadows mentioned by professor Kopal. How do you control your direction measurements, because it is not necessarily the same all over the plate for a long shadow?

KOPAL: This is the direction of the incident sunlight.

ARTHUR: That is right, but how do you know which direction this is?

KOPAL: If you have at least three control points, you can orientate. We have not done it yet, but this is the way it can be done in precision work.

ARTHUR: Yes, this means that in using the Pic-du-Midi plates which show the entire disk, you will have considerable variation from North Pole to South Pole.

KOPAL: The orientation can be inaccurate within a couple of minutes.

ARTHUR: Have you used the *Orthographic Atlas* in your work?

SUDBURY: Yes.

ARTHUR: With all due deference to my own work in the *Orthographic Atlas*, I am rather worried by seeing people put a weight on it that it will not bear. I assure you that the errors in some places are thicker than the width of a line. The errors of the Franz points in the limb region, on which it was based, would be about 002, and it was a question of guessing which control point was better than the other. And it certainly should not be used to control shadow measurements with low angles of illumination. I believe professor Kopal can tell you something about that – that the determination becomes rather sensitive with the position of the peak, and I am rather worried about that.

SUDBURY: The point with this determination is that one wishes to obtain a set of about 10 control points, preferably small craters within each of the fields. The purpose of consulting the *Orthographic Atlas*, in fact, was to obtain good identification.

GOUDAS: I would like to stress one point: namely, that it is necessary for the study of the figure of the Moon to have a homogeneous set of control points; and for the entire surface we need 432 points homogeneously distributed. This is a first-order system, of course, and covers the back side of the Moon; and, hence, 216 will be enough for the front side. Of course, we cannot go very close to the limb using the parallactic or stereoscopic technique; and, therefore, I think the number of 200 can be adopted. I suggest that one agency or two should undertake to propose, after a careful study, a set of points to be accepted by us, so that the different interested parties will then contribute separately towards the determination of the coordinates of the same 200 points using always sequential photography. This will hopefully disperse the confusion in the existing control systems, the evaluation of which is as difficult as the derivation of the systems.

WEIMER: Such a system including 40 points was proposed some years ago.

HOPMANN: The system was only used at Vienna, and I suggest that we adopt for the future the 196 points of ACIC and not the 40 points of dr. Weimer.

GOUDAS: Some control points are better for measurements than others. Therefore, certain characteristics will have to be used as criteria in the selection of the points to be adopted. For instance, irregular craters, or big craters, cannot be used. I was not aware of a previous effort to establish an International Control System, but if this is the case, then I suggest that the study be responsibly continued. It will be necessary to reject a few points, replace them by others in order to fulfil the homogeneity of the system, which is very essential, and then, say within a year, to conclude which system to adopt and improve upon.

WEIMER: Often in a given region you have no choice because there are no clearly identifiable features.

MILLS: I was going to ask, why only 200 points and not more or fewer?

ARTHUR: What is the magic about 200?

GOUDAS: This will provide one point per ten square degrees of the surface or so.

MILLS: Why ten square degrees?

GOUDAS: First, because this is a reasonable number of points to measure. Second, there is some magic in this number: namely, that in fitting harmonics you satisfy the orthogonality conditions with a small variance of 10% to 20% in the coefficients. If you have fewer points, you find that the coefficients of the harmonics expansion vary when you add more and more harmonics in the fit. If you have an infinite number of points, there is no change at all. If you have about the number of Schrutka or ACIC you find a 10% or 20% variance in the first four harmonics which include about 25 orthogonal functions. I think this is the right size of a first control system. If you have more points variance will become less, and therefore it is better to have 200, since this number is a compromise between the orthogonality requirement and the heavy task of computing points.

MEYER: Professor Kopal suggested that we have a committee to look into points and we had some other items that we wish to look over which have already been brought up during these present meetings. One, of course, is that those systems based on 4 or 5 plates differ from one another by the amount of difference which we are attempting to find, and it seems apparent that you have to go to either sequential or long exposure photography. Some of my colleagues have used, even in the last few years, librational differences of three degrees which indicate that they do not understand the problem. In addition, the terms discussed by dr. Goudas imply that those of us who work on approximate coordinates must have a homogeneous set of points.

GOUDAS: I would like to request from those who compute control systems to give in addition to the coordinates of the points the inside and outside elevation of the rim, because having the photograph, they are in the best position to do so. This is important because the rim must be subtracted from the height estimate in order to have the level which represents the true lunar surface.

KOPAL: May I propose an amendment to the motion and suggest that dr. Goudas be joined by colleagues from ACIC, AMS, etc.? If we assure membership from all major groups working in the field, this would facilitate, if not achieve, a large acceptance of the system.

CARDER: I'd like to echo all that professor Kopal said. Any participation by the ACIC lunar group would really be a Department of Defense effort and would include the AMS and perhaps a member of the army intelligence agency who is controlling this group, but speaking for the Air Force I wish to say that we shall be very happy to work with the scientific and academic community in such an effort that will aim finally to the endorsement of a control system by the I.A.U.

ARTHUR: There is something else that has not been mentioned yet, and this is uniformity in the publication of selenodetic results. I will not mention specific points, but sometimes I have to go through a paper three of four times to find out what particular things mean. We do not have the same notation for the selenodetic coordinates and some give latitudes and longitudes, some people give standard coordinates, etc. I must say one other thing: namely, that too much importance has been given to the derivation of these standard or spherical coordinates. The real work is in producing corrected photographic coordinates which anyone can use, and the emphasis should be placed on that. I am not in favour of impulse computation of the figure of the Moon. Let us face it, with modern techniques this is a simple job. It is not quite as simple to get the correct coordinates and I think the emphasis should be on that.

ECKHARDT: I just want to clarify a point regarding the term 1375 of Brown's lunar theory. Because of the long period of this term, the physical libration differs from the optical one by about 2% of opposite sign so the ephemeris does not carry this particular term of Brown.

DOWNIE: I would like to add my remarks to those of professor Minnaert and express the sincere thanks of the conference attendees to those who made this conference possible, Mahlon Hunt, the Air Force Research Laboratories, professor Kopal, the entire staff at Manchester here. Secondly, I would like to commend professor Kopal for solving the problem of long-range weather forecasting and thus being able to pick this week a year or so ago for the conference, unless he has solved the problem of weather control. So far, he has not let me in on his secret. So all I want to say is, thanks again professor Kopal; it was an extremely successful conference.

KOPAL: Thank you very much, colonel Downie; and thank you all ever so much for coming to Manchester from great distances and spending with us the past four days. As regards the weather control – you know, colonel, that I never keep any secrets from the Air Force (especially on a subject of so obvious strategic importance) – and so I should better explain. As has been known at least since the days of Agamemnon (that worthy ancestor of our dr. Goudas), the best way to beseech the clemency of the elements is to sacrifice (in an appropriate manner) to the gods the most beautiful maiden from our midst. In this connection, we thought first of course of Miss Ellen Finlay (see the frontispiece), but just had to give up the idea: for how would all our guests have arrived safely at Manchester, or (more important) departed again for their native shores without her efficient ministrations? Her sister Jean was our next candidate for the role of Iphigenia, but again without avail. For not only is she not under my jurisdiction, but the success with which she discharged (albeit with one exception) the duties of our honorary transport officer has rendered her truly indispensable. Mrs. Barbara Sudbury – another eminently suitable candidate for the sacrificial altar – could likewise not be thought of; for her husband would have turned the 43-inch Pic-du-Midi telescope against us in justifiable retaliation.

The group photograph appearing as a frontispiece to this volume reveals that, at this stage, we ran out of all suitable local candidates whose sacrifice could have propitiated the weather gods. I confess that, as a result, for several days before the meetings we watched the skies with some apprehension, and often feared the worst. That, instead, we were rewarded (and in Manchester, of all places) with a week of uninterrupted sunshine to share with our guests, should give us all a new measure of confidence in the advance of civilisation – at least since the days of Agamemnon and Iphigenia.

CONCLUDING SURVEY

M. G. J. MINNAERT

Professor Kopal has asked me "à l'improviste" to give a summary and some impressions about the interesting conference, to which we had the privilege to be present. Let me first say that I am not a specialist in the field of selenodesy and that my comments, therefore, might not always be to the point. I also apologise for inevitable incompleteness.

The enormous amount of research, devoted to lunar studies in the course of these last years, is really impressive. More especially we have admired the magnificent work done at Manchester, where, by a combination of ingeniosity and organisation, such a remarkable cooperation has been obtained with the Pic-du-Midi, with the Aeronautical Chart and Information Centre, with the U.S. Army Map Service, and with the Air Force Cambridge Research Laboratories.

The first day of this conference was almost entirely devoted to the mechanical properties of the Moon, as derived from the monumental work of Koziel on the lunar libration. This was based on careful heliometer measurements, made already a century ago, during a period of 40 years, and now discussed with the greatest care. The results establish in the first place that the "mechanical ellipticity" f of our satellite may definitively be considered to have the value 0.633, entirely different from the value 0.25 which would correspond to hydrostatical equilibrium. This led to interesting discussions concerning the singularities which are found in the solution of the equation of libration (Moutsoulas). Moreover, it appears that there is practically no "free libration" – a highly satisfactory result, since indeed any angular impulse, given to the Moon at the moment of its origin, would long have been damped out. From the theoretical point of view it still remains necessary to introduce in the equations the effects of elasticity and of friction in the lunar body.

After hearing the results of this investigation, completed by the collaborators of professor Koziel, one gets the impression that no essentially better results can be obtained by the classical methods; if a higher precision is really needed, it will have to be reached by entirely different means.

A quite different subject – the explanation of Cassini's second and third laws, was treated by professor Colombo. I could not help being impressed by his remark that astronomers nowadays should devote more time to celestial mechanics, a branch which has acquired new interest because of space research.

For a determination of the shape of the limb, several ingenious methods have been used, of which professor Rösch gave a lucid survey. The selenodesy of the other regions requires the determination of precise coordinates of lunar features. Such coordinates are not only basic for lunar cartography and for studies of the lunar shape (the "selenoid"); they will also allow a better determination of the Earth's radius, of the Moon's parallax, of the excentricity of the Moon's orbit, of latitude and longitude on Earth, of Chandler's motion, etc.

There was a general agreement that we ought to select a system of standard points, on the position of which all observatories should concentrate their efforts. A preliminary catalogue of such points was already earlier drawn up by Weimer. Once this network being well-established, all other features could have coordinates, referred to the standard-points. Coordinates on the Moon should be measured near full moon, in order to avoid shifts, due to lateral illumination (Davidson). Personal errors of measurement were investigated and found to be small (Hunt, Rackham). An important source of errors are the image displacements, due to the atmospheric refraction. The film of Carder and Meyer, showing the jumping images on successive exposures, made this point very convincing. Interesting was their suggestion to expose lunar photographs for a time of the order of 40 sec, on fine grain emulsions, in order to average out such irregular shifts, due to momentary and local seeing disturbances. Another possibility is of course to take many exposures and to compute mean values.

Some splendid examples of modern lunar cartography were displayed before us, e.g. the maps from the Aeronautical Chart and Information Center and the detailed isothermal maps of the Moon.

A first important result, to be drawn from such precise coordinate measurements, is the determination of the Moon's constants of rotation. Mr. Arthur has shown us how the long-focus refractor can be used for this purpose, and he has given new methods in order to improve the results.

Another aim of position measurements is the determination of absolute heights. The methods used were discussed by Mills and Davidson, Rackham, a.o. Up to now a decision concerning the general shape of the Moon cannot said to have been reached. There was a rather general feeling that we should not expect this shape to correspond necessarily to an ellipsoid, and that the present results suggest rather a spherical shape, to which here a slight extumescence, there a shallow excavation has been applied. The maria are not necessarily low, nor horizontal (Goudas).

It has certainly been a great experience for all of us to have heard on these subjects the comments of professor Hopmann, who has studied the Moon during a long life of devotion to science. His remarks filled us with admiration for the assiduity and cleverness of our predecessors, even though we are convinced that nowadays new methods have to be introduced.

Two such methods were discussed: the laser measurements and the lunar Orbiters. – The use of a laser beam, directed towards a point of the Moon, seems extremely difficult, because the returning light signal is drowned in the noise, while the energy which can be applied is limited by the material of the laser; only if by soft landing a corner reflector could first be deposited on the Moon, would the experience become useful (Orszag, Julian). – Much greater success may be expected already now from a lunar satellite. From Dr. Lorell's conclusions we remember that the position of such a satellite, provided with a corner reflector, could be determined with high precision: the distance may be obtained with a mean error of only 5 meters, the sight-line velocity with a precision of 3 mm/sec! It is true that such information does not extend to the back of the Moon, but one may feel confident that an orbit thus determined will be amply sufficient in order to find the potential distribution in the surroundings of the Moon and thus to obtain information about its moments of inertia and its interior structure. Dr. Veis and dr. Kaula complemented the discussion of the Orbiter method.

One step further: starting from a polar orbit, so well determined in space, we may hope to determine also the shape of the Moon, either by equipping the satellite with a range-finder which measures all the time and along different meridians its vertical distance to the lunar surface; or by determining this distance from stereoscopic photographs.

The last day we have been speaking about detailed relief measurements. We have heard how powerful the shadow method is, but we must keep in mind that several corrections have to be applied, so well elaborated by the Manchester astronomers. Determinations by the photometric method have the advantage, that they can be applied even if only one record is available provided spurious effects can be eliminated; the application of the isodensimeter may be of great use (Mrs. Rifaat). – It would be most interesting to test both methods simultaneously on the same, carefully selected object.

There have been also a few contributions on the relief of the lunar surface at the scale of centimeters or millimeters. Professor Hopmann showed us three wonderful panoramic views, obtained by the Soviet spacecraft Luna IX after its soft landing. In planning the Apollo mission, interesting research on the microstructure has been made, on which dr. Sasser reported.

In many problems of selenodesy, some measurements and calculations have to be repeated a great number of times, either because we need high precision, or because we have to investigate a considerable number of points on the Moon. For all such cases the use of digitised instruments and of electronic computers has become imperative (Nowicki, Mills, Davidson).

Finally I must mention in a few words the interesting reports about some observatories where lunar studies can be made. We listened to the description of the instruments at the famous Lowell Observatory and at the Cape Observatory; we heard about the great new possibilities for observations at Kottamia, at Hyderabad, at Kyoto. We congratulate our colleagues with their new institutes, which testify about the rapidly increasing interest in astronomy at all sides of the world. In carefully selected sites, surrounded by the desert or by the high mountain snow, they will help to furnish most valuable observations, increasing man's knowledge about the Universe.

In concluding this short survey, I am sure to express the general feeling of all here present, by congratulating those who have organised this conference, more especially the presidents professor Kopal and Mr. Hunt. We are grateful for the generous hospitality which we enjoyed here, and we highly appreciate the excellent planning and organisation of the conference by the leader and the staff of this Astronomical Department. Almost all of the work here discussed, is connected in some way with the activity of the Manchester group. You will certainly allow me to thank more in particular professor Kopal, who was not only a wonderful host, but who also all the time inspired our discussions.

INDEX OF NAMES

GPSR Compliance

The European Union's (EU) General Product Safety Regulation (GPSR) is a set of rules that requires consumer products to be safe and our obligations to ensure this.

If you have any concerns about our products, you can contact us on ProductSafety@springernature.com

In case Publisher is established outside the EU, the EU authorized representative is:

Springer Nature Customer Service Center GmbH
Europaplatz 3
69115 Heidelberg, Germany

Batch number: 10761697

Printed by Printforce, the Netherlands